Nanostructured Hard Materials

Scrivener Publishing
100 Cummings Center, Suite 541J
Beverly, MA 01915-6106

Publishers at Scrivener
Martin Scrivener (martin@scrivenerpublishing.com)
Phillip Carmical (pcarmical@scrivenerpublishing.com)

Nanostructured Hard Materials

Industrial Applications

Edited by

Inamuddin

Dept. of Applied Chemistry, Zakir Husain College of Engineering and Technology, Faculty of Engineering and Technology, Aligarh Muslim University, Aligarh, India

Tariq Altalhi

Dept. of Chemistry, College of Science, Taif University, Taif, Saudi Arabia

and

Mohammad Abu Jafar Mazumder

Chemistry Department, King Fahd University of Petroleum & Minerals, Dhahran, Saudi Arabia

Interdisciplinary Research Center for Advanced Materials, King Fahd University of Petroleum & Minerals, Dhahran, Saudi Arabia

Scrivener Publishing

WILEY

This edition first published 2026 by John Wiley & Sons, Inc., 111 River Street, Hoboken, NJ 07030, USA and Scrivener Publishing LLC, 100 Cummings Center, Suite 541J, Beverly, MA 01915, USA

For more information about Scrivener publications please visit www.scrivenerpublishing.com.

Wiley Global Headquarters
111 River Street, Hoboken, NJ 07030, USA

For details of our global editorial offices, customer services, and more information about Wiley products visit us at www.wiley.com.

The manufacturer's authorized representative according to the EU General Product Safety Regulation is Wiley-VCH GmbH, Boschstr. 12, 69469 Weinheim, Germany, e-mail: Product_Safety@wiley.com.

Library of Congress Cataloging-in-Publication Data

ISBN 9781394314140

Cover image: Generated with AI using Adobe Firefly
Cover design by Russell Richardson

Set in size of 11pt and Minion Pro by Manila Typesetting Company, Makati, Philippines

Printed in the USA

10 9 8 7 6 5 4 3 2 1

Contents

Preface **xvii**

1 Computational Modeling and Simulation: Predictive Modeling and Simulation of Nanostructured Hard Materials for Optimization and Design **1**
Puspendu Barik and Sasadhar Majhi
1.1 Introduction 2
1.2 Nanostructured Hard Materials 3
1.2.1 Metal Nanostructure 3
1.2.2 Semiconductor Nanostructure 4
1.2.3 Magnetic Nanostructure 4
1.2.4 Carbon-Based Nanostructure 6
1.2.5 Alloy 7
1.2.6 Composites 7
1.3 Modeling and Simulation of Nanostructured Hard Materials 8
1.3.1 Importance of Modeling and Simulation 8
1.3.2 Basic Categories of Modeling and Simulation 9
1.3.3 Simulation Strategies 11
1.4 The Outcome of Modeling and Simulation of Nanostructured Hard Materials 23
1.4.1 Insights into Growth Mechanisms 24
1.4.2 Prediction of Material Properties 25
1.4.3 Design of Novel Nanostructures 26
1.4.4 Defect Analysis 26
1.4.5 Validation and Interpretation of Experimental Results 27
1.4.6 Accelerating Material Discovery 27
1.4.7 Guiding Experimental Design 28
1.5 List of Popular Software for Nanostructured Hard Materials 28
1.5.1 MD Simulation 29
1.5.2 DFT Simulation 30
1.5.3 MC Simulation 31

1.5.4 Finite Element Method (FEM) 31
1.5.5 FDTD Software 32
1.6 Conclusion 32
References 33

2 Latest Developments for Ni-Based Wear Protective Surfaces: Coating Compositions and Methods 39
Onur Güler, Müslim Çelebi and Abdullah Hasan Karabacak
2.1 Introduction 40
2.2 Surface Protective Coatings for Wear-Resistant Materials 45
2.2.1 Traditional Ni-Based Wear-Resistant Coatings 57
2.2.1.1 Pure Ni-Based Coatings 57
2.2.1.2 Ni-P-Based Coatings 60
2.2.1.3 Ni-B-Based Coatings 65
2.3 Recent Progress on Ni-Based Wear-Resistant Coatings 69
2.4 Conclusion 95
2.5 Future Perspectives 97
References 99

3 Advances in Nanostructured Materials and Liquid Crystal Composites: Unveiling Structural Properties and Emerging Paradigms in Display Technologies 113
Driss Soubane and Mohamed El Garah
3.1 Introduction 114
3.2 Fundamentals of Nanostructured Materials 115
3.2.1 Nanostructured Material Classification 116
3.3 Fabrication of Nanostructures 119
3.3.1 Nanostructure Fabrication Techniques 120
3.4 Liquid Crystals: Basics and Properties 125
3.4.1 Liquid Crystal Classes 127
3.4.2 Principal Properties of the Liquid Crystal 129
3.5 Nanostructured Materials and Liquid Crystals Composites 131
3.5.1 Different Doping Materials in Binary Liquid Crystal Composites 134
3.5.2 Ternary Composite Materials for Display Applications 135
3.5.3 Beyond Binary and Ternary Composites 135
3.6 Pioneering Advances in Display Technology 136
3.7 Future Perspectives and Emerging Trends 136
3.8 Conclusion 137
Funding 138
References 138

4 Nanocomposites in Aerospace and Defense: Materials for Lightweight Structures, Thermal Management, and Ballistic Protection 149

Riyadh A. Al-Samarai and Yarub Al-Douri

4.1 Introduction 150
4.2 Advanced Ceramic Material Preparation 151
4.3 Methods for Creating Nanoceramics 152
4.3.1 Sol-Gel Chemistry for Nanoceramic 153
4.4 Recent Developments in General and Advanced Nanoceramics Fields 155
4.4.1 Nanoceramics Based on Silicate 157
4.4.2 Nanoceramics Based on Zirconium Dioxide 157
4.4.3 Nanoceramics Based on Titania 158
4.4.4 Nanoceramics Based on Borate 158
4.4.5 Nanoceramics Based on Silicon Carbide 159
4.4.6 Nanoceramics Based on Boron Carbide 159
4.5 Nanoceramics Applications 160
4.5.1 Electronics 160
4.5.2 Applications of Biomedical 161
4.5.3 Applications of Environmental 162
4.5.4 Applications in Space and Defense 162
4.5.5 Strategic Applications 163
4.6 Conclusion and Prospective Future of Nanoceramics 163
References 165

5 Cermet: Cutting Tool Materials for High Speed and Advancements in Nanocoating Technology 175

Riyadh A. Al-Samarai and Yarub Al-Douri

5.1 Introduction 176
5.2 Applications of High-Speed Machining 177
5.3 Machine Tools for HSM – Requirements 177
5.3.1 Hard Material Cutting 177
5.3.1.1 Hybrid Cutting 188
5.3.1.2 Chemical Vapor Deposition 190
5.3.1.3 Plasma Arc 191
5.3.1.4 Flame Synthesis 193
5.3.1.5 Laser Ablation 194
5.3.2 Method of Liquid Phase 194
5.3.2.1 Electrochemical 195
5.3.2.2 Precipitation 196
5.3.2.3 Sol–Gel 197

5.3.2.4 Hydrothermal 199
5.3.2.5 Microemulsion 200
5.4 Advancements in Nanocoating Technology 202
5.5 Application of Nanocoatings 203
Acknowledgements 203
References 204

6 Cermet: Composite Materials for Cutting Tools, Electrical Contacts, and Thermal Management 207
Riyadh A. Al-Samarai and Yarub Al-Douri
6.1 Introduction 208
6.2 Structural Formation 209
6.2.1 Nitride Titanium (TiN) 209
6.2.2 Carbonitride of Titanium (TiCN) 210
6.3 Coating of TiCN and TiN 211
6.4 Challenges and Restrictions of TiN and TiCN 212
6.5 Treatment Techniques 212
6.6 Factors that Improve Sintering Kinetics 213
6.7 Binderless 213
6.8 Prior Studies on Titanium Carbonate Nitride and Titanium Nitride without a Binder Used Sintering Techniques 215
6.9 Niobium Carbide 221
6.10 NbC Cermets 221
6.11 NbC-Co Cermets 222
6.12 NbC-Ni Cermets 222
6.13 Secondary Carbides' Impact 222
6.14 Mechanical Properties 223
6.15 Technical Developments 225
6.16 Concluded Remarks 229
References 230

7 Advancements in Nanocomposite Technology for Environmental Remediation: Innovations in Water Purification, Air Filtration, and Pollutant Capture 237
Ashish Mogra and Ranvir Singh Panwar
7.1 Introduction 238
7.2 Application of Nanocomposite in Environmental Remediation 240
7.2.1 Water Remediation 240
7.2.2 Air Remediation 240
7.2.3 Soil Remediation 240
7.2.4 Photocatalytic Degradation 241

7.2.5 Sensing and Detection 241
7.2.6 Oil Spill Cleanup 241
7.2.7 Enhanced Bioremediation 242
7.3 Objectives 242
7.4 Nanocomposites in Water Remediation 242
7.4.1 Removal of Heavy Metals 242
7.4.1.1 Adsorption 243
7.4.1.2 Catalysis 243
7.4.1.3 Magnetic Nanocomposites 244
7.4.2 Degradation of Organic Pollutants 244
7.4.2.1 Photocatalytic Degradation 245
7.4.2.2 Adsorption and Breakdown (Degradation) 245
7.4.3 Pathogen Elimination 246
7.4.3.1 Antibacterial Properties 246
7.4.3.2 Virus Removal 246
7.4.4 Desalination and Ion Removal 247
7.4.4.1 Selective Ion Exchange 247
7.4.4.2 Membrane Technology 248
7.4.5 Sensing and Monitoring 248
7.4.5.1 Nanocomposite-Based Sensors 249
7.4.5.2 Detection of Pollutants 249
7.5 Nanocomposites in Air Purification 250
7.5.1 Filtering of Particulate Matter 250
7.5.1.1 Nanocomposite Filters 250
7.5.1.2 Enhanced Filters 251
7.5.1.3 Electrostatic Capture 251
7.5.2 VOCs and Hazardous Gas Adsorption 252
7.5.2.1 Adsorption Capability 252
7.5.2.2 Selective Adsorption 252
7.5.3 Photocatalytic Applications 253
7.5.3.1 Photocatalytic Degradation 253
7.5.3.2 Self-Cleaning Surfaces 254
7.5.4 NOx and SOx Removal 254
7.5.5 Sensing and Monitoring of Air Quality 254
7.6 Nanocomposites in Soil Remediation 255
7.6.1 Heavy Metal Immobilization 255
7.6.1.1 Adsorption and Chelation 255
7.6.1.2 Phytoremediation Enhancement 256
7.6.2 Organic Contaminant Breakdown 257
7.6.2.1 Degradation of Pollutants 257
7.6.2.2 Sorption and Destruction 258

7.6.3 Stabilization of Contaminated Soil 258
7.6.4 Radioactive Waste Treatment 259
7.6.4.1 Radioactive Contaminant Removal 259
7.6.5 Sensing and Monitoring of Soil Quality 260
7.7 Challenges and Future Perspectives 260
7.7.1 Challenges 261
7.7.1.1 Environmental Impact 261
7.7.1.2 Health Risks 261
7.7.1.3 Regulatory Framework 261
7.7.1.4 Economic Viability 261
7.7.1.5 Technological Challenges 261
7.7.1.6 Public Perception 261
7.7.2 Future Perspectives 261
7.7.2.1 Sustainable Development 261
7.7.2.2 Advanced Functionalities 262
7.7.2.3 Smart Nanocomposites 262
7.7.2.4 Integrated Systems 262
7.7.2.5 Nanocomposite Recycling and Reuse 262
7.7.2.6 Advanced Sensing Technologies 262
7.7.2.7 Global Collaboration and Regulation 262
7.8 Conclusion 262
References 263

8 Nanoscale Coatings for Tribology: Applications, Challenges and Future Directives 269
Ranvir Singh Panwar and Ashish Mogra
8.1 Introduction 269
8.2 Overview of the Role of Tribology in Various Industries 272
8.2.1 Automotive Industry: Engine Component Coatings 272
8.2.2 Aerospace and Energy Industry: Steam and Wind Turbine Blade Coatings 274
8.2.3 Manufacturing Industry: Cutting Tool Coatings 275
8.2.4 Medical Devices: Orthopaedic Implant Coatings 276
8.2.5 Electronics: Hard Disk Drive Coatings 276
8.3 Challenges and Future Directives 276
8.3.1 Durability and Longevity 277
8.3.2 Uniformity and Consistency 277
8.3.3 Environmental, Sustainability and Health Concerns 277
8.3.4 Technical and Fabrication Challenges 277
8.3.5 Performance Under Extreme Conditions 277
8.3.6 Compatibility with Substrates 278

8.3.7 Characterisation and Testing 278
8.3.8 Cost and Economic Viability 278
8.4 Conclusion 280
References 281

9 Metal Borides: High-Temperature Materials for Aerospace, Automotive, Nuclear, and Electromagnetic Industries 283
Arun K. Chattopadhyay, Tuncay Simsek, Alican Yakın and Telem Simsek
9.1 Introduction and Overview 284
9.2 Production Methods of Metal Borides 287
9.3 Metal Borides Applications 291
9.3.1 Aerospace Applications of Metal Borides 294
9.3.2 Metal Borides Applications of Automotive Industries 299
9.3.3 Metal Borides Nuclear Applications 302
9.3.4 Electromagnetic Properties of Metal Borides 304
9.4 Conclusion 307
References 308

10 Hard Machining and Characteristics for Industrial Cutting Tools and Coating Deposition Applications 317
Riyadh A. Al-Samarai and Yarub Al-Douri
10.1 Introduction 318
10.2 Analysis of Hard Machining Research 322
10.3 Hard Machining Characteristics 323
10.3.1 Speed with Cutting Force 324
10.3.2 The Cutting Forces Major Axial (Thrust) Components 326
10.3.3 Power Invested in Creation of Surfaces 327
10.3.4 Requirements for Rigid Processing 328
10.4 Operations for Hard-Machining 328
10.4.1 Hard Turning 329
10.4.2 Reaming and Hard Boring 330
10.4.3 Milling of Hard 331
10.4.4 Broaching of Hard 332
10.4.5 Operations for Manufacturing Hard-Gear 333
10.4.6 Skiving 334
10.4.6.1 Turning of Hard 334
10.4.6.2 Skiving of Gear 334
10.5 Technology of Coating 336
10.5.1 Backdrop 336

10.5.2 Significance of Coating Layer 336
10.6 Future and Current Events 337
10.7 Conclusion 337
References 338

11 Metal Carbides: Tools and Wear-Resaretant Coatings for Machining, Drilling and Mining Operations 343
Riyadh A. Al-Samarai and Yarub Al-Douri
11.1 Introduction 344
11.2 Structural Formation 345
11.2.1 Nitride Titanium (TiN) 345
11.2.2 Carbonitride of Titanium (TiCN) 346
11.3 Coating of TiCN and TiN 347
11.4 Challenges and Restrictions of TiN and TiCN 348
11.5 Treatment Techniques 348
11.6 Factors That Improve Sintering Kinetics 349
11.7 Binderless 349
11.8 Prior Studies on Titanium Carbonate Nitride and Titanium Nitride without a Binder Used Sintering Techniques 351
11.9 Niobium Carbide 357
11.10 NbC Cermets 357
11.11 NbC-Co Cermets 357
11.12 NbC-Ni Cermets 358
11.13 Secondary Carbides' Impact 358
11.14 Mechanical Properties 358
11.15 Conclusion 361
References 362

12 Industrial Applications of Hard Materials: Wear-Resistant Coatings, Cutting Tools, and Advanced Manufacturing Techniques 369
Riyadh A. Al-Samarai and Yarub Al-Douri
12.1 Introduction 370
12.2 A Brief Analysis of Hard Machining Research 374
12.3 Characteristics of Hard Machining 376
12.3.1 Reduction in Cutting Force with Speed 376
12.3.2 The Cutting Forces Major Axial (Thrust) Components 377
12.3.3 Large Power Invested in Creation of Surfaces 379
12.3.4 Requirements for Rigid Processing 380

12.4 Operations for Hard-Machining 380
12.4.1 Hard Turning 380
12.4.2 Reaming and Hard Boring 382
12.4.3 Milling of Hard 383
12.4.4 Broaching of Hard 384
12.4.5 Operations for Manufacturing Hard-Gear 384
12.4.6 Skiving 386
12.4.6.1 Turning of Hard 386
12.4.6.2 Skiving of Gear 386
12.5 Coating Technology 388
12.5.1 Backdrop 388
12.5.2 Significance of Coating Layer 388
12.6 Concluding Remarks 389
References 390

13 Biomedical Applications of Hard Materials: Implants, Coatings and Drug Delivery Systems for Medical Devices 395
Fatema Tuz Zohera, Abul Kalam Azad and Madhusmruti Khandai
13.1 Introduction 396
13.2 Hard Materials for Biomedical Application 397
13.2.1 Drug-Delivery Systems 398
13.2.2 Tissue Engineering 398
13.2.3 Implantable Devices 398
13.2.4 Diagnostic Tools 399
13.3 Types of Implant Coatings 399
13.3.1 Biocompatible Coatings 399
13.3.1.1 Calcium Based Apatite Coatings 399
13.3.1.2 Bone Morphogenetic Protein Coatings (BMP) 400
13.3.1.3 RGD Peptide Based Coatings 400
13.3.1.4 Mg Based Coating 400
13.3.1.5 ZnO Based Coatings 401
13.3.1.6 TiO_2 Based Coatings 401
13.3.1.7 Carbon Based Coatings 401
13.3.1.8 TiN and CrN Based Coatings 402
13.3.2 Polymer Based Antimicrobial Coatings 402
13.3.2.1 Polymer Coatings 403
13.3.2.2 Coatings for Sustainable Antibiotic Release 403

13.3.2.3 The Properties of Coatings Systems and Their Resistivity to Corrosion 404
13.4 Ceramics 405
13.5 Commercially Available Multifunctional Coatings 405
13.6 Surface Modification 406
13.7 Conclusion 407
References 408

14 Antifouling Nano Filtration Membranes **415**
Divya D. Achari, Nandini A. Pattanashetti and Mahadevappa Y. Kariduraganavar
14.1 Introduction 415
14.2 Nanofiltration Membranes and Membrane Fouling 417
14.3 Types of Membrane Fouling 418
14.4 Mechanism of Fouling 419
14.5 Controlling Factors of Membrane Fouling 421
14.5.1 Effect of Surface Chemistry (Hydrophilicity/ Hydrophobicity) 421
14.5.2 Effect of Pore Size and Pore Size Distribution 422
14.5.3 Effect of Surface Structure 423
14.5.4 Effect of Surface Charge 423
14.5.5 Effect of Salts 423
14.5.6 Effect of pH 424
14.5.7 Effect of Operational Conditions 424
14.6 Antifouling Methods 425
14.6.1 Passive Antifouling Method 426
14.6.1.1 Fouling Resistance Method and Mechanism 426
14.6.1.2 Fouling Release Method and Mechanism 428
14.6.2 Active Antifouling Method 429
14.6.2.1 Off-Surface Antibacterial Mechanisms 430
14.6.2.2 On-Surface Antibacterial Mechanisms 431
14.7 Surface Modification of the Nanofiltration Membranes 432
14.7.1 Surface Modification by Surface Coating 432
14.7.2 Surface Modification *via* Surface Grafting 434
14.7.3 Surface Modification *via* Physical Blending 435
14.7.4 Surface Modification *via* Hydrophilic Materials 435
14.8 Conclusions and Future Perspectives 436
Acknowledgement 437
References 437

15 Nanocatalysts for Chemical Processes: Materials for Catalysis, Hydrogen Production, and Pollution Control 447
Shruti Mishra, Mustafa Aamir Hussain, Nisha V. Bora and Leena V. Bora
15.1 Introduction 448
15.2 Applications of Nanocatalysts 449
15.2.1 Biodiesel Production 449
15.2.2 Biomass Production Using Nanocatalysts 450
15.2.3 Nanocatalyst in Electrosynthesis 451
15.2.4 Air Pollution 451
15.2.4.1 Nanoadsorbents 452
15.2.4.2 Photocatalytic Nanocatalysts 453
15.2.5 Water Pollution 454
15.2.5.1 Carbon Nanotubes 454
15.2.5.2 Polymeric Nanoadsorbents 454
15.2.5.3 Zeolites 454
15.2.5.4 Metal Based Nanoadsorbent 455
15.2.6 Hydrogen Production 455
15.2.6.1 Hydrogen Production by Water Splitting Method 455
15.2.7 Hydrogen Production Using Spinel-Structured Nanocatalysts 458
15.2.7.1 Fabrication Process 460
15.3 Challenges and Future Perspectives 462
15.4 Conclusion 463
References 464

Index 471

Preface

The topic of this book exists within the broader context of nanotechnology's development and its impact across industries. Nanotechnology—the manipulation of matter at the nanoscale—has become a major field of study and application. It involves the design, synthesis, and characterization of materials and devices with unique properties at the nanoscale. A key focus of this book is nanostructured hard materials, which feature well-defined nanoscale structures and enhanced mechanical, electrical, optical, and chemical properties. These materials often take the form of nanoparticles, nanowires, or thin films, contributing to their unique performance.

Nanostructured hard materials have made a significant impact in various sectors. In electronics, they support device miniaturization, improve transistor performance, and enable wearable tech. In energy storage, they boost battery and supercapacitor efficiency. In solar energy, they improve light absorption and charge separation. In catalysis, their large surface areas and active sites enhance catalytic activity, crucial to industrial processes. In biomedicine, they enable targeted drug delivery and are key to tissue engineering scaffolds. Environmental engineering also benefits, with applications in water treatment, pollution control, and sustainable energy solutions.

This book explores industrial applications of nanostructured materials, covering synthesis and manufacturing techniques, characterization methods, properties, and sector-specific uses. It aims to be a comprehensive reference, combining theory and practice to encourage interdisciplinary collaboration and inspire further research.

Chapter 1 introduces computational modeling and simulation techniques to design nanostructured hard materials and optimize their performance. It covers predictive modeling and tools useful to materials scientists and engineers. Chapter 2 examines Ni-based wear-resistant surfaces with high hardness and self-lubricating oxide layers. It covers coating methods such as electroless plating, electroplating, laser cladding, and thermal spraying.

Chapter 3 explores nanostructures and liquid crystal composites for display technologies. It discusses fabrication techniques from VLS to self-assembly and analyzes doping effects in binary and trinary composites to enhance display performance. Chapter 4 reviews nanocomposites in aerospace and defense, highlighting their roles in lightweight structures, thermal management, and ballistic protection. It also discusses the emergence of nanoceramics in high-performance applications.

Chapter 5 introduces High-Speed Machining (HSM), explaining its definitions and distinguishing it from traditional processing. HSM is essential in die/mold production and other precision industries. Chapter 6 discusses cermets, particularly TiCN and TiN compounds, for their high-temperature resistance and performance on tool steels.

Chapter 7 covers nanocomposites in environmental remediation, focusing on their ability to remove contaminants, heavy metals, and microbes, and their role in improving water purification. Chapter 8 explores nanoscale coatings in tribology, with emphasis on nanoparticle treatments that enhance material properties and sustainability.

Chapter 9 reviews metal borides for high-temperature applications in space, automotive, nuclear, and electromagnetic industries. It discusses production methods and evaluates their properties and technological potential. Chapter 10 analyzes the evolution of tool coatings in modern grinding operations, discussing preparation methods and applications.

Chapter 11 examines wear-resistant coatings for machining, drilling, and mining. It assesses how TiN and TiCN coatings interact with various substrates to influence tribological performance. Chapter 12 discusses ultra-hard cutting tool materials, including PVD and CVD multilayer coatings, and their tribological performance in operations like milling and boring.

Chapter 13 focuses on biomedical applications of nano hard materials, particularly for targeted drug delivery and biomaterial development for regenerative medicine. Chapter 14 addresses antifouling nanofiltration membranes, discussing fouling mechanisms and strategies for improving membrane performance. Chapter 15 discusses nanocatalysts in biodiesel and biomass production, hydrogen electroproduction, and pollution remediation. It covers fabrication methods and future research directions.

We extend our thanks to everyone who contributed to this project and especially to Martin Scrivener and the team at Scrivener Publishing for their effort and guidance in bringing this work to fruition.

The Editors
July 2025

1

Computational Modeling and Simulation: Predictive Modeling and Simulation of Nanostructured Hard Materials for Optimization and Design

Puspendu Barik[1] and Sasadhar Majhi[2]*

[1]*Materials Science and Engineering Program, College of Arts and Sciences, American University of Sharjah, Sharjah, United Arab Emirates*
[2]*Department of Chemistry (UG & PG Dept.), Trivenidevi Bhalotia College, Raniganj, Kazi Nazrul University, West Bengal, India*

Abstract
This book chapter explores the application of computational modeling and simulation techniques in optimizing and designing nanostructured hard materials. The chapter delves into the intricacies of simulating the behavior of these materials to enhance their physical properties and performance, focusing on predictive modeling. The content concisely describes hard nanomaterials, a general idea of modeling and simulation with some essential strategies, and insights into leveraging computational tools for designing nanostructured hard materials with improved properties, offering a valuable resource for researchers and practitioners in materials science and engineering as well as those interested in computational modeling and simulation.

Keywords: Modeling and simulation, simulation strategies, computational chemistry, popular software, nanostructured hard materials, materials science impact

**Corresponding author*: sasadharmajhi@tdbcollege.ac.in

Inamuddin, Tariq Altalhi and Mohammad Abu Jafar Mazumder (eds.) Nanostructured Hard Materials: Industrial Applications, (1–38) © 2026 Scrivener Publishing LLC

1.1 Introduction

The study of hard nanostructured materials (HNM) has been a developing field for more than five decades after its first discovery in the 80's [1–3]. Nanomaterials are defined as substances possessing at least one dimension within the nanoscale range, usually ranging from 1 to 100 nanometers. They showcase distinctive characteristics such as innovative optical attributes, enhanced reactivity, and mechanical properties compared to their larger counterparts. These distinctive features arise from their diminutive size and elevated surface-to-volume ratio. A captivating field within the realm of chemistry is computational chemistry. This discipline harnesses computer simulations to address diverse chemical challenges, employing methodologies that play a crucial role in determining the structures and properties of intricate molecules and solids [4]. Computational modeling, a significant tool, assists in analyzing, processing, and managing the data to extract valuable information, and it is employed to study different physiochemical properties of nanomaterials containing shape, size, structure, surface, and porosity [5–8]. Molecular modeling, simulations, and computational approaches are handy tools in chemical engineering and the community as they can predict and explain experimentally noticed macromolecular structure, thermodynamics, and microscopic and macroscopic material properties [9, 10]. Investigative tools were furnished by theory, modeling, and simulation, which support nanotechnology [11]. Importantly, computational techniques are frequently applied to validate hypotheses that may not be available through classical experimentation.

In contemporary times, the incorporation of modeling and simulation holds substantial importance in the exploration of HNMs in interpreting experimental data, elucidating design principles, and making innovative property predictions for unfamiliar nanomaterial structures and dynamics, encompassing their interfaces across the quantum scale to the microscale. Thus, the theory and simulation approaches are now pivotal in chemistry, physics, and biology. However, to get a realistic and scientific understanding of the properties of the system without any misinterpretation, we must have good knowledge of designing simulations in studies of HNMs. For example, we must care about the design of a system at the nanoscale, choosing the most appropriate modeling and simulation method, using state-of-the-art parameters and accuracy, and accumulating enough statistics. In addition to these scientific aspects, it is also essential to consider the similarity or exact match of experimental findings and the reproducibility of the simulations. In the present book chapter, we have tried to describe a general introduction to hard nanomaterials, a basic idea of modeling and simulation with some essential strategies, and insights into leveraging computational tools for designing nanostructured hard materials with improved properties, offering a valuable resource for researchers and practitioners in materials science and engineering.

1.2 Nanostructured Hard Materials

Nanostructured materials involve a broad spectrum of materials engineered at the nanoscale, possessing unique properties and functionalities. At present, a significant portion of nanomaterials, such as transition metal chalcogenides, phosphides, nitrides, carbides, and boron nitride nanosheets, which fall under the category of two-dimensional nanomaterials, have garnered considerable interest due to their diverse appealing properties and applications, particularly in the fields of energy storage and conversion [12, 13]. However, the materials science community shows little interest in the transition metal borides, since nanomaterials are challenging to prepare [14]. Metal borides are applied in the industry, including superconductors (MgB_2), hard magnets ($Nd_2Fe_{14}B$), and hard materials (TiB_2) [15]. Conventional methods for synthesizing metal borides typically involve elevated temperatures and/or high pressures, requiring the combination of metal with boron, often achieved through arc melting at temperatures exceeding 2500°C. Research has explored alternative approaches for the low-temperature synthesis of nanocrystalline metal borides, specifically employing ion thermal flux synthesis in molten salts at temperatures below 1000°C [16]. Recent research invented a procedure for synthesizing nanocrystalline transition metal borides organized according to the Sn/$SnCl_2$ redox chemistry [14, 15].

HNMs are typically classified into several categories based on their composition. Here, we consider broad categories concerning HNMs.

1.2.1 Metal Nanostructure

Metal nanostructures emerge from diverse noble metals such as Au, Ag, Pt, Pd, and combinations thereof, along with metal oxides, exhibiting a size range of 1–100 nm. These nanostructures manifest in various shapes, including spheres, nanorods, nano cubes, nanospheres, and nanowires [17–19] and with different morphology (0D, 1D, 2D, and 3D). Due to their size-dependent behavior, these structures display exceptional properties and find applications in various fields such as electronics, catalysis, optics, and biomedical engineering. Surface plasmons are collective and coherent electron oscillations confined at the dielectric–metal interface, inducing optical properties hardly achievable in other optical materials. Tuning the aspect ratios of anisotropic nanocrystals is beneficial for specific applications.

1.2.2 Semiconductor Nanostructure

Nanostructures in the semiconductor realm, typically within the size range of 1-100 nanometers, exhibit distinct electronic transitions akin to isolated atoms and molecules, making them valuable for diverse applications, including solar energy conversion, optoelectronic devices, imaging, and detection [20]. Semiconductor nanostructures, including quantum dots, nanowires, and nanocrystals, have unique properties different from bulk due to quantum confinement effects that arise when the nanostructure size is comparable to the de Broglie wavelength of the charge carriers [21–23]. Contemporary comprehension of semiconductors is rooted in quantum physics, elucidating the dynamics of electrons and holes within a crystal structure and lattice. The quantum confinement effects observed in semiconductor nanostructures result in size-dependent optical and electronic properties, specifically size-dependent absorption and emission featuring discrete electronic transitions. In contrast to solid-state crystals and small-molecule chromophores and fluorophores, semiconductor nanostructures showcase distinctive optical properties influenced by their structure, composition, and size, proving beneficial for the development of photovoltaic and light-emitting devices. Additionally, the deliberate introduction of impurities or the application of electrical fields or light can modify the electrical properties of semiconductor materials, rendering them suitable for applications such as amplification, energy conversion, and switching [21, 23]. Intrinsic semiconductors are very pure semiconductor materials, i.e., having an equal number of holes and electrons, whereas, in extrinsic semiconductors, minute quantities of impurities (with trivalent or pentavalent dopant to obtain p-type or n-type, respectively) are doped to manipulate their conductivity and properties.

1.2.3 Magnetic Nanostructure

Magnetic nanostructures represent a captivating domain at the forefront of material science with their unique properties and transformative potential [24–26]. At the nanoscale, materials with magnetic properties showcase an array of behaviors that differ significantly from their bulk counterparts. From nanoparticles and nanowires to intricate thin films and composite structures, these nanostructures exhibit fascinating magnetic phenomena, often shaped by their size, shape, and arrangement. Magnetic nanostructures find applications in various domains, including data storage, biomedical applications (like magnetic imaging and targeted drug delivery), sensing technologies, spintronics, and magnetic refrigeration, owing to their unique magnetic characteristics. Through precise engineering and manipulation of size, shape, composition, and orientation, magnetic properties like coercivity, remanence, and magnetic susceptibility can be tailored, fine-tuning material behavior for specific applications. The exploration of nanomagnetism primarily focuses on creating innovative magnetic nanostructures and unravelling the connections between structure and magnetic properties. The synthesis of magnetic nanostructures involves the fabrication of nanoparticles, nanoclusters, nanowires, multilayers, and thin films.

Hard magnetic nanostructured materials, e.g., FePt, Nd-Fe-B, SmCo, have received more attention presently since hard magnetic nanostructures are applied for recording media, bonded magnet construction, and microelectromechanical system utilization [27]. Magnetism, whether attractive or repulsive, typically arises from the movement of electrons within the atoms of a substance. The magnetic moment associated with an electron is derived from its angular momentum. Following the Bohr model, the orbital angular momentum of an electron is quantized, occurring in integral multiples of $h/2\pi$. The magnetic moment (μ) is calculated based on the current and area of the loop resulting from the electron's orbit around the nucleus within the atom. The concept of the Bohr magneton ($\mu_B = eh/4\pi m_e = 9.27 \times 10^{-24}\ JT^{-1}$)embodies the notion from the earlier quantum theory that an electron in its ground state possesses non-zero angular momentum. Hard magnetic nanostructures contain high magnetic anisotropy, while low anisotropy is the critical feature of soft magnetic nanostructures; magnetic anisotropy illustrates how direction influences the magnetic property of a material [28]. Shape anisotropy and crystalline anisotropy are the two main components of magnetic anisotropy. Various attractive magnetic phenomena were exhibited by magnetic nanostructures, including extra anisotropy contributions. Besides, magnetic nanoparticles exhibit high saturation field, high field irreversibility, and superparamagnetism, whereas multilayer thin films and nanorods show high saturation magnetization. Studies have disclosed that single-domain superparamagnetic nanoparticles are crucial in new biomedical uses, including targeted drug delivery and gene delivery [29]. Ferrofluids (magnetic fluids), organic liquid or water-based colloidal suspensions of ferromagnetic nanoparticles, including magnetite (Fe_3O_4) and Co-ferrite ($CoFe_2O_4$), can be applied in microfluidic devices [30]. Molecular nanomagnets, also called single-molecule magnets, which are intermediate between classical and quantum magnets, have received more attention due to their exciting merit, including the vital crowd toward quantum computers [31].

1.2.4 Carbon-Based Nanostructure

Carbon-based nanostructures from various carbon allotropes are materials of carbon atoms arranged in various configurations on the nanometer scale, with unique properties that are useful for electronics, optoelectronics, energy storage, electrochemical applications, and a diverse range of biological specialties [21, 32, 33]. Common carbon-based are carbon nanotubes (CNT), carbon nanofibers, carbon nanohorns, fullerenes, nanodiamonds (ND), graphene, graphene quantum dot (GQD), graphene oxide (GO), and carbon quantum dots. These carbon-based nanostructures can be arranged in 0D such as carbon nanodots, GQD, 1D such as CNTs, 2D such as graphene, and 3D such as nanodiamond materials. Fullerenes (C_{60}, C_{70}, C_{76}, C_{82}, and C_{84}) (0D nanostructure), a carbon nano allotrope with hybridization between sp^2 and sp^3, have a soccer ball-shaped molecular structure of carbon atoms arranged in a spherical shape with unique electronic properties making it useful in electronics and energy storage. Carbon nanodots (0D), a new class of carbon-based nanostructures, are quasi-spherical nanoparticles with a 2–10 nm diameter, comprising almost sp^3-hybridized mostly amorphous carbon. Carbon nanodots' size, shape, and structural arrangements yield unique optical properties, such as strong photoluminescence, good photostability, relatively low toxicity, and good biocompatibility, making them useful in bioimaging, sensing, and optoelectronics. Crystalline GQDs are comprised of sp^2-hybridized carbon atoms and show nonzero band gaps (can be easily altered by dimension, morphology, geometry, and edges), promising for electronic, optical, and electrochemical applications. Carbon nanotubes (CNTs) are cylindrical nanostructures formed by rolling up single-layered graphene sheets, with carbon atoms arranged in a hexagonal lattice. Depending on the number of graphene layers, CNTs can be categorized as either single-walled (SWNTs) or multi-walled (MWNTs). They exhibit high electrical conductivity, mechanical strength, and thermal stability, rendering them valuable in diverse electronic and energy storage applications. Another noteworthy carbon-based nanostructure is graphene (2D), consisting of a single-layer arrangement of carbon atoms in a 2D honeycomb lattice. This structure imparts exceptional electronic characteristics, attributed to the quantum hall effect and massless Dirac fermions. 3D-ND, mainly composed of sp^3 carbon, has recently attracted much attention in various fields, like bio-electrochemistry, electroanalytical chemistry, and electrocatalysis. With progress on these frontages of carbon-based nanostructures, carbon nanomaterials are precisely optimistic in numerous industrial, biomedical, and environmental applications.

1.2.5 Alloy

Alloy nanostructures, combining different metallic elements at the nanoscale with enhanced mechanical, electrical, magnetic, and catalytic properties compared to their bulk counterparts, represent a fascinating realm of materials science, having unique properties and functionalities [34–36]. The size, composition, and arrangement of atoms in alloy nanostructures significantly impact their properties, which can be tailored through controlled synthesis of these structures. Engineers and scientists can alter their properties for specific applications with increased strength, improved corrosion resistance, superior conductivity, or catalytic efficiency. Nowadays, their applications extend to various fields, from biomedical devices and electronics to energy storage and catalysis, promising them for next-generation technology advancements in areas such as sensors, drug delivery systems, high-performance computing, and renewable energy devices. However, the precise control and fabrication of alloy nanostructures face challenges in consistently achieving desired properties. Despite these challenges, ongoing research and advancements continue to unlock the potential of alloy nanostructures, driving innovation in materials science and engineering. Controlling defects designed stronger crystalline alloys usually to retard dislocation motion. Interface, line, point, and volume defects are types of these defects [37]. Materials gain strength through various mechanisms such as grain boundary, solid solution, precipitate, and dislocation. The essential attributes of structural metallic materials often include both high ductility and high strength, a combination particularly crucial for industrially prevalent aluminum-based alloys [38]. A novel hierarchical nanostructured aluminum alloy has recently been developed. This alloy consists of aluminum nanograins surrounded by nano-sized metallic glass shells. It exhibits an exceptional ultrahigh yield strength of 1.7 GPa under compression of 1.2 GPa. Additionally, it demonstrates remarkable plasticity, with plasticity higher than 70% under compression and plasticity of 15% under tension [38].

1.2.6 Composites

Nanocomposites of nanostructures represent an innovative class of materials with tailored properties, offering immense potential for advancing technology and addressing challenges in various industries by harnessing the unique characteristics of nanomaterials within composite structures [39–41]. Nanocomposites refer to materials with two or more distinct phases, where at least one phase must be at the nanometer scale and exhibit enhanced or novel properties compared to its components. For example, incorporating nanoparticles into a polymer matrix can improve mechanical strength, stiffness, electrical conductivity, thermal properties, optical properties, and chemical reactivity beyond what either component could achieve alone. Another example is that carbon nanotube composites in aerospace can offer lightweight yet strong materials for aircraft components. Advances in manufacturing techniques and material engineering, by predicting unique properties by computational technique continually improve the quality and scalability of nanocomposites.

1.3 Modeling and Simulation of Nanostructured Hard Materials

1.3.1 Importance of Modeling and Simulation

Modeling and simulation have turned out to be essential tools in nanotechnology, particularly for studying the behavior of HNMs. The exploration of highly desirable applications in electronics, catalysis, and medicine is facilitated by the dynamic role of modeling and simulation in the study of HNMs, which possess distinct properties and behaviors deviating from their bulk counterparts. One of the critical motivations why modeling and simulation are essential for HNMs is that experimental characterization techniques often struggle to thoroughly understand their structure and behavior at the atomic level due to having complex shapes, defects, and surface structures. However, computational models and simulations provide insights into the atomic arrangement, stability, and dynamics of HNMs, helping to bridge the gap between theory and experiment. Furthermore, modeling and simulation allow scientists to explore the properties of HNMs under different conditions that may be difficult or impossible to achieve experimentally. Therefore, knowledge about the HNMs with numerous conditions is crucial for designing and optimizing HNMs for specific applications, as it enables researchers to predict and understand how these materials will perform under real-world conditions. Modeling and simulation deliver a cost-effective and time-efficient solution to screen and evaluate a wide range of HNM compositions, sizes, and shapes instead of synthesizing and characterizing numerous samples, predicting the properties, and narrowing down the most promising candidates for further experimental investigation. In general, the essential tools for studying HNMs are modeling and simulation, providing a comprehensive grasp of structure-property relationships. These tools enable the exploration of HNMs under diverse conditions, facilitating the design and optimization of materials for practical applications. As nanotechnology advances, the importance of modeling and simulation in this field will only be nurtured, contributing to the development of state-of-the-art HNM-based technologies and devices.

1.3.2 Basic Categories of Modeling and Simulation

Modeling and simulating NHMs is vital for understanding their behaviors and designing innovative applications. There are four primary modeling and simulation techniques categories: atomistic simulation, continuum models, quantum mechanical simulation, and multiscale modeling. Each category has its strengths and limitations, inducing the choice of approach based on the specific properties and phenomena in nanostructures. However, integrating these models offers a more comprehensive understanding of nanostructure.

Atomistic Simulations

The atomistic models focus on individual atoms and corresponding interactions to simulate nanostructures to study materials' behavior, properties, and interactions. Molecular dynamics (MD), Monte Carlo simulations, and *ab initio* simulations are the most common atomistic modeling. MD tracks atomic motion over time, revealing dynamic behaviors, while Monte Carlo simulates probabilistic events, providing statistical insights into nanostructure properties. *Ab initio* simulations engage quantum mechanical principles without empirical parameters to estimate accurate insights into materials' electronic structures, energy levels, and chemical properties. Therefore, these simulations provide insights into materials' dynamics, structure, and thermodynamics.

Continuum Models

Continuum models are mathematical and computational approaches to describe the behavior of materials or nanostructures, considering them as continuous media and treating them as having uniform properties across space. The equations of continuum mechanics describe the phenomena at larger scales. Finite element analysis (FEA) and continuum mechanics methods fall under this category, suitable for studying macroscopic behavior or collective properties. Though continuum models provide valuable insights into macroscopic behaviors, continuum models might not capture atomic-scale phenomena in materials. Consequently, these models are often used alongside atomistic or other higher-resolution models to understand materials and their behaviors across different scales comprehensively.

Quantum Mechanical (QM) Simulations

QM simulations for nanostructures consider the quantum nature of nanostructures essential for accurately describing phenomena at the nanoscale or atomic scale, empowering the development of nanoelectronics, nanomedicine, energy materials, and more. These simulations provide a deep understanding of quantum effects to reveal the discrete nature of matter and interactions at atomic and molecular levels, controlling the design and optimization of nanostructured materials with tailored functionalities for diverse applications. Density functional theory (DFT) and tight-binding models are examples. Critical aspects of QM simulations for nanostructures include: (i) *Accuracy in the atomic-scale picture* - HNMs exhibit unique properties influenced by their size, shape, and surface properties. QM simulations completely understand quantum confinement, surface effects, and size-dependent properties; (ii) *DFT for HNMs* - DFT calculates electronic properties, providing insights into electronic structure, energy levels, and charge distributions. This method is precious for predicting and understanding the properties of nanoparticles, nanowires, and quantum dots; (iii) *Designing novel NHMs* - QM simulations guide the design of novel nanomaterials with tailored properties *via* external factors, such as doping, surface functionalization, or changing the size and shape, which impact their electronic, optical, and catalytic properties. Simulating HNMs at the quantum level is computationally demanding due to the complexity and number of particles involved that determine the challenge, approximations, accuracy, and computational efficiency.

Multiscale Modeling

HNMs exhibit behaviors across various length and time scales, from the atomic and molecular level to the mesoscale and beyond. Multiscale modeling is a powerful approach, and it integrates different modeling approaches, allowing a comprehensive understanding of nanostructures and capturing both their atomic-level details and macroscopic behaviors within a unified framework. The main advantage of multiscale modeling is that computational resources can be optimized by employing higher-resolution models where necessary using coarse-grained models at larger scales, and it produces a balance between accuracy and computational efficiency and enables simulations of larger systems for longer timescales. Understanding the emergent properties of NHMs can be achieved by studying the collective behavior of atoms and molecules. Therefore, it helps predict and optimize the performance of nanomaterials for specific applications in fields like electronics, medicine, and energy. In recent years, advances in computational power, algorithms, and simulation techniques have continually enhanced multiscale modeling capabilities for nanostructures.

1.3.3 Simulation Strategies

Researchers must choose a perfect simulation technique or combination of various techniques for their particular interest in the material and characteristics needed over an overwhelming number of simulation techniques and software. However, one must consider the critical considerations for selecting a simulation method - properties of interest, length-scales, and timescales involved, the level of programming versus graphical user interfaces, and computational resource requirements.

Quantum Mechanics - Approximations and Approaches

The principles of quantum mechanics govern the behavior of particles at the atomic scales, making it vital in understanding and simulating HNMs. However, exact quantum calculations for large nanostructures are often computationally unaffordable. Hence, various approximations and approaches are imposed for efficient simulations and reliable outcomes representing the experimental findings. Each method/approximation has strengths and limitations concerning computational efficiency, accuracy, and the specific aspects of nanostructures they address. Researchers often combine multiple approaches or develop hybrid methods to overcome limitations and obtain a more comprehensive understanding of nanostructured materials and their properties. As computational power advances, so does the potential for more accurate simulations of complex nanostructures using quantum mechanics.

Born–Oppenheimer (BO) approximation

The BO approximation, a renowned concept in molecular dynamics and named after its original contributors, Max Born and Robert Oppenheimer from the early era of quantum mechanics, aids in elucidating the wave functions of electrons and nuclei within a molecule. This approximation streamlines the resolution of the Schrödinger equation by assuming that it is feasible to separate the electronic and nuclear motions in molecules. In other words, the uncertainty regarding the positions of the nuclei can be disregarded when deducing the wave function for the electrons. This approximation is justified by the significant difference in mass between electrons and nuclei, allowing the assumption that nuclei are nearly stationary relative to electron motion. According to the BO approximation, the wavefunction of a molecule is considered the product of electronic and nuclear wavefunctions. This approximation is based on two key assumptions: (i) the electronic wavefunction depends on the nuclear positions but not on their velocities, and (ii) the motion of nuclei (such as rotation or vibration) experiences a smeared potential from the fast electrons, with electrons reacting instantaneously to any changes in nuclei. The assumption of fixed nuclear positions relative to electrons significantly simplifies the scenario. Disregarding nuclear motion makes it impractical for exploring molecular dynamics beyond the ground state. However, the BO approximation proves valuable in quantum chemistry and molecular spectroscopy, facilitating faster computation of molecular wavefunctions and other properties, particularly for large molecules. In elementary mathematics, the nonrelativistic Hamiltonian (H) for a molecule can be expressed as the sum of five terms. These include the kinetic energy of the nuclei ($T^N(\vec{R})$) and electrons ($T^e(\vec{r})$), the potential energy arising from electron-electron repulsions ($V^e(\vec{r})$), the potential energy resulting from nucleus-nucleus repulsions ($V^N(\vec{R})$), and the potential energy associated with the attraction of electrons by the nuclei, summed over all electrons and nuclei ($V^{N-e}(\vec{R},\vec{r})$).

$$H = T^e(\vec{r}) + T^N(\vec{R}) + V^e(\vec{r}) + V^N(\vec{R}) + V^{N-e}(\vec{R},\vec{r})$$

The BO approximation allows for the independent calculation of the electron and nuclei Hamiltonians. Initially, an electronic Hamiltonian ($H^{electron}$) can be employed in the Schrödinger equation to depict the motion of electrons, disregarding the kinetic energy term for the nuclei.

$$H^{electron} \cdot \Psi^e(\vec{R},\vec{r}) = E^f(\vec{R}) \cdot \Psi^e(\vec{R},\vec{r})$$

The mentioned equation yields the effective nuclear potential function ($E^f(\vec{R})$), depicting the potential energy surface for the system. The effective potential for the nuclear Hamiltonian, $H^N = T^N(\vec{R}) + E^f(\vec{R})$, can be determined. This Hamiltonian is then applicable in the Schrödinger equation for nuclear motion, capturing the vibrational, rotational, and translational states of nuclei and enabling predictions of the vibrational spectra of molecules.

Hartree–Fock (HF) approximation

The HF approximation is a foundational technique in quantum mechanics employed to approximate the behavior of a many-body quantum system, with a particular focus on comprehending the electronic structure of atoms, molcculcs, and solids [42]. The HF approximation, also known as the molecular orbital approximation, plays a crucial role in quantum chemistry. It serves as an iterative computational method employed to solve the time-independent Schrödinger equation for many-body electronic systems, which are described by the many-body Hamiltonian. This approximation elucidates the motion of atomic and subatomic particles. The HF approximation seeks to determine optimal single-electron wavefunctions (orbitals) that characterize the electron behavior in a many-electron system. It assumes that each electron moves independently within an average field generated by all other electrons, without directly considering their interactions. This simplification makes the problem mathematically tractable. However, HF struggles when the correlation energy is in a strong regime. HF method involves three iterative approaches - (i) guessing the initial wavefunctions of the electrons, (ii) solving the HF equations to calculate an effective potential for each electron based on the average electron distribution from all other electrons, and (iii) adjusting the wavefunctions using the updated potential and iterating until a self-consistent solution is reached, where the wavefunctions no longer change significantly.

Thomas–Fermi theory

The Thomas-Fermi (TF) theory, also known as the 'statistical theory,' represents a semi-empirical and statistical mechanical model applied to elucidate the characteristics of many-electron systems. It specifically addresses systems where the electrons do not interact with each other, and their motion is considered random. This theory is often employed to study particle systems, such as electrons in a metal or atoms in a gas, when exposed to an external potential. The theory was developed independently by Llewellyn Thomas and Enrico Fermi in the 1920s and 1930s [43]. While Schrödinger's equation provides precise density and energy information, it becomes challenging to handle for large atoms or molecules with numerous electrons. A notable prediction from the Thomas-Fermi theory is the occurrence of the "Thomas-Fermi screening" effect. In this phenomenon, electrons within a metal or other conducting material respond to an external electric field by generating a "screening" charge that counteracts the applied field. This effect is important in understanding the behavior of metals and other conducting materials in various applications, including electrical circuits and electronic devices. The TF theory has found extensive application in the examination of diverse systems, encompassing metals, semiconductors, and various materials like superconductors and superfluids. This theory is adept at detailing the electronic structure of atoms and molecules. Notably, the TF theory has proven successful in predicting various properties of atoms and solids, including ionization energies, atomic radii, and bulk properties such as the equation of state. However, it fails to accurately describe certain phenomena that require a more detailed quantum mechanical treatment, such as fine structure splitting and chemical bonding.

A starting point for the TF theory is the expression of TF energy functional (E_ρ) for a molecule with K nuclei of charges ($Z_i > 0$) at fixed locations, $\mathbb{R}_i \in \mathbb{R}^3 (i = 1, \ldots., K)$.

The TF theory commences with the formulation of the TF energy functional (E_ρ) for a molecule featuring K nuclei with positive charges ($Z_i > 0$) positioned at fixed locations, denoted as, $\mathbb{R}_i \in \mathbb{R}^3 (i = 1, \ldots., K)$ and it can be expressed as

$$E_\rho = \frac{3\hbar^2\gamma}{10m}(3\pi^2)^{2/3}\int \rho(\boldsymbol{x})^{5/3} d^3\boldsymbol{x} - \int V(\boldsymbol{x})\rho(\boldsymbol{x})d^3\boldsymbol{x} + \frac{1}{2}\iint \frac{\rho(\boldsymbol{x})\rho(\boldsymbol{y})}{|\boldsymbol{x}-\boldsymbol{y}|} d^3\boldsymbol{x}d^3\boldsymbol{y} + U$$

Where, $\gamma = (3\pi^2)^{2/3}\hbar^2/2m$, $\rho(x) \geq 0$ and $\int \rho = N$. Here, $V(x)$ (Coulomb potential) and U (the nuclear-nuclear repulsion) can be expressed as

$$V(\boldsymbol{x}) = \sum_{j=1}^{K} Z_j \left|\boldsymbol{x} - \boldsymbol{R}_j\right|^{-1}$$

$$U = \sum_{1 \leq i < j!l < K} Z_i Z_j \left|\boldsymbol{R}_i - \boldsymbol{R}_j\right|^{-1}$$

In the given TF functional equation, the initial term accounts for the minimum quantum-mechanical kinetic energy required by N fermionic electrons to generate an electron density $\rho(\boldsymbol{x})$. The subsequent term involves the attractive interaction, mediated by the Coulomb potential (V), between the N electrons and the K nuclei. The third term serves as an approximate representation of the electron-electron repulsive energy.

The TF theory accurately yields the electron density in an actual atom as $Z \rightarrow \infty$, with the Thomas-Fermi density aligning with the observed quantum mechanical density.

Hohenberg–Kohn theorems

The Hohenberg-Kohn theorems represent foundational principles within the realm of DFT, a cornerstone of quantum mechanics employed to elucidate the electronic structure of atoms, molecules, and solids. These theorems, formulated by Pierre Hohenberg and Walter Kohn in 1964, play a crucial role in comprehending the connections between electron density and the electronic properties of a system [44]. There are two key theorems within the Hohenberg-Kohn framework [42]: (i) The first Hohenberg-Kohn theorem posits that the external potential of a quantum system is uniquely determined by its ground-state electron density. In other words, the ground-state electron density encapsulates all the information necessary to deduce the external potential confining the electrons in the system. This theorem guarantees a unique correspondence between the potential and the density, ensuring that distinct external potentials cannot produce an identical ground-state density. (ii) The second Hohenberg-Kohn theorem asserts the existence of a variational principle for the ground-state energy of a system. It states that the true ground-state energy is the minimum of a functional dependent on the electron density. While the precise form of this function, often referred to as the "universal function," remains unknown, the theorem guarantees that the true ground-state energy can be determined by minimizing this function with respect to the electron density. These theorems laid the groundwork for DFT, providing the theoretical foundation for understanding the quantum mechanical behavior of electrons in a system without requiring explicit solutions to the many-body Schrödinger equation, which becomes computationally infeasible for larger systems. While these theorems elucidate the principles underpinning DFT, the practical implementation relies on approximations for the exchange-correlation function, representing the interactions between electrons. Nevertheless, the Hohenberg-Kohn theorems remain a pivotal contribution to quantum mechanics, enabling efficient and accurate computational studies of the electronic structure and properties of diverse materials at the atomic and molecular scales.

Kohn–Sham formalism

The Kohn-Sham equations represent a set of mathematical expressions, specifically the noninteracting Schrödinger equation or a Schrödinger-like equation describing a fictitious system known as the "Kohn–Sham system."

These equations are applied in physics and quantum chemistry, particularly within DFT, to simplify the intricate task of understanding the behavior of electrons in atoms and molecules. Widely utilized in atomic and condensed matter physics, as well as quantum chemistry, the Kohn-Sham formalism is a prominent method for determining electronic structure. To grasp this formalism, one can consider the foundation laid by the two theorems of Hohenberg and Kohn (HK), which underpin the Kohn-Sham density functional theory (KS-DFT). The core principle of DFT posits that the precise calculation of ground and excited state properties in the presence of an external field, $\boldsymbol{F}_{ext}(\boldsymbol{r}) = -\nabla v(\boldsymbol{r})$, for a many-electron system can be achieved from the ground state density, $\rho(\boldsymbol{r})$. In KS-DFT, the energy is expressed as a function of the ground state density, $\rho(\boldsymbol{r})$, capturing a system where the kinetic energy of noninteracting Fermions is precisely treated. This approach fundamentally diverges from the field description in quantal density functional theory. The first HK theorem asserts that the external potential energy, $v(\boldsymbol{r})$, is uniquely determined by the ground state density $\rho(\boldsymbol{r})$. The second theorem states that the ground state energy functional, $E(\rho)$, can determine $\rho(\boldsymbol{r})$ alone from the variational principle.

Molecular Dynamics (MD)

MD simulation stands as a potent computational technique employed to investigate the temporal evolution, behavior, and interactions of atoms and molecules, grounded in the principles of Newtonian mechanics [45, 46]. Offering a comprehensive comprehension of the dynamic attributes of systems on an atomic scale, MD simulation enables scientists to explore diverse phenomena like chemical reactions, protein folding, and material properties. These simulations furnish details about time evolution and kinematic properties, providing valuable insights into the structure, thermodynamics, and kinetics of intricate systems by numerically solving equations of motion that may be challenging to observe experimentally. The particles interact with each other through forces determined by a potential energy function, which describes the potential energy surface of the system. In a MD simulation, the positions and velocities of the atoms and molecules are calculated using the forces that act between them at each time step. Typically commencing with an initial configuration of the system, the simulation involves the numerical solution of equations of motion to generate a trajectory of the system over time. MD simulations have become essential in many fields, including chemistry, physics, materials science, and biology. MD simulation is beneficial for studying the behavior of complex systems, such as proteins, lipids, and polymers, where experimental techniques may be difficult or impossible to apply. Some of the critical applications of molecular dynamics simulations include (i) the three-dimensional structure prediction of molecules and biological macromolecules, (ii) the accurate prediction of thermodynamic properties of molecules and materials, particularly effective in sampling configurational space, (iii) effective for the study of the dynamics and conformational changes of molecules and biological macromolecules, such as the folding of proteins and the binding of ligands to proteins, (iv) study the properties of materials, such as their mechanical properties, thermal conductivity, and diffusion coefficients, (v) study the binding of drugs to their target proteins and to design new drugs with desired properties.

The MD simulation facilitate the comprehension and observation of various fundamental processes by replicating the continuous motion of atoms driven by interatomic forces, represented as the negative derivative of potential energy (U) concerning distance. The principles of molecular mechanics provide insight into potential energy and, by extension, the system's potential energy. The primary form of potential energy in a force field is represented by the following equation [46]:

$$U = E_{bonded} + E_{non-bonded} = E_{bonds} + E_{angles} + E_{torsions} + E_{vdW} + E_{Coulomb}$$

The initial three terms account for the impacts of bonds, angles, and torsions among covalently linked atoms, ensuring a chemically valid structure. The last two terms pertain to nonbonding interactions, specifically van der Waals interatomic forces (e.g., the Lennard-Jones potential function) and Coulomb's long-range electrostatic forces. The motion of atoms is determined by Newton's second law, as expressed in the following equation.

$$F_i = m_i \cdot \frac{d^2 r_i(t)}{dt^2} = \frac{\partial U(r)}{\partial r_i}$$

Where, $r_i(t)$ is the position vector of the ith particle at time t with a mass of m_i and an applied external force of F_i. Nevertheless, the selection of a force field constitutes a crucial determinant of the simulation results. Ongoing advancements in computational capabilities enable simulations to be conducted at shorter time scales, specifically up to a few femtoseconds. Out of numerous computational approaches, MD simulation proves to be an efficient method for examining spatiotemporal structural alterations in material systems on atomic length scales and within femtosecond time frames [45].

Density Functional Theory (DFT)

DFT stands as a potent computational approach employed in quantum mechanics and condensed matter physics to explore the electronic structure of atoms, molecules, and solids. In contrast to the Hartree-Fock method, which emphasizes wavefunctions, DFT aims to characterize the ground state properties of a system by focusing on the electron density instead of the many-body wavefunction. At the heart of DFT lies the Hohenberg-Kohn theorem, declaring that the ground-state electron density uniquely determines both the system's ground-state wavefunction and energy. DFT simplifies the intricate many-body problem by utilizing an auxiliary non-interacting system of fictitious particles (Kohn-Sham particles) with an equivalent density to that of the real system. Solving the Kohn-Sham equations allows for determining the ground-state properties by treating the exchange-correlation effects as an effective potential. In DFT, the crucial challenge is describing the exchange and correlation interactions between electrons, usually done through an exchange-correlation function, representing the effects beyond the classical Coulomb interactions.

The overall electronic energy can be expressed through the following equation, with each term contingent on the total electron density, *n*.

$$E_{total} = E_T + E_V + E_J + E_X + E_C$$

The kinetic energy (E_T) of the electrons, the potential energy (E_V) arising from the electron-nuclear attraction due to Coulombic forces with the nuclear center, E_J signifies the potential energy attributed to electron-electron repulsion arising from the average Coulomb repulsion between pairs of electrons. This term does not account for the correlated motion of electrons. Additionally, E_X stands for the exchange energy of the electrons, and E_C denotes the total correlation energy of the electrons. The Kohn–Sham equation constitutes the Schrödinger equation for a system of noninteracting electrons that collectively generate the same density as the original system. As a result, it is capable of calculating the kinetic energy of each electron in terms of Kohn–Sham orbitals, akin to the methodology employed in Hartree-Fock theory. Various mathematical functions can be employed to express the electron density. For instance, the Gaussian basis function is capable of representing the electron density, particularly when considering a single atom-centered scenario. This is often applied in original DFT calculations that utilize plane-wave functions as basis sets.

The primary advantages of DFT are (i) efficiency, i.e., compared to many-body methods, DFT is more computationally feasible for larger systems;

(ii) accuracy, i.e., DFT provides reasonable accuracy for predicting various properties like total energies and electronic structures, and some molecular properties, and (iii) applicability. Due to its capacity to strike a balance between accuracy and computational cost, DFT finds extensive application in fundamental domains of physics, chemistry, and materials science. It is commonly used for predicting molecular structures, elucidating reaction mechanisms, and determining the electronic and optical properties of materials, among other applications. Typically, DFT is perfect for ground-state properties (remember, it is a theory of the ground-state), including cohesive energies, equilibrium crystal structures, elastic constants, charge densities, vibrational properties, and qualitative (most of the time quantitative) prediction of band topology.

Monte Carlo (MC)

The MC simulation, also known as the MC Method or a multiple probability simulation, is a mathematical technique used to estimate the potential outcomes of uncertain events through probabilistic sampling. This computational method, commonly employed in various fields such as nanostructure simulation, is a powerful tool for modeling and analyzing complex systems based on probabilistic sampling. MC simulation is a stochastic method that relies on repeated random experiments. The calculated results are statistical averages of all the random and repeated samplings whose variance strongly depends on the number of samplings. In the context of nanostructures, MC methods are valuable for studying the behavior and properties of materials at the atomic or molecular level. MC methods use statistical sampling to estimate properties such as thermodynamic quantities (e.g., free energy, entropy), which are essential for understanding the behavior of nanostructures. MC simulations can model the structure and properties of NHMs by simulating the positions and interactions of atoms or molecules. These are particularly useful in understanding the behavior of materials at nanoscale dimensions, where quantum effects and surface phenomena play significant roles. MC methods can explore the energy landscape of a nanostructure by randomly moving atoms or molecules and accepting or rejecting these moves based on energy changes, finding low-energy configurations, or optimizing the structure of NHMs. MC simulations can study nanostructure phase transitions and phase diagrams under different conditions (e.g., temperature, pressure) to provide insights into how materials transition between different phases at the nanoscale. MC methods can investigate the effects of defects, surface roughness, or impurities on mechanical, electronic, or optical properties. Specialized variants like Quantum Monte Carlo (QMC) are used to handle the quantum nature of particles in nanostructures more accurately. QMC methods can provide highly accurate solutions for the electronic structure of materials. MC simulations are computationally intensive, especially for complex systems with numerous particles. Still, they offer valuable insights into the behavior and properties of nanostructures, complementing experimental investigations. Their ability to simulate atomic-level details makes them essential tools for understanding and designing nanoscale materials.

First-principle methods

First-principles methods in nanostructure simulations refer to computational techniques that directly solve the quantum mechanical equations governing the behavior of atoms and molecules from fundamental physical principles without the need for empirical parameters. Here are some standard first-principles methods for nanostructure simulations – (i) *DFT*: As mentioned earlier, DFT is a widely used first-principles method that calculates the electronic structure of materials based on the electron density for studying the properties of nanostructures, such as their electronic, magnetic, and structural properties, (ii) *Ab Initio Molecular Dynamics*: It combines DFT calculations with classical molecular dynamics simulations to study the nanostructures at finite temperatures, providing insights into their dynamical behavior, phase transitions, and reactions, (iii) *Wavefunction-based Methods*: These methods are computationally expensive and offer high accuracy by directly solving the Schrödinger equation for small systems using coupled cluster theory or configuration interaction like small nanostructures or benchmarking calculations, (iv) *Tight-Binding Approaches*: Tight-binding methods, not as computationally expensive as DFT, offer a compromise between accuracy and computational cost to simplified Hamiltonians to describe the electronic structure of nanostructures, (v) *Many-Body Perturbation Theory*: This method enhances the precision of Density Functional Theory (DFT) by incorporating various methods such as the GW approximation or Bethe-Salpeter equation. These approaches account for many-body effects, enabling the prediction of electronic excitations and optical properties with increased accuracy. Overall, these first-principles methods allow researchers to investigate various aspects of nanostructures, including electronic structure (e.g., band structures, the density of states, and charge distributions), structural optimization (i.e., determining the most stable geometries and predicting atomic arrangements), mechanical properties (e.g., elastic constants, mechanical stability, and deformation behavior), chemical reactivity (i.e., studying reactions on surfaces, adsorption energies, and catalytic properties), and optical and electronic properties (i.e., assessing optical transitions, energy levels, and conductivity). While they can be computationally demanding, advancements in algorithms and high-performance computing have expanded their applications in exploring nanomaterials, aiding in designing and developing innovative nanostructured materials for various technological applications.

Finite Element Method (FEM)

The FEM proves to be a robust technique for numerically addressing differential equations encountered in engineering and applied science. The fundamental concept of FEM involves dividing any system into numerous simple geometric elements, such as triangles or quadrilaterals, known as finite elements, through a specific spatial discretization. Following this, the concept of weighted residuals is utilized to estimate the solution within each finite element domain, ensuring the smooth continuity of dependent variables and their first derivatives as transition occurs one element to the next. This method enables the conversion of complex partial differential equations in challenging problems into sets of ordinary differential equations over time. It is particularly effective for systems characterized by irregular geometries and steep gradients.

Finite-difference time-domain (FDTD)

The FDTD method, also known as Yee's method, stands out as a straightforward full-wave technique for solving a diverse array of problems in electromagnetics and photonics. This approach, both in its conceptual framework and in its application across time and space domains, proves highly efficient in tackling various challenges. These challenges include phenomena like scattering from metallic objects and dielectrics, as well as addressing issues related to antennas, microstrip circuits, and electromagnetic absorption in materials spanning from the nanoscale to the microscale. FDTD serves as a numerical analysis technique specifically designed for modeling computational electrodynamics, providing approximate solutions for electromagnetics and photonics problems. With its time-domain capability, FDTD possesses the unique ability to cover a broad frequency range through a single simulation. Furthermore, FDTD can tackle frequency domain problems by leveraging Fourier transforms to calculate complex Poynting vectors and determine the transmission/reflection of light. Employing the solution of time-dependent Maxwell's curl equations in non-magnetic materials, FDTD proves to be a versatile approach in addressing a wide range of electromagnetic phenomena:

$$\frac{\partial \vec{D}}{\partial t} = \nabla \times \vec{H}$$

$$\vec{D}(\omega) = \varepsilon_0 \varepsilon_r(\omega) E(\omega)$$

$$\frac{\partial \vec{H}}{\partial t} = -\frac{1}{\mu_0} \nabla \times \vec{H}$$

Where, H, D, and E are the magnetic, displacement, and electric fields, respectively. The $\varepsilon_r(\omega)$ (complex relative dielectric constant), where $\varepsilon_r(\omega) = n^2$, and n signifies the refractive index. Through the FDTD technique, the computational analysis computes the electric (E) and magnetic (H) fields across the entire computational domain over time. This capability enables the modeling of a diverse range of linear and nonlinear dielectric and magnetic materials. Yee introduced the numerical translation into a time-stepping algorithm to solve vector electromagnetic field problems on a rectilinear grid. According to the above equations, it is visible that the electric field is curling around the magnetic flux, and the magnetic fields are curling around the electric flux, which is the basic idea of the Yee algorithm.

FDTD computational solutions for Maxwell's equations have gained popularity in material science owing to various advantages over alternative techniques. These include the absence of the need for matrix inversion, high accuracy, and robustness in addressing electromagnetic wave interaction problems. As a time-domain technique, FDTD eliminates the necessity for calculating structure-dependent Green functions. Additionally, it solely requires parallel-processing computer architectures and offers straightforward visualization of electromagnetic wave propagation through materials. Hundreds of simulation tools implement FDTD algorithms, e.g., ANSY's Lumerical, CST Studio Suite, Opti FDTD, OmniSim, etc. In their recent review, Teixeira *et al.*, provide a thorough description of the FDTD technique, addressing ongoing challenges and exploring opportunities to enhance its capabilities for current and future applications. The comprehensive discussion encompasses a wide range of technological areas, including optics, plasmonic, metamaterials, biomedicine, geophysics, and quantum applications [47].

Semi-Empirical Methods

Semi-empirical methods balance accuracy and computational cost, utilizing experimental data to parameterize theoretical models and thereby reducing the computational complexity with a reasonable level of accuracy for nanostructure simulations. These methods are beneficial for studying large systems that are computationally challenging compared to rigorous quantum mechanical methods. Researchers often choose semi-empirical methods based on the specific properties of interest, the size of the system being studied, and the balance between computational cost and desired accuracy for their simulations of nanostructures in various fields of nanoscience, such as modeling large molecular systems, understanding electronic properties of nanomaterials, exploring catalytic reactions on nanostructures, and predicting the behavior of complex nanoscale systems. Some prominent semi-empirical methods are usually employed in nanostructure simulations: (i) *Empirical Tight Binding (ETB)* - provides an approximation of the quantum mechanical behavior of electrons within a solid. This is achieved by accounting for interactions between atoms through a set of empirical parameters, which are adjusted to replicate experimental data or higher-level quantum mechanical calculations. ETB models prove to be computationally efficient and are widely employed for investigating electronic structures and properties of diverse nanostructures. (ii) *Extended Hückel Method* - built upon the Hückel molecular orbital theory, this approach incorporates supplementary terms to accommodate electron-electron interactions and overlap integrals between atomic orbitals. This extension strikes a balance between accuracy and computational cost, making it well-suited for studying the electronic structures of molecular systems and nanostructures. (iii) *Parameterized Tight-Binding Models* – these models describe the interactions between atoms in a material using adjustable parameters derived by fitting experimental data or higher-level quantum mechanical calculations and allow for simulations of more extensive systems than first-principles methods, making them valuable for studying nanomaterials' electronic, optical, and mechanical properties. (iv) *Semi-Empirical Quantum Chemistry Methods* - methods like PM3, AM1, and MNDO are semi-empirical quantum chemistry approaches used to predict molecular structures, energies, and properties of molecules and nanostructures and simplify the computational load using semi-empirical Hamiltonians and empirical parameters derived from experimental data.

1.4 The Outcome of Modeling and Simulation of Nanostructured Hard Materials

HNMs display distinctive characteristics attributed to their compact size and extensive surface area, rendering them highly promising for a multitude of applications. Understanding their growth mechanisms and structural properties is crucial for optimizing their synthesis and tailoring their characteristics. The following sub-sections delve into the outcomes of modeling and simulating nanostructured materials to study their growth, structure, and other optimizations, with some notable examples in recent years. It summarizes the outcomes of modeling and simulating HNMs, highlighting their significance in understanding growth mechanisms and structural properties, aiming to guide further research and applications in this field. Modeling and simulating HNMs for studying growth, structure, and other optimizations yielded invaluable insights into fundamental processes governing their formation and properties. The outcomes enhance our understanding and offer opportunities for designing novel materials with tailored properties for various applications. The synergy between computational simulations and experimental studies is pivotal in advancing the field of nanostructured materials. Continued research into advanced computational techniques and experimental validation should be encouraged. Additionally, interdisciplinary collaborations between materials scientists, physicists, and computational experts can propel the field toward innovative applications and discoveries. Modeling and simulating nanostructured materials to study their growth, structure, and other optimizations can yield several significant outcomes [48]:

1.4.1 Insights into Growth Mechanisms

Simulation helps in understanding the fundamental processes governing the growth of nanostructures, such as nucleation, diffusion, aggregation, and epitaxial growth. This knowledge can contribute to controlling and optimizing the synthesis of nanostructured materials for specific applications. Conventional growth methods like molecular beam epitaxy (MBE), chemical vapor deposition (CVD), and its derivative, atomic layer deposition (ALD), are capable of generating a diverse array of structures including multilayers, porous structures, nanoclusters, and nanowires, suitable for applications in nanoelectronics and various fields. In this context, the MC technique proves valuable in studying growth kinetics, allowing for the analysis of nanostructure formation mechanisms at the atomic scale. The duration of model processes in the MC technique corresponds to real times, providing insights into the real-time evolution of the growth processes. In their work, Neizvestny *et al.*, elucidated the process of silicon nanocluster formation within silicon dioxide layers. Additionally, they employed the Monte Carlo (MC) simulation technique to study the growth of nanowhiskers made of silicon (Si) and gallium arsenide (GaAs) using *SilSim3D* software [49]. Gocalinska *et al.*, conducted a systematic investigation into the morphology of InP formation on AlInAs during metalorganic vapor-phase epitaxy (MOVPE). Their study involved the utilization of a kinetic model, which accounted for the deposition and decomposition of polyatomic precursors, along with the subsequent adatom kinetics [50]. DFT and MC methods are employed to forecast atomistic interface structures, particularly in the construction of semiconductor/oxide and metal/insulator interfaces. These predictions mimic the growth process through techniques like ALD or CVD, utilizing kinetic MC and other methodologies [48, 51–53]. The transformation of an amorphous oxide film into a crystalline oxide film at semiconductor/oxide and metal/insulator interfaces is significantly impacted by critical factors such as oxide-film thickness, substrate orientation, temperature, and the metal–oxygen system. This variability poses a challenge in simulating these structures accurately [54].

1.4.2 Prediction of Material Properties

Simulations enable the prediction of various material properties like mechanical strength, electrical conductivity, thermal behavior, and optical properties based on their structure and composition. This helps in designing materials with desired functionalities. In their recent review, Chandel *et al.*, discuss how the MD simulation technique is adept at exploring the physical movements of atoms and molecules. This is achieved through the numerical integration of Newton's equations of motion, providing a dynamic perspective of the studied system. The review particularly emphasizes static, buckling, and vibration analyses applied to various nanostructures, such as nanorods, nanobeams, nanoplates, and nanoshells [55]. In their comprehensive work, Hao *et al.*, provide a detailed account of the significant role played by theoretical calculations in the development of advanced high-performance thermoelectric materials [56]. They explored DFT-based calculations to assess the relative stability involved in the formation of two-phase nanostructures and the band alignments of both simple and compositionally alloyed nanostructures. Additionally, the MD simulation technique was employed to calculate the lattice thermal conductivity.

Furthermore, DFT, in conjunction with cluster expansion and MC methods, proves useful in gaining additional insights into carrier mobilities and the quaternary phase diagram [56]. MD simulation can predict specific heat and thermal conductivity of nanofluids containing HNMs, like TiO_2, Al_2O_3, CuO, Si_3N_4, and many more [57]. Bora *et al.*, in their review, explained that the integration of conventional simulation domains with innovative pathways can yield precise representations of nanomaterial properties, behavior, and phenomena [11]. Modern DFT, based on the first principles, can predict material properties accurately [58]. The stability of the HNMs is also crucial along with material properties and can be predicted by DFT and a combination of various atomistic modeling and simulations [58–60]. Hou *et al.*, did an atomistic simulation of low-dimensional nanostructures of Si to predict extreme-scale supercomputing [61]. Yao *et al.*, employed the notion of complex networks to simulate carrier transport processes within assemblies of nanostructures, encompassing various geometries and morphologies [62]. Castin *et al.*, detailed enhanced atomistic Monte Carlo (MC) models utilizing neural networks trained with ab initio data to investigate binary FeCu and FeCr alloys [63].

1.4.3 Design of Novel Nanostructures

Simulation allows researchers to explore and design new nanostructures with unique properties that might not exist. It opens possibilities for innovative materials tailored to specific electronics, medicine, and catalysis applications. Choi *et al.*, employed the Nanomaterial Modeler within CHARMM-GUI to create diverse nanomaterial models and corresponding topologies. Through MD simulations, they gained insights into the structure, dynamics, and underlying mechanisms of complex systems containing nanomaterials [64]. The coherent quantum effect is gaining significance in addressing the heat dissipation bottleneck in semiconductor nanoelectronics, especially as the characteristic size approaches a few nanometers. The quantum mechanical model can be applied to study the phonon-phonon scattering, which influences thermal resistance or conductance in thin films of semiconductors [65]. Quantum mechanical modeling enables the prediction of magnon-phonon scattering heat transport, specifically addressing quantum heat transport within magnetic nanostructures across three-dimensional ferromagnetic/nonmagnetic interfaces [66].

1.4.4 Defect Analysis

Studying nanostructures at an atomic level through simulation can provide insights into defects, interfaces, and surface properties. This knowledge is crucial in enhancing the stability and performance of nanostructured materials. Wang *et al.*, conducted a molecular dynamics study to clarify the impact of nanostructure, surface interference, and oxygen adsorption on the oxidation behavior, adhesion, and friction characteristics of ultrathin amorphous carbon films [45]. Atomistic models aid in simulating the crystal structure, microstructural features (including grain boundaries, dislocations, point defects, and morphology), as well as the structure of binary nanoparticle superlattices [59]. Sinclair *et al.*, revealed new features of GO structure with atomistic simulations based on theoretical and experimental evidence [67]. Niekiel and colleagues utilized atomistic simulations employing semi-empirical many-body potentials to quantitatively compare experimental data. Their aim was to gain a deeper understanding of the defect structure present in Ag nanowires with a fivefold twinning arrangement [68].

1.4.5 Validation and Interpretation of Experimental Results

Simulations complement experimental studies by providing a theoretical framework to interpret experimental data. They help in validating hypotheses and understanding complex experimental observations. Sayle *et al.*, developed fully atomistic models and employed a simulation approach to create atomistic models of nanomaterials, contributing to the systematic construction of nanoparticle arrays. They investigated nanoparticle arrays featuring cubic, tetragonal, and hexagonal symmetries, incorporating primitive, face-centered cubic, and body-centered cubic nanoparticles. The DL_POLY code was utilized to execute all molecular dynamics simulations [59]. Utilizing atomistic simulations that integrate semi-empirical many-body potentials along with X-ray diffraction (XRD) experiments, it is possible to forecast the internal structure of fivefold twinned Ag nanowires [68].

1.4.6 Accelerating Material Discovery

Computational modeling accelerates the discovery of new nanostructured materials by predicting their properties before synthesis, thereby reducing the trial-and-error approach in experimental laboratories. A comprehensive comprehension of the impact of variations in local densities of states (LDOS) on the magnetic properties of HNMs, along with an understanding of the size-dependent alterations in interatomic distances, local spin magnetic moments, and charge distributions per site and orbital, is essential for the purposeful design of HNMs with specific magnetic properties [58]. In a recent publication, Zheng and colleagues outlined an integrated approach that combines experimental and computational methodologies to anticipate the enzyme-like catalytic capabilities of metal nanomaterials and alloys. This framework holds significance for their medical applications, particularly in expediting the research and development of antibacterial nanomedicine [69]. Song and co-authors conducted atomistic simulations to investigate the crystallographic orientation relationships involved in heterogeneous nucleation. They employed a comprehensive polycrystalline system and utilized MD simulation for their study [70]. The Materials Project (www.materialsproject.org) serves as a central component of the

Materials Genome Initiative, aiming to expedite materials innovation through high-throughput computing methods. Its purpose is to unveil the properties of all documented inorganic materials [71].

1.4.7 Guiding Experimental Design

The design and optimization of nanostructures often involve a combination of experimental and computational techniques. Simulation and computational methodologies play a pivotal role in shaping experimental design by offering insights into the behavior and properties of nanostructures at the atomic and molecular scales. Simulation results can guide experimental design by suggesting parameters and conditions for synthesizing specific nanostructures with desired properties. By integrating computational techniques like MD, DFT, MC, FEM, artificial intelligence (AI), and machine learning (ML), and their combinations, researchers can guide experimental design by predicting and understanding the properties, behaviors, and performance of nanostructures before conducting physical experiments to save time and resources and lead to more informed experimental strategies. Liu and colleagues conducted a computational analysis to examine spatial nanostructures and their response to tensile strain. This investigation aims to inform the precise design of nanostructures for applications in soft strain sensors, electromagnetic shielding materials, and other electronic devices in the future [72]. Quantum mechanical methods, MD, MC, and continuum models are computational tools usually used in nanocarrier-based formulation design in drug delivery systems. The simulation of targeted drug delivery systems foresees a quantitative description of HNMs transportation and delivery processes in the dynamic biological system [73].

1.5 List of Popular Software for Nanostructured Hard Materials

Simulating HNMs often involves complex calculations and simulations that require specialized software. As of the update in January 2024, several popular software tools exist for nanostructure material simulation. Keep in mind that the popularity and availability of software may change over time. However, it is essential to consider specific research needs, computational resources, and the type of simulations intended to be performed before choosing a software tool. Additionally, computational materials science is dynamic, and new tools with modifications may emerge regularly. Always check for the latest developments and user reviews when selecting simulation software. We incorporate popular software or packages (freeware and shareware) in each category.

1.5.1 MD Simulation

MD simulation software enables researchers to model the behavior of molecules and materials at atomic and molecular levels over time. Here are some notable MD simulation software to name a few used extensively in scientific research:

- LAMMPS (Large-scale Atomic/Molecular Massively Parallel Simulator): Highly versatile and widely used for simulating molecular dynamics for its scalability and flexibility, accommodating various interatomic potentials and simulation methods.
- GROMACS (GROningen MAchine for Chemical Simulations): Popular in biochemistry, but also widely applied to simulate the behavior of nanoparticles, polymers, and other materials for its efficiency and performance on high-performance computing (HPC) systems.
- NAMD (NAnoscale Molecular Dynamics): Specializes in simulating large biomolecular systems but is also employed for simulating nanomaterials due to its scalability and efficient parallelization.
- CHARMM (Chemistry at HARvard Molecular Mechanics): Originally designed for simulating biomolecules, CHARMM is also used for modeling various nanomaterials and macromolecules, offering diverse force fields and simulation techniques.
- AMBER (Assisted Model Building with Energy Refinement): Primarily used in biochemistry, AMBER has modules for simulating biomolecules but also applies to nanomaterials research through force field modeling.
- OpenMM: Known for its speed and flexibility, OpenMM is designed to perform molecular simulations on CPUs and GPUs for diverse systems, including nanomaterials and biomolecules.
- DL_POLY: This package is used for classical molecular dynamics simulations and is suitable for simulating materials, liquids, and biomolecules.

1.5.2 DFT Simulation

DFT stands as an extensively employed computational method in materials science and quantum chemistry for simulating the electronic structure of atoms and molecules. Several notable software packages, among others, are utilized for conducting DFT simulations:

- VASP (Vienna Ab initio Simulation Package): VASP is a top-rated and powerful software for performing DFT calculations in modeling solid-state materials and surfaces and is widely used for studying nanomaterials and their properties.
- Quantum ESPRESSO: This comprises a comprehensive set of programs designed for electronic structure calculations and materials modeling using DFT. The purpose is to simulate the properties of nanomaterials, surfaces, and interfaces.
- Gaussian: While Gaussian is primarily known for its capabilities in computational chemistry, it also includes modules for DFT calculations for studying molecules, reactions, and the electronic properties of nanomaterials.
- CRYSTAL: CRYSTAL is a versatile software program designed for the ab initio investigation of crystalline solids based on Hartree-Fock or Kohn-Sham DFT for accurate calculations of the electronic structure of nanostructures.
- CASTEP: This software is explicitly tailored for modeling materials through DFT to compute the structural, electronic, and energetic properties of solids, surfaces, and nanostructures.
- WIEN2k: Particularly suitable for studying the electronic structure of solids, WIEN2k is used for calculations based on DFT, especially for systems with strong electron correlations.
- ABINIT: An open-source package for performing electronic structure calculations using DFT for various materials, including nanomaterials.
- CRYSTAL: Primarily focused on periodic systems, CRYSTAL studies crystalline solids and periodic systems, providing accurate results for materials containing transition metals.
- SIESTA: A versatile software package used for simulating materials and nanostructures in handling large systems and is often used for molecular electronics and surface science studies.

- CP2K: While CP2K is versatile and covers various simulation methods, it includes DFT functionality and is used for studying condensed-phase systems, surfaces, and interfaces.

1.5.3 MC Simulation

MC simulation is a computational technique used in various fields to model the behavior of complex systems through repeated random sampling. Here are some popular software tools used for Monte Carlo simulations:

- MATLAB: While not exclusively designed for Monte Carlo simulations, MATLAB offers a robust environment with various toolboxes (like Statistics and Machine Learning Toolbox) that can be used to implement Monte Carlo simulations.
- GAMS (General Algebraic Modeling System): Widely used in optimization and modeling, GAMS can be employed for Monte Carlo simulations involving mathematical modeling and optimization problems.
- AnyLogic: Known for its simulation modeling capabilities, AnyLogic enables Monte Carlo simulations for various applications, including logistics, manufacturing, and social systems.
- Crystal Ball: This software, integrated with Microsoft Excel, provides tools for Monte Carlo simulations in financial, engineering, and risk analysis applications.

1.5.4 Finite Element Method (FEM)

When it comes to simulating nanostructures using the FEM, there are several software tools available to handle the unique challenges and characteristics of nanoscale simulations.

- COMSOL Multiphysics: COMSOL is a versatile simulation software that supports finite element analysis for various applications, including nanostructures for simulating physical phenomena such as heat transfer, electromagnetics, fluid flow, and structural mechanics.
- ANSYS: ANSYS is a comprehensive simulation software that covers a broad spectrum of engineering applications, including structural, thermal, and fluid dynamics simulations,

which can be adapted for nanostructure simulations using appropriate modeling techniques.

1.5.5 FDTD Software

The FDTD method is widely used for simulating electromagnetic wave propagation, interaction, and scattering in various structures and materials. Here are some popular software packages used for FDTD simulations:

- Lumerical FDTD Solutions: A user-friendly interface and powerful simulation capabilities make it attractive for simulating various optical phenomena and photonic devices.
- Meep (or MIT Electromagnetic Equation Propagation): Open-source software for simulating electromagnetic systems. Meep allows users to model a variety of optical and electromagnetic processes.
- CST Studio Suite: While CST has various modules for different electromagnetic simulations, its FDTD module is used for modeling electromagnetic behavior, particularly in complex structures and materials.
- RSoft FullWAVE: A comprehensive simulation tool for photonic devices and components. It utilizes FDTD methods to analyze and design optical systems and waveguides.

1.6 Conclusion

In conclusion, this chapter of the book emphasizes the significance and the multifaceted realm of computational modeling and simulation in advancing the optimization and design of nanostructured hard materials. By providing a concise overview of predictive modeling techniques, the chapter offers a valuable resource for researchers and engineers seeking to enhance the performance of HNMs. The insights gained from simulations contribute to a deeper understanding of the intricate dynamics at the nanoscale, guiding the design process for improved properties and functionalities. The chapter has highlighted the significance of predictive modeling, elucidating how simulations contribute to a nuanced understanding of material responses under various conditions. This understanding, in turn, empowers the optimization process, allowing for the tailoring of HNMs to meet specific performance criteria. From mechanical strength to thermal conductivity, the ability to predict and fine-tune these properties through computational models marks a paradigm shift in materials design. By embracing these tools, researchers can navigate the complexities of materials design with unprecedented precision, shaping the future landscape of advanced materials and technologies.

Integrating computational tools in materials science is pivotal for achieving innovative breakthroughs in developing HNMs as technology evolves. Looking ahead, ongoing advancements in simulation methodologies hold promising prospects for further refining and accelerating the optimization and design processes in this critical field. Looking forward, the collaborative efforts of material scientists, computational experts, and technologists are crucial for harnessing the full potential of predictive modeling and simulation. As computational power advances and simulations become increasingly sophisticated, the field stands at the cusp of transformative discoveries. The ongoing synergy between theory and experiment, facilitated by computational models, promises to propel nanostructured hard materials into uncharted territories, unlocking new frontiers of performance and functionality.

Furthermore, integrating artificial intelligence and machine learning into computational simulations introduces a dynamic dimension to the optimization process. As algorithms learn and adapt from vast datasets, the potential for discovering novel material configurations and properties becomes even more promising. This synergy between computational modeling and emerging technologies opens avenues for innovative breakthroughs, accelerating the pace of material design and discovery.

References

1. Brus, L.E., On the development of bulk optical properties in small semiconductor crystallites. *J. Lumin.*, 31–32, 381, 1984.
2. Brus, L., Zero-Dimensional "Excitons" in Semiconductor Clusters. *IEEE J. Quantum Electron.*, 22, 1909, 1986.
3. Brus, L., Electronic wave functions in semiconductor clusters: Experiment and theory. *J. Phys. Chem.*, 90, 2555, 1986.
4. Clark, T., *A Handbook of Computational Chemistry: A Practical Guide to Chemical Structure and Energy Calculations*, Wiley Interscience, John Wiley, New York, 1985.
5. Majhi, S., *Applications of nanoparticles in organic synthesis under ultrasonication, in Nanoparticles in Green Organic Synthesis*, pp. 279–315, Elsevier, Amsterdam, 2023.
6. Rajput, A., Shevalkar, G., Pardeshi, K., Pingale, P., Computational nanoscience and technology. *OpenNano*, 12, 100147, 2023.

7. Powers, K.W., Carpinone, P.L., Siebein, K.N., Characterization of Nanomaterials for Toxicological Studies, in: *Methods in Molecular Biology*, vol. 926, pp. 13–32, Humana Press, Totowa, NJ, 2012.
8. Silva, L.B., Ferreira, E.F.B., Maryam, Espejo-Román, J.M., Costa, G.V., Cruz, J.V., Kimani, N.M., Costa, J.S., Bittencourt, J.A.H.M., Cruz, J.N., Campos, J.M., Santos, C.B.R., Galantamine Based Novel Acetylcholinesterase Enzyme Inhibitors: A Molecular Modeling Design Approach. *Molecules*, 28, 1035, 2023.
9. Gartner, T.E. and Jayaraman, A., Modeling and Simulations of Polymers: A Roadmap. *Macromolecules*, 52, 755, 2019.
10. Grajciar, L., Heard, C.J., Bondarenko, A.A., Polynski, M.V., Meeprasert, J., Pidko, E.A., Nachtigall, P., Towards *operando* computational modeling in heterogeneous catalysis. *Chem. Soc. Rev.*, 47, 8307, 2018.
11. Bora, T., Dousse, A., Sharma, K., Sarma, K., Baev, A., Hornyak, G.L., Dasgupta, G., Modeling nanomaterial physical properties: theory and simulation. *Int. J. Smart Nano Mater.*, 10, 116, 2019.
12. Fan, Z., Huang, X., Tan, C., Zhang, H., Thin metal nanostructures: synthesis, properties and applications. *Chem. Sci.*, 6, 95, 2015.
13. Dinh, K.N., Liang, Q., Du, C.-F., Zhao, J., Tok, A.I.Y., Mao, H., Yan, Q., Nanostructured metallic transition metal carbides, nitrides, phosphides, and borides for energy storage and conversion. *Nano Today*, 25, 99, 2019.
14. Jothi, P.R., Yubuta, K., Fokwa, B.P.T., A Simple, General Synthetic Route toward Nanoscale Transition Metal Borides. *Adv. Mater.*, 30, 1704181, 2018.
15. Akopov, G., Yeung, M.T., Kaner, R.B., Rediscovering the Crystal Chemistry of Borides. *Adv. Mater.*, 29, 1604506, 2017.
16. Gouget, G., Beaunier, P., Portehault, D., Sanchez, C., New route toward nanosized crystalline metal borides with tuneable stoichiometry and variable morphologies. *Faraday Discuss.*, 191, 511, 2016.
17. Sharma, Y. and Kapoor, P., Noble metal nanostructures for various applications. *Mater. Today Proc.*, 2023. https://doi.org/10.1016/j.matpr.2023.03.628.
18. Yu, S., Zhang, C., Yang, H., Two-Dimensional Metal Nanostructures: From Theoretical Understanding to Experiment. *Chem. Rev.*, 123, 3443, 2023.
19. Jamal, F., Rafique, A., Moeen, S., Haider, J., Nabgan, W., Haider, A., Imran, M., Nazir, G., Alhassan, M., Ikram, M., Khan, Q., Ali, G., Khan, M., Ahmad, W., Maqbool, M., Review of Metal Sulfide Nanostructures and their Applications. *ACS Appl. Nano Mater.*, 6, 7077, 2023.
20. Smith, A.M. and Nie, S., Semiconductor Nanocrystals: Structure, Properties, and Band Gap Engineering. *Acc. Chem. Res.*, 43, 190, 2010.
21. Baig, N., Kammakakam, I., Falath, W., Nanomaterials: a review of synthesis methods, properties, recent progress, and challenges. *Mater. Adv.*, 2, 1821, 2021.
22. McDonald, M.P., Chatterjee, R., Si, J., Jankó, B., Kuno, M., Dimensional crossover in semiconductor nanostructures. *Nat. Commun.*, 7, 12726, 2016.

23. Terna, A.D., Elemike, E.E., Mbonu, J.I., Osafile, O.E., Ezeani, R.O., The future of semiconductors nanoparticles: Synthesis, properties and applications. *Mater. Sci. Eng., B*, 272, 115363, 2021.
24. Wang, S., Xu, J., Li, W., Sun, S., Gao, S., Hou, Y., Magnetic Nanostructures: Rational Design and Fabrication Strategies toward Diverse Applications. *Chem. Rev.*, 122, 5411, 2022.
25. Ali, A., Shah, T., Ullah, R., Zhou, P., Guo, M., Ovais, M., Tan, Z., Rui, Y., Review on Recent Progress in Magnetic Nanoparticles: Synthesis, Characterization, and Diverse Applications. *Front. Chem.*, 9, 629054, 2021.
26. Fernández-Pacheco, A., Streubel, R., Fruchart, O., Hertel, R., Fischer, P., Cowburn, R.P., Three-dimensional nanomagnetism. *Nat. Commun.*, 8, 15756, 2017.
27. Shao, Z. and Ren, S., Rare-earth-free magnetically hard ferrous materials. *Nanoscale Adv.*, 2, 4341, 2020.
28. Ma, Z., Mohapatra, J., Wei, K., Liu, J.P., Sun, S., Magnetic Nanoparticles: Synthesis, Anisotropy, and Applications. *Chem. Rev.*, 123, 3904, 2023.
29. Materón, E.M., Miyazaki, C.M., Carr, O., Joshi, N., Picciani, P.H.S., Dalmaschio, C.J., Davis, F., Shimizu, F.M., Magnetic nanoparticles in biomedical applications: A review. *Appl. Surf. Sci. Adv.*, 6, 100163, 2021.
30. Oehlsen, O., Cervantes-Ramírez, S.I., Cervantes-Avilés, P., Medina-Velo, I.A., Approaches on Ferrofluid Synthesis and Applications: Current Status and Future Perspectives. *ACS Omega*, 7, 3134, 2022.
31. Chiesa, A., Macaluso, E., Petiziol, F., Wimberger, S., Santini, P., Carretta, S., Molecular Nanomagnets as Qubits with Embedded Quantum-Error Correction. *J. Phys. Chem. Lett.*, 11, 8610, 2020.
32. Segawa, Y., Ito, H., Itami, K., Structurally uniform and atomically precise carbon nanostructures. *Nat. Rev. Mater.*, 1, 15002, 2016.
33. Farmand, M., Jahanpeyma, F., Gholaminejad, A., Azimzadeh, M., Malaei, F., Shoaie, N., Carbon nanostructures: a comprehensive review of potential applications and toxic effects. *3 Biotech*, 12, 159, 2022.
34. Sun, L.G., Wu, G., Wang, Q., Lu, J., Nanostructural metallic materials: Structures and mechanical properties. *Mater. Today*, 38, 114, 2020.
35. Li, X., Lu, L., Li, J., Zhang, X., Gao, H., Mechanical properties and deformation mechanisms of gradient nanostructured metals and alloys. *Nat. Rev. Mater.*, 5, 706, 2020.
36. de Almeida, R.B.M., Barbosa, D.B., do Bomfim, M.R., Amparo, J.A.O., Andrade, B.S., Costa, S.L., Campos, J.M., Cruz, J.N., Santos, C.B.R., Leite, F.H.A., Botura, M.B., Identification of a Novel Dual Inhibitor of Acetylcholinesterase and Butyrylcholinesterase: *In Vitro* and *In Silico* Studies. *Pharmaceuticals*, 16, 95, 2023.
37. George, E.P., Raabe, D., Ritchie, R.O., High-entropy alloys. *Nat. Rev. Mater.*, 4, 515, 2019.
38. Wu, G., Liu, C., Sun, L., Wang, Q., Sun, B., Han, B., Kai, J.-J., Luan, J., Liu, C.T., Cao, K., Lu, Y., Cheng, L., Lu, J., Hierarchical nanostructured aluminum

alloy with ultrahigh strength and large plasticity. *Nat. Commun.*, 10, 5099, 2019.

39. Nepal, D., Kang, S., Adstedt, K.M., Kanhaiya, K., Bockstaller, M.R., Brinson, L.C., Buehler, M.J., Coveney, P.V., Dayal, K., El-Awady, J.A., Henderson, L.C., Kaplan, D.L., Keten, S., Kotov, N.A., Schatz, G.C., Vignolini, S., Vollrath, F., Wang, Y., Yakobson, B.I., Tsukruk, V.V., Heinz, H., Hierarchically structured bioinspired nanocomposites. *Nat. Mater.*, 22, 18, 2023.
40. Omanović-Mikličanin, E., Badnjević, A., Kazlagić, A., Hajlovac, M., Nanocomposites: a brief review. *Health Technol. (Berl)*, 10, 51, 2020.
41. de Sousa dos Santos, E.L.V., Cruz, J.N., da Costa, G.V., de Sá, E.M.F., da Silva, A.K.P., Fernandes, C.P., de Faria Mota Oliveira, A.E.M., Duarte, J.L., Bezerra, R.M., Tavares, J.F., da Costa, T.S., dos Anjos Ferreira, R.M., dos Santos, C.B.R., Souto, R.N.P., Essential Oil of Ocimum basilicum against Aedes aegypti and Culex quinquefasciatus: Larvicidal Activity of a Nanoemulsion and *In Silico* Study. *Separations*, 11, 97, 2024.
42. Blinder, S.M., Introduction to the Hartree-Fock method, in: *Mathematical Physics in Theoretical Chemistry*, pp. 1–30, Elsevier, Amsterdam, 2019.
43. Spruch, L., Pedagogic notes on Thomas-Fermi theory (and on some improvements): atoms, stars, and the stability of bulk matter. *Rev. Mod. Phys.*, 63, 151, 1991.
44. Hohenberg, P. and Kohn, W., Inhomogeneous Electron Gas. *Phys. Rev.*, 136, B864, 1964.
45. Wang, S. and Komvopoulos, K., A molecular dynamics study of the oxidation mechanism, nanostructure evolution, and friction characteristics of ultrathin amorphous carbon films in vacuum and oxygen atmosphere. *Sci. Rep.*, 11, 3914, 2021.
46. Padhi, A.K., Janežič, M., Zhang, K.Y.J., Molecular dynamics simulations: Principles, methods, and applications in protein conformational dynamics, in: *Advances in Protein Molecular and Structural Biology Methods*, pp. 439–454, Elsevier, Amsterdam, 2022.
47. Teixeira, F.L., Sarris, C., Zhang, Y., Na, D.-Y., Berenger, J.-P., Su, Y., Okoniewski, M., Chew, W.C., Backman, V., Simpson, J.J., Finite-difference time-domain methods. *Nat. Rev. Methods Primers*, 3, 75, 2023.
48. van der Giessen, E., Schultz, P.A., Bertin, N., Bulatov, V.V., Cai, W., Csányi, G., Foiles, S.M., Geers, M.G.D., González, C., Hütter, M., Kim, W.K., Kochmann, D.M., LLorca, J., Mattsson, A.E., Rottler, J., Shluger, A., Sills, R.B., Steinbach, I., Strachan, A., Tadmor, E.B., Roadmap on multiscale materials modeling. *Model Simul. Mat. Sci. Eng.*, 28, 043001, 2020.
49. Neizvestny, I.G. and Shwartz, N.L., Monte Carlo Simulation of Semiconductor Nanostructure Growth, in: *Advances in Semiconductor Nanostructures*, pp. 345–364, Elsevier, Amsterdam, 2017.
50. Gocalinska, A.M., Mura, E.E., Manganaro, M., Juska, G., Dimastrodonato, V., Thomas, K., Zangwill, A., Vvedensky, D.D., Pelucchi, E., Early stages of InP nanostructure formation on AlInAs. *Phys. Rev. B.*, 101, 165310, 2020.

51. Dkhissi, A., Mazaleyrat, G., Estève, A., Rouhani, M.D., Nucleation and growth of atomic layer deposition of HfO2 gate dielectric layers on silicon oxide: a multiscale modelling investigation. *PCCP*, 11, 3701, 2009.
52. Zeng, Z., Hansen, M.H., Greeley, J.P., Rossmeisl, J., Björketun, M.E., Ab Initio Thermodynamic Modeling of Electrified Metal–Oxide Interfaces: Consistent Treatment of Electronic and Ionic Chemical Potentials. *J. Phys. Chem. C*, 118, 22663, 2014.
53. Netzer, F.P. and Fortunelli, A. (Eds.), *Oxide Materials at the Two-Dimensional Limit*, Springer International Publishing, Cham, 2016.
54. Reichel, F., Jeurgens, L.P.H., Mittemeijer, E.J., The thermodynamic stability of amorphous oxide overgrowths on metals. *Acta Mater.*, 56, 659, 2008.
55. Chandel, V.S., Wang, G., Talha, M., Advances in modelling and analysis of nano structures: a review. *Nanotechnol. Rev.*, 9, 230, 2020.
56. Hao, S., Dravid, V.P., Kanatzidis, M.G., Wolverton, C., Computational strategies for design and discovery of nanostructured thermoelectrics. *npj Comput. Mater.*, 5, 58, 2019.
57. Mohebbi, A., Prediction of specific heat and thermal conductivity of nanofluids by a combined equilibrium and non-equilibrium molecular dynamics simulation. *J. Mol. Liq.*, 175, 51, 2012.
58. Makkar, P. and Ghosh, N.N., A review on the use of DFT for the prediction of the properties of nanomaterials. *RSC Adv.*, 11, 27897, 2021.
59. Sayle, D.C., Seal, S., Wang, Z., Mangili, B.C., Price, D.W., Karakoti, A.S., Kuchibhatla, S.V.T.N., Hao, Q., Möbus, G., Xu, X., Sayle, T.X.T., Mapping Nanostructure: A Systematic Enumeration of Nanomaterials by Assembling Nanobuilding Blocks at Crystallographic Positions. *ACS Nano*, 2, 1237, 2008.
60. Koch, W. and Holthausen, M.C., *A Chemist's Guide to Density Functional Theory*, Wiley, Germany, 2001.
61. Hou, C., Zhu, A., Zhang, S., Zhao, M., Ye, Y., Xu, J., Ge, W., Atomistic simulation of low-dimensional nanostructures toward extreme-scale supercomputing. *CCF Trans. High Perform. Comput.*, 5, 3, 2023.
62. Yao, H., Hsieh, Y.-P., Kong, J., Hofmann, M., Modelling electrical conduction in nanostructure assemblies through complex networks. *Nat. Mater.*, 19, 745, 2020.
63. Castin, N., Messina, L., Domain, C., Pasianot, R.C., Olsson, P., Improved atomistic Monte Carlo models based on *ab-initio* -trained neural networks: Application to FeCu and FeCr alloys. *Phys. Rev. B*, 95, 214117, 2017.
64. Choi, Y.K., Kern, N.R., Kim, S., Kanhaiya, K., Afshar, Y., Jeon, S.H., Jo, S., Brooks, B.R., Lee, J., Tadmor, E.B., Heinz, H., Im, W., CHARMM-GUI Nanomaterial Modeler for Modeling and Simulation of Nanomaterial Systems. *J. Chem. Theory Comput.*, 18, 479, 2022.
65. Guo, Y., Bescond, M., Zhang, Z., Luisier, M., Nomura, M., Volz, S., Quantum mechanical modeling of anharmonic phonon-phonon scattering in nanostructures. *Phys. Rev. B*, 102, 195412, 2020.

66. Pan, H., Tang, L.-M., Chen, K.-Q., Quantum mechanical modeling of magnon-phonon scattering heat transport across three-dimensional ferromagnetic/nonmagnetic interfaces. *Phys. Rev. B*, 105, 064401, 2022.
67. Sinclair, R.C. and Coveney, P.V., Modeling Nanostructure in Graphene Oxide: Inhomogeneity and the Percolation Threshold. *J. Chem. Inf. Model.*, 59, 2741, 2019.
68. Niekiel, F., Bitzek, E., Spiecker, E., Combining Atomistic Simulation and X-ray Diffraction for the Characterization of Nanostructures: A Case Study on Fivefold Twinned Nanowires. *ACS Nano*, 8, 1629, 2014.
69. Zheng, J.-J., Wang, X., Li, Z., Shen, X., Wei, G., Xia, P., Zhou, Y.-G., Wei, H., Gao, X., Integrated Computational and Experimental Framework for Inverse Screening of Candidate Antibacterial Nanomedicine. *ACS Nano*, 18, 1531, 2024.
70. Song, H. and Hoyt, J.J., An atomistic simulation study of the crystallographic orientation relationships during the austenite to ferrite transformation in pure Fe. *Model Simul. Mat. Sci. Eng.*, 23, 085012, 2015.
71. Jain, A., Ong, S.P., Hautier, G., Chen, W., Richards, W.D., Dacek, S., Cholia, S., Gunter, D., Skinner, D., Ceder, G., Persson, K.A., Commentary: The Materials Project: A materials genome approach to accelerating materials innovation. *APL Mater.*, 1, 011002, 2013.
72. Liu, J., Zhao, F., Tao, Q., Cao, J., Yu, Y., Zhang, X., Visualized simulation for the nanostructure design of flexible strain sensors: from a numerical model to experimental verification. *Mater. Horiz.*, 6, 1892, 2019.
73. T, A., Narayan, R., Shenoy, P.A., Nayak, U.Y., Computational modeling for the design and development of nano based drug delivery systems. *J. Mol. Liq.*, 368, 120596, 2022.

2

Latest Developments for Ni-Based Wear Protective Surfaces: Coating Compositions and Methods

Onur Güler*, Müslim Çelebi and Abdullah Hasan Karabacak

Karadeniz Technical University, Metallurgical and Materials Engineering, Trabzon, Turkey

Abstract
Nanoscale tribology and wear resistance explores cutting-edge materials and techniques for minimizing friction, enhancing lubrication, and creating durable, wear-resistant surfaces at the nanoscale. Ni-based wear protective surfaces enhance nanoscale tribology and wear resistance through multiple mechanisms. They exhibit high hardness and low surface roughness, minimizing adhesion and abrasion during sliding. Moreover, Ni-based alloys form self-lubricating oxide layers, reducing friction and wear by acting as solid lubricants. These attributes reduce friction, minimize material loss, and extend component lifespans, making Ni-based wear protective surfaces a vital tool in enhancing wear resistance and optimizing the performance of various engineering systems at the nanoscale. In recent years, Ni-based wear-resistant surfaces have been developed with the incorporation of various additive materials to achieve tribological perfection. These advanced surfaces suitable for industrial applications are obtained through coating methods such as electroless coating, electroplating, laser cladding, thermal spraying, CVD, and PVD. Consequently, these Ni-based wear-resistant surfaces with excellent characteristics form the basis for cost-effective, high-performance products and serve the purpose of enhancing wear resistance in industrial applications. In line with this, in this book chapter, we will discuss the compositions and methods used for cost-performance products in the context of Ni-based wear-resistant surfaces that keep pace with recent advancements and technology.

**Corresponding author*: onurguler@ktu.edu.tr

Inamuddin, Tariq Altalhi and Mohammad Abu Jafar Mazumder (eds.) Nanostructured Hard Materials: Industrial Applications, (39–112) © 2026 Scrivener Publishing LLC

Keywords: Ni-based surfaces, nanotribology, wear resistance, coating methods, tribological properties, latest technologies

2.1 Introduction

Tribology is the scientific investigation and engineering field that concentrates on the analysis of wear, friction, and lubrication occurring between interacting surfaces. It involves comprehending the interactions between surfaces in relative motion, understanding the mechanisms underlying friction and wear, and developing strategies to reduce or mitigate these occurrences. Since tribology plays a crucial role in many industrial fields and in everyday life, its importance cannot be overstated [1]. In the industrial sector, tribology is crucial for optimizing machine performance, reducing energy consumption, and extending the lifespan of mechanical components [2, 3]. It is a cornerstone of automotive engineering, ensuring efficient engines and smooth-running transmissions [4]. In healthcare, tribology is fundamental for the design of prosthetic joints and orthopedic implants, enhancing the quality of life for millions [5]. Moreover, it impacts environmental sustainability by reducing energy waste and material consumption. In essence, tribology is the unsung hero that keeps the world moving, contributing to technological advancements, economic growth, and the overall well-being of society [6, 7]. Tribology's significance extends beyond the confines of specific fields, permeating diverse sectors. It forms the bedrock of innovations in transportation, manufacturing, and healthcare, with potential applications ranging from space exploration to nanotechnology. Its role in improving efficiency, reducing maintenance costs, and minimizing environmental impact underscores its global importance. Tribology research continues to push boundaries, addressing emerging challenges in sustainability and energy conservation [8, 9]. In a world increasingly focused on efficiency and environmental responsibility, tribology remains an essential discipline at the heart of progress, shaping the future of technology and industry. When considering the potential energy savings that could occur in various applications as seen in Figure 2.1, friction and wear related to the tribology is evident that there is no area of study that should be overlooked [10].

Nano tribology is a specialized branch of tribology focused on studying friction, wear, and lubrication at the nanoscale. It entails analyzing surfaces and materials with dimensions on the order of nanometers, equivalent to one billionth of a meter. It delves into the fundamental understanding of interatomic and intermolecular interactions at these extremely small

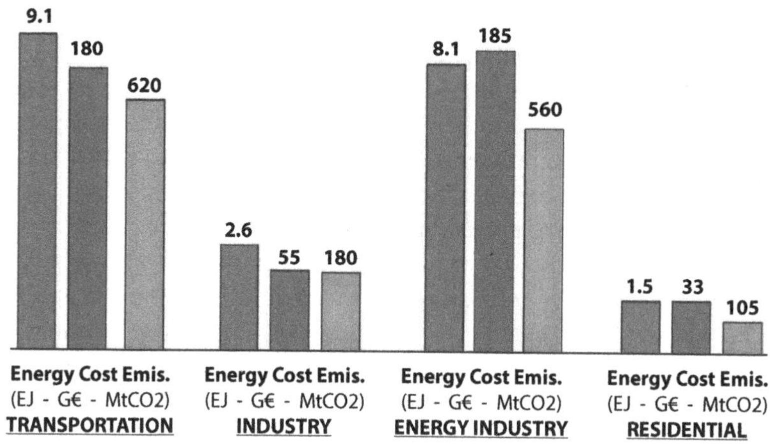

Figure 2.1 The likely yearly reductions in worldwide energy consumption, expenses, and carbon dioxide emissions because of the widespread adoption of advanced tribology practices ([10], with permission from Elsevier).

scales, where conventional tribology principles may not apply directly. Nano tribology plays a critical role in various fields, particularly in nanotechnology, materials science, and the development of advanced lubricants and coatings. The importance of nano tribology lies in its ability to address challenges and opportunities associated with miniaturization and the manipulation of materials at the nanoscale. In nanotechnology, where the behavior of materials can differ significantly from their macroscopic counterparts, nanotribology helps engineers and scientists design and optimize nanoscale devices and systems, such as microelectromechanical systems (MEMS) and nanoelectromechanical systems (NEMS) [11–13]. It is essential to ensure the reliability and functionality of these devices by controlling friction and wear at the nanoscale. Furthermore, nanotribology has implications for improving the performance of various industrial processes and products. It aids in the development of advanced lubricants and coatings that can reduce friction and wear in precision machinery, leading to energy savings, longer component lifespans, and reduced maintenance costs [14]. Nano-tribological research also contributes to the understanding of biological systems, such as protein interactions and cell mechanics, which has implications for fields like biotechnology and medicine [15]. Nanotribology, an intricate realm of scientific exploration, delves into the enigmatic dynamics of minute sliding interactions, where the paramount protagonists are surface forces and their confrontation with externally applied forces. This esoteric discipline is driven by dual compelling rationales, both unveiling profound insights. Primarily, it kindles curiosity about friction within diminutive systems, wherein the diminishment of scale amplifies the ascendancy of surface forces, eventually rendering them comparable in magnitude to their externally imposed counterparts. Here, the diminutive dimensions serve as an aperture into a world where traditional frictional paradigms undergo metamorphosis. Secondarily, nanotribology embarks on the intricate journey of deciphering the intricate choreography of microscopic encounters between surface asperities. These microscopic skirmishes are, in essence, the architects of macroscopic frictional phenomena. In the nuanced scrutiny of singular asperity interludes, nanotribology aspires to bestow upon us the essential building blocks for the edifice of fundamental frictional comprehension [16–20]. Molecular dynamics simulation, an intricate orchestration of computational prowess, orchestrates a mesmerizing dance at the infinitesimal atomic scale as seen in Figure 2.2 [21]. In one mesmerizing act, it unveils the intricate choreography of contact, where an amorphous carbon tip, akin to a celestial wanderer, encounters the diamond sample's pristine expanse. This cosmic encounter, a ballet of atoms, reveals the delicate interplay of forces and the transient bonds that form and dissolve in the realm of nanoscale friction (Figure 2.2a). In another spellbinding performance, it delves into the enigmatic world of the tribo-layer, a clandestine realm that emerges between

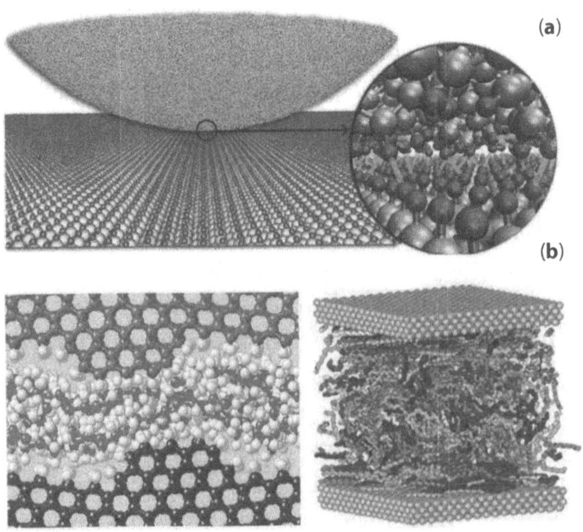

Figure 2.2 Molecular dynamics simulation on the atomic level. (a) Interaction between an irregular carbon apex and a crystalline diamond specimen, (b) Lubricious stratum amidst two gliding interfaces ([21] with permission from Elsevier).

two sliding surfaces. This interstitial tableau, a symphony of molecular interactions, weaves an intricate tapestry of transient structures and intricate chemical reactions, shedding light on the clandestine secrets of tribological phenomena (Figure 2.2b).

Coatings are crucial in the creation of surfaces resistant to wear, especially in the field of nanotribology, where interactions take place on the nanometer scale. These nanoscale coatings are engineered to mitigate friction and wear, enhancing the performance and durability of various components and devices. Nanotribology, which delves into intermolecular interactions and surface phenomena at the atomic and molecular levels, demands innovative solutions to minimize wear and friction [22, 23]. Coatings excel in this regard by providing a protective barrier that can withstand extreme conditions. In the automotive industry, for example, nanotribological coatings are applied to engine components, such as pistons and camshafts, reducing friction and increasing fuel efficiency. These coatings can also be found in cutting tools and bearings, significantly extending their lifespan, and improving machining processes [24, 25]. Additionally, in the semiconductor industry, where precision is paramount, coatings on systems ensure the smooth operation of devices, preventing friction and wear [26, 27]. Moreover, the medical field benefits from wear-resistant coatings in prosthetic joints, reducing friction and enhancing the longevity of implants [28]. In electronics, these coatings enable the development of smaller, more efficient devices by reducing friction between moving parts [29]. The aerospace sector utilizes them in aircraft engines, where high temperatures and extreme conditions necessitate robust protection against wear [30]. In summary, nanotribological coatings are instrumental in creating wear-resistant surfaces that drive innovation and reliability across various industries. They enable the design of longer-lasting, more efficient devices, contributing to improved performance, reduced maintenance costs, and increased sustainability in the modern world.

As shown in Figure 2.3a, coatings are of great importance in obtaining wear-protective surfaces, as wear can occur in all areas used, from macro-sized to nano-sized materials [31]. As seen in Figure 2.3b, with the increase of nano WC particles in surface coatings, wear mechanism shows less adhesion, as an example of an even simple coating type beneficial to wear protection [32].

Moreover, considering the global industrial coatings market size (Figure 2.4), it is clearly understood that the need for coatings for surface protection processes in the industry is increasing day by day.

Numerous articles in the literature have delved into the surface protective application methods of hard coatings, their investigation in terms

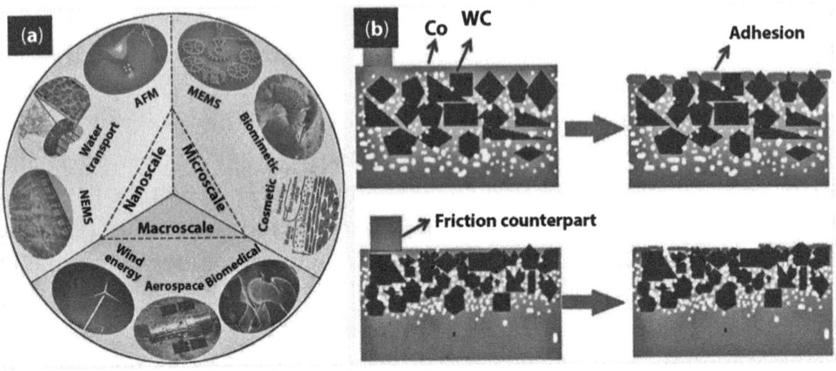

Figure 2.3 The importance of tribology and coatings; (a) diversity of tribological working area ([31], licensed by CC-BY 4.0) and surface protective mechanism thanks to the coating ([32], licensed by CC-BY 4.0).

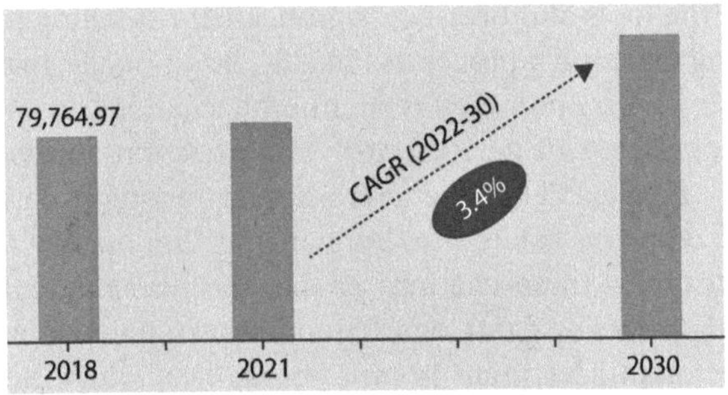

Figure 2.4 Global spending on industrial coatings (USD million) [180].

of tribological properties, and the development of coatings that exhibit self-lubricating characteristics at high temperatures, tailored to specific application needs [33–35]. Additionally, there is extensive research focusing on the synergy between coatings and wear-resistant polymers [36] or metals [37], enhancing their resistance to wear. Given the swift progression of wear-resistant materials and their significant influence on energy usage, it is crucial to present an extensive examination of the recent developments in materials incorporating coatings based on Nickel. The aim of this chapter is to offer an in-depth examination, concentrating on the recent progress in wear-resistant materials, especially those incorporating coatings based on Nickel. It encompasses novel design approaches, elucidates key properties, and explores a wide array of applications for these materials. The spotlight is placed on the unique combination of properties that Ni-based coatings bring to the table, making them instrumental in addressing the ever-growing demands for wear-resistant materials in various industrial sectors.

Here in the introductory section, we briefly touched upon the significance of tribology in industry and the methods used to achieve wear-resistant surfaces at the nanotribological scale. We also highlight the importance of Ni-based coatings in this context. Moving on to Section 2.2, we delve into surface protection techniques and compositions commonly applied for the purpose of wear resistance. We'll specifically discuss the advantages of Ni-based coatings over other types, typical strategies for surface engineering in the pursuit of wear-resistant materials, and novel design concepts and their components that have emerged in recent years, keeping pace with technological advancements. Section 2.3 will be dedicated to a comprehensive exploration of how the tribological properties of Ni-based coatings have evolved through recent developments. We'll draw from the literature to elucidate the nanotribology-oriented wear mechanisms, emphasizing the diversity of substrate materials onto which these coatings are applied, including various metals and plastics. In Section 2.4, we will explore a diverse range of uses for Ni-based coatings with wear resistance, including applications in aerospace components, automotive elements, wind turbines, micro-/nano-electromechanical systems (MEMS/NEMS), atomic force microscopy (AFM), and biomedical devices. Section 2.5 will wrap up the review, delving into the challenges and future prospects of this pivotal field, presenting a thorough summary of the recent progress and potential pathways in the domain of wear-resistant materials incorporating Ni-based coatings.

2.2 Surface Protective Coatings for Wear-Resistant Materials

Wear-resistant surface protection coatings, as discussed in the literature, refer to specialized coatings or layers applied to the surfaces of materials or components to enhance their resistance to wear, friction, and corrosion. These coatings are crucial for various industries due to their ability to extend the lifespan and improve the performance of critical components and machinery. Certainly, there are several types of anti-wear surface protection coatings used in various industries. These coatings are formulated with the purpose of minimizing friction, preventing wear, and improving the longevity of components and equipment. The following provides a detailed overview of some common types:

Hard Coatings

Hard Chromium Coatings: Chromium is electroplated onto the substrate to create a hard, wear-resistant surface. It is commonly used in applications like hydraulic rods, piston rings, and cutting tools. As can be seen in Figure 2.5 as an example from the literature, while strip abrasion constitutes the primary mechanism of wear, a phenomenon of material transfer was observed at the apex of the pins. The pins lacking a coating and those subjected to chrome plating exhibited varying levels of strip material transfer, reflecting their distinct frictional properties. Interestingly, the chrome-plated pins demonstrated a considerably lesser amount of attached strip material in comparison to the uncoated ones. Given the uniformity of the test conditions, it is plausible to attribute this behavior exclusively to the variance in the tribochemical interactions within the contacting pairs [38].

Ceramic Coatings: Ceramic materials like titanium nitride (TiN), titanium carbide (TiC), and aluminum oxide (Al_2O_3) are deposited as coatings. These are known for their hardness, corrosion and tribological resistance, and high-temperature stability. In a study [39], it was determined that TiN surface protective coatings offer superior tribological properties than chrome coating as seen in Figure 2.6a. Additionally, the heightened resistance to wear can be attributed to the formation of a reinforcing phase composed of *in-situ* synthesized TiC-Al_2O_3 on the copper surface as can be understood in Figure 2.6b [40]. This process of enhancing strength entails forming a sturdy layer on the surface of copper by synthesizing TiC and Al_2O_3 within the material. This newly formed phase acts as a protective

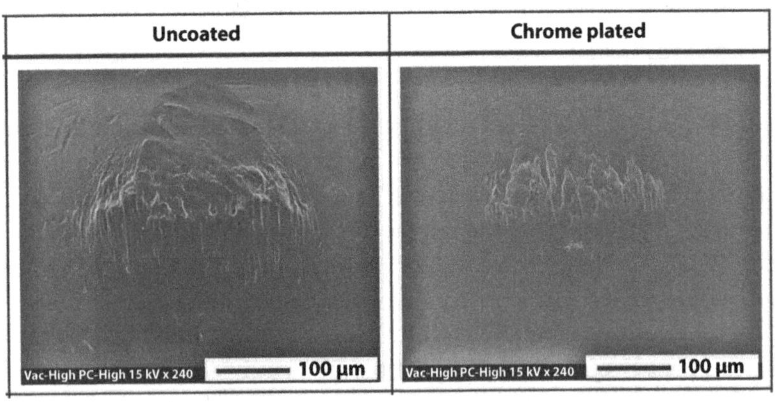

Figure 2.5 The scratched surfaces of pins uncoated and coated with hard chromium ([38], licensed by CC-BY-4.0).

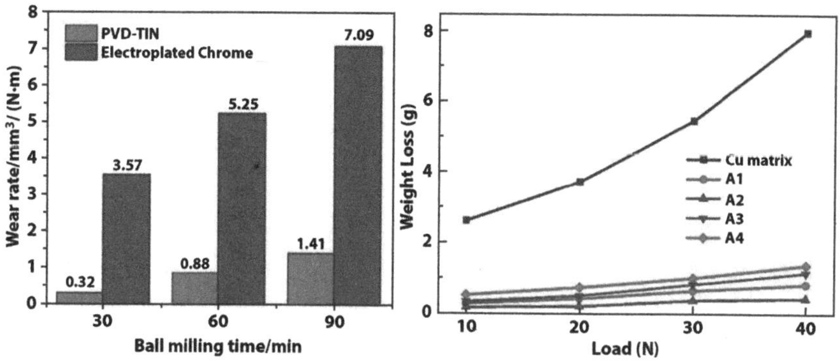

Figure 2.6 Wear test results for different times of sliding effect; (a) TiN and Cr plating ([39] licensed by CC-BY-4.0), (b) TiC/Al_2O_3 coatings in different contents of Al_2O_3 described as A1-A4 ([40], with permission from Elsevier).

barrier, bolstering the copper's wear resistance and prolonging its durability in the face of abrasive forces.

Diamond-Like Carbon (DLC) Coatings: DLC coatings blend the hardness of diamond with the flexibility of graphite. They are used in various applications, including automotive components and medical devices. In a sample obtained from the literature, DLC coated aluminum materials exhibited significantly lower wear track, friction coefficient, and wear loss values compared to uncoated materials, as shown in Table 2.1 [41]. The underlying mechanism behind this observation can be explained as follows: The DLC coating, composed of amorphous carbon with properties resembling those of natural diamond, provides exceptional hardness, smoothness, and a low coefficient of friction. Upon application to aluminum surfaces, the DLC coating functions as a protective barrier, diminishing the direct interaction between the aluminum and abrasive or opposing surfaces. As a result, the DLC-coated aluminum materials experience

Table 2.1 Wear test results of pure aluminum and DLC-coated aluminum ([41], licensed by CC-BY-4.0).

Material	Wear loss (m3)	Wear track (μm)	Coefficient of friction (μ)
Aluminum substrate	4.85 × 10–9	1500	0.95
DLC coated substrate	1.8 × 10–11	190	0.01

reduced wear, lower frictional forces, and minimized wear loss. This is due to the DLC's inherent resistance to abrasion and its ability to maintain a smooth, low-friction interface, ultimately enhancing the overall performance and durability of the material.

Polymer Coatings

Polytetrafluoroethylene (PTFE): PTFE coatings provide excellent lubricity and low friction properties. They are used in industries where reduced friction and non-stick properties are crucial, such as food processing and mold release applications. As seen in Figure 2.7, the friction coefficient for Si_3N_4/TiC ceramics coated with PTFE exhibited a conspicuous reduction when compared to their uncoated counterparts under identical conditions. The PTFE coating, known for its exceptional low-friction properties and non-stick nature, forms a lubricious layer on the surface of Si_3N_4/TiC ceramics. This layer serves as a barrier that diminishes direct contact between the

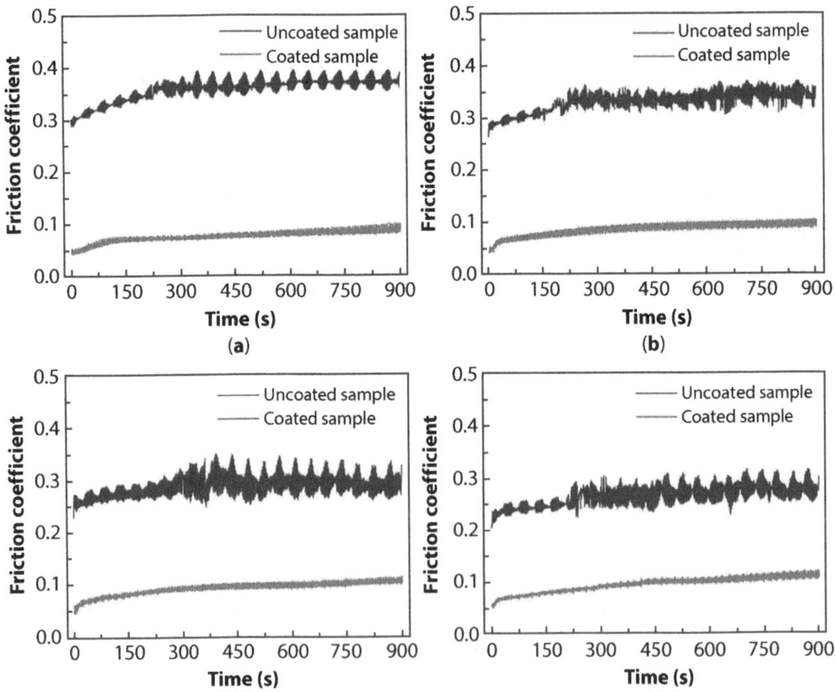

Figure 2.7 Friction coefficient profiles at various applied loads: (a) 20 Newtons; (b) 40 Newtons; (c) 60 Newtons; (d) 80 Newtons ([42], with permission from Elsevier).

ceramic material and the opposing surface, thereby substantially decreasing the frictional forces produced during relative motion [42].

Polymer Matrix Composites: These coatings consist of a polymer matrix reinforced with hard particles like ceramics or metals. They offer a balance between wear resistance and toughness and find applications in various mechanical components. In a study, following the wear tests, the surface of the samples was examined using a confocal microscope. Corresponding to that study, Figure 2.8 illustrates the worn surfaces of the reference epoxy coating and 0.3% graphene oxide reinforced epoxy coating. Notably, all samples exhibited abrasion grooves on their surfaces and the abrasive wear was primarily attributed to micro-cutting. The most significant material loss occurred on the peaks of surface waviness, resulting in a notable flattening of these areas. In these peak regions, the grooves were wider and deeper, owing to the higher pressure exerted by the abrasive wheel compared to the valleys of the waviness. The reference coating EP0 was particularly affected by abrasive wear, displaying even deeper and larger groove areas as seen in Figure 2.8a. However, in the case of the coating containing 0.3% graphene oxide, dark areas were observed, predominantly in the sample where GO was introduced. It is noteworthy that graphene oxide is recognized as a lubricant, and due to its uniform distribution on the surface, it has the potential to reduce the coefficient of friction in powder coatings [43].

Metal-Based Coatings

Nickel-Based Coatings: Coatings based on nickel, such as electroless nickel (EN), are recognized for their resistance to wear, resistance to corrosion, and their capacity to form robust bonds with substrates. They can be

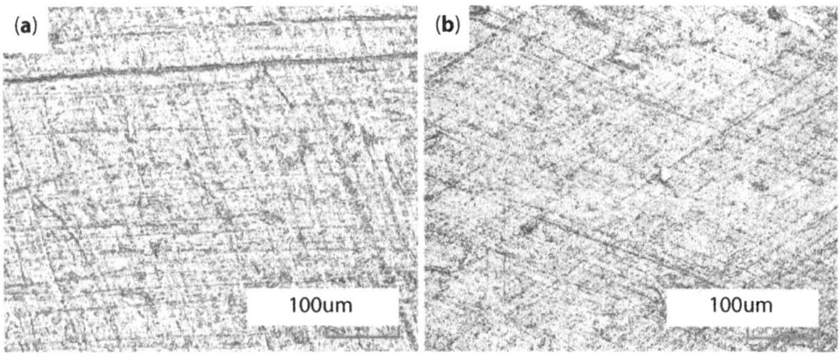

Figure 2.8 Wear surfaces: (a) epoxy coating, (b) 0.3% graphene oxide reinforced epoxy coating fabricated by extruded process ([43], licensed by CC-BY-4.0).

self-lubricating and are used in industries like aerospace and automotive. If we were to provide examples from the literature regarding these specialized coatings, one can refer to the findings presented in Figure 2.9. In this context, it becomes apparent that the thickness of the coating is crucial in determining erosion rates. In particular, the most substantial coating, with a thickness of 70 μm (Ni-P-NiTi), showed the highest erosion rate, while the slimmest coating, measuring 12 μm (Ni-P), demonstrated the least erosion rate. Upon analyzing the four coatings, regardless of their compositional variations, a noticeable pattern emerges: the erosion rates rise with an increase in coating thickness. The connection between erosion rate and coating thickness is explained by the existence of residual stresses. The coating process induces compressive residual stresses through deformation, leading to the generation of tensile stresses within the subsurface. This cold working process, in turn, leads to significant alterations in the material's mechanical properties. The fundamental principle underpinning this phenomenon finds practical application in various industries as a mechanical treatment method aimed at enhancing wear resistance, a process commonly referred to as shot peening ([44], licensed by CC-BY-4.0).

Copper-Nickel Alloys: These alloys offer good corrosion resistance and wear resistance. They are often used in marine applications, such as ship propellers. For instance, Ni-P-Cu coatings are specialized coating materials used to protect steel surfaces. The development of these coatings aims to boost the corrosion resistance and wear resistance of steel simultaneously. As evident from a study example in Figure 2.10, polarization curves are indicative of the corrosion resistance of a steel surface as a result of an electrochemical experiment. Ni-P-Cu coatings create a barrier against corrosion by covering the steel surface. This barrier prevents direct contact with the steel surface and slows down the progression of

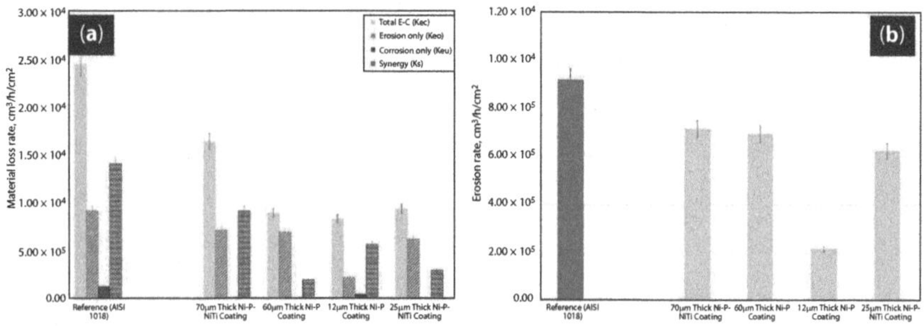

Figure 2.9 The values of (a) erosion and corrosion losses and (b) only erosion loss for the test time of 6 h ([44], licensed by CC-BY-4.0).

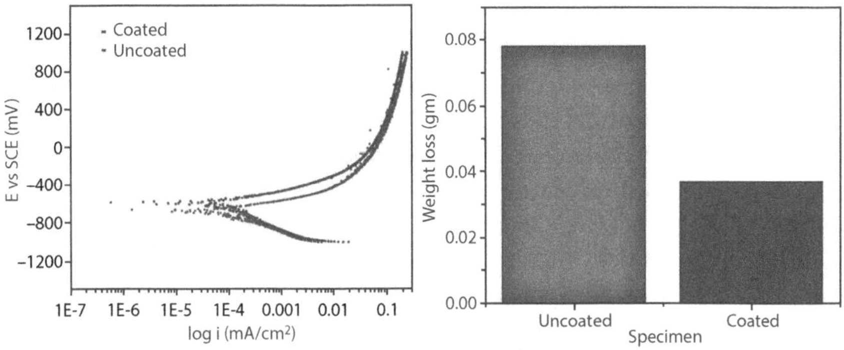

Figure 2.10 The polarization curves and the weight loss caused by corrosion and wear, respectively, on both coated and uncoated samples ([45], with permission from Elsevier).

corrosion. Consequently, the coated steel surface is exposed to less corrosion, leading to a leftward shift in the polarization curve, indicating improved corrosion resistance. Furthermore, Ni-P-Cu coatings impart additional durability and hardness to the steel surface. This provides better protection against mechanical effects such as friction, impact, and wear. Consequently, coated steel surfaces experience less wear, resulting in reduced wear loss [45].

Bronze Coatings: Bronze coatings contain copper and tin and provide good wear resistance. They are used in applications like bearings and bushings. Figure 2.11 illustrates the results of a study conducted in the literature [46], comparing the wear loss values of different bronze coating types applied to steel materials under dry conditions with the wear losses

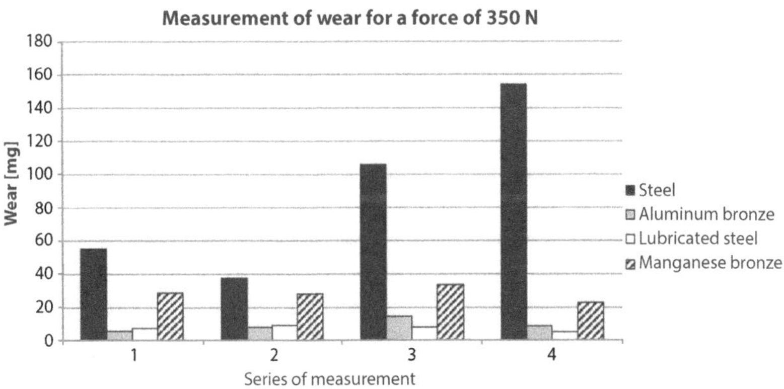

Figure 2.11 Wear loss values of substrate steel and different bronze coatings in dry conditions and in only lubricated steel ([46], with permission from Elsevier).

exhibited by uncoated steel under lubricated conditions. As observed, the wear tests conducted under dry conditions with bronze-aluminum coatings have yielded nearly identical wear loss values to those obtained from the uncoated tests under lubricated conditions. This finding suggests that the application of bronze-aluminum coatings on steel in dry conditions provides wear resistance comparable to uncoated steel in lubricated conditions. It highlights the effectiveness of these coatings in reducing wear and friction, potentially making them a cost-effective and environmentally friendly alternative for applications where lubrication is not feasible or desired. The remarkable similarity in wear loss values between these two scenarios suggests that the bronze-aluminum coatings are highly effective at reducing wear, even in dry conditions. This can be attributed to the superior wear-resistant properties of bronze coatings, which serve as a protective barrier and minimize both adhesive and abrasive wear.

Composite Coatings

Metal-Matrix Composites: These coatings combine a metal matrix with hard particles like carbides or borides. They provide improved resistance to wear and are frequently utilized in cutting tools and components of industrial machinery. In a thorough examination centered on the wear attributes of metal matrix composite (MMC) coatings, a notable study [47] was undertaken to evaluate the wear performance of two different compositions: one with 5% tungsten carbide (WC) content and another with 10%. The coatings were meticulously fabricated at a controlled deposition rate of 400 mm·min^{-1}, and their wear behavior was meticulously scrutinized *via* ball-on-disc dry sliding tests performed under ambient room temperature conditions. Illustrated in Figure 2.12 are detailed depictions of surface topographies and cross-sectional profiles corresponding to the regions that underwent wear during the experimental trials. It is readily discernible that the wear severity is significantly exacerbated in the coating characterized

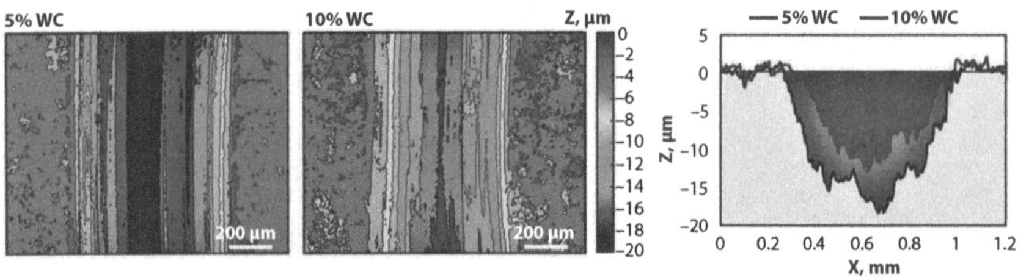

Figure 2.12 The effect of WC content in the wear track depth of WC-Co coatings ([47], licensed by CC BY-NC-ND 4.0).

by the 5% WC composition. This observation unequivocally suggests that an increased volumetric fraction of tungsten carbide (as embodied by the 10% WC coating) yields a commensurately reduced wear scar. In the realm of MMC coatings, an intricate wear mechanism comes into play. Tungsten carbide particles within the composite serve as robust fortifications, effectively impeding the matrix material's propensity to erode and substantially mitigating material loss during the sliding test. This protective effect is attributable to the hardness and abrasive resistance inherent to tungsten carbide, which counteracts the abrasive forces and abrasion-induced wear typically encountered in such tribological scenarios. Consequently, these findings underscore the pivotal role played by tungsten carbide content in enhancing the wear resistance and durability of MMC coatings, shedding valuable light on the underlying wear mechanisms at work in such advanced materials systems.

Ceramic-Matrix Composites: These coatings incorporate ceramic particles in a matrix material, providing a combination of hardness and

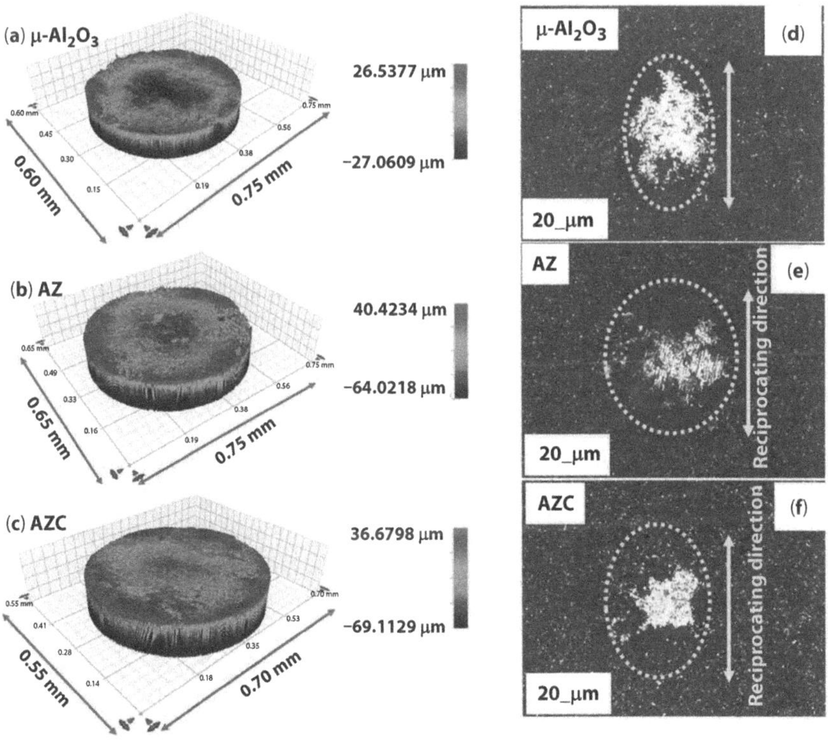

Figure 2.13 Surface profile of wear scars ([48], with permission from Elsevier).

toughness. They find applications in high-stress environments. If we take an example from the literature [48], as shown in Figure 2.13, surfaces coated with only Al_2O_3 exhibit more deformation, while ceramic-based composite coatings demonstrate lower wear deformation and a wider wear track width. This can be observed in the 3D topography and SEM images of worn surfaces coated with Al_2O_3–ZrO_2 (AZ), Al_2O_3–ZrO_2-CNT (AZC), and μ-Al_2O_3 materials. The wear mechanism in this context can be interpreted as follows: Surfaces coated solely with Al_2O_3 experience more deformation, indicating that the Al_2O_3 coating is likely to deform and wear away more easily under the given conditions. This could be due to the properties of Al_2O_3 itself, such as its hardness, toughness, and adhesion to the substrate material. Ceramic-based composite coatings, such as Al_2O_3–ZrO_2 and Al_2O_3–ZrO_2-CNT, exhibit lower wear deformation and wider wear tracks. This suggests that these composite coatings offer superior wear resistance. The incorporation of materials like ZrO_2 and carbon nanotubes (CNT) is anticipated to enhance the hardness, toughness, and overall resistance to wear in the coating. ZrO_2, for instance, is known for its high toughness and resistance to fracture, which can contribute to reduced deformation and wider wear track.

Thermal Sprayed Coatings

HVOF (High-Velocity Oxygen Fuel) Coatings: These coatings are applied using a high-velocity flame to deposit metals, ceramics, or carbides. They provide outstanding resistance to wear and find applications across diverse industries, such as aerospace and power generation. Praveen and Arjunan [49], carried out a study on the influence of different HVOF ceramic coatings on the wear performance of steel. High-temperature erosion is a significant issue affecting various industries, leading to increased maintenance periods, time costs, and reduced efficiency. In many cases, particularly within the power generation sector, oxidation and erosion are the primary factors contributing to material failure. However, the most severe scenarios result from the combined effects of erosion and oxidation. To gain insight into the interplay between these two processes, it is essential to comprehensively understand both the oxidation characteristics and high-temperature erosion tendencies of the considered coatings. Consequently, Figure 2.14 illustrates the erosion wear rate of NiCrSiB coatings reinforced with Al_2O_3, WC-Co, and n-Al_2O_3.

Each type of anti-wear surface protection coating has its advantages and is chosen based on specific application requirements. These coatings are essential for reducing friction, preventing wear, and extending the life of critical components in industries ranging from manufacturing and

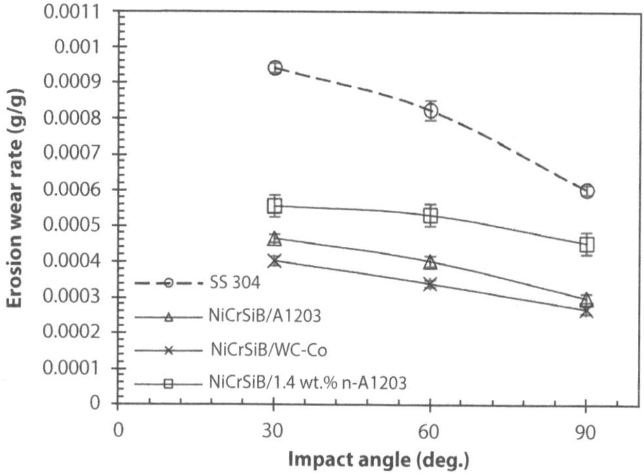

Figure 2.14 Wear rate values of the samples coated with different concentrations at 750°C ([49], licensed by CC BY-NC-ND 4.0).

automotive to aerospace and medical devices but Ni-based coatings are often preferred over other coatings such as hard chromium, ceramic, DLC, PTFE coatings in various industrial applications due to several scientifically proven advantages. Here are some reasons why Ni-based coatings are favored:

Corrosion Resistance: Ni-based coatings offer excellent corrosion resistance, protecting the underlying material from chemical reactions and environmental factors. This makes them particularly suitable for applications in aggressive or corrosive environments [50].

Wear Resistance: Ni-based coatings exhibit high wear resistance, ensuring durability and extended service life for components subjected to abrasive conditions. This resistance to wear is attributed to the hardness and toughness of the Ni-based materials [51].

Adhesion Strength: Ni-based coatings typically adhere well to various substrate materials, ensuring that they stay in place and provide consistent protection even under stress and dynamic conditions [52].

Temperature Resistance: Ni-based coatings possess the ability to endure high temperatures, rendering them appropriate for applications where exposure to extreme heat is a consideration. This property is crucial in industries like aerospace and manufacturing [53].

Versatility: Ni-based coatings can be tailored to specific application requirements through variations in composition [54], deposition methods [55], and post-processing treatments [56]. This versatility allows them to address a wide range of wear and corrosion challenges.

Low Friction: Ni-based coatings can be engineered to have low friction properties, reducing energy consumption, and enhancing the efficiency of moving components. This is especially important in applications like automotive engines [57].

Smooth Surface Finish: Ni-based coatings can be applied to achieve a smooth and uniform surface finish, which can be

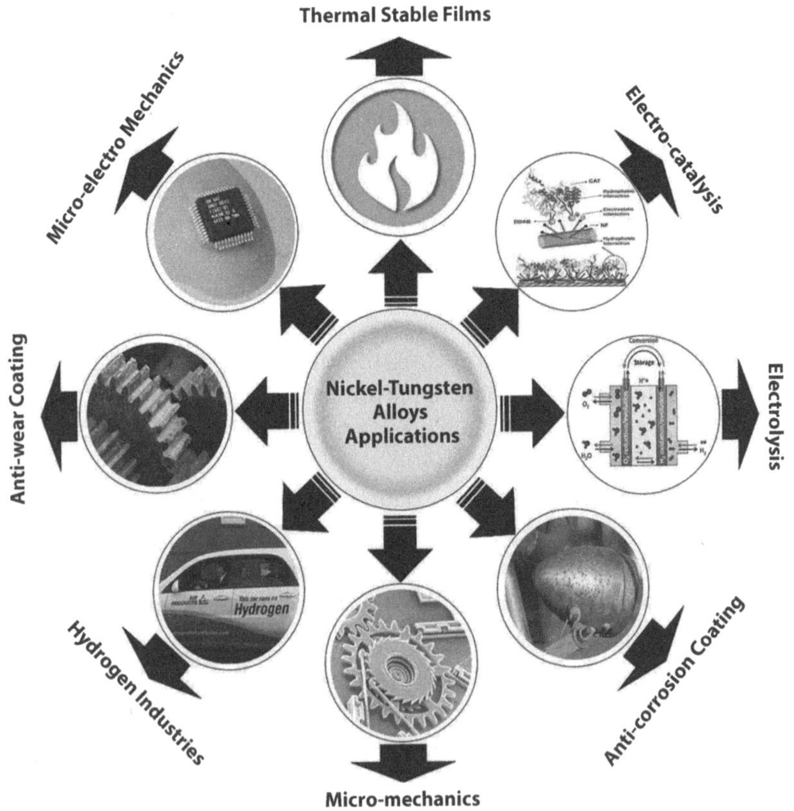

Figure 2.15 Variety of application areas of Ni-based coatings ([59] with permission from Elsevier).

essential in industries where precise tolerances and reduced friction are critical [58].

The effectiveness of Ni-based coatings in preference can be attributed to the combination of these advantages, which are backed by scientific research and practical applications. Additionally, ongoing research and development efforts continue to improve Ni-based coating technologies, making them even more appealing for various industries as seen in Figure 2.15 [59].

2.2.1 Traditional Ni-Based Wear-Resistant Coatings

2.2.1.1 Pure Ni-Based Coatings

Nickel coatings serve as a corrosion-resistant shield on various metals, such as steel and zinc, safeguarding them against deterioration due to corrosion, erosion, and abrasion. It's important to highlight that in this context, nickel is primarily utilized as an intermediate layer in decorative finishes, frequently in conjunction with chromium coatings. These nickel coatings are typically created through electrodeposition using Watts-type solutions, which contain a blend of nickel sulfate, boric acid, and nickel chloride. Under normal plating conditions, around 98% of the cathodic current is dedicated to the nickel deposition, with the remaining portion contributing to proton discharge. Alternatively, electroless deposition techniques can be utilized, employing reducing agents such as sodium borohydride or sodium hypophosphite to generate deposits of nickel-boron and nickel-phosphorus alloys [60]. Comparatively, electrodeposits are generally acknowledged to offer superior corrosion resistance, although electroless deposits have the advantage of easily covering intricate and internal surfaces. Nickel coatings demonstrate corrosion resistance in numerous chemical settings [61]. Nevertheless, it is crucial to refrain from subjecting nickel coatings to sulfur dioxide, chlorine, sodium hypochlorite, salts, and nitric acid that contain mercury or silver. To provide effective corrosion protection, nickel coatings may need to reach thicknesses of about 120 μm, and in cases requiring machining, even thicker layers are necessary. In decorative applications, the use of a dual-layer nickel coating offers a two-fold advantage, enhancing both aesthetics and corrosion protection. This method involves a sulfur-free nickel undercoat acting as the cathode to the bright nickel topcoat. Consequently, corrosive attacks primarily target the surface, significantly delaying penetration to the underlying substrate metal. It's important to recognize that applying nickel coatings may impact various material properties. Notably, nickel-coated steels with a tensile strength exceedingly approximately 420 MN mm^{-2} tend to exhibit reduced fatigue strengths [60]. Extensive literature is available for exploring different engineering aspects of nickel coatings, machining, encompassing plating, post-treatment techniques for promoting adhesion, and a comprehensive discussion of various properties associated with electrodeposited nickel coatings. Nickel coatings have witnessed increased adoption of thermal spray applications following electroplating and electroless plating processes. These thermal spray methods encompass Twin Wire-Arc Spray (TWA) [62], High-Velocity Oxy-Fuel (HVOF) [63], and Cold Spray techniques [64], each offering unique advantages and operating principles. TWA is recognized for its efficiency in creating dense and well-adhered nickel coatings. It utilizes two electrically charged wires, typically one being pure nickel, which are melted by an electric arc. The molten material is propelled onto the substrate surface, where it solidifies to form a coating. TWA coatings exhibit low porosity and excellent bond strength, making them suitable for corrosion protection and wear resistance [65]. HVOF is renowned for its ability to produce high-density nickel coatings with reduced porosity and improved adhesion. During this procedure, a fuel gas is burned in the presence of oxygen to generate a high-velocity, high-temperature gas stream, propelling the molten nickel particles onto the substrate. HVOF coatings offer exceptional corrosion resistance and wear properties, making them suitable for critical applications in various industries [66]. Cold Spray stands out for its unique cold, supersonic particle deposition process. Unlike other thermal spray methods, it does not involve the melting of the material. Instead, solid nickel particles are accelerated to supersonic speeds and impact the substrate, causing mechanical bonding without significant thermal degradation. Cold Spray coatings are highly dense, maintain the material's original properties, and exhibit exceptional adhesion, making them ideal for sensitive substrates and complex geometries. Regarding the scientific principles that underlie these techniques, they all involve the kinetic energy of particles, contributing to the creation of coatings. Solid or molten nickel particles are expelled onto the substrate at significant velocity. Upon impact, these particles undergo deformation and stick to the surface, creating a cohesive and protective coating [67].

As a result of their study involving electrolytic pure Ni coatings on Cu substrates, Aniekwe and Utigard [68], examined the oxidation resistance at high temperatures and observed that pure Nickel coatings were highly effective in forming oxidation layers. Their research indicates that pure Nickel coatings create a very thin oxide layer even at high temperatures, with a thickness of less than 10 times. This finding suggests that pure

Nickel coatings offer effective resistance against oxidation at high temperatures, making them a viable option for protecting copper substrates and other underlying materials from oxidation. Such coatings can be utilized to prevent or delay oxidation in applications exposed to elevated temperatures, such as industrial equipment or electronic components. Lee and his colleagues, on the other hand, have achieved electroplated Ni-C composite coatings on copper (Cu) substrates by incorporating nano-diamond powders into the Ni coating, as part of their efforts to improve pure Ni coatings. They found that as the carbon (C) content increased, the coating thickness decreased. This phenomenon was attributed to the barrier effect of particle reinforcement on the reduction of Ni ions. Furthermore, they observed that pulse current was more effective than direct current and resulted in increased surface hardness. They determined that a stirring speed below 300 rpm, an electrolyte temperature between 30-50°C, and high C content were the most suitable parameters for Ni-C electroplated coatings, yielding the highest hardness values [69], for tribological applications. It should be noted that higher stirring speeds and temperatures may hinder the homogeneous reduction of Ni ions and C particles onto the surface. The application of pure Ni coatings to different substrates can be achieved through various methods, each yielding different results. Laser cladding represents a technique employed to enhance the corrosion resistance of Hastelloy-N alloy within a molten salt setting [70]. The pure Ni coating applied through laser cladding exhibited favorable metallurgical bonding with the substrate [70]. Another method is electrochemical co-deposition, where graphene platelets (GPL) are co-deposited with nickel to form composite coatings. These GPL/Ni composite coatings exhibited better corrosion resistance and higher microhardness compared to pure Ni coatings [71].

In addition to these methods, thermal spraying has been employed to deposit an Al/Ni composite coating on a pure Ti substrate [72]. This technique involves the deposition of an intermetallic NiAl layer *in situ*, which provides high-temperature oxidation resistance [72]. Cold spraying is an alternative technique employed for depositing pure Ni coatings. A research investigation into low-pressure cold-sprayed (LPCS) coatings involved the examination of the microstructure, compactness, and mechanical characteristics of Ni coatings deposited through LPCS [73]. Electrodeposition is a commonly used method for applying Ni coatings, particularly for decorative purposes. However, it can also be used to deposit Ni matrix composite coatings with various boride particles [74]. These composite coatings have the capability to enhance tribological characteristics, including reducing the coefficient of friction, in comparison to coatings made entirely of electrodeposited pure Ni [75]. The outcomes derived from the application of pure nickel coatings onto diverse substrates differ based on the particular application and methodology employed. For instance, in a research on the corrosion performance of Ni-Co alloy coatings in a marine atmosphere, the study explored pure Ni coatings and Co-Ni alloys with varied compositions. The corrosion behavior of these coatings was evaluated, providing insights into their performance in a marine environment [76]. Electroless pure Ni-based coatings have been extensively studied and applied to different substrates, offering unique advantages and results. The use of electroless nickel coatings has gained popularity because of their capacity to deliver enhanced wear resistance, hardness, and protection against corrosion. They can be applied to various substrates, including difficult-to-plate materials, providing an eco-friendly solution [77].

Overall, the application of pure Ni-based coatings to different substrates using various methods offers opportunities to enhance corrosion resistance, mechanical properties, and tribological performance. These coatings can be tailored to specific applications and environments, providing a versatile solution for a range of industries.

2.2.1.2 Ni-P-Based Coatings

Ni-P coatings, a form of electroless nickel plating, are applied to impart corrosion resistance, wear resistance, and other advantageous properties to a range of substrates. Ni-P coatings were first developed in the early 20th century as an alternative to traditional electroplating processes, which required the use of electrical current. Electroless nickel plating, including Ni-P, was first patented in the 1940s. Since then, it has become a widely adopted surface-finishing technique for various industrial applications. In this coating type, the substrate is thoroughly cleaned and treated to remove any contaminants, oxides, or scale. The substrate is immersed in a solution that contains a sensitizing agent, which promotes the adhesion of the coating. The substrate is submerged in a solution of electroless nickel, comprising a reservoir of nickel ions and a substance that acts as a reducing agent. The nickel ions deposit on the sensitized surface without the need for an electrical current. To create a Ni-P coating, a phosphorus-containing compound is introduced into the bath alongside the nickel source. The amount of phosphorus in the coating can be adjusted to influence properties like hardness, corrosion resistance, and solderability. Regulating the deposition rate and ensuring the quality of the coating depends significantly on the parameters within the plating bath. While Buchtik *et al.* [78], observed an almost 2-fold increase in the hardness values of heat-treated coatings, as shown in Figure 2.16, they highlighted that a rise in the P ratio results in

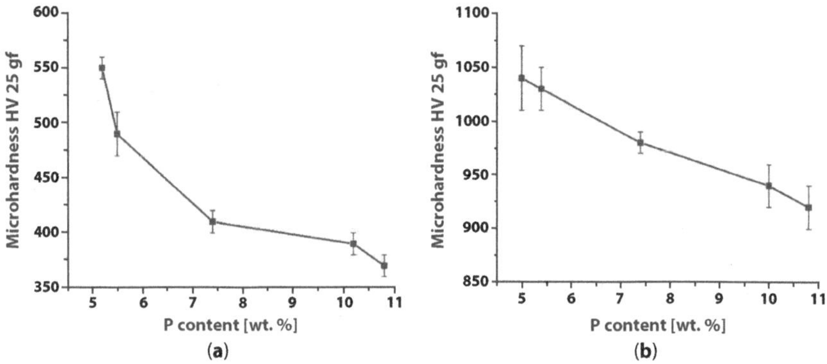

Figure 2.16 Change in hardness according to P ratio; (a) Ni-P coating, (b) heat-treated Ni-P coating ([78], licensed by CC BY 4.0).

a reduction of hardness values because of the transformation of the Ni-P structure into an amorphous state.

The specific chemicals and additives in the electroless nickel bath can be modified to achieve the desired coating characteristics. Ni-P coatings offer excellent protection against corrosion, making them valuable in applications exposed to harsh environments. The enhanced hardness and lubricating properties of the Ni-P coating lead to improved wear resistance and reduced friction. Ni-P coatings provide uniform coverage, even on complex shapes and internal surfaces. The inclusion of phosphorus can enhance the solderability of components coated with Ni-P. Ni-P coatings can be customized to attain a polished, reflective surface or a matte appearance [79–81]. The application of Ni-P coatings is versatile, extending to a range of materials such as metals (e.g., steel [82], aluminum [83], copper [84]), plastics [85], and ceramics [86]. They are particularly useful for components exposed to harsh conditions, such as those used in the aerospace, automotive, and oil and gas industries. Ni-P coatings are highly effective at preventing corrosion, extending the lifespan of coated components. These coatings reduce friction and wear, making them suitable for moving parts and components subject to abrasion. Ni-P coatings can be tailored to suit specific applications by adjusting the phosphorus content [87], and deposition conditions [88]. They provide consistent coverage, even in complex geometries. Ni-P coatings are conducive to soldering, benefiting the electronics industry [89]. Ni-P coatings are used on components like aircraft landing gear, hydraulic systems, and engine parts for their corrosion resistance and wear protection. They find application in automotive components such as pistons, valves, and transmission parts to enhance durability.

Ni-P coatings are used in printed circuit boards and connectors to improve solderability and reduce wear. In this industry, Ni-P coatings are applied to valves, pumps, and drilling equipment to withstand harsh conditions and prevent corrosion. Ni-P coatings protect equipment used in aggressive chemical environments. In summary, Ni-P coatings have a rich history and a versatile range of applications across various industries. The capacity to offer corrosion resistance, wear resistance, and other favorable attributes, coupled with their versatility across various substrates, positions them as a valuable option for improving the durability and performance of components in demanding conditions [90].

Conversely, composite coatings based on Ni-P represent sophisticated surface modifications, integrating electroless nickel plating with the addition of various materials or nanoparticles to augment the properties and functionalities of the coating. These composite coatings are designed to improve specific characteristics beyond what traditional Ni-P coatings can offer. Composite coatings of Ni-P generally consist of two primary constituents: Electroless Nickel Matrix; the base of the composite is electroless nickel, often a high-phosphorus version, which provides excellent corrosion resistance, wear resistance, and uniform coverage. Incorporated Materials: to create a composite, other materials or nanoparticles are added to the electroless nickel matrix. These materials are chosen based on the desired enhancement of specific properties. Ni-P-based composite coatings offer several advantages, depending on the nature of the incorporated materials: Incorporating hard particles, such as silicon carbide, boron nitride, or diamond, can substantially enhance the coating's hardness and wear resistance [91–94]. This makes these coatings suitable for applications where abrasion and wear are critical concerns. Materials like PTFE (polytetrafluoroethylene) [95], or graphite [96], can be added to reduce friction and provide self-lubricating properties, making them ideal for components that require smooth operation.

When Tsongas and his team [97], coated steel with electroless Ni-P and SiC added Ni-P and analyzed the outcomes, they observed notable enhancements in hardness and wear resistance, as depicted in Figure 2.17. Applying these coatings and subjecting them to heat treatment resulted in significant improvements in the mechanical properties of the steel substrate. The observed increases in hardness are particularly important for applications involving mechanical loads or abrasive forces. The protective layers formed by these Ni-P and SiC added Ni-P coatings enhanced the material's resistance to deformation and wear. This is especially beneficial in sectors like automotive, aerospace, and manufacturing, where materials frequently encounter demanding operational conditions.

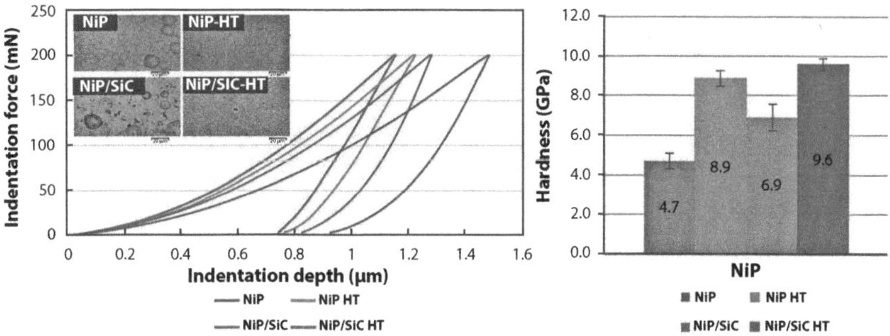

Figure 2.17 Hardness investigation and values of the Ni-P and Ni-P/SiC coatings with and without heat treatment ([97], licensed by CC-BY 4.0).

The fabrication of Ni-P coatings involves electroless/electrolytic plating, a process known for its ability to uniformly coat complex geometries. The key fabrication parameters and coating mechanisms are crucial aspects to consider in understanding the scientific underpinnings of Ni-P coatings. The plating bath for Ni-P coatings typically contains nickel salts, reducing agents, complexing agents, and phosphorus sources. The precise composition is critical for achieving the desired properties. Changes in the composition of the plating solution can affect the phosphorus levels in the coating, thereby influencing both corrosion resistance and hardness. It is crucial to keep the pH within a designated range in the plating solution to manage deposition rates and achieve consistent coating thickness. Temperature is equally vital, impacting reaction kinetics and the microstructure of the deposited coating. Elevated temperatures typically result in higher phosphorus content and enhanced corrosion resistance. Surface preparation of the substrate is a key parameter in Ni-P coating fabrication. Proper cleaning, activation, and pre-treatment of the substrate surface promote adhesion and uniform coating deposition. The duration of the electroless plating process, or deposition time, influences the thickness of the Ni-P coating. Extended deposition durations can result in the formation of thicker coatings, influencing characteristics such as hardness and resistance to wear [98–100].

Genova and colleagues [101], investigated the influence of the Ni source, represented by the $NiSO_4$ ratio, and the complexing agent, citric acid amount, on the phosphorus content within Ni-P coatings and the deposition rate. This examination is depicted in Figure 2.18. The study delves into the intricate relationships between the amount of $NiSO_4$ used as the Ni source, the concentration of citric acid as the complexing agent, and the resulting phosphorus content and deposition rate of the Ni-P

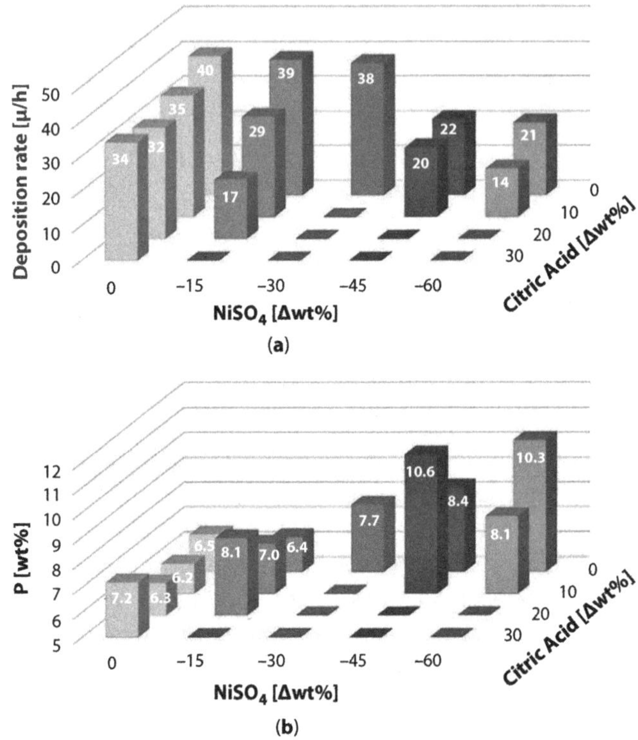

Figure 2.18 $NiSO_4$ and citric acid content effect on P content and deposition rate of Ni-P coating ([101], licensed by CC-BY 4.0).

coating. The combination with a concentration of +10 wt% citric acid and -45 wt% $NiSO_4$ stands out as the most advantageous, demonstrating superior outcomes in both plating rates of about 30 μm/h and P of 10.5 wt%. This highlights the intricate interdependence among the various reactants, emphasizing the impossibility of isolating their effects. The characteristics of the system are significantly impacted by the mutual interactions between each constituent. While the quantity of $NiSO_4$ has a noticeable effect on the simultaneous deposition of phosphorus, achieving an ideal equilibrium involving plating rate and bath stability necessitates concurrent optimization of the complexing agent's quantity.

The electroless plating of Ni-P coatings is an autocatalytic process, wherein the deposited coating serves as a catalyst for additional deposition. This self-sustaining mechanism ensures even coverage on intricate surfaces. Phosphorus is incorporated into the Ni-P coating during the electroless plating process, and the phosphorus content can be adjusted by controlling the composition of the plating solution. Higher phosphorus content contributes to increased corrosion resistance but may result in reduced hardness. Ni-P coatings can exhibit an amorphous or nanocrystalline structure, depending on the deposition conditions. The absence of a well-defined crystal structure contributes to the wear properties and hardness of the coating. The presence of phosphorus in Ni-P coatings is crucial for corrosion protection. A higher phosphorus content results in enhanced corrosion resistance, rendering Ni-P coatings suitable for use in aggressive environments. Therefore, Ni-P coatings, fabricated through electroless plating, offer a versatile solution for enhancing the properties of substrates in various industrial settings. By understanding and optimizing fabrication parameters such as plating bath composition, pH, temperature, substrate preparation, and deposition time, manufacturers can tailor Ni-P coatings to meet specific application requirements. The coating mechanisms involving autocatalytic deposition, phosphorus incorporation, and the formation of amorphous structures contribute to the unique properties that make Ni-P coatings valuable for corrosion preservation and wear resistance in diverse industries.

2.2.1.3 Ni-B-Based Coatings

Nickel-boron (Ni-B) based coatings represent a class of advanced surface treatments that have gained prominence in various industrial applications due to their unique properties and advantages. To comprehensively understand Ni-B coatings, it is essential to delve into their history, significance, usage areas, composition, fabrication methods, and key parameters, while also exploring the underlying coating mechanisms. The origin of Ni-B coatings dates back to the mid-20th century, as researchers aimed to improve the performance of materials in challenging environments [102]. Over the years, advancements in deposition techniques and alloy formulations have led to the optimization of Ni-B coatings, making them indispensable in a variety of industrial applications. Ni-B coatings typically consist of a nickel matrix with dispersed boron particles. The boron content can vary, and the coating's properties are influenced by the concentration of boron [103]. The primary usage areas of Ni-B coatings span across diverse industries, including aerospace, automotive, oil and gas, electronics, and tooling. Their applications range from protecting components against corrosion and wear to providing a low-friction surface for moving parts [104]. Understanding the coating mechanisms is crucial for grasping the functionality of Ni-B coatings. The deposition process involves the formation of a metallic matrix with embedded boron particles, and the coating mechanisms can be categorized into several key aspects: Ni-B coatings are commonly applied through an electroless plating method, wherein nickel ions undergo autocatalytic reduction in the presence of a reducing agent containing boron [105]. This process ensures uniform coating thickness on complex geometries and provides excellent adhesion to various substrates. During the deposition process, boron diffuses into the nickel matrix, forming a solid solution. As the coating cools, boron precipitates as hard, wear-resistant particles within the nickel matrix. This dispersion of boron contributes to the enhanced mechanical properties of the coating [106]. Ni-B coatings can exhibit amorphous or nanocrystalline structures, which further enhance their hardness and wear resistance [107]. The absence of a well-defined crystal structure in the coating contributes to improved mechanical properties and resistance to crack propagation. The incorporation of boron into the nickel matrix results in solid solution strengthening. The smaller boron atoms occupy interstitial positions in the nickel lattice, leading to increased hardness and improved mechanical properties [108]. Several methods are employed for fabricating Ni-B coatings, with electroless or electrodeposited plating being the most widely used. The key steps in the fabrication process include surface preparation, activation, bath preparation, deposition, and post-treatment. The success of the coating depends on the optimization of parameters such as bath composition, temperature, deposition time, and pH [109–111]. The characteristics of Ni-B coatings rely significantly on diverse parameters employed during the fabrication process. The temperature during deposition is crucial in influencing the structure and characteristics of the coating. Higher temperatures can lead to the formation of nanocrystalline structures, enhancing hardness [112]. The deposition rate, uniformity, and properties of the resulting coating are influenced by the pH and composition of the plating bath. Maintaining optimal conditions is essential for achieving high-quality coatings [113]. Surface preparation of the substrate is vital for ensuring good adhesion of the Ni-B coating [114]. Widely recognized for their outstanding hardness, resistance to wear, and corrosion protection, electroless Ni-B coatings can be elevated through the integration of supplementary materials, thus evolving them into composite coatings. The addition of extra elements enables customization of the properties of these coatings to fulfill specific application needs, promoting the advancement of advanced and multifunctional surface treatments. The incorporation of additional materials into electroless Ni-B coatings is driven by the aim to improve mechanical properties, tribological performance, and resistance to corrosion [115–119]. Commonly used additives include W [120], SiC

[121], Mo [122], and various nanoparticles. These additions serve as reinforcing agents, imparting unique attributes to the composite coatings.

Mukhopadhyay and colleagues [123], explored the influence of incorporating W and Mo on the hardness measurements of coatings based on Ni-B. They emphasized that the addition of W resulted in an improvement of hardness by more than 10%, while the presence of Mo led to a decrease in hardness, as shown in Table 2.2. The study emphasized that coatings subjected to heat treatment at 350°C for one hour exhibited a notably increased microhardness, ascribed to the phenomenon of precipitation hardening. The addition of W led to solid solution strengthening, enhancing the hardness of electroless Ni-B-W coatings. Conversely, Ni-B-Mo coatings displayed a lower microhardness compared to other coatings. This was ascribed to the coarse-grained structure of Ni-B-Mo coatings, possibly arising from increased growth stresses associated with Mo co-deposition. Significantly, following the heat treatment of Ni-B-Mo coatings (450°C), there was no substantial change in microhardness, indicating enhanced thermal stability of the deposits. The noted elevated microhardness in heat-treated Ni-B and its ternary counterparts is consistent with findings in the current body of literature.

Matik [124], investigated the advancement in surface properties of Fe-based materials produced through the powder metallurgy method with Ni-B coatings and explored how this development contributed to material characteristics such as friction, wear, hardness, and corrosion (Figure 2.19). According to the results, the wear rate in uncoated samples was determined to be three times greater than that of Ni-B coated ones, with the friction coefficient also exhibiting higher values. Furthermore, hardness values increased by more than six times owing to the hard Ni-B coating, and the loss rate resulting from corrosion decreased by more than tenfold. The columnar morphology of Ni-B coatings was noted to reduce the contact surface area under wear conditions, while oxide-based lubricants

Table 2.2 Hardness values of the samples versus heat treatment temperature and coating composition ([123] licensed by CC-BY 4.0).

Sample	Hardness (HV100)	
	Heat treating (350°C)	**Heat treating (450°C)**
Ni–B	1060 ± 20	×
Ni–B–Mo	738 ± 20	690 ± 20
Ni–B–W	1181 ± 20	×

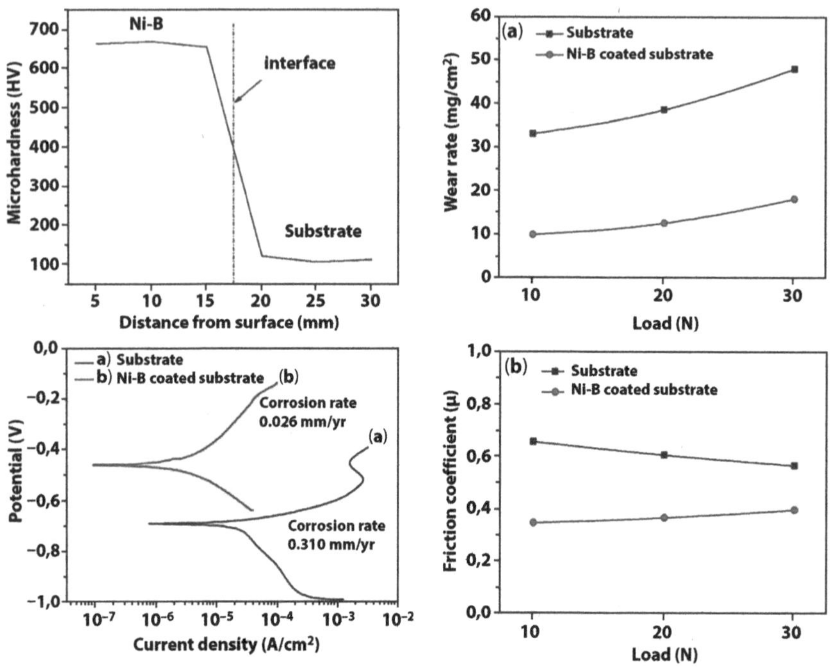

Figure 2.19 Hardness, polarization curves, wear rate and friction coefficient values of the samples ([124], licensed by CC-BY 4.0).

settling into voids on the rough surface were identified to reduce the friction coefficient. The enhanced corrosion resistance observed in samples coated with Ni-B can be attributed to the protective belongings of the Ni-B layer, acting as a barrier against corrosive agents. The Ni-B coating effectively prevents direct contact between the underlying Fe-based material and corrosive elements, thereby reducing the corrosion rate. The effectiveness of this protective capability is notably heightened by the Ni-B coating's consistent and compact arrangement. The hard Ni-B layer serves as a protective shield, offering resistance against mechanical wear and abrasion. The coating's ability to withstand deformation and maintain its hardness under external forces contributes to the overall durability and longevity of the coated material.

Das and Sahoo [125], conducted research into how coating parameters affect the properties of coatings obtained through the electroless Ni-B coating process. They utilized the Taguchi L27 array method in their study, considering parameters such as bath temperature (A), the ratio of the reducing agent (B), the ratio of the nickel source (C), and annealing temperature (D). They found that subjecting Ni-B coatings to heat treatment

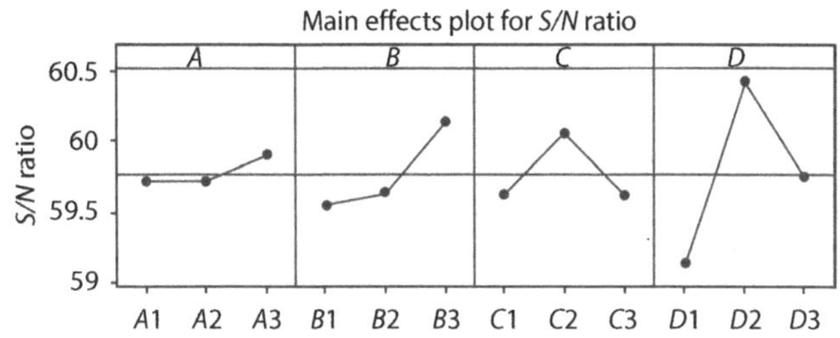

Figure 2.20 The impact of key factors on the signal-to-noise ratio ([125], licensed by CC-BY 3.0).

induces a transition from an amorphous structure to a crystalline one. The optimized condition was identified as A3B3C2D2, as shown in Figure 2.20, resulting in approximately 19% higher hardness compared to the first condition. ANOVA results indicated that the annealing temperature significantly impacted the hardness value of the Ni-B coating. Additionally, the concentration of the reducing agent had a noteworthy effect, contributing to 65% of the hardness variation. There was a moderate level of interaction among all the factors. The coating displayed a range of boron content, varying from 5.72% to 7.46%.

In summary, Ni-B-based coatings represent a technologically advanced solution for enhancing the properties of various materials. Their unique combination of hardness, wear resistance, and corrosion resistance makes them invaluable in industries where reliability and durability are paramount. The coating mechanisms involving electroless deposition, diffusion, precipitation, amorphous structure, and solid solution strengthening contribute to the exceptional performance of Ni-B coatings. As technology continues to advance, further refinements in fabrication techniques and coating formulations are expected, opening up new possibilities for Ni-B coatings in cutting-edge applications.

2.3 Recent Progress on Ni-Based Wear-Resistant Coatings

In recent times, Ni-based coatings have undergone diverse advancements through the application of various coating methods onto different types of substrates and composite processes, leading to the progressive diversification and enhancement of material properties. The utilization of Ni-based coatings has evolved significantly, with innovative techniques and compositional modifications contributing to the refinement of material characteristics. Various coating methods, such as thermal spraying, electroplating, and chemical vapor deposition, have been employed to apply Ni-based coatings onto substrates with distinct properties, thereby enabling tailored solutions for specific applications. Additionally, advancements in composite processes involving Ni-based materials have played a pivotal role in achieving superior material performance. By making strategic compositional adjustments, researchers have effectively enhanced the mechanical, thermal, and corrosion resistance properties of Ni-based coatings, broadening their suitability across various industries. This scientific progress reflects a comprehensive effort to harness the full potential of Ni-based coatings, paving the way for multifaceted improvements in material functionality and adaptability [126–132].

Various methods have been explored to minimize issues such as low coating homogeneity, uneven coating thickness, and interfacial weakness. This scientific pursuit involves systematic experimentation with different techniques aimed at mitigating the challenges associated with coating processes. The optimization efforts focus on enhancing the uniformity of coatings, addressing issues related to thickness variations, and strengthening interfacial bonds. Through the systematic evaluation of diverse methodologies, researchers strive to minimize the detrimental effects of low coating homogeneity, thickness irregularities, and interfacial weaknesses. This scientific approach not only underscores the commitment to overcoming challenges in coating applications but also contributes to the advancement of knowledge and innovation in materials science.

Yu *et al.* [133], have recently applied nickel-based alloy coatings on copper metals using the jet electrodeposition method. In this context, they assessed the wear properties of Ni, Ni-Fe, Ni-W, and Ni-Fe-W coatings. While they determined the hardness of these coatings as 350, 459, 526, and 507 HV, respectively, after heat treatment at 800°C, these values increased to 504, 599, 654, and 632 HV. Furthermore, they stated that Ni-Fe-W alloy coatings displayed the smallest wear track, as illustrated in Figure 2.21. The improvement observed in Ni-Fe and Ni-W alloy coatings was attributed to the appearance of a solid solution facilitated by the alloy element. This solid solution played a role in strengthening the structure oriented toward the underlying substrate. Despite Ni-W alloy coating having the highest hardness value, they noted that abrasive wear dominated in such coatings, leading to increased wear. In Ni-Fe-W coatings, the contribution of Fe resulted in more durable surfaces, where adhesive abrasive wear mechanisms played a combined role. Moreover, the Ni-Fe-W coating exhibited increased hardness, and hard phases such as WO_3 and agglomerated Fe_2O_3

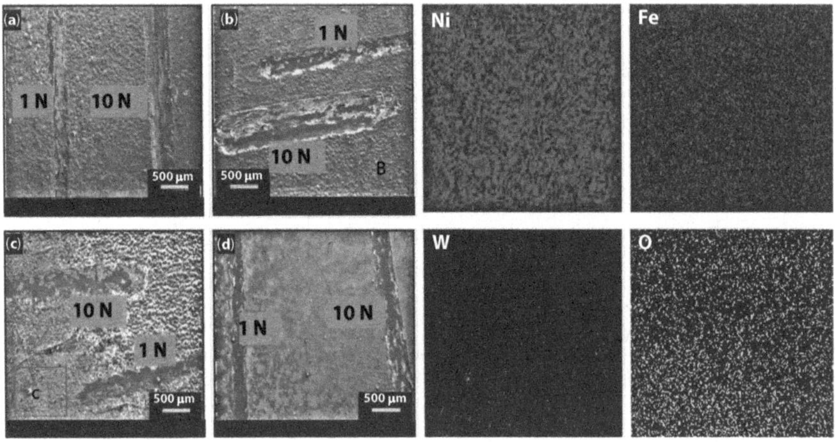

Figure 2.21 Wear traces exhibited by Ni-based alloy coatings deposited using jet electrodeposition with varying compositions, along with elemental analysis of these wear traces: (a) Ni, (b) Ni-W, (c) Ni-Fe and (d) Ni-Fe-W ([133], licensed by CC-BY 4.0).

appeared following heat treatment at 800°C. Ni-Fe-W alloy coatings exhibited lower friction coefficients, higher wear resistance, and increased oxidation resistance at elevated temperatures.

In conclusion, this study highlights the efficacy of the jet electrodeposition method in applying diverse nickel-based alloy coatings, emphasizing its importance and critical relevance for advancing materials with tailored properties for specific industrial requirements.

In another study, a Ni-P coating layer was applied to particle-shaped morphologies to shape these structures, and bulk materials with Ni-P interfaces were produced by electroless coating followed by hot pressing. In this context, Güler [134], applied Ni-P-CS (candle soot) coatings to copper powders, followed by hot pressing to produce copper composites with Ni-P-CS interfaces (Figure 2.22). According to the obtained results, composites with Ni-P-CS interfaces exhibited more than 10 times less wear loss compared to pure Cu materials (Figure 2.23). Clearly demarcating the Ni-P and Ni-P-CS layers at interfaces reveals the establishment of a strong connection in the interface, as depicted in Figure 2.22. During wear, the Ni-P-CS stored around each Cu particle continuously resisted abrasive effects through the functionality of both a hard and lubricating layer, ensuring effective protection (Figure 2.23).

The electroless coating technique has been effectively employed to modify the particle morphology of Ni-based coatings, showcasing its ability to customize coatings for specific particle structures. This versatility holds

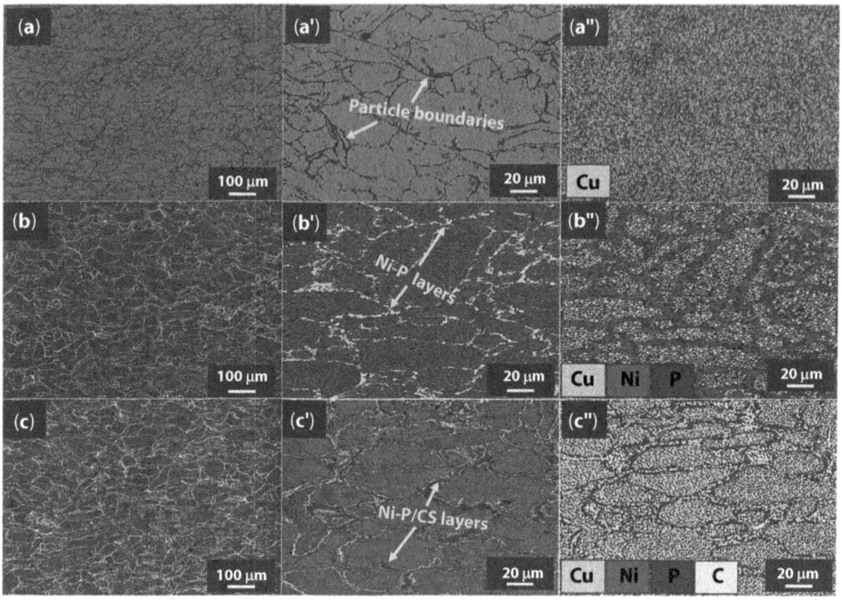

Figure 2.22 SEM investigation of hot-pressed samples of (a) pure Cu, (b) Ni-P coated and (c) Ni-P-CS coated Cu particles ([134], with permission from Elsevier).

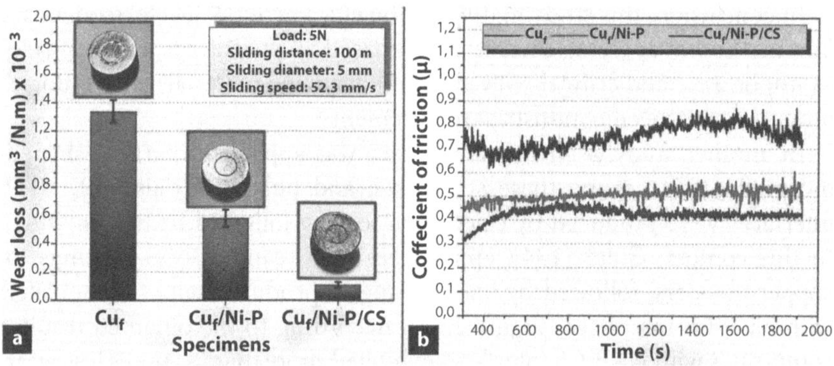

Figure 2.23 Wear outcomes of hot-pressed samples containing coated Cu particles; (a) wear rate and (b) coefficient of friction values ([134], with permission from Elsevier).

critical implications for various applications where the unique characteristics of Ni-based coatings on particle morphologies are essential. The ability to apply electroless coating to Ni-based coatings not only showcases its adaptability but also suggests its potential significance in critical applications that demand precise control over coating characteristics for enhanced performance and functionality.

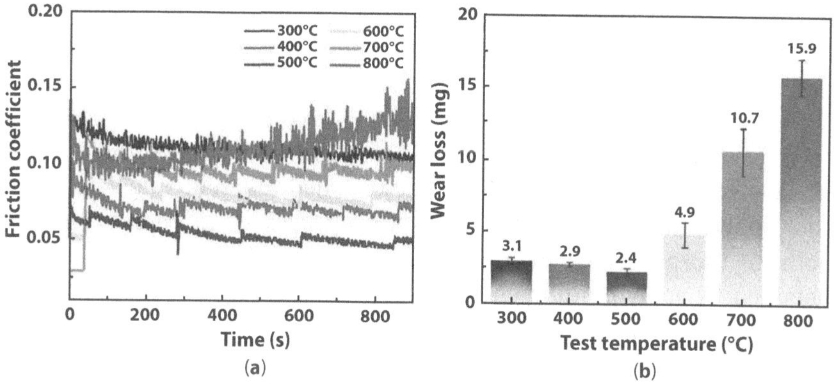

Figure 2.24 The results obtained during wear tests conducted at different temperatures for the coating; (a) coefficient of friction (b) wear loss values ([135], with permission from Elsevier).

Wu and colleagues [135], applied nickel-based coatings on a copper substrate using the plasma cladding method. Achieving an extraordinary coating thickness of 2mm, they utilized Ni60A powder (comprising Cr: 15.0, Fe: 8.0, Si: 4.0, B: 3.5, C: 0.8, and Ni in balance) as the coating element. Wear tests were conducted under high temperatures, up to 600°C, making it a critical and highly interesting study for coppery blast furnace tuyere applications. The results revealed that, at 400°C, the coatings exhibited wear resistance 17 times greater than that of pure copper. Up to 600°C, the protective interface formed by the NiO oxide layer reduced wear losses. However, beyond 600°C, an increase in wear losses was observed due to the softening of the coating and peeling of the oxide layer from the surface (Figure 2.24). The heightened wear resistance can be ascribed to the coating's microstructure, comprising a solid solution of γ-(Cu, Fe, Ni), reinforcing phases (Cr23C6 and CrB), an intermetallic compound of Fe3Ni, and interdendritic phases of Ni3Si.

This method may be crucial for applications where a high coating thickness is required. Achieving an exceptional coating thickness of 2mm, the plasma cladding technique employed by Wu and the team [135], holds significant importance for specific applications. The substantial thickness not only enhances the protective capabilities of the coating but also addresses the requirements of applications demanding robust and durable coatings. This is particularly relevant in industries such as coppery blast furnace tuyere applications, where the coating's ability to withstand wear and high temperatures is of utmost importance. The notable thickness achieved through this method contributes to the longevity and resilience of the coating, making it well-suited for scenarios where thicker coatings play a pivotal role in ensuring extended service life and enhanced performance.

Tan and his team [136], applied Ni-W-TiO_2 composite coatings on steel surfaces using the pulse electrodeposition method. However, unlike traditional electrodeposition processes, they applied different values of ultrasonic power during the electrodeposition process. The improvement in both the morphological development and wear resistance of the coatings obtained by applying different ultrasonic power values was emphasized. In accordance with the findings presented in Figure 2.25, a rise in coating thickness was noted at an ultrasonic power of 50 W, resulting in a coating that was 6 μm thicker compared to the untreated one. It was determined that when no ultrasonic power was applied or when it exceeded 50 W, TiO_2 particles agglomerated, and they hindered the reduction of ions during Ni-W coating. Therefore, the optimum ultrasonic power value was reported to be 50 W.

When examining the morphological development, it was observed from the obtained results that this development directly influences the wear resistance of the coatings. Upon reviewing the wear loss values exhibited by

Figure 2.25 The morphology of NiW-TiO_2 coatings obtained by different ultrasonic powers; (a) 0 W, (b) 25 W, (c) 50 W and (d) 75 W ([136], with permission from Elsevier).

Table 2.3 Wear rate values of the coatings fabricated under different ultrasonic powers during the pulse electrodeposition [136].

Ultrasonic power (W)	Wear rate at room temperature	Wear rate at 300°C
0	~11.4×10–6 mm3/N·m	~10.4×10–6 mm3/N·m
25	~9.5×10–6 mm3/N·m	~11.35×10–6 mm3/N·m
50	8.6×10–6 mm3/N·m	7.4×10–6 mm3/N·m
75	~12×10–6 mm3/N·m	~14.5×10–6 mm3/N·m

the coatings in Table 2.3, it was found that the addition of 50 W ultrasonic power contributed to the homogenization of TiO_2 particle distribution within the Ni-W matrix, resulting in the lowest wear loss values. Moreover, the results indicate that the coating generated at 50 W displayed a more intricate and compact microstructure, characterized by a crystallite size of 14 nm. This outcome implies that the application of 50 W ultrasonic power led to the formation of a coating characterized by small and closely packed crystalline entities. This refined microstructure is beneficial for improving wear resistance as it offers a stronger and more cohesive surface, decreasing the chances of material loss and wear under abrasive conditions. The compact arrangement of crystallites contributes to the coating's durability and ability to withstand mechanical stresses, thereby reinforcing its overall wear resistance performance.

As a result, the application of ultrasonic power in the fabrication of Ni-based coatings is notably crucial, as indicated by the obtained results. The fine and closely packed crystalline entities formed under the influence of ultrasonic power create a more robust surface, minimizing material loss and improving the coating's ability to withstand abrasive conditions. In summary, the incorporation of ultrasonic power in the electrodeposition process for Ni-based coatings effectively addresses challenges related to particle distribution, microstructural development, agglomeration, and wear resistance, leading to coatings with enhanced quality and performance. Therefore, the incorporation of ultrasonic power, especially at the specified levels, emerges as a crucial parameter in the fabrication of Ni-based coatings, offering a pathway to achieve coatings with superior microstructural characteristics and enhanced wear resistance properties.

The research undertaken by Zhong *et al.* [137], emphasizes the influence of modifications on coatings, specifically in the case of Ni-W-P and unalloyed Ni-W-P coating improved with graphene. Traditional application of these coatings to surfaces has been a common practice, but the introduction of graphene/graphene oxide particles, such as furan epoxide graphene (FGO), as modifying agents has been found to significantly impact wear resistance, as illustrated in Figure 2.26. As depicted in the figure, unreinforced Ni-W-P coatings exhibit the widest groove width and depth among the tested coatings. On the other hand, G and GO-enhanced coatings create less distinct wear tracks, and the depth of these tracks is considerably reduced compared to unalloyed coatings. This results in improved friction properties, showcasing better performance. It is noteworthy that FGO-enhanced coatings induce the shallowest wear tracks, and the worn region is determined to be the narrowest among all coatings. The study attributes this effect to the modification process facilitated by FGO, which promotes a more even distribution within the Ni-W-P coating structure. This improved distribution is suggested to reduce surface irregularities, consequently minimizing the possibility of agglomeration on the coating surface. In conclusion, the utilization of FGO in the modification process is noted to enhance the wear properties of Ni-W-P coatings. The enhanced distribution of FGO within the coating structure decreases roughness, thereby improving wear resistance.

The findings suggest that addressing wear issues in coatings could benefit from the pre-modification of additive particles before their incorporation. Pre-modifying the contributing particles, as exemplified by the case of reinforcement particles, appears to offer a promising solution. This process enables a more effective dispersion of the modified particles within the

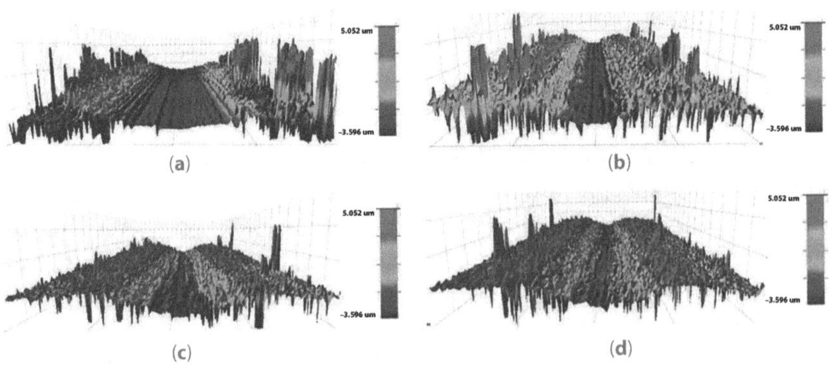

Figure 2.26 3D morphology of the worn surfaces; (a) Ni-W-P, (b) Ni-W-P-G, (c) Ni-W-P-GO and (d) Ni-W-P-FGO ([137], with permission from Elsevier).

coating structures. Consequently, it not only mitigates wear problems but also contributes to reducing surface irregularities, as evidenced by the shallower wear tracks and narrower worn regions observed in enhanced wear-resistant coatings compared to unmodified ones. Therefore, it is recommended that, in tackling wear challenges, consideration should be given to modifying additive particles before their application in Ni-based coatings.

The use of Ni-based coatings on polymer surfaces has recently attracted considerable interest, broadening the research horizon. In scenarios where wear resistance plays a vital role, Ni-P coatings have been utilized to advance the tribological characteristics of specific polymers. In this context, Güler and colleagues [138], achieved a twofold reduction in wear loss by electroplating Ni-P coatings onto mechanically reinforced Novolac matrix composites compared to coating-free samples. They highlighted that Ni-P coatings containing 2%-8% P exhibit a nanocrystalline structure, significantly increasing the surface hardness and wear resistance. Considering that the analysis of wear debris in nano-tribological studies is an effective method for determining wear mechanisms, their examination of wear debris, as shown in Figure 2.27, revealed that the debris generated from Ni-P-coated surfaces consists of submicron particles. Especially on the robust surfaces coated with hard Ni-P, they noted that, during friction, abrasive particles of Ni-P interact with the surface and detach from the matrix material on the paths of wear. The presence of Fe, Ni, and O elements in the region is apparent in Figure 2.27d and Figure 2.27f. The robust Ni-P layer facilitates the migration of iron from the abrasive steel surface, showing an iron content approximately four times higher than that of uncoated surfaces, which is around 13%. Given the small size of abrasive particles, debris size, and restricted iron transfer, it is confirmed that abrasion is the primary wear mechanism in Ni-P coated composites.

The understanding derived from this is that Ni-based coatings can be made preferable for tribological applications not only on specific types of materials but also on polymer surfaces. Ni-based coatings can extend their application beyond certain material types to include polymer surfaces, making them a preferred choice for tribological applications. Addressing the difficulties in rendering polymer surfaces resistant to wear poses a significant challenge. The versatility of Ni-based coatings in this context could provide a solution by offering a protective layer that enhances wear resistance on polymer surfaces. This adaptability underscores the potential of Ni-based coatings to mitigate wear-related challenges across a broader spectrum of materials, including polymers.

An interesting method for tribological modification of another Ni-based coating was pursued by Liu and his team [139], where LGC and CLGC

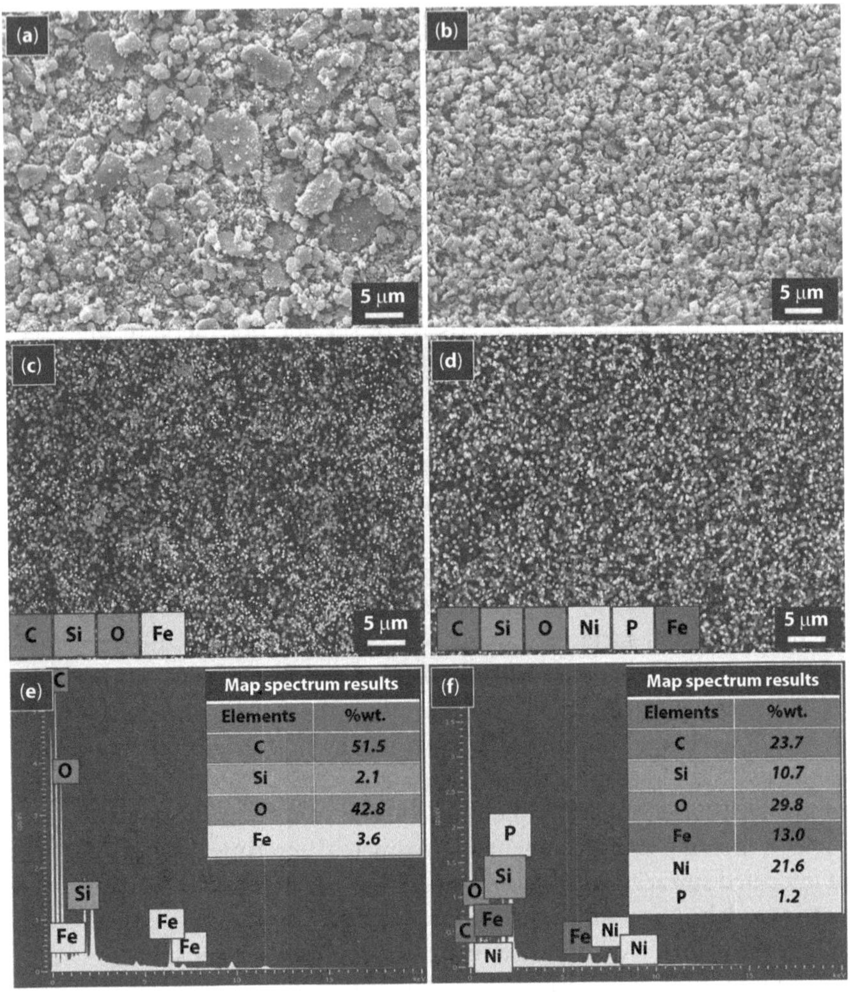

Figure 2.27 SEM and EDS findings of debris for (a-c-e) uncoated samples and (b-d-e) Ni-P coated samples ([138], with permission from Elsevier).

coatings were applied onto Cu substrates. Unlike the traditional LGC method, a novel approach utilizes cold spraying of the NiCu alloy onto the interface between the substrate and coating. This serves to mitigate the adverse effects arising from the high reflectivity of the Cu substrate and minimizes the mismatch in the coefficient of thermal expansion between the substrate and coating. Additionally, the quality of the coating was investigated by altering the production parameter of the NiCu layer applied through cold spraying, specifically the gas pressure. Increasing the pressure of the carrier gas resulted in a shorter spraying distance, consequently increasing particle velocity toward the substrate. Consequently, the compression and work-hardening interactions among particles heightened, resulting in an elevation of both the quality and the deposition rate of the coating. Subsequently, after achieving surface improvement with the cold spraying method under optimal parameters, analyses of the wear resistance of the coatings obtained by creating the composite gradient revealed improvements in wear resistance, as demonstrated in Figure 2.28. This improvement is particularly noticeable in the pre-coating wear resistance achieved in the CLGC coating.

Based on this result, it can be asserted that when the laser cladding method is employed for applying Ni-based coatings onto Cu substrates, the use of cold spraying as an intermediate layer eliminates the high reflectivity problem, resulting in coatings with a well-established interface. This statement encapsulates a significant finding.

Consequently, the tribological characteristics of Ni-based coatings are impacted by diverse factors associated with coating composition, application techniques, process parameters, and other variables. The specific Ni-based structures used in the coating can significantly impact tribological properties. Different Ni alloys offer varying hardness, wear resistance, and frictional characteristics. The addition of elements such as tungsten, cobalt, or boron can alter the coating's hardness, adhesion, and tribological behavior. Coating application methods, and method parameters are also critical to obtain the enhanced tribological features of the coatings. The microstructure characteristics of the coating, encompassing factors such as grain size and orientation, play a pivotal role in influencing hardness, friction, and resistance to wear. The crystallographic structure of the coating, such as the presence of certain phases, can influence tribological behavior. The surface quality of both the coating and substrate can influence friction

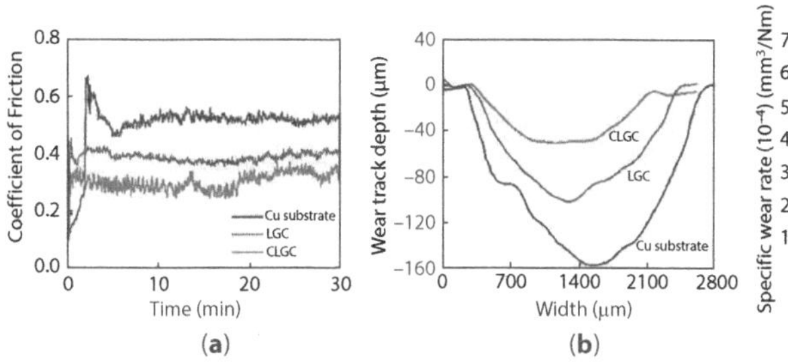

Figure 2.28 Wear results of samples; (a) coefficient of friction, (b) wear track depth and (c) wear rate values; *LGC: Laser Cladded Gradient Coating and CLGC: Cold Spray/Laser Cladding Composite Gradient Coating* ([139], licensed by CC-BY 4.0).

and wear properties. Surfaces with rough textures may undergo increased abrasive wear. Post-deposition heat treatment can alter the microstructure and hardness of the coating, impacting its tribological performance. The bonding between the substrate and coating, impacted by surface preparation and deposition conditions, is essential for ensuring wear resistance. Interface quality, including the presence of interlayers or transition zones, can affect the overall tribological behavior. Understanding and optimizing these factors are essential for tailoring Ni-based coatings to specific tribological requirements in various applications.

Tribological Properties of Ni-based Wear-resistant Coatings

Tribology is the scientific investigation of surfaces in motion relative to each other, encompassing the analysis of friction, wear, and lubrication. In industrial applications, nickel-based wear-resistant coatings play a critical role in enhancing tribological properties, contributing to increased durability and reduced friction in diverse mechanical systems [140]. Nickel-based coatings stand out for their ability to significantly minimize friction between sliding surfaces, thereby enhancing energy efficiency [141]. Beyond this, these coatings demonstrate impressive wear resistance, functioning as protective layers that withstand abrasive wear and shield the substrate material from the external environment. Nickel-based coatings demonstrate remarkable effectiveness in withstanding different types of wear, such as adhesive, abrasive, and fatigue wear, due to their effective combination of high hardness and toughness [142]. Moreover, these coatings often offer corrosion resistance, acting as a barrier against corrosive substances. Depending on the application, some nickel-based coatings may incorporate lubricating properties, further minimizing friction and wear. Notably, the temperature resistance of nickel-based coatings renders them suitable for challenging environments, such as those encountered in manufacturing processes or aerospace applications [140]. Overall, the tribological properties of these coatings are integral in enhancing the performance and longevity of components across various industries. Ongoing research endeavors aim to improve nickel-based coatings by exploring new compositions and deposition techniques to meet evolving industrial demands [143–145].

The following outlines key aspects of the tribological properties exhibited by nickel-based wear-resistant coatings:

Friction Reduction: Critical to enhancing wear resistance in nickel-based coatings is the reduction of friction [146]. Friction, characterized as the impediment to relative movement between two contacting surfaces, holds a crucial position in tribology and wear resistance. Efforts to minimize friction are effective in mitigating wear and surface damage resulting from the rubbing or sliding of materials. The substantial hardness inherent in nickel-based coatings is a key contributor to friction reduction. Materials with high hardness exhibit enhanced resistance to deformation and surface damage, establishing a protective barrier between interacting surfaces [140, 147]. As a result, when two surfaces undergo sliding motion, the robust nickel-based coating effectively resists wear and tear, thereby reducing the likelihood of material loss and degradation [148].

High Hardness and Toughness: The intrinsic hardness of nickel-based coatings stands as a pivotal attribute defining their success in various tribological applications. While nickel itself possesses inherent hardness, its application as a coating further enhances the wear resistance of underlying substrates. Several factors contribute to the hardness of these coatings, including the deposition process and the integration of alloying elements [149]. Hardness, a critical characteristic, significantly influences the coatings' ability to withstand friction, resist wear, and provide effective surface protection. The composition of nickel-based coatings plays a crucial role in determining hardness, with electroless nickel-phosphorus (Ni-P) coatings demonstrating a noteworthy correlation between phosphorus content and hardness [138, 140]. Higher phosphorus content generally results in increased hardness, allowing for customization based on specific application requirements. The incorporation of alloying elements, such as boron or silicon, contributes to the development of a sophisticated microstructure, further enhancing hardness and overall tribological performance. Microstructural properties, including grain size and distribution, are vital contributors to hardness in nickel-based coatings. A fine-grained microstructure, often associated with higher hardness, strengthens the coating along grain boundaries [150]. Post-deposition heat treatment processes, like annealing, offer a means to optimize the crystallographic structure, establishing a balance between hardness and other mechanical properties. Enhanced hardness equips the coating to resist abrasive wear, preventing deformation and surface damage during frictional contact [151]. Industries such as manufacturing, automotive, and aerospace benefit from nickel-based coatings due to their ability to maintain structural integrity under harsh conditions. While hardness is fundamental, the complementary property of toughness plays a crucial role in resisting crack propagation and increasing overall durability. In tribological scenarios involving dynamic loads, impacts, and cyclic stresses, toughness becomes particularly critical. Well-designed nickel alloys exhibit toughness that prevents premature coating failure, especially in applications with varying mechanical forces. This capability ensures that the coating can absorb energy and deform plastically before succumbing to fractures or spalling. The synergy between hardness and toughness in nickel-based coatings offers a comprehensive solution to wear-related challenges. The hard surface resists abrasive wear, while toughness prevents catastrophic failures, extending the coating's lifespan. This synergistic effect proves advantageous in intricate applications like cutting tools, where the coating faces a complex interaction of force and wear. The delicate balance between these properties enables nickel-based coatings to consistently deliver reliable performance in demanding tribological environments [140, 152–153].

Lubrication Enhancement: Effective lubrication is crucial in influencing the tribological characteristics of Ni-based coatings, providing an added dimension to their overall performance. It holds significant importance in reducing friction, minimizing wear, and overall enhancing the efficiency of tribological systems. Lubricants can be integrated into the coatings themselves or applied externally, providing crucial support in optimizing the performance of Ni-based coatings. Specifically, Ni-based coatings can be customized to include solid lubricants like molybdenum disulfide or graphite. These solid lubricants play a role in establishing a low-friction interface between surfaces, generating a defensive layer that minimizes direct contact between the coating and mating surface. Consequently, wear and friction are mitigated. Some Ni-based coatings also inherently exhibit lubricating properties, facilitating smooth operation in diverse tribological environments. Efficient lubrication not only minimizes wear on Ni-based coatings by averting direct metal-to-metal contact but also extends their service life. This becomes particularly crucial in applications involving sliding or rolling motion, where wear-induced material loss could lead to component failure. The synergy between lubricity and the inherent wear resistance of Ni-based coatings results in prolonged service life and reduced maintenance requirements across various industrial settings. Moreover, lubrication in Ni-based coatings contributes to temperature regulation during friction processes. By reducing the heat generated through friction, lubrication plays a critical role in maintaining the stability of the coating and preventing degradation caused by thermal factors. This feature proves especially advantageous in high-speed applications or environments characterized by elevated temperatures. In essence, the integration of effective lubrication not only enhances the tribological performance of Nickel-based coatings but also contributes to their resilience in challenging operational conditions [154–156].

Wear mechanisms of Ni-based Wear Resistance Coatings

Different tribotests, employing various contact configurations such as ball-on-disk, pin-on-disc, and block-on-ring, were utilized to assess the wear and friction characteristics of Nickel-based coatings. These configurations are commonly favored in tribological testing. Numerous factors influence tribological properties, encompassing coating type, composition, parameters, method of application, wear type and parameters, as well as the substrate material's type and properties. Coatings applied to surfaces in tribology-based applications serve the dual purpose of imparting smoothness and hardness, ultimately mitigating friction and wear. Among these, friction emerges as a crucial element that influences the effectiveness of mechanical assemblies with sliding surface contact and frequently plays a substantial role in wear, thereby restricting the service life of devices. The primary goal of applying a coating is to reduce friction, and the low friction force observed in Ni coatings is ascribed to the low surface energy of the Ni-based coating surface. This feature diminishes adhesion forces between surface and wear particles [129, 148].

Alloying nickel-based (Ni-based) alloys improves their wear properties by introducing hard phases and enhancing solid solution strengthening. Elements like chromium, molybdenum, and tungsten form hard carbides or intermetallic compounds, increasing the alloy's hardness and wear resistance [157]. From a nanotribology perspective, this enhancement in resistance to wear is primarily due to the altered interaction at the nano scale between the alloy surface and the counterface. The presence of harder phases reduces adhesion and abrasion, two primary wear mechanisms at this scale. These harder phases resist the mechanical removal of material (abrasion) and reduce material transfer between contacting surfaces (adhesion). Additionally, some alloying elements can lead to the formation of protective tribofilms and modify surface roughness, further influencing wear behavior. The overall effect is a significant enhancement in the wear resistance of Ni-based alloys, making them suitable for demanding applications where resistance to surface degradation is crucial [158–160].

In their research, Yu and colleagues [146], investigated the impact of enhancing nickel-based coatings with iron (Fe) and tungsten (W) through a jet electrodeposition technique. Following the application of these coatings, the specimens were heat-treated at 800°C for a duration of 30 minutes. The study measured the friction coefficients over time for four distinct coatings (Ni, Ni-W, Ni-Fe, and Ni-Fe-W) under the stresses of 1 N and 10 N, as depicted in Figure 2.29. The findings revealed a significant reduction in friction at lower loads due to the alloying process, with the Ni-W-Fe coating exhibiting the most notable decrease in friction coefficient (Figure 2.31-a). Under a heavier load of 10 N, the Ni-W coating displayed elevated friction coefficients, a consequence of tungsten's contribution to increased hardness and subsequent abrasive wear (Figure 2.31-b). Notably, the Ni-Fe-W coating continued to show the lowest friction coefficients at this higher load, highlighting that the integration of Fe and W not only fortifies the mechanical strength of the nickel coating but also promotes the development of a durable tribofilm. This film effectively maintains its

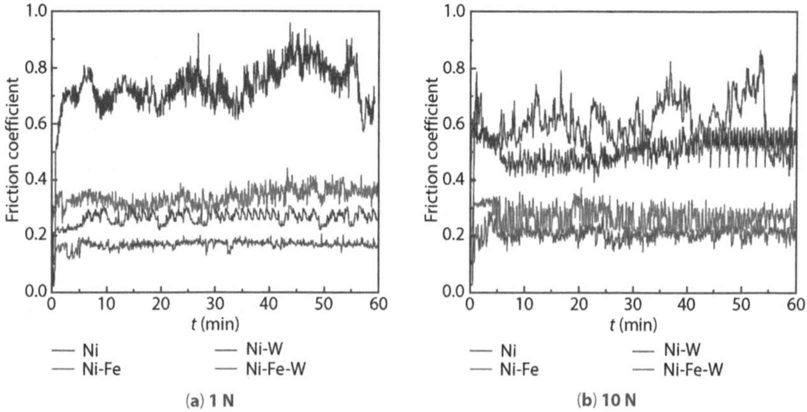

Figure 2.29 The variation of friction coefficient values ([146]; licensed by CC-BY 4.0).

protective qualities under the strain of greater loads, indicating a beneficial composite effect that boosts the coating's wear resistance.

The incorporation of hard ceramic particles like SiC, B4C, TiC, Al2O3, WC into Ni-based coatings significantly alters wear mechanisms by enhancing the coatings' hardness and resistance to abrasive wear [31]. These ceramics, being much harder than metallic counterparts, act as barriers to deformation and cutting actions posed by abrasive particles, reducing material removal rates. Additionally, they contribute to the reduction of adhesive wear by limiting direct metal-to-metal contact and by providing a thermal barrier that retains the coating's integrity at high temperatures, ensuring the wear resistance is maintained even under thermal cycling [161]. Moreover, the embedded ceramics can potentially deflect or blunt the abrasive particles, which further contributes to the longevity of the coating. These particles also generate a load-bearing effect, shifting the load from the softer Ni matrix to the harder ceramic particles. This action reduces the contact stress on the Ni-based coating, consequently alleviating wear [162]. The addition of lubricious additives such as graphene, h-BN, graphite, MoS2, CNTs, and GO to Ni-based coatings generally modifies the wear mechanisms by creating a self-lubricating effect that reduces friction between contact surfaces. These additives have the ability to create a lubricating layer as a result of the wear process. This layer functions as a solid lubricant, reducing the extent of metals contact and consequently lowering the rate of wear. This layer, often referred to as a tribofilm, can be continuously replenished during operation, which helps to sustain the coating's low friction over time. The overall result is a reduction in the coefficient of friction, which leads to a decrease in both adhesive and abrasive wear. Additionally, the presence of these lubricious additives in the coating can distribute the applied load more evenly across the surface, reducing the stress on any single point and thereby diminishing the likelihood of wear and surface fatigue [163, 164].

Yan *et al.* [165], studied the tribological characteristics of Ni-GO/SiC coatings deposited on 2218 alloy through the electroplating method. They tested three different coatings: Ni-GO, Ni-SiC, and Ni-GO/SiC, coded as S2, S3, and S4, respectively, with the uncoated aluminum base material coded as S1. The findings, as depicted in Figure 2.30, demonstrate that the sample coated with Ni-GO (S2) displayed the least coefficient of friction. This suggests that GO is effective in establishing a protective nano-tribolayer, which diminishes wear by minimizing surface contact. The Ni-SiC sample (S3) displayed the highest friction values, suggesting that while SiC particles enhance abrasion resistance, they do not significantly reduce friction compared to GO. Nonetheless, the wear rate examination in Figure

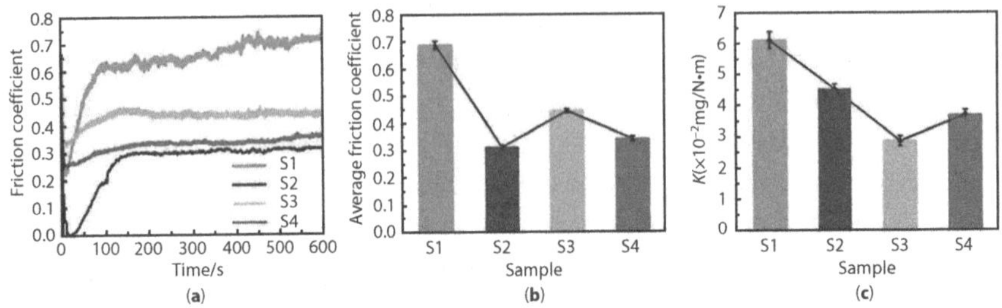

Figure 2.30 Changes in Friction Coefficient and Wear Rate: (a) Friction Coefficient over Time; (b) Mean Friction Coefficient; (c) wear rate ([165]; licensed by CC- BY 4.0).

2.30-c indicates that the Ni-SiC coating exhibited the most minimal wear rate, attributed to the superior barrier function of SiC against wear. The Ni-GO/SiC combined coating strikes a balance, offering moderate friction reduction and improved abrasion resistance, underscoring the importance of selecting suitable coating compositions for specific service conditions.

According to their findings, each coating shows a different wear mechanism. Figure 2.31 presents SEM images and 3D morphologies of worn surfaces, revealing distinct wear mechanisms for each sample. Sample S1 (uncoated 2218AlA) shows severe adhesive, abrasive, and mild fatigue wear, characterized by water wavy marks, plastic deformation, and deep grooves with a wear depth of 22.998 μm. Sample S2 (Ni-GO coating) exhibits mild adhesive and abrasive wear with a smoother, flatter surface and shallow grooves (3.299 μm depth), indicating effective lubrication by GO nanosheets. Sample S3 (Ni-SiC coating) demonstrates severe abrasive wear with deeper, wider, and more evenly distributed grooves (7.691 μm depth), and less plastic deformation due to higher microhardness. Sample S4, which features a coating of Ni-GO/SiC, exhibits a combination of mild abrasive and adhesive wear characteristics. This amalgamation resembles the wear traits observed in both S2 and S3. As a consequence, it produces a moderately rough surface with a wear depth measuring 6.972 μm. In general, the S4 specimen is deemed more effective in reducing friction and enhancing wear resistance, attributed to the synergistic effects of SiC particles and GO nanosheets.

In a separate investigation, Zhang *et al.* [166], applied Ni-P and Ni-P-CS coatings onto low carbon steel using the electroless method, aiming to discern discrepancies in their respective wear behaviors. Upon scrutinizing the surface images post-wear test in Figure 2.32, a discernible similarity emerges in the wear characteristics of both coatings, with the prevailing mechanisms being adhesive wear complemented by a minor abrasive wear

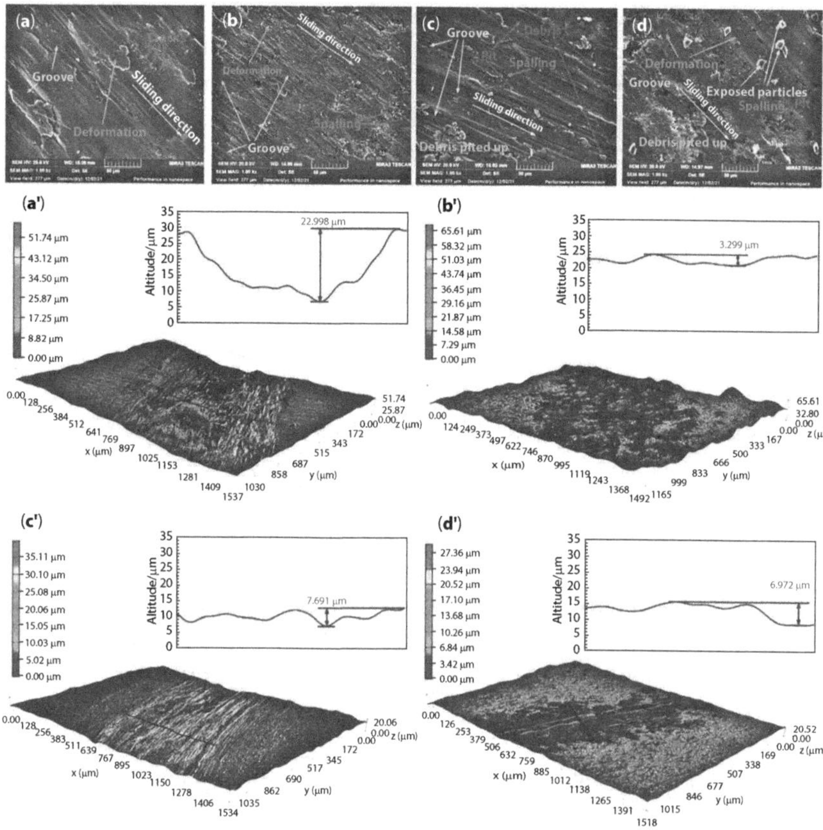

Figure 2.31 Analyze SEM images and 3D morphologies of worn surfaces: (a, a') S1, (b, b') S2, (c, c') S3, and (d, d') S4 samples ([165]; licensed by CC- BY 4.0).

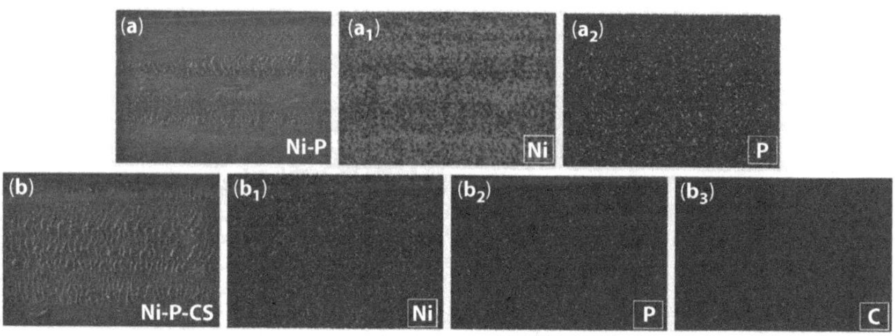

Figure 2.32 SEM images depicting the Ni-P and Ni-P-CS composite coating, along with elemental mappings (a1, a2) for Ni-P and (b1, b2, b3) for Ni-P-CS, were acquired through EDS analysis following friction and wear ([166]; licensed by CC- BY 4.0).

element. The Ni-P coating demonstrates a mean coefficient of friction of approximately 0.55, accompanied by a wear loss value totaling 25 mg. In sharp contrast, the Ni-P-CS coating, which integrates candle soot as a reinforcing lubricant, displays a unique behavior, showcasing an average coefficient of friction around 0.35 and a notably decreased wear loss, amounting to only 1 mg. This discrepancy underscores the potential of incorporating candle soot as a lubricating reinforcement to elicit a distinct coating mechanism. The detailed mechanism analysis is shown schematically in Figure 2.33. In the initial stage (A), the coating undergoes cracking and peeling due to the substrate material's hardness surpassing that of the coating. Candle soot acts as a solid lubricant, reducing the friction coefficient and force of the system during friction. As the coating experiences wear, compressed candle soot within the coating fills the wear gap, resulting in a smoother surface characterized by improved lubricity. This contributes to the stabilization of the friction coefficient (C).

The tribological properties of Ni-based wear-resistant coatings can be modified by the treatment processes applied to the substrate material. Wu *et al.* [167], investigated the influence of preheating treatments, spanning between 500°C-800°C, on the tribological characteristics of a Cu substrate coated with Ni60 alloy using the plasma cladding method. The impact of the Ni coating and the preheating duration on wear loss and friction force is illustrated in Figure 2.34. The incorporation of the nickel coating is evidently linked to an important reduction in both the wear loss and friction coefficient values. Furthermore, the preheating process has also been observed to markedly influence the tribological properties. The optimum preheating temperature for the best tribological performance is identified to be 500°C. Moreover, the influence of preheating temperature and the Ni60 coating on the wear mechanism is depicted in Figure 2.35. Pure copper's worn surfaces exhibit clear tearing, microcracks, and shallow grooves,

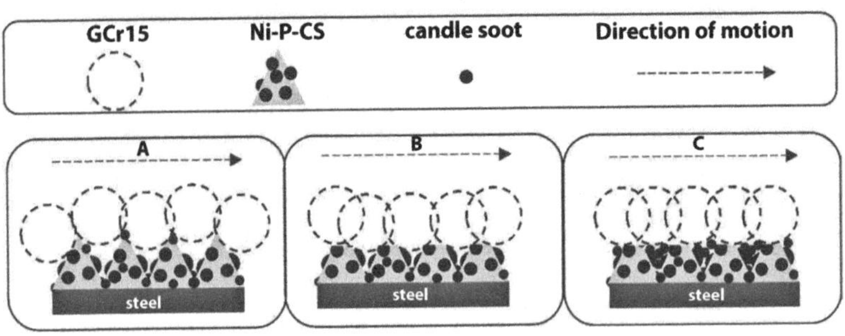

Figure 2.33 A diagram illustrating the friction mechanism of the coating ([166]; licensed by CC- BY 4.0).

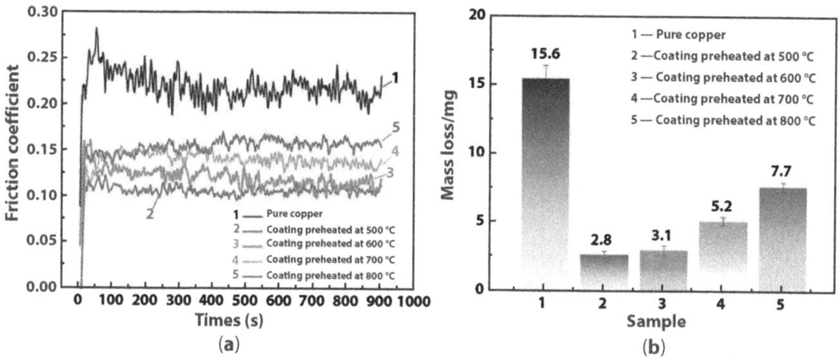

Figure 2.34 Assess the outcomes of the wear tests performed on the given samples, paying particular attention to (a) the friction coefficient and (b) the mass loss attributed to wear ([167], licensed by CC- BY 4.0).

indicating a predominant adhesive wear mechanism (Figure 2.35-a and b). At 600°C, the sample coated with Ni60 alloy, after preheating, displays only a few scratches and minor tearing, suggesting that the primary wear mechanism is primarily abrasive (Figure 2.35-c and d). However, at 800°C, the fabricated coating's worn surface exhibits increased tearing, indicating a rise in adhesive wear. Additionally, an important number of tracks (Figures 2.35-e and f) suggest the coexistence of both abrasive and adhesive wear mechanisms. In conclusion, the applied pre-treatment and the Ni60 coating have distinct effects on the wear mechanisms of the Cu substrate material.

As mentioned earlier, wear-resistant coatings based on Ni are employed on various substrates to improve their longevity and performance. Common substrates include steel and stainless steel, which are chosen for their strength and durability; aluminum alloys, favored in aerospace and automotive industries for their lightweight properties; and titanium alloys, known for their high strength-to-weight ratio and corrosion resistance. Copper alloys are also used, especially in electrical and thermal applications, due to their conductivity. In specialized cases, ceramics and even plastics or composites can serve as substrates, particularly when weight and basic wear resistance are crucial factors. For example, Saroha *et al.* [168], employed the electrodeposition method to coat ABS, resulting in a significant enhancement in wear resistance. The untreated ABS displayed a wear rate of 0.871 mm3/m, while the ABS coated with nickel exhibited a considerably reduced wear rate of 0.039 mm3/m. The notable improvement in wear performance is attributed to the enhanced hardness and mechanical properties imparted by the nickel coating.

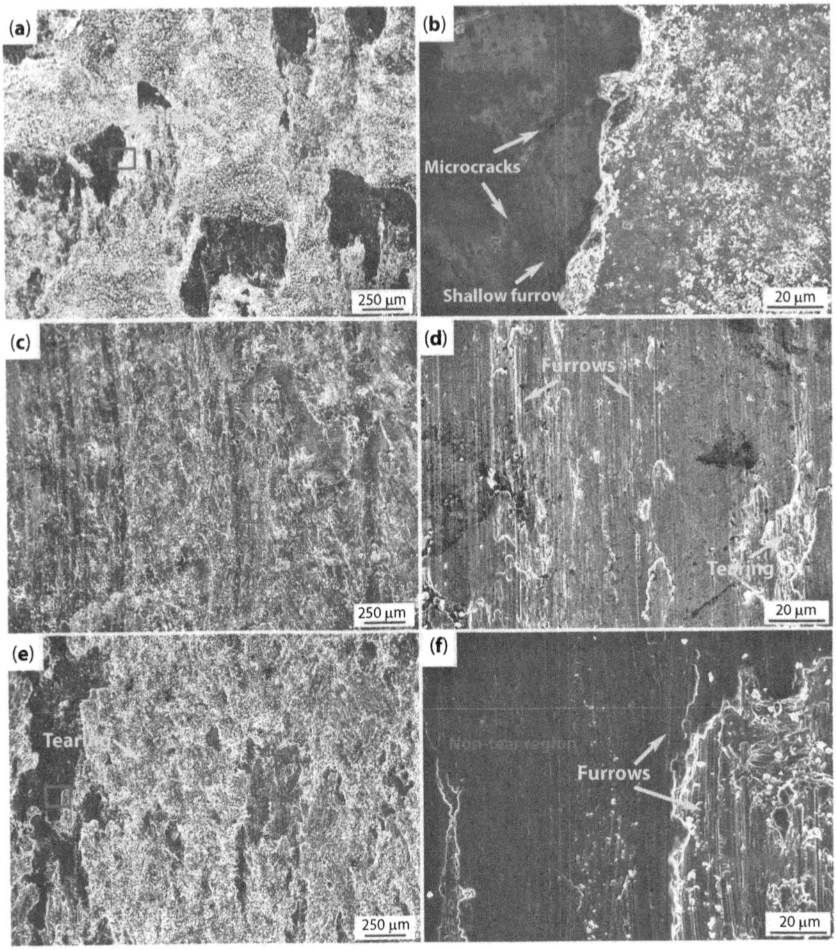

Figure 2.35 Analyze the surface morphologies of worn samples, highlighting the characteristics of (a–b) untreated Cu, (c–d) Ni60 A coating (600°C), and (e–f) Ni60 A coating (800°C) ([167], licensed by CC- BY 4.0).

Hence, the influence of Ni-based coatings on the wear resistance and mechanisms of the underlying materials they are applied to is significant, and this impact is highly contingent on various factors. The extent of this effect is notably influenced by parameters such as the composition of the coating, the inherent properties of the substrate, the specific type of coating employed, and other relevant variables. The nuanced interplay of these factors underscores the complexity in understanding and optimizing the performance of Ni-based coatings in enhancing wear resistance for diverse applications.

Applications of Ni-based Wear-resistant Coatings

Ni-based coatings are recognized as versatile solutions in different sectors, bestowing numerous desirable attributes upon metal surfaces upon application. These coatings are well-known for their outstanding characteristics, such as exceptional wear resistance, elevated hardness, a low coefficient of friction, resistance to corrosion, and efficient electrical conductivity. The utilization of nickel-based coatings spans a wide spectrum of applications, with their effectiveness influenced by both the production method employed and the specific composition of the coating. The diverse range of properties exhibited by these coatings contributes to their widespread adoption in different fields. Furthermore, the applications of nickel-based alloys are significantly enriched by the distinctive characteristics introduced by the 2nd and 3rd components, further enhancing their adaptability and performance in various industrial contexts.

Nickel-tungsten (Ni-W) coatings stand out as preferred choices in various applications, particularly in cutting tools and metalworking tools. Celebrated for their outstanding features, these coatings provide a compelling blend of elevated hardness, remarkable wear resistance, and resistance to corrosion. Their application in cutting operations proves to be highly advantageous, significantly enhancing tool longevity by providing enhanced wear resistance. As micromechanical technology continues to advance, there is a growing demand for superior mechanical and physical properties. In response to this, nickel-tungsten alloys have emerged as an optimal solution, showcasing an ideal blend of exceptional hardness and wear resistance properties. This makes them particularly well-suited for the fabrication of microelectromechanical structures and microdevices. In specialized applications like electrical contacts, the need for an effective corrosion and diffusion barrier layer is paramount. Nanocrystalline Ni-W coatings play a pivotal role in improving corrosion resistance, minimizing wear on contacts, and substantially reducing the formation of intermetallic compounds resulting from diffusion at interfaces. These unique attributes position these coatings as highly effective barriers in electronic and electrical contacts. Furthermore, in scenarios requiring a harmonious balance of electrical conductivity and wear resistance, Ni-W-GO (Graphene Oxide) coatings emerge as a versatile option. Their utilization adds an extra layer of adaptability, making them suitable for applications demanding both electrical performance and durability [168–170].

Ni-B coatings play a pivotal role across diverse industries, offering a multitude of exceptional attributes that contribute to their widespread preference. Renowned for their outstanding solderability, lubricity, high hardness, and remarkable mechanical wear resistance, these coatings find indispensable applications in critical sectors such as the oil and gas pipelines, chemical industry, nuclear energy facilities, defense technology, automotive components, electronic devices, and hydrogen production. The distinct advantages of electroless nickel coatings reduced by borohydride are particularly noteworthy, mainly due to their outstanding hardness and inherent resistance to mechanical wear. In comparison to tool steel and hard chrome coatings, electroless Ni-B coatings stand out as leaders in wear resistance. This makes them indispensable for heavy-duty conditions and industrial applications with high wear potential, where their performance stands out. An added value of electroless Ni-B coatings lies in their ability to replace gold in the electronics industry, showcasing their versatility and cost-effectiveness. The columnar structure inherent in these coatings ensures an effective retention of lubricants, proving particularly advantageous under adhesive wear conditions. Solderability is another area where electroless Ni-B coatings excel, outperforming electroless Ni-P coatings due to their thin oxide layers. Furthermore, their critical role in preventing copper diffusion, especially in copper interconnect technologies, underscores their significance. The magnetic properties, low resistance, and electrocatalytic activity of electroless Ni-B coatings contribute to their widespread adoption in various industrial applications. Consequently, these coatings remain in constant demand, driven by their broad potential for use in industrial settings, unique properties, and exceptional durability [102, 171].

Ni-B-CeO2 coatings emerge as a valuable solution for enhancing the performance of catalytic converters, particularly in the automotive industry. Catalytic converters play a crucial role in vehicles by transforming exhaust gases into less harmful components, contributing to environmental sustainability. The coatings applied to these converters are instrumental in reducing harmful emissions by augmenting catalytic activity. The incorporation of cerium oxide (CeO2) in Ni-B-CeO2 coatings proves to be a key factor in elevating catalytic properties. Cerium oxide, as a component, actively engages in redox reactions within catalytic converters, enhancing the efficiency of oxidation-drying cycles. This, in turn, aids in the purification of engine exhaust gases, making the process more effective. The unique features of Ni-B-CeO2 coatings make them a more efficient and durable choice for catalytic converters employed in automotive exhaust systems [172].

Ni-B-Si3N4 coatings represent a specialized category of coatings tailored for applications in the surface coating of cutting tools, drills, router bits, and similar precision cutting instruments. These coatings are pivotal in prolonging the lifespan of cutting tools by substantially enhancing wear resistance, thereby ultimately enhancing overall machining performance. Their prevalence is particularly notable in the metal processing industry, where the demand for robust wear-resistant coatings is high. In challenging conditions, such as high-speed machining (HSM), the importance of wear resistance becomes paramount. Ni-B-Si3N4 coatings excel in these scenarios due to their exceptional high-temperature resistance, a quality attributed to the presence of silicon nitride (Si3N4). This key component enables these coatings to optimize cutting tool performance with anti-wear properties, making them well-suited for demanding metal cutting and shaping applications. The preference for Ni-B-Si3N4 coatings in the metal processing industry underscores their effectiveness in extending the life of cutting tools, contributing to enhanced efficiency and cost-effectiveness in machining operations. As a result, these coatings continue to be a favored choice for applications where wear resistance and durability are critical factors [173, 174].

Ni-B-Al2O3 coatings find application in enhancing the performance of critical structural elements, including engine cylinders, valves, accessories, and electronic systems. These coatings serve the purpose of augmenting hardness, wear resistance, and corrosion resistance in these components, contributing to their overall durability and longevity [175].

Ni-P coatings play a crucial role in the construction of offshore wind turbine blades, where they are deposited on Glass Fiber Reinforced Plastic (GFRP) substrates. This application serves multiple purposes, including providing resistance to damage caused by lightning strikes, adding conductivity, and enhancing wear-corrosion resistance. Furthermore, electroless nickel coatings, especially Ni-P, are preferred for enhancing the wear and wear-corrosion characteristics of Carbon Fiber Reinforced Plastic (CFRP) materials. CFRP finds widespread application in the space, aviation, and automotive sectors owing to its lightweight benefits. In marine applications, the suitability of Ultra High Molecular Weight Polyethylene (UHMWPE) engineering plastics paired with electroless Ni-P coatings has been identified. In particular, under seawater lubrication conditions, UHMWPE engineering plastics exhibit the least coefficient of friction when interacting with electroless Ni-P coatings. This property makes them ideal for various marine applications. Moreover, the applicability of electroless nickel coatings in the maritime and arms industry has been examined. It has been emphasized that "dry" coatings, particularly in dusty environments, should be preferred. These coatings not only provide low friction but also offer corrosion protection and durability, making them valuable for weapons operating in challenging and abrasive conditions [176, 177].

Exploring the utilization of Ni-P-PTFE composite coatings to enhance aluminum alloy efficiency in lightweight external gear pump bearing regions has been a focal point of investigations. When Ni-P-Si3N4 coatings are applied to the surfaces of aluminum piston materials, they demonstrate moderate friction and wear resistance. However, notable improvements in wear resistance have been observed in iron-based bearings designed for water-lubricated applications. Another notable variation of coating, Ni-P-Graphite-SiC, displays outstanding wear resistance and a reduced coefficient of friction. This combination makes it well-suited for diverse applications, including mold materials, automobile parts, and various other fields. The distinct properties of each coating highlight their potential in addressing specific challenges and optimizing performance in targeted applications [140, 178, 179].

The applicability of Ni plating in a commercial brake valve assembly has been demonstrated, revealing comparable performance to hard chrome platings. On the contrary, electroless plating exhibits minimal wear and moderate friction throughout the observed sliding distance. As a result, electroless Ni plating presents itself as a viable substitute for chrome coatings, providing a more environmentally friendly option given the well-known environmental toxicity linked to chrome platings. Among the diverse metallization methods available, Ni-P coating stands out as a preferred choice for producing metal-coated fabrics. This selection is based on its appealing advantages, including coating uniformity, excellent conductivity, the capability to metalize non-conductors, and overall flexibility. In the context of EMI shielding fabrics, Ni-P coating is employed to enhance wear resistance. Moreover, introducing SiO_2 particles into the coating proves to be effective in enhancing the abrasion resistance of the fabric [140].

In conclusion, Ni-based coatings have demonstrated their versatility and indispensability across a myriad of industries. These coatings, whether nickel-tungsten, Ni-B, Ni-B-CeO_2, or Ni-P variants, offer unique properties that cater to specific needs in cutting tools, automotive catalytic converters, marine applications, and various other critical components. The outstanding wear resistance, corrosion resistance, and hardness displayed by these coatings make a substantial contribution to the durability and efficiency of the coated materials. Moreover, their diverse applications underscore the adaptability of Ni-based coatings in addressing challenges and optimizing performance in targeted industrial sectors. As a result, these coatings remain at the forefront of material enhancement, playing a pivotal role in advancing technology, improving durability, and promoting environmentally friendly alternatives.

2.4 Conclusion

From the comprehensive overview provided, it can be concluded that nanotribology and the application of nanoscale coatings are pivotal in advancing the performance, efficiency, and durability of materials across a broad spectrum of industries. By delving into friction, wear, and lubrication at the nanometer scale, nanotribology offers innovative solutions to challenges posed by miniaturization and the unique behavior of materials at such scales. The deployment of Ni-based coatings, in particular, exemplifies the field's capacity to enhance wear resistance and reduce friction, contributing significantly to sectors such as automotive, aerospace, electronics, and medicine. These coatings not only extend the lifespan of components and devices but also support the development of more sustainable and energy-efficient technologies. In essence, the exploration of nanotribology and wear-resistant coatings encapsulates a critical area of scientific inquiry, driving forward technological innovation and offering promising avenues for future advancements.

The discourse has meticulously navigated through the diverse array of coatings, including hard chromium, ceramic, diamond-like carbon (DLC), polymer-based, and metal-based coatings, each tailored to mitigate wear, friction, and corrosion, thereby enhancing component longevity and performance. Specifically, hard chromium coatings have been spotlighted for their efficacy in reducing material transfer and wear, as evidenced by the diminished strip material adhesion on chrome-plated pins. Ceramic coatings, particularly titanium nitride (TiN), have demonstrated superior tribological properties, with the *in-situ* synthesis of TiC-Al2O3 reinforcing phases on copper surfaces markedly augmenting wear resistance. DLC coatings, with their unique amalgamation of diamond hardness and graphite flexibility, have shown significant reductions in wear track depth, friction coefficient, and wear loss on aluminum substrates.

Polymer coatings, notably PTFE, have been highlighted for their exceptional lubricity, substantially lowering friction coefficients on Si3N4/TiC ceramics. Polymer matrix composites reinforced with graphene oxide have revealed a pronounced reduction in abrasive wear, attributed to graphene oxide's lubricating properties. Among metal-based coatings, nickel-based coatings stand out for their comprehensive benefits, including superior corrosion and wear resistance, strong adhesion, temperature resilience, and low friction, making them a preferred choice in demanding environments.

The technological progress and utilization of Ni-based coatings, including pure Ni, Ni-P, and Ni-B coatings, have played a significant role in progressing surface engineering to achieve improved wear resistance. Pure Ni coatings, primarily applied through electrodeposition and electroless deposition, are distinguished by their corrosion resistance and adaptability for decorative finishes, offering protection against a wide spectrum of chemical environments. The introduction of thermal spray techniques like High-Velocity Oxy-Fuel, Twin Wire-Arc Spray, and Cold Spray has further diversified the application spectrum of Ni coatings, ensuring dense and well-adhered layers suitable for combating wear and corrosion.

Developed as a substitute for conventional electroplating, Ni-P coatings utilize electroless nickel plating to provide exceptional wear resistance and perfect corrosion. The versatility of Ni-P coatings is underscored by their ability to be tailored for specific applications through adjustments in phosphorus content, influencing their hardness and corrosion resistance. Innovations such as the inclusion of hard particles and self-lubricating materials into Ni-P based composite coatings have led to notable enhancements in mechanical properties, demonstrating significant improvements in wear resistance and hardness.

Ni-B coatings emerge as a class of advanced surface treatments with a unique blend of hardness, wear resistance, and reduced friction, attributed to the incorporation of boron. The adaptability of Ni-B coatings to various deposition methods, including electroless plating and innovative thermal spray techniques, highlights their role in addressing industrial challenges related to harsh environments and demanding operational conditions.

The integration of alloying elements and ceramic particles into Ni-based coatings has been a pivotal advancement, significantly improving the coatings' mechanical and functional properties. These enhancements include increased hardness, wear resistance, and thermal stability, making Ni-based coatings indispensable in high-demand sectors such as aerospace, automotive, and manufacturing. The application of innovative approaches, such as the jet electrodeposition method and the incorporation of ultrasonic power during electrodeposition, has led to coatings with superior wear properties and morphological development, showcasing the potential for tailored material solutions.

Alloying elements such as phosphor (P), chromium (Cr), molybdenum (Mo), tungsten (W), and boron (B) into Ni-based coatings has been shown to significantly influence wear mechanisms. Cr and W, for example, are instrumental in forming hard carbides within the Ni matrix, contributing to an increased resistance to abrasive wear. This is particularly evident in environments where hard particles or rough surfaces are in contact, necessitating a robust barrier against material removal. Mo, on the other hand, enhances the coating's resistance to adhesive wear by reducing metal-to-metal contact and thus mitigating material transfer between sliding surfaces. Boron's contribution to the formation of an amorphous structure in coatings like Ni-B enhances wear resistance by providing a uniform, hard surface that lacks grain boundaries, thereby reducing avenues for crack propagation and material removal.

The wear mechanisms prevalent in Ni-based coatings are directly influenced by the specific composition and the operational environment. For instance, coatings reinforced with hard carbides (Cr, Mo, W) predominantly resist abrasive wear due to their enhanced hardness, making them suitable for applications involving high contact stress and abrasive particles. In contrast, coatings with elements that improve lubricity or form protective layers, such as boron, are more adept at reducing adhesive wear, making them ideal for applications where sliding contact is prominent.

Moreover, incorporating solid lubricants such as molybdenum disulfide or graphite into Ni-based coatings imparts self-lubricating properties, leading to a substantial reduction in friction and wear in sliding contacts. This enhancement is crucial in applications where maintaining low friction is paramount to the system's efficiency and longevity.

The dynamic field of Ni-based coatings is characterized by ongoing innovation, with research efforts focused on optimizing coating compositions, application methods, and performance in diverse environmental conditions. As the demand for materials with enhanced durability and specialized functionalities grows, Ni-based coatings will undoubtedly play a crucial role in meeting these needs, driving forward the frontiers of material science and engineering.

2.5 Future Perspectives

As we navigate the evolving landscape of material s science and surface engineering, the future of Ni-based wear protective surfaces holds promising prospects with ongoing advancements in coating compositions and application methods. Researchers and industry experts are actively exploring innovative formulations to further enhance the performance characteristics of Ni-based coatings. The incorporation of advanced nanomaterials, such as graphene oxide, and the exploration of novel alloy combinations are areas of intense investigation. These developments aim to push the boundaries of hardness, wear resistance, and corrosion protection, ensuring that Ni-based coatings continue to meet the ever-growing demands of diverse industrial applications.

Furthermore, the optimization of application methods, including precision deposition techniques and tailored surface treatments, is a key focus for future research. The goal is to achieve even more uniform and durable coatings, addressing specific challenges in intricate components and critical machinery. The integration of artificial intelligence and machine learning algorithms in coating processes may offer unprecedented control and customization, allowing for the fine-tuning of coating properties based on real-time performance feedback.

- The integration of complex composite materials, such as Ni-P/PTFE/TiO2/ZrO2, and the utilization of diverse composite particles like SiC, CeO2, TiO2, Al2O3, or CNTs, will play a pivotal role in addressing specific needs for wear resistance and corrosion protection. These materials hold the potential to generate next-generation composite coatings with tailored properties, ensuring enhanced durability and performance in challenging environments.
- The use of cutting-edge nanoparticles such as TiNi or CN in the Ni-based coating matrix is positioned to develop the hardness, strength, and overall wear resistance and corrosion of Ni-based coatings. This strategic integration of new nanoparticles aligns with the pursuit of achieving maximum protection and durability at cost-effective levels, meeting the evolving requirements of diverse industries.
- For the electroless Ni coating, the quest for increased equipment productivity, optimal bath efficiency, and the development of new plating technologies remain focal points for research. The application of composite materials and the reduction of chemical usage are essential considerations. Exploring new opportunities and refining the implementation of Electroless coatings in corrosion prevention will drive innovation in the coming years.
- Nanocomposite coatings, enriched with nanoparticles like B4C, are expected to be a focus for future investigations. These coatings possess distinctive features, including elevated hardness, superior wear resistance, appropriate mechanical properties, effective corrosion resistance, and satisfactory thermal stability. Exploring strategies to integrate a greater quantity of nanoparticles, with the potential to result in coatings that are exceptionally hard and resistant to corrosion, will be a significant focus.
- The investigation into improved thermal stability in nanocrystalline Ni-W and Ni-B alloys entails exploring the

inclusion of rare metal elements. The study of heavily nano twinned Ni-W/B structures for thermal stability and the creation of Ni-W/B films as diffusion barriers for high-temperature coatings represents crucial focal points for future research. These progressions are especially pertinent for implementations in the vanes and blades of first and second-stage turbines.

- To enhance diverse properties, it is suggested to concurrently incorporate trivalent components like iron, copper, tin, molybdenum, or nanoparticles in binary Ni-based coatings.
- The rapid cooling and solidification during laser cladding induce structural and performance differences at the particle-matrix interface compared to processes like sintering and casting. Optimizing the process and assessing material compatibility are crucial for enhancing the composite coating's overall properties, particularly at the interface.

In conclusion, the future of Ni-based wear protective surfaces is dynamic and multifaceted, driven by advancements in materials science, nanotechnology, and sustainable practices. The ongoing exploration of tailored compositions and innovative methods positions Ni-based coatings as integral components in the evolving landscape of wear protection, ensuring they meet the diverse and demanding needs of modern industries.

References

1. Czichos, H. and Winer, W.O., Tribology: A Systems Approach to the Science and Technology of Friction, Lubrication and Wear. *ASME. J. of Lubrication Tech.*, 100, 4, 513–514, 1978.
2. Tzanakis, I., Hadfield, M., Thomas, B., Noya, S.M., Henshaw, I., Austen, S., Future perspectives on sustainable tribology. *Renew. Sustain. Energy Rev.*, 16, 4126–4140, 2012.
3. Shah, R., Woydt, M., Huq, N., Rosenkranz, A., Tribology meets sustainability. *Ind. Lubr. Tribol.*, 73, 430–435, 2020.
4. Das, S. and Das, S., Applications of Tribology on Engine Performance, in: *Energy, Environ. Sustain.*, pp. 307–325, 2019.
5. Moghadasi, K., Mohd Isa, M.S., Ariffin, M.A., Mohd Jamil, M.Z., Raja, S., Wu, B., Yamani, M., Bin Muhamad, M.R., Yusof, F., Jamaludin, M.F., Bin Ab Karim, M.S., Binti Abdul Razak, B., Bin Yusoff, N., A review on biomedical implant materials and the effect of friction stir based techniques on their

mechanical and tribological properties. *J. Mater. Res. Technol.*, 17, 1054–1121, 2022.
6. Ciulli, E., Tribology and Sustainable Development Goals, in: *Mech. Mach. Sci.*, vol. 108 MMS, pp. 438–447, 2022.
7. Assenova, E., Majstorovic, V., Vencl, A., Kandeva, M., Green tribology and quality of life. *Int. Conv. Qual. 2012*, vol. 40, pp. 32–38, 2012.
8. Shah, R., Martini, A., Woydt, M., Wong, H., Green tribology. *Tribol. Ind.*, 42, 592–596, 2020.
9. Ciulli, E., Tribology and Industry: From the Origins to 4.0. *Front. Mech. Eng.*, 5, 2019.
10. Holmberg, K. and Erdemir, A., The impact of tribology on energy use and CO2 emission globally and in combustion engine and electric cars. *Tribol. Int.*, 135, 389–396, 2019.
11. Bîrleanu, C., Pustan, M., Şerdean, F., Merie, V., AFM nanotribomechanical characterization of thin films for MEMS applications. *Micromachines*, 13, 23, 2022.
12. de Sousa dos Santos, E.L.V., Cruz, J.N., da Costa, G.V., de Sá, E.M.F., da Silva, A.K.P., Fernandes, C.P., de Faria Mota Oliveira, A.E.M., Duarte, J.L., Bezerra, R.M., Tavares, J.F., da Costa, T.S., dos Anjos Ferreira, R.M., dos Santos, C.B.R., Souto, R.N.P., Essential Oil of Ocimum basilicum against Aedes aegypti and Culex quinquefasciatus: Larvicidal Activity of a Nanoemulsion and In Silico Study. *Separations*, 11, 2024.
13. Sadeghi, B., Cavaliere, P., Shabani, A., Pruncu, C.I., Lamberti, L., Nano-scale wear: A critical review on its measuring methods and parameters affecting nano-tribology. *Proc. Inst. Mech. Eng. Part J. J. Eng. Tribol.*, 238, 125–155, 2024.
14. Sahoo, P. and Das, S.K., Tribology—A Tool for Mechanical and Industrial Engineering. *Materials Forming, Machining and Tribology*, pp. 1–37, 2022.
15. Singh, H. and Kaur, K., Role of nanotechnology in research fields: Medical sciences, military & tribology- A review on recent advancements, grand challenges and perspectives. *Mater. Today Proc.*, 2023.
16. Dahlan, N.A., Thiha, A., Ibrahim, F., Milić, L., Muniandy, S., Jamaluddin, N.F., Petrović, B., Kojić, S., Stojanović, G.M., Role of Nanomaterials in the Fabrication of bioNEMS/MEMS for Biomedical Applications and towards Pioneering Food Waste Utilisation. *Nanomaterials*, 12, 4025, 2022.
17. Montero, L., Allaoui, S., Hivet, G., Characterisation of the mesoscopic and macroscopic friction behaviours of glass plain weave reinforcement. *Compos. Part A Appl. Sci. Manuf.*, 95, 257–266, 2017.
18. Johnson, K.L., The contribution of micro/nano-tribology to the interpretation of dry friction. *Proc. Inst. Mech. Eng. Part C J. Mech. Eng.*, 214, 11–22, 2000.
19. Bhushan, B., Nanotribology and nanomechanics. *Wear*, 259, 1507–1531, 2005.
20. Sarfraz, M.H., Zubair, M., Aslam, B., Ashraf, A., Siddique, M.H., Hayat, S., Cruz, J.N., Muzammil, S., Khurshid, M., Sarfraz, M.F., Hashem, A., Dawoud, T.M., Avila-Quezada, G.D., Abd_Allah, E.F., Comparative analysis of phyto-fabricated chitosan, copper oxide, and chitosan-based CuO nanoparticles:

antibacterial potential against Acinetobacter baumannii isolates and anticancer activity against HepG2 cell lines. *Front. Microbiol.*, 14, 1188743, 2023.
21. Srivastava, I., Kotia, A., Ghosh, S.K., Ali, M.K.A., Recent advances of molecular dynamics simulations in nanotribology. *J. Mol. Liq.*, 335, 116154, 2021.
22. Breakspear, S., Noecker, B., Popescu, C., AFM for Hair Surface Characterisation, in: *Surf. Sci. Adhes. Cosmet.*, pp. 215–243, 2021.
23. Scorzoni, C., Goldoni, G., De Renzi, V., The Gecko® Approach to Friction: A Novel Teaching Learning Sequence, in: *Challenges Phys. Educ.*, pp. 273–285, 2021.
24. Kałużny, J., Waligórski, M., Szymański, G.M., Merkisz, J., Różański, J., Nowicki, M., Al Karawi, M., Kempa, K., Reducing friction and engine vibrations with trace amounts of carbon nanotubes in the lubricating oil. *Tribol. Int.*, 151, 106484, 2020.
25. Tung, S.C. and McMillan, M.L., Automotive tribology overview of current advances and challenges for the future. *Tribol. Int.*, 37, 517–536, 2004.
26. Bhushan, B., Nanotribology and nanomechanics of MEMS/NEMS and BioMEMS/BioNEMS materials and devices. *Microelectron. Eng.*, 84, 387–412, 2007.
27. Kim, S.H., Asay, D.B., Dugger, M.T., Nanotribology and MEMS. *Nano Today*, 2, 22–29, 2007.
28. Ching, H.A., Choudhury, D., Nine, M.J., Abu Osman, N.A., Effects of surface coating on reducing friction and wear of orthopaedic implants. *Sci. Technol. Adv. Mater.*, 15, 14402, 2014.
29. Sayfidinov, K., Cezan, S.D., Baytekin, B., Baytekin, H.T., Minimizing friction, wear, and energy losses by eliminating contact charging. *Sci. Adv.*, 4, 3808–3824, 2018.
30. S, A., Krishnamurthy, S., Kandasubramanian, B., Kumar B, P., Probing into atomically thin layered nano-materials protective coating for aerospace and strategic defence application – A review. *J. Alloys Compd.*, 968, 172203, 2023.
31. Zhai, W., Bai, L., Zhou, R., Fan, X., Kang, G., Liu, Y., Zhou, K., Recent Progress on Wear-Resistant Materials: Designs, Properties, and Applications. *Adv. Sci.*, 8, 2003739, 2021.
32. Gu, Y., Xia, K., Wu, D., Mou, J., Zheng, S., Technical characteristics and wear-resistant mechanism of nano coatings: A review. *Coatings*, 10, 233, 2020.
33. Houdková, S., Zahálka, F., Kašparová, M., Berger, L.M., Comparative study of thermally sprayed coatings under different types of wear conditions for hard chromium replacement. *Tribol. Lett.*, 43, 139–154, 2011.
34. Kesavan, D. and Kamaraj, M., The microstructure and high temperature wear performance of a nickel base hardfaced coating. *Surf. Coat. Technol.*, 204, 4034–4043, 2010.
35. Yu, M., Liu, F., Du, F., Synthesis and properties of a green and self-cleaning hard protective coating. *Prog. Org. Coat.*, 94, 34–40, 2016.

36. Polanec, B., Zupanič, F., Bončina, T., Tašner, F., Glodež, S., Experimental investigation of the wear behaviour of coated polymer gears. *Polymers (Basel)*, 13, 3588, 2021.
37. Song, W.-M., Yang, G.-R., Lu, J.-J., Hao, Y., Ma, Y., Microstructure and wear behaviour of Ni-based surface coating on copper substrate. *Wear*, 262, 868–875, 2007.
38. Mekicha, M.A., de Rooij, M.B., Matthews, D.T.A., Pelletier, C., Jacobs, L., Schipper, D.J., The effect of hard chrome plating on iron fines formation. *Tribol. Int.*, 142, 106003, 2020.
39. Pan, P., Gao, J., Si, C., Yao, Q., Guo, Z., Zhang, Y., Tribological Properties of TiN Coating on Cotton Picker Spindle. *Coatings*, 13, 959, 2023.
40. Yang, F., Qin, Q., Shi, T., Lu, X., Chen, C., Guo, Z., Volinsky, A.A., *In-situ* synthesis of TiC-Al2O3 coating on copper surface. *Surf. Coat. Technol.*, 373, 65–74, 2019.
41. Natrayan, L., Gaur, P., Merneedi, A., Kaliappan, S., Patil, P.P., Sivaprakash, V., Chewaka, M.D., Investigation of Tribological Behaviour on DLC Coatings for AA5051 using DC Sputtering. *Adsorpt. Sci. Technol.*, 2022, 2022.
42. Song, W., Wang, S., Lu, Y., Zhang, X., Xia, Z., Friction behavior of PTFE-coated Si3N4/TiC ceramics fabricated by spray technique under dry friction. *Ceram. Int.*, 47, 7487–7496, 2021.
43. Radziejewska, J. and Grzelka, J., Effect of concentration GO and diamond wax and method of introducing additives on morphology and properties of epoxy powder coating. *Polym. Test.*, 117, 107866, 2023.
44. Jensen, R., Farhat, Z., Islam, M.A., Jarjoura, G., Erosion–Corrosion of Novel Electroless Ni-P-NiTi Composite Coating. *Corros. Mater. Degrad.*, 4, 120–141, 2023.
45. Bhattacharyya, S., Mohanty, D., Kumar, P., Kalyan Das, S., Sahoo, P., Pal, S.K., Chakraborty, S., A corrosion and tribo-failure analysis of Ni-P-Cu coated mild steel (AISI-1040) at varied copper concentration. *Eng. Fail. Anal.*, 146, 107063, 2023.
46. Gamon, W. and Aniołek, K., Examination of the sliding wear of bronze coatings on railway buffer heads. *Wear*, 448–449, 203235, 2020.
47. Ostolaza, M., Arrizubieta, J.I., Lamikiz, A., Ukar, E., Study of the flexural behaviour and bonding strength of WC-Co metal matrix composite coatings produced by Laser Directed Energy Deposition. *Surf. Coat. Technol.*, 463, 129538, 2023.
48. Kushwaha, S., Bhadauria, A., Bajpai, S., Tiwari, A., Pandey, K.K., Keshri, A.K., Balani, K., Mechanical, microstructural, and fretting wear behaviour of Al2O3–ZrO2-CNT based bimodal composite coatings. *Wear*, 532–533, 205127, 2023.
49. Praveen, A.S. and Arjunan, A., High-temperature oxidation and erosion of HVOF sprayed NiCrSiB/Al2O3 and NiCrSiB/WC–Co coatings. *Appl. Surf. Sci. Adv.*, 7, 100191, 2022.

50. Kumar, M., Kant, S., Kumar, S., Corrosion behavior of wire arc sprayed Ni-based coatings in extreme environment. *Mater. Res. Express*, 6, 106427, 2019.
51. Sui, Y., Yang, F., Qin, G., Ao, Z., Liu, Y., Wang, Y., Microstructure and wear resistance of laser-cladded Ni-based composite coatings on downhole tools. *J. Mater. Process. Technol.*, 252, 217–224, 2018.
52. Zhan, X., Liu, Y., Yi, P., Feng, W., Feng, Z., Jin, Y., Effect of Substrate Surface Texture Shapes on the Adhesion of Plasma-Sprayed Ni-Based Coatings. *J. Therm. Spray Technol.*, 30, 270–284, 2021.
53. Grilli, M.L., Valerini, D., Slobozeanu, A.E., Postolnyi, B.O., Balos, S., Rizzo, A., Piticescu, R.R., Critical raw materials saving by protective coatings under extreme conditions: A review of last trends in alloys and coatings for aerospace engine applications. *Materials (Basel)*, 14, 1656, 2021.
54. Raghavendra, C.R., Basavarajappa, S., Sogalad, I., Kumar, S., A review on Ni based nano composite coatings. *Mater. Today Proc.*, 39, 6–16, 2020.
55. Sadeghimeresht, E., Markocsan, N., Nylén, P., A Comparative Study on Ni-Based Coatings Prepared by HVAF, HVOF, and APS Methods for Corrosion Protection Applications. *J. Therm. Spray Technol.*, 25, 1604–1616, 2016.
56. Skulev, H., Malinov, S., Basheer, P.A.M., Sha, W., Modifications of phases, microstructure and hardness of Ni-based alloy plasma coatings due to thermal treatment. *Surf. Coat. Technol.*, 185, 18–29, 2004.
57. Wang, X., Zhu, L., Zhou, Z., Liu, G., Liu, E., Zeng, Z., Wu, X., Tribological Properties of WC-Reinforced Ni-Based Coatings Under Different Lubricating Conditions. *J. Therm. Spray Technol.*, 24, 1323–1332, 2015.
58. Wu, B., Xu, B.-S., Zhang, B., Dong, S.Y., The effects of parameters on the mechanical properties of Ni-based coatings prepared by automatic brush plating technology. *Surf. Coat. Technol.*, 201, 5758–5765, 2007.
59. Allahyarzadeh, M.H., Aliofkhazraei, M., Rezvanian, A.R., Torabinejad, V., Sabour Rouhaghdam, A.R., Ni-W electrodeposited coatings: Characterization, properties and applications. *Surf. Coat. Technol.*, 307, 978–1010, 2016.
60. Taylor, S.R., Coatings for Corrosion Protection: Metallic, in: *Encycl. Mater. Sci. Technol.*, pp. 1–5, 2001.
61. Mévrel, R., State of the art on high-temperature corrosion-resistant coatings. *Mater. Sci. Eng., A*, 120–121, 13–24, 1989.
62. Fournier, J., Miousse, D., Legoux, J.G., Wire-arc sprayed nickel based coating for hydrogen evolution reaction in alkaline solutions. *Int. J. Hydrogen Energy*, 24, 519–528, 1999.
63. Cabral Miramontes, J., Pedraza Basulto, G.K., Gaona Tiburcio, C., Zambrano Robledo, P.D.C., Poblano Salas, C.A., Almeraya Calderón, F., Coatings characterization of Ni-based alloy applied by HVOF. *Aircr. Eng. Aerosp. Technol.*, 90, 336–343, 2018.

64. Luo, X.T., Li, Y.J., Li, C.X., Yang, G.J., Li, C.J., Effect of spray conditions on deposition behavior and microstructure of cold sprayed Ni coatings sprayed with a porous electrolytic Ni powder. *Surf. Coat. Technol.*, 289, 85–93, 2016.
65. Parameters, K., Thermal Spray Coatings – Keepsake. *Metall. Ceram. Prot. Coatings*, pp. 261–289, 2007.
66. Luo, X., Smith, G.M., Sampath, S., On the Interplay Between Adhesion Strength and Tensile Properties of Thermal Spray Coated Laminates—Part I: High Velocity Thermal Spray Coatings. *J. Therm. Spray Technol.*, 27, 296–307, 2018.
67. Alidokht, S.A., Vo, P., Yue, S., Chromik, R.R., Cold Spray Deposition of Ni and WC-Reinforced Ni Matrix Composite Coatings. *J. Therm. Spray Technol.*, 26, 1908–1921, 2017.
68. Aniekwe, U.V. and Utigard, T.A., High-Temperature Oxidation of Nickel-Plated Copper vs Pure Copper. *Can. Metall. Q.*, 38, 277–281, 1999.
69. Lee, W.H., Tang, S.C., Chung, K.C., Effects of direct current and pulse-plating on the co-deposition of nickel and nanometer diamond powder. *Surf. Coat. Technol.*, 120–121, 607–611, 1999.
70. Zhu, H., Li, B., Chen, M., Qiu, C., Tang, Z., Improvement of corrosion resistance of Hastelloy-N alloy in LiF-NaF-KF molten salt by laser cladding pure metallic coatings. *Coatings*, 8, 322, 2018.
71. Jiang, K., Li, J., Liu, J., Electrochemical codeposition of graphene platelets and nickel for improved corrosion resistant properties. *RSC Adv.*, 4, 36245–36252, 2014.
72. Jia, Q., Li, D., Li, S., Zhang, Z., Zhang, N., High-temperature oxidation resistance of NiAl intermetallic formed in Situ by thermal spraying. *Coatings*, 8, 292, 2018.
73. Koivuluoto, H., Lagerbom, J., Kylmälahti, M., Vuoristo, P., Microstructure and mechanical properties of low-pressure cold-sprayed (LPCS) coatings. *J. Therm. Spray Technol.*, 17, 721–727, 2008.
74. Dordsheikh Torkamani, A., Velashjerdi, M., Abbas, A., Bolourchi, M., Maji, P., Electrodeposition of Nickel matrix composite coatings *via* various Boride particles: A review. *J. Compos. Compd.*, 3, 91–98, 2021.
75. Góral, A., Lityńska-Dobrzyńska, L., Kot, M., Effect of Surface Roughness and Structure Features on Tribological Properties of Electrodeposited Nanocrystalline Ni and Ni/Al2O3 Coatings. *J. Mater. Eng. Perform.*, 26, 2118–2128, 2017.
76. Sharifi, K. and Ghorbani, M., Corrosion behaviour of Ni-Co alloy coatings at Kish Island (marine) atmosphere. *Bull. Mater. Sci.*, 37, 713–719, 2014.
77. Sudagar, J., Lian, J., Sha, W., Electroless nickel, alloy, composite and nano coatings - A critical review. *J. Alloys Compd.*, 571, 183–204, 2013.
78. Buchtík, M., Krystýnová, M., Másilko, J., Wasserbauer, J., The effect of heat treatment on properties of Ni-P coatings deposited on a AZ91 magnesium alloy. *Coatings*, 9, 461, 2019.

79. Ţălu, S., Stach, S., Méndez, A., Trejo, G., Ţălu, M., Multifractal Characterization of Nanostructure Surfaces of Electrodeposited Ni-P Coatings. *J. Electrochem. Soc.*, 161, D44–D47, 2014.
80. Ahmadi Ashtiani, A., Faraji, S., Amjad Iranagh, S., Faraji, A.H., The study of electroless Ni–P alloys with different complexing agents on Ck45 steel substrate. *Arabian J. Chem.*, 10, S1541–S1545, 2017.
81. Omar, R., Oraby, E., Abdelrhman, Y., Aboraia, M., Effect of glycine as a complex agent on the surface and corrosion properties of Ni-P and Ni-P/Al2O3 electroless coating. *Anti-Corros. Methods Mater.*, 67, 593–603, 2020.
82. Gu, C., Lian, J., Li, G., Niu, L., Jiang, Z., High corrosion-resistant Ni-P/Ni/Ni-P multilayer coatings on steel. *Surf. Coat. Technol.*, 197, 61–67, 2005.
83. Wang, Y., Guan, L., He, Z., Zhang, S., Singh, H., Hayat, M.D., Yao, C., Influence of pretreatments on physicochemical properties of Ni-P coatings electrodeposited on aluminum alloy. *Mater. Des.*, 197, 109233, 2021.
84. Cui, C., Du, H., Liu, H., Xiong, T., Corrosion behavior of the electroless Ni-P coating on the pore walls of the lotus-type porous copper. *Corros. Sci.*, 162, 108202, 2020.
85. Su, Y., Zhou, B., Liu, L., Lian, J., Li, G., Electromagnetic shielding and corrosion resistance of electroless Ni-P and Ni-P-Cu coatings on polymer/carbon fiber composites. *Polym. Compos.*, 36, 923–930, 2015.
86. Severin, J.W., Hokke, R., De With, G., Adhesion of electrolessly deposited Ni(P) layers on alumina ceramic. I. Mechanical properties. *J. Appl. Phys.*, 75, 3402–3413, 1994.
87. Lin, J.D. and Chou, C.-T., The influence of phosphorus content on the microstructure and specific capacitance of etched electroless Ni-P coatings. *Surf. Coat. Technol.*, 368, 126–137, 2019.
88. Wu, Y., Zhang, Z., Xu, K., Lu, J., Wang, A., Dai, X., Zhu, H., A study on the formation conditions of amorphous nickel-phosphorus (Ni-P) alloy by laser-assisted electrodeposition. *Appl. Surf. Sci.*, 535, 147707, 2021.
89. Alam, M.O., Chan, Y.C., Hung, K.C., Reliability study of the electroless Ni-P layer against solder alloy. *Microelectron. Reliab.*, 42, 1065–1073, 2002.
90. Li, L., Wang, J., Xiao, J., Yan, J., Fan, H., Sun, L., Xue, L., Tang, Z., Time-dependent corrosion behavior of electroless Ni–P coating in H2S/Cl– environment. *Int. J. Hydrogen Energy*, 46, 11849–11864, 2021.
91. Dong, D., Chen, X.H., Xiao, W.T., Yang, G.B., Zhang, P.Y., Preparation and properties of electroless Ni-P-SiO 2 composite coatings. *Appl. Surf. Sci.*, 255, 7051–7055, 2009.
92. Chen, W., Gao, W., He, Y., A novel electroless plating of Ni-P-TiO2 nano-composite coatings. *Surf. Coat. Technol.*, 204, 2493–2498, 2010.
93. Li, Z.H., Wang, X.Q., Wang, M., Wang, F.F., Ge, H.L., Preparation and tribological properties of the carbon nanotubes-Ni-P composite coating. *Tribol. Int.*, 39, 953–957, 2006.

94. Xu, H., Yang, Z., Li, M.K., Shi, Y.L., Huang, Y., Li, H.L., Synthesis and properties of electroless Ni-P-Nanometer Diamond composite coatings. *Surf. Coat. Technol.*, 191, 161–165, 2005.
95. Lee, M., Park, J., Son, K., Kim, D., Kim, K., Kang, M., Electroless Ni-P-PTFE Composite Plating with Rapid Deposition and High PTFE Concentration through a Two-Step Process. *Coatings*, 12, 1199, 2022.
96. Liu, S., Bian, X., Liu, J., Yang, C., Zhao, X., Fan, J., Zhang, K., Bai, Y., Xu, H., Liu, Y., Guo, T., Structure and properties of Ni-P-graphite (Cg)-TiO2 composite coating. *Surf. Eng.*, 31, 420–426, 2015.
97. Tsongas, K., Tzetzis, D., Karantzalis, A., Banias, G., Exarchos, D., Ahmadkhaniha, D., Zanella, C., Matikas, T., Bochtis, D., Microstructural, surface topology and nanomechanical characterization of electrodeposited Ni-P/SiC nanocomposite coatings. *Appl. Sci.*, 9, 2901, 2019.
98. Mousavi Anijdan, S.H., Sabzi, M., Zadeh, M.R., Farzam, M., The effect of electroless bath parameters and heat treatment on the properties of Ni-P and Ni-P-Cu composite coatings. *Mater. Res.*, 21, e20170973, 2018.
99. Ashassi-Sorkhabi, H. and Rafizadeh, S.H., Effect of coating time and heat treatment on structures and corrosion characteristics of electroless Ni-P alloy deposits. *Surf. Coat. Technol.*, 176, 318–326, 2004.
100. Moniruzzaman, M. and Roy, S., Effect of pH on electroless Ni-P coating of conductive and non-conductive materials. *Int. J. Automot. Mech. Eng.*, 4, 481–489, 2011.
101. Genova, V., Paglia, L., Pulci, G., Pedrizzetti, G., Pranzetti, A., Romanelli, M., Marra, F., Medium and High Phosphorous Ni-P Coatings Obtained *via* an Electroless Approach: Optimization of Solution Formulation and Characterization of Coatings. *Coatings*, 13, 1490, 2023.
102. Barati, Q. and Hadavi, S.M.M., Electroless Ni-B and composite coatings: A critical review on formation mechanism, properties, applications and future trends. *Surf. Interfaces*, 21, 100702, 2020.
103. López, J.R., Méndez, P.F., Pérez-Bueno, J.J., Trejo, G., Stremsdoerfer, G., Meas, Y., The effect of boron content, crystal structure, crystal size on the hardness and the corrosion resistance of electrodeposited Ni-B coatings. *Int. J. Electrochem. Sci.*, 11, 4231–4244, 2016.
104. Shakoor, R.A., Kahraman, R., Gao, W., Wang, Y., Synthesis, characterization and applications of electroless Ni-B coatings-A review. *Int. J. Electrochem. Sci.*, 11, 2486–2512, 2016.
105. Oraon, B., Majumdar, G., Ghosh, B., Improving hardness of electroless Ni-B coatings using optimized deposition conditions and annealing. *Mater. Des.*, 29, 1412–1418, 2008.
106. Barman, M., Barman, T.K., Sahoo, P., Effect of borohydride concentration on tribological and mechanical behavior of electroless Ni-B coatings. *Mater. Res. Express*, 6, 126575, 2019.
107. Pal, S. and Jayaram, V., Effect of microstructure on the hardness and dry sliding behavior of electroless Ni–B coating. *Materialia*, 4, 47–64, 2018.

108. Nemane, V. and Chatterjee, S., Evaluation of microstructural, mechanical, and tribological characteristics of Ni-B-W-SiC electroless composite coatings involving multi-pass scratch test. *Mater. Charact.*, 180, 111414, 2021.
109. Oraon, B., Majumdar, G., Ghosh, B., Parametric optimization and prediction of electroless Ni-B deposition. *Mater. Des.*, 28, 2138–2147, 2007.
110. Anik, M., Körpe, E., Şen, E., Effect of coating bath composition on the properties of electroless nickel-boron films. *Surf. Coat. Technol.*, 202, 1718–1727, 2008.
111. Hamid, Z.A., Hassan, H.B., Attyia, A.M., Influence of deposition temperature and heat treatment on the performance of electroless Ni-B films. *Surf. Coat. Technol.*, 205, 2348–2354, 2010.
112. Duru, E., Doğan, F., Uysal, M., Akbulut, H., Aslan, S., Optimization of Ni-B coating bath and effect of DMAB concentration on hardness and wear. *Surf. Interfaces*, 22, 100880, 2021.
113. Guerra, L., Echeverría, F., Correa, E., Effect of pH on Dimethylamine Borane Reduced Electroless Nickel Deposits on AISI/SAE 1045 Steel Surface. *Met. Mater. Int.*, 26, 773–782, 2020.
114. Vijayanand, M. and Elansezhian, R., Effect of different pretreatments and heat treatment on wear properties of electroless Ni-B Coatings on 7075-T6 Aluminum Alloy. *Procedia Eng.*, 97, 1707–1717, 2014.
115. Shakoor, R.A., Waware, U.S., Kahraman, R., Popelka, A., Yusuf, M.M., Corrosion behavior of electrodeposited Ni-B coatings modified with SiO2 particles. *Int. J. Electrochem. Sci.*, 12, 4384–4391, 2017.
116. Shakoor, R.A., Kahraman, R., Waware, U.S., Wang, Y., Gao, W., Synthesis and properties of electrodeposited Ni-B-CeO2 composite coatings. *Mater. Des.*, 59, 421–429, 2014.
117. Sheu, H.H., Wang, Q.Y., Lee, Y.W., Effects of Boron Carbide Particle Content on Corrosion Resistance and Wear Resistance of Ni-B/B4C Composite Coatings. *Int. J. Electrochem. Sci.*, 17, 221166, 2022.
118. Wang, Y., Wang, S.J., Shu, X., Gao, W., Lu, W., Yan, B., Preparation and property of sol-enhanced Ni-B-TiO2 nano-composite coatings. *J. Alloys Compd.*, 617, 472–478, 2014.
119. Shakoor, R.A. and Bahgat Radwan, A., Remarkable improvement in properties of NI-B coatings by the addition of mixed oxide nanoparticles. *Mater. Sci. Technol. Conf. Exhib. 2015, MS T 2015*, vol. 1, pp. 171–178, 2015.
120. Eraslan, S. and Ürgen, M., Oxidation behavior of electroless Ni-P, Ni-B and Ni-W-B coatings deposited on steel substrates. *Surf. Coat. Technol.*, 265, 46–52, 2015.
121. Salari Mehr, M., Akbari, A., Damerchi, E., Electrodeposited Ni-B/SiC micro- and nano-composite coatings: A comparative study. *J. Alloys Compd.*, 782, 477–487, 2019.
122. Mukhopadhyay, A., Barman, T.K., Sahoo, P., Tribological characteristics of electroless Ni-B-Mo coatings at different operating temperatures. *Surf. Rev. Lett.*, 26, 2019.

123. Mukhopadhyay, A., Barman, T.K., Sahoo, P., Davim, J.P., Comparative study of tribological behavior of electroless Ni-B, Ni-B-Mo, and Ni-B-W coatings at room and high temperatures. *Lubricants*, 6, 67, 2018.
124. Matik, U., Improvement of surface properties of iron based powder metal compacts by electroless NiB coating. *J. Fac. Eng. Archit. Gazi Univ.*, 33, 1603–1610, 2018.
125. Das, S.K. and Sahoo, P., Influence of process parameters on microhardness of electroless Ni-B coatings. *Adv. Mech. Eng.*, 2012, 2012.
126. Dushik, V.V., Ruban, E.A., Drovosekov, A.B., Shaporenkov, A.A., Rozhanskiy, N.V., Synergetic Effect in Ni–P–W and W–C Multilayer Coating Systems Obtained by Chemical-Catalytic Metallization and Chemical-Vapor Deposition. *J. Surf. Investig.*, 17, 1364–1371, 2023.
127. Nazari, H., Barati Darband, G., Arefinia, R., A review on electroless Ni–P nanocomposite coatings: effect of hard, soft, and synergistic nanoparticles. *J. Mater. Sci.*, 58, 4292–4358, 2023.
128. Chintada, V.B., Gurugubelli, T.R., Tamtam, M.R., Koutavarapu, R., Advancements in Nickel-Phosphate/Boron Based Electroless Composite Coatings: A Comprehensive Review of Mechanical Properties and Recent Developments. *Materials (Basel)*, 16, 6116, 2023.
129. Liu, Y., Li, Z., Li, G., Tang, L., Friction and wear behavior of Ni-based alloy coatings with different amount of WC–TiC ceramic particles. *J. Mater. Sci.*, 58, 1116–1126, 2023.
130. Wang, J., Zhou, J., Zhang, T., Meng, X., Li, P., Huang, S., Zhu, H., Ultrasonic-Induced Grain Refinement in Laser Cladding Nickel-Based Superalloy Reinforced by WC Particles. *Coatings*, 13, 151, 2023.
131. Wang, Y., Weng, C., Deng, Z., Sun, H., Jiang, B., Fabrication and performance of nickel-based composite mold inserts for micro-injection molding. *Appl. Surf. Sci.*, 615, 156417, 2023.
132. Hüner, B., Demir, N., Kaya, M.F., Ni-Pt coating on graphene based 3D printed electrodes for hydrogen evolution reactions in alkaline media. *Fuel*, 331, 125971, 2023.
133. Yu, J., Wang, Y., Zhao, X., Li, Q., Qiao, Q., Zhao, J., Zhai, S., Wear Resistance of Ni-Based Alloy Coatings. *Adv. Mater. Sci. Eng.*, 2019, 2019.
134. Güler, O., The effect of candle-soot reinforced electroless Ni-P particle interface on tribological performance in copper-based materials produced by flake powder metallurgy method: Transfer of candle-soot from waste to solid-lubricant. *Colloids Surf., A*, 673, 131858, 2023.
135. Wu, W., Zhang, C., Wang, R., Zhang, Y., Lu, X., A high temperature wear-resistant Ni-based alloy coating for coppery blast furnace tuyere application. *Surf. Coat. Technol.*, 464, 129550, 2023.
136. Tan, S., Algül, H., Kiliçaslan, E., Alp, A., Akbulut, H., Uysal, M., The effect of ultrasonic power on high temperature wear and corrosion resistance for Ni based alloy composite coatings. *Colloids Surf., A*, 656, 130345, 2023.

137. Zhong, J., Zhang, S., He, Y., Fan, Y., Li, Z., Yan, L., Zhou, H., Cheng, X., Song, J., Li, H., Study on the use of furan epoxide modified graphene to enhance the corrosion resistance and wear resistance of electroless Ni-W-P coatings. *Surf. Coat. Technol.*, 473, 129946, 2023.
138. Güler, O., Kocaman, M., Akçay, S.B., Çuvalcı, H., Varol, T., Structure and tribological properties of composites based on a Novolac matrix with Ni–P electrolytic coating. *Surf. Coat. Technol.*, 451, 129025, 2022.
139. Liu, S., Sun, Y., Zhai, P., Fan, P., Zhang, Y., Li, M., Fang, J., Wu, R., Cai, Z., Microstructure and Properties of Nickel-Based Gradient Coatings Prepared Using Cold Spraying Combined with Laser Cladding Methods. *Materials (Basel)*, 16, 1627, 2023.
140. Sahoo, P. and Das, S.K., Tribology of electroless nickel coatings - A review. *Mater. Des.*, 32, 1760–1775, 2011.
141. Rosenkranz, A., Costa, H.L., Baykara, M.Z., Martini, A., Synergetic effects of surface texturing and solid lubricants to tailor friction and wear – A review. *Tribol. Int.*, 155, 106792, 2021.
142. Zhu, D., Zhang, X., Ding, H., Tool wear characteristics in machining of nickel-based superalloys. *Int. J. Mach. Tools Manuf.*, 64, 60–77, 2013.
143. Abdeen, D.H., El Hachach, M., Koc, M., Atieh, M.A., A review on the corrosion behaviour of nanocoatings on metallic substrates. *Materials (Basel)*, 12, 210, 2019.
144. Galedari, S.A., Mahdavi, A., Azarmi, F., Huang, Y., McDonald, A., A Comprehensive Review of Corrosion Resistance of Thermally-Sprayed and Thermally-Diffused Protective Coatings on Steel Structures. *J. Therm. Spray Technol.*, 28, 645–677, 2019.
145. Majeed, M.N., Yousif, Q.A., Bedair, M.A., Study of the Corrosion of Nickel-Chromium Alloy in an Acidic Solution Protected by Nickel Nanoparticles. *ACS Omega*, 06, 48, 2022.
146. Yu, J., Wang, Y., Zhao, X., Li, Q., Qiao, Q., Zhao, J., Zhai, S., Wear Resistance of Ni-Based Alloy Coatings. *Adv. Mater. Sci. Eng.*, 2019, 1–7, 2019.
147. Bhutta, M.U. and Khan, Z.A., Wear and friction performance evaluation of nickel based nanocomposite coatings under refrigerant lubrication. *Tribol. Int.*, 148, 106312, 2020.
148. Wood, R.J.K. and Lu, P., Coatings and Surface Modification of Alloys for Tribo-Corrosion Applications. *Coatings*, 14, 99, 2024.
149. Kundu, S., Sahoo, P., Das, S.K., Optimization studies on electroless nickel coatings: A review. *Int. J. Manuf. Mater. Mech. Eng.*, 4, 1–25, 2014.
150. Priyadarshi, P., Katiyar, P.K., Maurya, R., A review on mechanical, tribological and electrochemical performance of ceramic particle-reinforced Ni-based electrodeposited composite coatings. *J. Mater. Sci.*, 57, 19179–19211, 2022.
151. Beltowska-Lehman, E., Bigos, A., Szczerba, M.J., Janusz-Skuza, M., Maj, L., Debski, A., Wiazania, G., Kot, M., Heat treatment of ultrasonic electrodeposited Ni-W/ZrO2 nanocomposites. *Surf. Coat. Technol.*, 393, 125779, 2020.

152. Jungk, J.M., Michael, J.R., Prasad, S.V., The role of substrate plasticity on the tribological behavior of diamond-like nanocomposite coatings. *Acta Mater.*, 56, 1956–1966, 2008.
153. Gheisari, R. and Polycarpou, A.A., Tribology of Selected Hard Coatings for Oil and Gas Applications Up to 450°C. *Tribol. Lett.*, 67, 1–16, 2019.
154. Liu, N., Liu, Q., Li, Z., Bai, Y., Sun, Y.W., Li, Z.D., Bao, M.Y., Zhan, H., Guo, D.G., Ma, Y.S., Tribological behavior of plasma-sprayed metal based solid self-lubricating coatings under heavy load. *Wear*, 486–487, 204108, 2021.
155. Chen, J., An, Y., Liu, G., Chen, G., Zhao, X., Jia, L., Tribological Performance and Thermal Stability of a Novel Cold Sprayed Nanostructured Ni-based Lubrication Coating. *J. Therm. Spray Technol.*, 31, 1702–1711, 2022.
156. Kocaman, M., Akçay, S.B., Güler, O., Çuvalcı, H., Varol, T., Çuvalcı, O., The effect of recycled zinc powder content on the properties of novel novolac matrix hybrid composites. *Tribol. Int.*, 188, 108814, 2023.
157. Ezugwu, E.O., Wang, Z.M., Machado, A.R., The machinability of nickel-based alloys: A review. *J. Mater. Process. Technol.*, 86, 1–16, 1998.
158. Behera, A. and Sahoo, A.K., Wear behaviour of Ni based superalloy: A review. *Mater. Today Proc.*, 33, 5638–5642, 2020.
159. Marenych, O. and Kostryzhev, A., Strengthening mechanisms in nickel-copper alloys: A review. *Metals (Basel)*, 10, 1–18, 2020.
160. Cheng, J., Li, F., Zhu, S., Hao, J., Yang, J., Li, W., Liu, W., High temperature tribological properties of a nickel-alloy-based solid-lubricating composite: Effect of surface tribo-chemistry, counterpart and mechanical properties. *Wear*, 386–387, 39–48, 2017.
161. Li, S., Huang, K., Zhang, Z., Zheng, C., Li, M., Wang, L., Wu, K., Tan, H., Yi, X., Wear mechanisms and micro-evaluation of WC + TiC particle-reinforced Ni-based composite coatings fabricated by laser cladding. *Mater. Charact.*, 197, 112699, 2023.
162. Azim, S., Gangopadhyay, S., Mahapatra, S.S., Mittal, R.K., Singh, R.K., Role of PVD coating on wear and surface integrity during environment-friendly micro-drilling of Ni-based superalloy. *J. Clean. Prod.*, 272, 122741, 2020.
163. Sreekumar, R., Nair, A.S., S.S, S., Recent trends and developments in two-dimensional materials based electrodeposited nickel nanocomposite coatings. *FlatChem*, 36, 100434, 2022.
164. Zhang, Y. and Chromik, R.R., Tribology of Self-Lubricating Metal Matrix Composites, in: *Self-Lubricating Compos. Second Ed.*, pp. 31–71, 2022.
165. Yan, Y., Lu, L., Huo, Y., Dry Sliding Tribological Behaviors of Electrodeposited Ni-GO/SiC Composite Coating on the 2218 Aluminum Alloy. *Materials (Basel)*, 15, 2834, 2022.
166. Zhang, X., Wang, H., Liu, G., Friction and Wear of Electroless Ni-P-CS Composite Coating. *Metals (Basel)*, 13, 315, 2023.
167. Wu, W., Ye, S., Wang, R., Zhang, C., Zhang, Y., Lu, X., Microstructure and wear behavior of plasma cladded Ni-based alloy coating on copper under

different preheating temperature. *J. Mater. Res. Technol.*, 23, 1609–1617, 2023.
168. Saroha, V., Pabla, B.S., Bhogal, S.S., Characterization of ABS for enhancement of mechanical properties. *Int. J. Innov. Technol. Explor. Eng.*, 8, 2164–2167, 2019.
169. Fan, Y., He, Y., Luo, P., Shi, T., Chen, X., Pulse Current Electrodeposition and Properties of Ni-W-GO Composite Coatings. *J. Electrochem. Soc.*, 163, D68–D73, 2016.
170. Zhang, X.H., Li, X.X., Liu, W.J., Fan, Y.Q., Chen, H., Liang, T.X., Preparation and tribological behavior of electrodeposited Ni–W–GO composite coatings. *Rare Met.*, 38, 695–703, 2019.
171. Srinivasan, K.N., Meenakshi, R., Santhi, A., Thangavelu, P.R., John, S., Studies on development of electroless Ni-B bath for corrosion resistance and wear resistance applications. *Surf. Eng.*, 26, 153–158, 2010.
172. Gültekin, D., Duru, E., Akbulut, H., Improved wear behaviors of lead-free electroless Ni[sbnd]B and Ni-B/CeO2 composite coatings. *Surf. Coat. Technol.*, 422, 127525, 2021.
173. Mirhosseini, S.S. and Mahboubi, F., Effect of plasma nitriding on tribological properties of nickelboron-nanodiamond electroless coatings. *Surf. Coat. Technol.*, 435, 128216, 2022.
174. Krishnaveni, K., Sankara Narayanan, T.S.N., Seshadri, S.K., Electrodeposited Ni-B-Si3N4 composite coating: Preparation and evaluation of its characteristic properties. *J. Alloys Compd.*, 466, 412–420, 2008.
175. Aslan, S. and Duru, E., Microstructure and Wear Properties of Electrodeposited Ni-B-Al2O3 Composite Coating on Low Carbon Steel at Elevated Temperature. *J. Mater. Eng. Perform.*, 31, 1693–1704, 2022.
176. Panja, B., Das, S.K., Sahoo, P., Tribological Behavior of Electroless Ni–P Coating in Brine Environment. *J. Inst. Eng. Ser. D*, 95, 153–159, 2014.
177. Nisa, Z.U., Chuan, L.K., Guan, B.H., Ayub, S., Ahmad, F., Anti-Wear and Anti-Erosive Properties of Polymers and Their Hybrid Composites: A Critical Review of Findings and Needs. *Nanomaterials*, 12, 2194, 2022.
178. Gadhari, P. and Sahoo, P., Electroless nickel-phosphorus composite coatings: A review. *Int. J. Manuf. Mater. Mech. Eng.*, 6, 14–50, 2016.
179. Pancrecious, J.K., Ulaeto, S.B., Ramya, R., Rajan, T.P.D., Pai, B.C., Metallic composite coatings by electroless technique–a critical review. *Int. Mater. Rev.*, 63, 488–512, 2018.
180. Grand View Research Report, Industrial Coatings Market, 2018, https://www.grandviewresearch.com/industry-analysis/powder-coatings-market-analysis

3

Advances in Nanostructured Materials and Liquid Crystal Composites: Unveiling Structural Properties and Emerging Paradigms in Display Technologies

Driss Soubane[1,2*] and Mohamed El Garah[3†]

[1]Institut National de la Recherche Scientifique - Centre Énergie, Matériaux et Télécommunications (INRS-EMT), Varennes, Québec, Canada
[2]Cadi Ayyad University - Multi-Disciplinary Faculty of Safi, Department of Physics, Energy = Materials × Computation × Confinement (ε =mc²) EBCG-Lab, Safi, Morocco
[3]LRC, LASMIS, Antenne de Nogent – 52, Pôle Technologique de Sud – Champagne, Nogent, France

Abstract
In this chapter, we provide a comprehensive exploration of recent advancements in nanostructured materials doped liquid crystal composites, emphasizing their applications in displays. Beginning with a brief introduction, we systematically delve into the fundamentals of nanostructured materials, categorizing them into 0D, 1D, and 2D structures. Detailed discussions on fabrication techniques, including lithography, VLS/SLS, PVD, CVD, ALD, and self-assembly, establish a thorough foundation. The core of the chapter revolves around the interaction between nanostructures and liquid crystals. Various doping materials, such as titania, Au and Ag quantum dots, silicon dioxide quantum dots, and others, are examined in binary liquid crystal composites. The exploration extends to ternary composite materials, incorporating photochromic/photoresponsive polymers and chiral dopants for display applications. Beyond binary and ternary composites, we outline emerging trends and perspectives in this dynamic field, encompassing the

**Corresponding author*: driss.soubane@inrs.ca
†*Corresponding author*: mohamed.el_garah@utt.fr
Driss Soubane: ORCID: https://orcid.org/0000-0002-8873-3799
Mohamed El Garah: ORCID: https://orcid.org/0000-0003-2217-9681

Inamuddin, Tariq Altalhi and Mohammad Abu Jafar Mazumder (eds.) Nanostructured Hard Materials: Industrial Applications, (113–148) © 2026 Scrivener Publishing LLC

intrinsic properties of liquid crystals, including their classes, phases, and principal characteristics. Pioneering advances in display technology are spotlighted, offering insights into innovative designs that enhance display performance. We conclude by laying the groundwork for future developments, emphasizing the potential for continued advancements in the synergy between nanostructured materials and liquid crystals for displays.

Keywords: Nanostructured materials, liquid crystal composites, display technology, doping materials, ternary composites, pioneering advances, fabrication techniques, emerging trends

3.1 Introduction

The term *nano* originates from the Latin word *nanus*, signifying a dwarf being. It specifically denotes a billionth of a meter, i.e., 10^{-9} m. In a general sense, an object with a size less than 100 nm is considered the threshold to distinguish between the bulk and nanoscale world [1]. It is essential to note that the 100 nm upper limit is not a scientifically rigorous barrier separating classical from quantum object behavior; rather, it is a recommendation by the International Organization for Standardization. The scientifically recognized threshold is the Bohr radius, which theoretically coincides with the manifestation of quantum effects. To be morc precise, the concept of volume-specific surface area was introduced a decade ago. A nano-object is defined as having a surface-to-volume ratio equal to or higher than 60 m^2/cm^3 [2].

In fact, the utilization of nanotechnology dates back centuries. Examples such as the 4th-century Lycurgus Cup, which changes color due to the presence of quantum dots made of silver and gold [3], and Saladin's Damascus sword, crafted with carbon nanotubes [4], to defeat the Crusaders, are remarkable illustrations of advanced ancient civilizations. Unfortunately, the techniques involved in their creation have been mysteriously lost over time.

At the 1959th annual meeting held by the American Physical Society, R. Feynman delivered his famous speech entitled 'There is plenty of room at the bottom: an invitation to enter a new field of physics' [5], paving the way to manipulate single atoms. He challenged physicists, offering a $1000 prize to the first successor to fabricate a miniaturized motor and/or to include Encyclopedia Britannica on the head of a pen. W. McLellan from Caltech and T. Newmann at Stanford University separately fulfilled the two challenges in 1960 and 1985, respectively [6, 7].

3.2 Fundamentals of Nanostructured Materials

The impact of nanotechnology on society is indisputably evident in improving individual lifestyles, encompassing energy savings, enhanced computer performance, and advancements in health welfare [8–10]. Nanotechnology's small scale allows for direct use in human blood for medical purposes, aiding in disease diagnosis and making surgeries more localized and less damaging. This downsizing aligns with the pace of the second law of thermodynamics. Additionally, nanotechnology contributes to environmental preservation, unlocking the way for a clean energy transition by integrating nanostructures into renewable energy technologies, such as photovoltaic solar cells [11–14], photo-catalysts [15, 16], and nano-capacitors for energy storage [17–19].

Human progress, spanning from the Stone Age with the invention of the potter's wheel to the present, has seen phases like the Bronze Age, Iron Age, and the acquisition of the know-how to produce glass. The 18th century marked the industrial revolution, and the 20th century witnessed the information technology era, leading to the revolutionary rise of artificial intelligence (AI) driven by the maturation of nanotechnology [20]. This advancement significantly contributed to enhancing storage capacity and expediting data processing, paving the way for the potential development of quantum computers [21]. The concept of Kurzweil's singularity [22], appears both aspirational and daunting as we contemplate the future trajectory of technology.

Nanotechnology involves feature sizes less than 100 nm. To better visualize this scale, consider the size comparison between the sun and a human being, analogous to comparing the same human being to a DNA diameter of about 2.5 nm. It encompasses the manipulation, study, and application of nanoscale materials and devices, spanning scientific disciplines such as physics, chemistry, biology, engineering, and materials.

Furthermore, IBM designed a computer that was 50 feet in size, and through the progression outlined in Gordon Moore's second law [23], we now have computers at a much smaller actual size. In 1974, Professor Norio Taniguchi of Tokyo University coined the term *nanotechnology* [24]. His pioneering work in this field was later recognized with a lifetime achievement award in 1999 by The European Society for Precision Engineering and Nanotechnology.

This chapter initiates with a discussion of nanostructured materials, including classification and fabrication techniques, fundamentals of liquid crystal, and composite materials. The latter part delves into future perspectives and emerging trends.

3.2.1 Nanostructured Material Classification

Based on the above discussion, distinctions among nanomaterials are made based on the number of nanoscaled dimensions of a bulk material. Nanostructured materials are classified into four types based on their sizes, as presented in Figure 3.1: 0D, 1D, 2D, and 3D. Materials with dimensions ranging from 1 to 100 nm fall into the category of nanomaterials.

Nanomaterials are further classified according to their spatial dimensions:

- Zero-dimensional (0D): Nanomaterials with three dimensions smaller than 100 nm.
- One-dimensional (1D): Where one dimension is larger than the other two.
- Two-dimensional (2D): Where two dimensions are smaller than the third.

Three-dimensional (3D) nanomaterials are not confined to the nanoscale in any dimension. The classification of this type of nanostructured materials is presented in Figure 3.1.

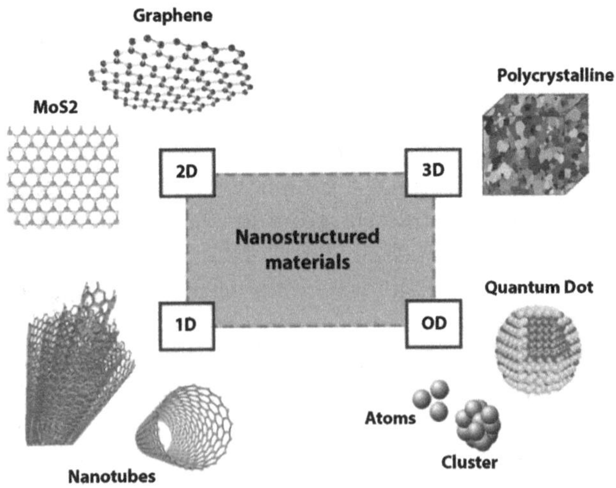

Figure 3.1 Classification of nanostructured materials according to their sizes.

a. 0D Nanostructures

In the 0D nanostructure category, all three dimensions are quantum-confined, indicating that their dimensions fall within the range with an upper limit of 100 nm. They take the form of clusters or spheres that appear as point-like particles, including quantum dots, nanodots, nanocrystals, nanoparticles, etc. Generally, 0D nanostructured materials can be crystalline or amorphous, exhibiting various shapes [25].

A particularly interesting 0D structure is Fullerene, also known as Buckminsterfullerene, which forms a polygon with sixty carbon atoms. In 1985, the group led by Kroto [26], successfully fabricated it. Its characteristic shape is that of a buckyball, a sphere. Other forms, such as C70 and C84, extend the applications of this unique material, making it valuable in nanotechnology, medicine, and other realms [27–29]. Quantum dots, prepared with atoms from groups II-VI or III-V, are also known as fluorescent semiconductive nanoparticles. The most well-known ones are CdSe, CdTe, and InP, which can be considered as potential candidates for use in biosensing and therapy, for example. Different nanodots have been recently reported in the literature [8].

b. 1D Nanostructures

In this case, two out of the three dimensions are on the nanometer scale, with only one dimension free from quantum confinement, leading to its classification as a 1D nanostructure. This category includes nanowires, nanotubes (if the wire is hollow), nanorods [30], nanoneedles, whiskers, fibers, fibrils, etc. Fibers and nanowires are presumed to be longer than whiskers and nanorods (Figure 3.2). This class of nanostructures exhibits unique electronic and thermal behavior of energy carriers, making them potential candidates for applications in optoelectronics, mechanics, and interconnecting devices, all of which have been thoroughly investigated.

It is worth noting that in the early definition, whiskers and fibers were not strictly considered to be in the nanometric size range; microscaled

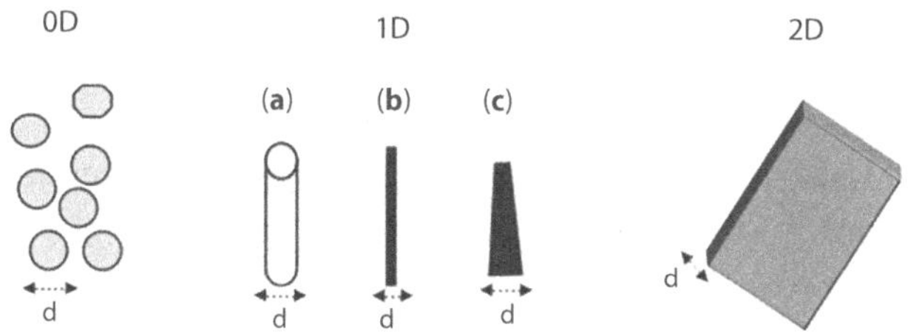

Figure 3.2 Typical down-scaled nanostructures classes: 0D: ball, 1D: (a) wire, (b) tube, (c) rod, 2D: plate with thickness < 100 nm where 0 < d < 100 nm.

material was also accepted in this category. Nevertheless, the diameter upper limit for nanowires and nanorods is set to a few hundred nanometers. Nanorods can take several forms: columns, pillars, or rods. They can be grown using various techniques, including, for example, physical vapor deposition (PVD) and chemical vapor deposition (CVD), etc. [31–33]. Carbon nanotubes are the most well-known class of 1D nanomaterials, formed by carbon atoms with a honeycomb structure and a tubular shape. Two categories are reported: single-walled carbon nanotubes and multi-walled carbon nanotubes. They are characterized by interesting properties and can be exploited for many applications [34–37]. Other 1D nanomaterials include nanowires, where the most relevant are polymeric or metallic.

c. 2D Nanostructures

In the 2D nanostructure classification, the nanometric dimension is along only one dimension, forming a plane-like shape. The straightforward approach is to control the layer thickness to a nanoscale to achieve such a nanostructure. Various works have been undertaken to obtain these structures, including nanoplates, nanosheets, nanowalls, nanodisks, etc.

Graphene [38], is one of the famous 2D structures, consisting of a single layer of carbon atoms crystallized in a hexagonal Bravais lattice, making it a zero-gap semiconductor. This unique material exhibits exceptional mobility, thermal conductivity, and unique mechanical features. Graphene was first exfoliated by Andre Geim and Konstantin Novoselov in 2004 [39], awarded the Nobel Prize in physics. Graphene oxide is a derivative of Graphene that has also been reported. Furthermore, various elements other than carbon can also form 2D nanostructured materials. For example, transition metal dichalcogenides (TMDs) are 2D materials having a formula of MX_2 (M=transition metal and X=chalcogen element) exhibiting a hexagonal plane structure. They are characterized by three different phases: 1T, 2H, and 3R. The nanosheets, with 2H phase, are stable and have good optical properties (strong absorption and emission) in visible regions. Borophene is also another type of 2D nanomaterial recently investigated in materials science and nanotechnology field. It has unique physicochemical properties and can be exploited to develop potential applications mainly in chemical sensors [40].

Nanotechnology focuses on producing nanoscale systems with practical applications in everyday life. Nanostructures, exhibiting one, two, or three dimensions at the nanoscale, are respectively termed quantum-well, quantum-wire, quantum-rod, or quantum-tube for a hollowed wire, and quantum dots.

Applications of nanotechnology are pervasive across various industrial domains, ranging from computer-based electronics and fuel cell electrodes to photonic applications, including laser compounds. Furthermore, nanotechnology finds applications in various areas, including cosmetic products [41–43], sports equipment [44], vehicle supply [45], and the pharmaceutical industry [46].

3.3 Fabrication of Nanostructures

In general, nanofabrication methods are divided into two categories: top-down and bottom-up, depending on the processes involved in creating nanostructures. The top-down approach involves the controlled use of tools to reduce the size of nanoscale devices to the desired shape [47, 48], obtaining tiny structures displaying different physical and chemical properties. This method relies mainly on lithography and techniques such as cutting, etching, or grinding (Figure 3.3). Various techniques are employed for this purpose, including all types of lithography, physical vapor deposition, and chemical vapor deposition [49] (originating in 1960 at AT & Bell Laboratory).

Although the top-down approach has played a crucial role in the fabrication of nanostructured materials, it faces limitations, especially when the desired material requires a size less than 100 nm. These limitations include high manufacturing costs, imperfections in the processed materials, and long manufacturing times, particularly in lithography patterning with short wavelengths, where X-rays cannot achieve resolutions finer than 10 nm.

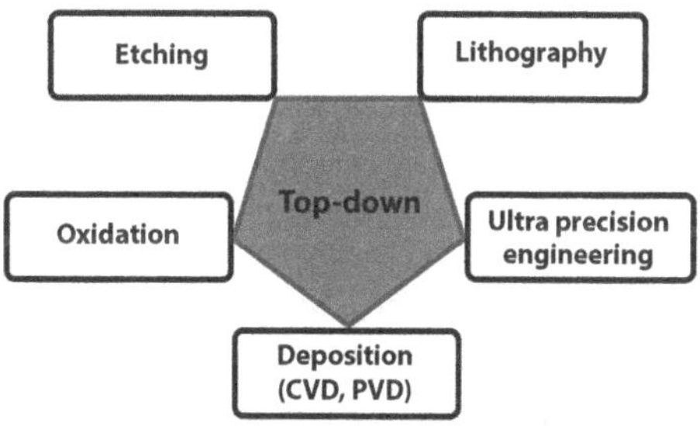

Figure 3.3 Illustration of various top-down deposition techniques for nanostructures. The diagram highlights diverse methods, including lithography, sputtering, and etching, representing key top-down approaches employed in the controlled deposition of nanostructures for advanced material engineering.

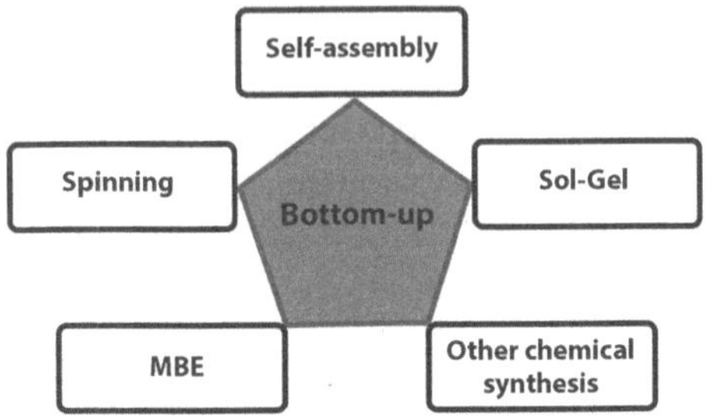

Figure 3.4 Visualization of bottom-up nanostructure synthesis techniques. The diagram showcases diverse bottom-up methodologies, such as chemical synthesis, sol-gel, and self-assembly, elucidating the controlled fabrication of nanostructures at the molecular or atomic level.

The milling of metallic and ceramic materials involves the generation of heat, necessitating the use of cryogenic liquids. However, the resulting nanomaterials are not monodispersed and lack shape homogeneity. Commonly employed techniques for this purpose include plasma etching, anodization, and lithography.

Conversely, the bottom-up approach focuses on manufacturing nanostructured materials from atoms or molecules. In this approach, the controlled assembly of atoms leads to the formation of desired nanostructures with nanometric sizes. Two methods are commonly used in the bottom-up approach to fabricate nanostructures: gas-phase synthesis and liquid-phase synthesis. The most well-known is the self-assembly method, aiming to construct nano-objects and involving chemical synthesis, self-organization, or positional assembly. Numerous techniques inducing self-assembled arranging, such as sol-gel, spinning methods, and molecular beam epitaxy (MBE) (Figure 3.4), are equally employed [50]. The obtained nanostructures are of high quality, enabling uniform dispersion and better control of the shape.

Compared to the bottom-up approach, the top-down approach is reputed to result in nanostructures with lower stoichiometry but higher defect density and less long-range ordering. However, it is more consistent with mass production [51].

3.3.1 Nanostructure Fabrication Techniques

a. Lithography
Lithography, a precision patterning technique for thin layers, involves a meticulous process. Various methods have been employed for patterning

nanostructured materials. It commences with applying a photoresist coating onto a substrate, covering it with a carefully designed photomask. Controlled exposure of light onto the photoresist transfers intricate patterns from the mask, marking the inception of this sophisticated method. The process advances to the etching stage, selectively removing material from the substrate. Subsequently, the photoresist is removed, leaving behind the meticulously patterned thin layer [52]. Different lithography modalities, distinguished by their radiation sources, contribute to this refined technique. Optical lithography, introduced in the 1960s for patterning integrated electronic circuits, uses light (400 nm) to achieve a resolution of 1 μm. Electron beam lithography, commonly used for manufacturing integrated circuits, involves exposing the surface to a beam with small dimensions. Nanoimprint lithography, based on mechanical embossing, is relevant for manufacturing advanced nanostructures with higher resolutions than conventional lithography. Other reported lithography methods include X-ray lithography and extreme ultraviolet lithography, where the pattern dimensions are intricately determined by the wavelength of the employed radiation source.

b. Vapor Liquid Solid (VLS)/Solution Liquid Solid (SLS)

The Vapor Liquid Solid (VLS) technique, proposed in the 1960s by Wagner and Ellis, gained widespread recognition in the 1990s as a bottom-up approach to obtaining nanowires with nanometric dimensions. The VLS method consists of three main steps illustrated in Figure 3.5. The target material for nanowires is evaporated on a liquid metallic catalyst droplet, such as gold. The mixing of the liquid and vapor guides the growth of nanowires and determines their diameter. As it solidifies, the nanowire grows taller as long as the precursor is available unless the growth process is intentionally interrupted by breaking specified criteria.

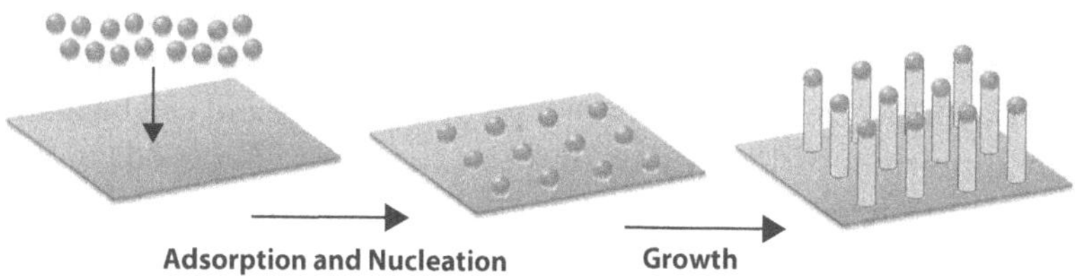

Figure 3.5 Illustration showcasing the sequential steps of the vapor-liquid-solid process in the growth of semiconductor nanowires. This mechanism involves depositing vapor-phase material onto a liquid catalyst droplet, leading to the solidification of nanowires with precise dimensions and superior crystalline quality.

On the other hand, in the SLS approach, the initial stage of nanowire elaboration involves a solution-phase reaction. The choice between the VLS or SLS approaches depends on the targeted properties of the nanowires and the synthesis requirements. This method has been instrumental in fabricating silicon (Si) and germanium (Ge) nanowires [53–56], for example.

c. Physical Vapor Deposition

The physical method in question, aptly named, does not involve chemical reactions. However, growth takes place at the atomic scale, allowing for classification into two primary processes: evaporation and sputtering. In the former, such as thermal evaporation, the source is heated to form the film. The material is initially heated in a vacuum chamber to create vapor pressure. Thermal energy propels the flow of this vapor, leading to the formation of a film on the substrate. In the latter, the process involves the evaporation of the target, with an inert gas (often argon) to form plasma. Rapid cooling of the gas then causes condensation, resulting in the formation of a film on the substrate. The PVD process is mainly used for the synthesis of pure metals, alloys, and ceramics (oxides, nitrides, and carbides).

Following the logic outlined above, Molecular Beam Epitaxy, evaporation, pulsed laser deposition (PLD), cathodic arc deposition, and sputtering are sub-classes of the physical vapor deposition approach. MBE involves the evaporation of atoms or molecules in a high vacuum environment, resulting in the formation of high-quality nanostructures tightly adherent to substrates and offering more atomic precision control. MBE excels in precision and control at the atomic level, particularly in the fabrication of nanostructured semiconductors. In contrast, the evaporation technique is less atomically precise, relying on thermal heating to form vapor, achieving a higher growth rate. Cathodic arc deposition is also a technique reputed for manufacturing functional films. An electrical arc can be performed to evaporate the target, leading to its deposition on the substrate. Pulsed Laser Deposition is a versatile technique that facilitates the growth of a wide range of nanostructures [57]. It relies on using a laser excimer source to generate a plasma plume through target ablation. Generally, films produced by PVD techniques can have high performance and can be made in an environmentally friendly way. They can, therefore, offer good durability, resistance to corrosion, and oxidation. In Figure 3.6, we present the different PVD methods.

d. Chemical Vapor Deposition

In the manufacturing of nanomaterials by the CVD technique, various competing processes are involved, including the transport of reactants,

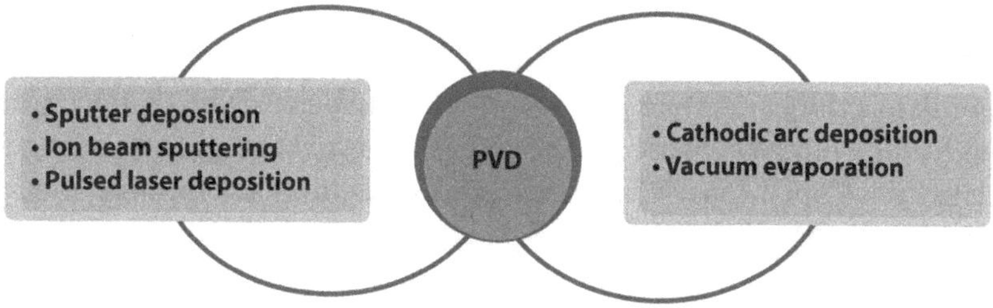

Figure 3.6 Overview of PVD processes family - nanostructure deposition techniques.

adsorption and desorption of precursors, nucleation, and coating growth. Various gas elements then chemically react to form a vapor phase, leading to the deposition of a solid complex compound on a substrate. Other reaction products must be gaseous so that they can be removed from the reactor. This type of deposition is carried out at variable pressure, with energy systematically supplied to promote the reactions.

Depending on the type of energy used to activate the reaction, several CVD methods are available. In Thermal CVD, the temperature of the substrate provides the energy needed to activate the reaction. Metal Organic Chemical Vapor Deposition (MOCVD) employs metalorganic precursors containing bonds characteristic of the materials to be deposited. Due to their thermal instability, they can decompose and react at lower temperatures. Plasma-Enhanced Chemical Vapor Deposition (PECVD) employs plasma to boost reactions. The species are produced either in the core of the gas phase or on the surface of the substrate. This process offers deposition at low temperatures, ranging from 25 to 400°C. The deposition rates in this technique are higher than those of conventional CVD.

Additionally, there are other techniques derived from CVD. Figure 3.7 summarizes the CVD processes family.

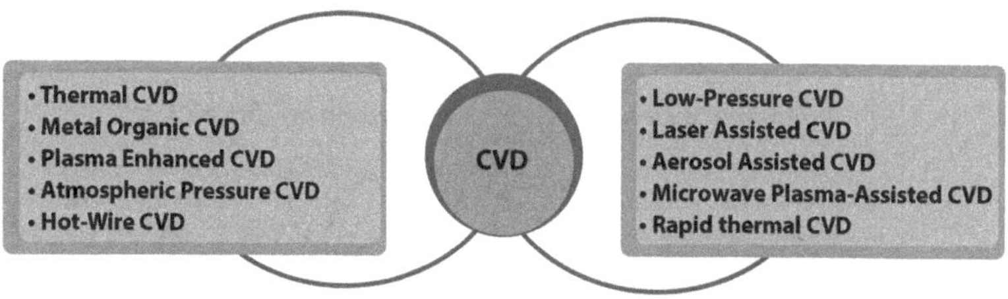

Figure 3.7 Overview of CVD processes family-nanostructure deposition techniques.

e. Atomic Layer Deposition

Atomic Layer Deposition (ALD) can be considered a version of chemical vapor deposition, and more illustratively, it is associated with surface reactions leading to vapor-phase self-assembly. The ALD method was relatively recently developed [58, 59]. Its uniqueness lies in its exceptional control over growth, allowing for precise atom-by-atom or molecule-by-molecule assembly, as its name suggests. The precursors are introduced into the reaction chamber to form the desired films. They are pulsed alternately, one at a time, and separated by an inert gas to avoid any gas-phase reaction.

The ALD mechanism leads to self-limiting growth, resulting in perfect conformity and uniformity of film thickness (Figure 3.8). After the gas has been introduced, saturation is reached when the entire surface of the substrate is covered with a monolayer of this gas. The remaining gas is pumped off while a second gas is introduced. The second gas adsorbs covalently to the first monolayer. The process can be repeated several times to achieve the desired coating thickness. This makes it a preferred method for controlling thickness and facilely fabricating 2D nanostructures.

f. Self-Assembly

The self-assembly of nanostructures is a process in which atoms or molecules spontaneously organize themselves into ordered structures that can reach a scale of a few hundred nanometers (Figure 3.9). This technique is based on the use of natural forces such as attraction, chemical bonding, electrostatic, or capillary forces to spontaneously build the desired nanostructure. It is also perceived as an advanced version of CVD. The uniqueness of this approach is showcased in high precision through building nanostructures at the atomic scale, enabling the accurate assembly of single atoms and outstanding control of the fabrication rate.

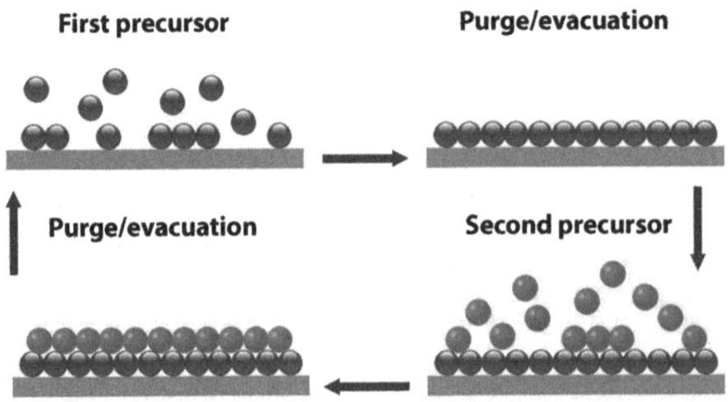

Figure 3.8 Atomic layer deposition steps-precise nanostructure growth through sequential surface reactions.

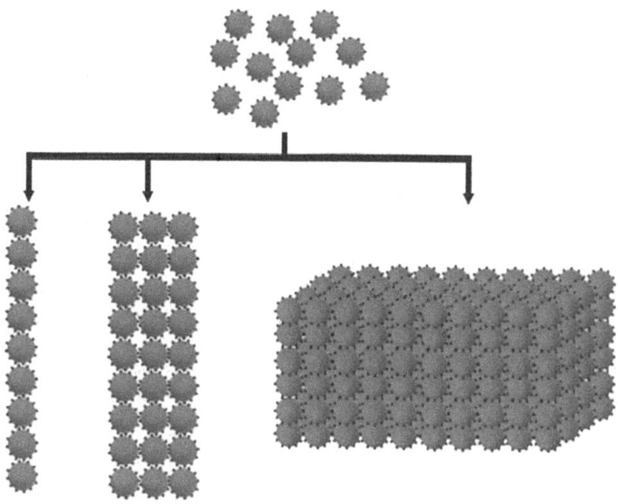

Figure 3.9 Self-assembly process of the building blocks in progress.

This can be achieved, for example, by immersing a substrate into a solution of an active surfactant in an organic solvent [60, 61]. The spontaneous fabrication of nanostructured materials from building blocks is of significant scientific interest. By studying the self-assembly of these nano-objects, researchers can gain vital insights into the behavior of complex systems. Self-assembly is a fundamental principle of the promising bottom-up approach to nanofabrication, playing a central role in the field of nanotechnology, where potential applications can be realized. To date, this approach has been widely used to develop nano-electronic devices, drugs, biosensors, and efficient photovoltaic devices.

3.4 Liquid Crystals: Basics and Properties

The accidental adventure story of the liquid crystalline phase began in 1888 with F. Reinitzer [62], who first discovered liquid crystals. He associated two distinct melting points with the first-discovered liquid crystal, cholesteryl benzoate. It undergoes an initial transition from a solid into a smoggy liquid upon gradual heating and then transforms into a transparent liquid at higher temperatures. Otto Lehmann was the first to introduce the term *liquid crystal* (LC) and mesophase, observing that the substance is birefringent and fluid. Due to their unique characteristics, scientists have since endeavored to replicate nature and create artificial liquid crystals. It was not until the end of the 1960s that the first practical application of liquid crystals was introduced, with the appearance of the first liquid crystal displays. The difference between liquid crystals and other states-liquid, solid, and gas-is that, while the substance remains liquid, its properties are anisotropic and can change depending on direction. The liquid crystal phase is a state of matter between a dilute gas and a crystalline.

As the term 'liquid crystal' suggests, it describes a substance that combines features of both a liquid, with fluidity, optoelectronic, and magnetic anisotropic properties (Figure 3.10). Moreover, it presents a repetitive lattice akin to a crystal, shaped by specific criteria. In this sense, a liquid crystal represents an intermediate, stable thermodynamic phase with the inherent fluidity of a liquid and solid-like optical characteristics. In simpler terms, upon closer examination, a liquid crystal appears liquid-like on a macroscopic scale while revealing intricate microscopic levels of order [63, 64].

Molecular crystalline solids crystallography terminology extends to liquid crystalline solids, where the individual components, named mesogens, serve as the basic building blocks of liquid crystals with weaker bonds compared to primary bonds. Depending on the internal degree of freedom, we distinguish between nematic and smectic phases, among others. The sub-classification arises from two parameters: transitional order and directional order. Thermotropic transitions occur due to temperature variations, while lyotropic transitions are controlled by changing solvent concentrations.

Based on these specific characteristics and applications, we can classify them into lyotropic phases found in cell membranes and proteins, and thermotropic phases with three sub-classes: cholesteric or chiral liquid

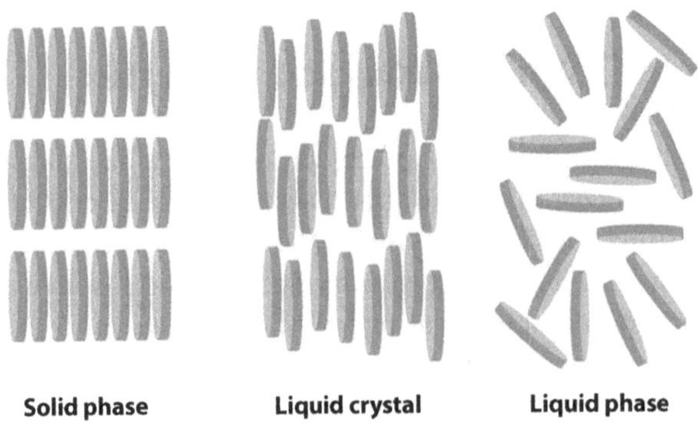

Figure 3.10 Liquid crystalline mesophase highlighting its position between the frontier of isotropic liquid phase and solid ordered phase.

crystals composed of asymmetric cholesterol, nematic, and smectic. Each sub-class has specific applications within a certain temperature range.

Unlike in a higher entropy liquid phase where molecules are in disorder, in the LC state, mesogens prefer to align along a director axis. In the solid molecular state, molecules have a privileged orientation with an infinitesimal transitional degree of freedom. This distinguishing feature of LCs is why the terms *mesogenic state* and *liquid crystal* are used interchangeably.

When defining positional order and orientational order, we essentially describe the basis in a lattice. In the case of LCs, this pertains to the molecules, or more precisely, the mesogens. Their mass centers have specific positions, and their axes clearly orient towards privileged directions. These criteria are crucial in distinguishing among different phases, as detailed in the next section.

3.4.1 Liquid Crystal Classes

Liquid crystals, as mentioned earlier, exhibit characteristics of both liquids and solids, with molecules ordered similarly to those in a crystal. For a molecule to be classified as a liquid crystal, it must meet three specific criteria: flexible ends, exhibiting rigidity at their centers, and displaying elongation exceeding their width [65–68]. These criteria ensure the unique properties that distinguish liquid crystals from other states of matter. Now, let us explore some common liquid crystal classes (Figure 3.11):

a. Lyotropic Phase

Lyotropic is a term composed of *lyo* and *tropic*, where the former means *solvent*, and the latter means *forms*. The lyotropic phase is formed when

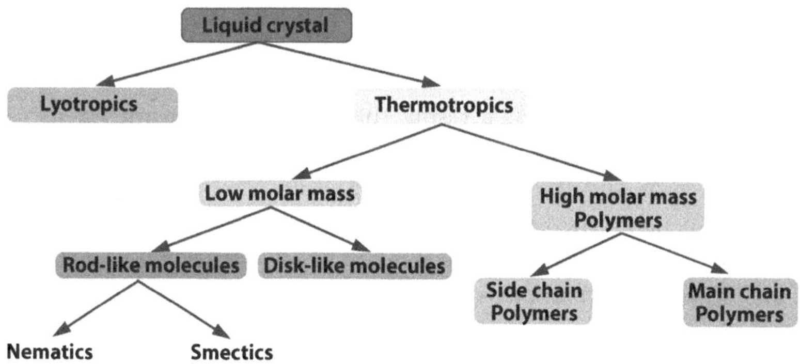

Figure 3.11 Comprehensive classification of liquid crystals.

amphiphilic molecules are dissolved in a solvent. Two factors, temperature and component concentrations are the main determinants that imply phase transitions in liquid crystals. This LC phase is composed of a combination of solvent molecules, hydrates, and mesogens. Depending on the rod-like geometrical shapes, either cone or cylinder, of the hydration and solvation, a peculiar cylindrical-like shape in a sequence of polar and non-polar films generates a laminar phase. When a lyotropic mesogen is dissolved in a solvent, mesophases are formed depending on several parameters, such as the size of the head group, the number of aliphatic chains of the amphiphile and the polarity of the solvent. Lyotropic liquid crystals can be divided into three different phases. The lamellar phase consists of bilayers of amphiphilic molecules adjacent to the aqueous phase. The thickness of the bilayer increases with increasing solvent concentration. Hexagonal phase can be formed from cylinders that aggregate to form two-dimensional structures. These cylinders are themselves formed from surfactant micelles. The third phase of Lyotropic LCs is cubic where three structures can be formed: the body-centered cubic, gyroid-cubic and double diamond lattice phase.

b. Thermotropic Phase

The term thermotropic is made up of two words: "thermo" and "tropic", meaning "temperature" and "shape" respectively. This explains why molecules that change shape as a function of temperature represent a class of thermotropic liquid crystals. When liquid crystals are heated, they lose long-range order. The phase transition in this case is then temperature oriented. Thus, during a heating process with increasing temperature, the first emerging phase is the smectic phase. With further heating, a transition occurs into the nematic phase and then into the isotropic liquid.

i. Smectic Phase

   The liquid crystal phase is associated with the preservation of positional order. This occurs during the heating process, which may disrupt the orientational orientation by changing the temperature; as a result, the arrangement of mesogens differs, allowing the distinction between smectic C when the director and the plane are titled at an arbitrary angle and smectic A when this angle is specifically 90°.

   It is noteworthy that the smectic phase occurs at temperatures lower than those of the cholesteric and nematic phases.

ii. Nematic Phase

   It is specifically characterized by a lack of positional order, even though this phase showcases long-range orientational order. Various parameters can modify this phase structure, including electromagnetic light or surface mechanical treatment.

iii. Cholesteric Phase
Also known as chiral nematic liquid crystal, it is characterized by a lack of positional order but the presence of orientational order. The director form peculiarly takes the shape of a helix. The cholesteric phase is sensitive to pitch, where pitch represents the repeat distance along the director. Hence, electromagnetic fields and state variables, including pressure and temperature, significantly affect the pitch.

This class of liquid crystals has two distinct sub-classes: intrinsic and extrinsic. It becomes intrinsic when we introduce a chiral amphiphilic molecule, such as soap, into a nematic lyotropic liquid phase. In contrast, it becomes extrinsic when a chiral non-amphiphilic molecule is used. These liquid crystals find applications in various industrial contexts, including display systems, sensors, fabric synthesis, temperature indicators, and more.

3.4.2 Principal Properties of the Liquid Crystal

a. Optical Properties
Liquid crystals, under the influence of polarized electromagnetic waves, exhibit variable refractive indices based on molecular alignment, emphasizing their anisotropic characteristics. Temperature variations induce changes in optical properties, allowing LCs to diffuse and transmit electromagnetic waves based on molecular organization. LCs can also showcase nonlinear optics through the Kerr effect. These unique features make LCs suitable for displays, with bistability prompting consideration for use in on-off states as memories [69, 70].

b. Structural properties
To investigate the structural patterns and phases of materials, X-rays prove to be a suitable tool. X-ray radiation provides valuable insights into the molecular organization within LCs, offering details on their order and the distribution of molecular positions.

Moreover, X-rays play a crucial role in distinguishing between different LC phases, such as smectic, nematic, and cholesteric, shedding light on transitions between these phases. The orientation of molecules and the inter-distance between film layers can be precisely determined with the aid of X-rays, helping identify irregularities in molecular structure, defects, and disorder within the material.

Real-time characterization using X-rays is instrumental in unraveling the impact of external factors, such as temperature variations and the influence of electromagnetic waves on the structure and behavior of LCs. This comprehensive analysis provided by X-ray techniques enhances our understanding of the intricate features and dynamic behavior of LCs.

Transmission Electron Microscopy (TEM) represents a powerful imaging technique that utilizes electrons, exhibiting a shorter wavelength compared to light, allowing for superior resolution. However, applying TEM to LC samples introduces specific challenges, primarily stemming from the high vacuum conditions required for optimal imaging. To circumvent this issue and ensure a faithful representation of the native state of LCs, special precautions must be taken during sample preparation. Cryogenic TEM (Cryo-TEM) emerges as a pivotal technique in this context, preserving the initial microstructure and offering more accurate insights into their features.

c. Mechanical Properties

Liquid crystals exhibit unique mechanical properties influenced by mesogen positions, orientations, and phase transitions. These characteristics can be effectively investigated through various techniques:

i. Rheology
   Rheological studies enable the exploration of deformation dynamics under applied force, unveiling the plasticity, elasticity, and viscosity of liquid crystals.
ii. Atomic Force Microscopy
   Atomic force microscope (AFM) facilitates the visualization of surface topography, offering insights into the elasticity of liquid crystals and the structure at the nanoscale.
iii. Dynamic Mechanical Analysis
   Dynamic mechnical analysis (DMA) provides a temporal evolution of viscoelasticity in liquid crystals, allowing for a comprehensive understanding of its variations with changes in temperature.
iv. Acoustic Techniques
   Measurement of sound velocity across the material serves as an effective means to reveal mechanical properties such as elasticity in liquid crystals.

By employing these techniques, researchers gain valuable insights into the mechanical behavior of liquid crystals, shedding more light on their unique anisotropic properties and responses to external influences.

3.5 Nanostructured Materials and Liquid Crystals Composites

The dispersion of nanomaterials in a liquid crystal matrix is a technique essential for the development of cost-effective and environmentally friendly approaches. The goal is to enhance various physical parameters of these materials, with carbon dots and liquid crystals composites emerging as more cost-effective and eco-friendly alternatives. These composites, including carbon dots, have shown potential for applications in low-energy scenarios [71].

Neha *et al.*, conducted a review on the impact of carbon dots on liquid crystal properties, focusing on their dispersion in nematic and ferroelectric liquid crystals. The study revealed that the use of carbon dots as a dopant significantly enhanced the electro-optical and dielectric properties of liquid crystals [71].

Additionally, nanotubes, nanorods, and nanowires have gained attention as functional materials in nanotechnology. Incorporating these nano-objects into liquid crystals has led to the development of innovative materials. Singh *et al.*, studied the electro-optical properties of a composite formed by multi-walled carbon nanotubes (MWCNTs) and commercial crystal liquid E7, showing a significant improvement in device parameters with the addition of MWCNTs [72–74].

In recent years, the integration of nanostructures with liquid crystals has garnered significant attention, as evidenced by numerous relevant publications [8, 9, 20]. The incorporation of nanomaterials into LCs has proven instrumental in fine-tuning optoelectrical properties [75–78].

Several key parameters of LCs, including phase shift temperature, elasticity, viscosity, optical constants, and optical conductivity, can be effectively tuned by doping them with nanoparticles. Extensive research has been dedicated to optimizing displays by enhancing the electro-optical parameters of LC-nanoparticle composites [75, 77–83]. These composites fall into three main categories: nanoparticle segregation or dispersion, nanoparticle nature (ferroelectric, semiconducting, or metallic), and nanoparticle shape.

Among these, the class involving nanoparticles dispersed in nematic LCs warrants special attention. Achieving the stability of such composites necessitates a robust interaction between LCs and nanoparticles. This is achieved through strategies such as coating nanoparticles with organic ligands [84], or adjusting the size, concentration, chemical content, or temperature of quantum dots. The intricate dynamics of interaction between nanoparticles and LCs are shaped by the surface capping of nanoparticles using organic ligands. Ligands play a crucial role in determining how LCs anchor onto the nanoparticle surface [85].

The mutual influence between nanoparticles and LC hosts holds promising applications. Extensive investigations into the effects of nanoparticle doping on the alignment, electro-optical, dielectric, and physico-chemical properties of LCs hosts reveal significant enhancements. Nanoparticle doping often results in improved electro-optical performance, including reduced operating voltages, enhanced switching speeds, higher contrast ratios, and improved LCs alignment - crucial factors for display applications.

The improved electro-optical performance is attributed to various factors, including the molecular ordering or disordering of the LC host, the ion-capturing abilities of nanoparticles, alterations in viscoelastic properties, and dielectric anisotropy. Key metrics for evaluating the performance of nanoparticle/LC systems in display applications include the contrast ratio, threshold voltage, and response time. These parameters are expressed mathematically, taking into account crucial variables such as the splay elastic constant, dielectric anisotropy, rotational viscosity, cell thickness, and applied voltage. These mathematical expressions provide essential insights for assessing the effectiveness of nanoparticle/LC composites in practical display scenarios.

Contrast Ratio

The contrast ratio C_R is typically a function of the maximum and minimum transmittance values of the display. It is defined as:

$$C_R = \frac{T_{max}}{T_{Min}} \tag{3.1}$$

where T_{max} is the maximum transmittance and T_{min} is the minimum transmittance.

Threshold Voltage

The threshold voltage V_{Th} for a liquid crystal display can be expressed as:

$$V_{Th} = \sqrt{\frac{\pi^2 \kappa}{\epsilon_0 \Delta\epsilon}} \tag{3.2}$$

where: κ is the splay elastic constant, ϵ_0 is the permittivity of free space and Δϵ is the dielectric anisotropy.

Response Time

The response time τ is often related to the rotational viscosity, dielectric anisotropy, applied voltage, and cell thickness. It can be approximated by:

$$\tau = \frac{\gamma d^2}{\pi^2 \kappa (V - V_{Th})} \tag{3.3}$$

where γ is the rotational viscosity, d is the cell thickness and V is the applied voltage.

In effect, the splay elastic constant κ describes the elastic resistance to splay deformation (i.e., the bending of liquid crystal molecules) and is crucial in determining the threshold voltage and response time. The dielectric anisotropy Δϵ is the difference in dielectric constants parallel and perpendicular to the liquid crystal director. A higher dielectric anisotropy typically lowers the threshold voltage. The rotational viscosity γ represents the resistance to the rotational motion of liquid crystal molecules, with higher rotational viscosity resulting in slower response times. The cell thickness d also plays a significant role; thicker cells can lead to higher threshold voltages and slower response times. The applied voltage V across the liquid crystal cell influences the alignment of liquid crystal molecules, thereby affecting the display's optical properties. By controlling these parameters and understanding their corresponding equations, one can optimize the design and performance of nanoparticle/LC composite systems in display applications.

Quantum dots play a complex role as dopants in liquid crystals. Whether non-mesogenic or pro-mesogenic, their introduction disrupts LCs molecular orientation, leading to a concentration-dependent destabilization of the LCs phase. Typically, this results in a lower Nematic-to-Isotropic phase transition temperature to the isotropic phase in liquid crystal systems and a blurred differential scanning calorimetry thermogram peak. The impact of nanoparticles on

LCs is less predictable than that of simple dopants, with various nanoparticles often causing a reduction in the nematic to isotropic phase transition temperature and LCs orientational ordering [85–88]. The intricate interplay between nanoparticles and LCs influences their thermal and ordering characteristics.

Binary component fabrication is easily achieved by dispersing nanoparticles in LCs to alter the composite's substantial physical properties. This process does not require sophisticated chemical techniques for implementation.

3.5.1 Different Doping Materials in Binary Liquid Crystal Composites

The use of different doping materials in binary liquid crystal composites offers a diverse range of options for enhancing performance, each with its unique characteristics. One notable material is Titania Quantum Dots, known for their environmentally friendly and cost-effective nature. Doping nematic LCs with titania 0D particles has demonstrated improved physical properties and enhanced system performance [8]. This inclusion has proven effective in reducing image sticking and flickering in display devices, thereby contributing to better luminescent features in such applications.

Another significant class of materials includes Au and Ag Quantum Dots. The dispersion of nanosized particles of Au and Ag in LCs colloid has been a prominent area of study [75, 78, 89–91], primarily due to the relative ease in preparing these nanoparticles. Numerous investigations have showcased their excellent performance in the fabrication of display devices [92–97]. Doping LCs with Au/Ag nano-sized particles has been demonstrated to enhance the photoluminescence of LCs, opening avenues for utilization in LCs display applications [98–101].

Silicon Dioxide Quantum Dots have garnered interest in SiO_2-LCs composites for more than three decades [102], exhibiting applications, especially in bistable LCs displays. Extensive research has been devoted to this system [75, 103–111]. Silicon dioxide, also known as aerosil nano-sized particles, exhibits surface specificity, featuring hydrophobic and hydrophilic aspects. Hydrogen bonds bridge LCs molecules and silicon dioxide dots.

Additionally, other materials, such as Diamond, zinc oxide, and CeO_2 nanoparticles, have been intensively studied nanoparticles [112–123].

Ferroelectric and Ferromagnetic 0D Materials represent another crucial category. The introduction of ferroelectric nano-sized particles, such as $Sn_2P_2S_6$, into LCs systems in 2003 marked a significant development [124, 125]. Their utilization in LCs systems has contributed to the improvement of dielectric anisotropy and a reduction in the potential difference threshold. Following this breakthrough, various other 0D particles, including strontium titanate nanoparticles [126], $BaTiO_3$ [127, 128], $Sn_2P_2S_6$ [124, 129], and $Bi_4Ti_3O_{12}$ [130], have been explored. Research confirms that ferroelectric 0D materials enhance the electro-optical parameters of LC matrices, reducing the threshold for the Freedericksz transition and influencing temperature phase shifts [131–133]. It is worth noting some inverse effects of incorporating ferroelectric 0D materials into LCs host [134, 135].

Ferromagnetic 0D materials, possessing a permanent magnetic dipole moment, stand in contrast to LCs molecules, which are diamagnetic with a tiny magnetic susceptibility. Consequently, the introduction of ferromagnetic 0D particles in LCs induces a strong response to magnetic fields. Researchers have studied $CeCoFe_2O_4$ nano-sized particles within nematic LC composites [136].

3.5.2 Ternary Composite Materials for Display Applications

Ternary composite materials in display applications mark a paradigm shift by incorporating three distinct elements into the classic LC-nanoparticle system to enhance performance [137]. This shift involves adding organic components, such as photo-responsive polymers or photochromic compounds [137–140]. The integration of LCs, nanoparticles, and photochromic molecules introduces increased complexity, necessitating a delicate equilibrium and optimal parameters for advanced display device fabrication [141].

The amalgamation of LCs, nanoparticles, and photochromic molecules enables innovative smart display technology, allowing the tuning of optical properties under accommodating ambient conditions. When exposed to electromagnetic waves, the photochromic molecules undergo reversible color shifts alongside changes in transparency [137]. This unique feature finds application in creating energy-efficient display devices, showcasing the multifaceted potential of ternary composites in advancing display technologies [137].

Additionally, a method for switching and fine-tuning optical effects in the ternary system involves using chiral dopants. The distinctive spiral structure of cholesteric LCs results in quasi-elliptical polarization with a specific affinity for particular electromagnetic wavelengths, while rejecting others. This property is applied in reflective display technology, providing energy-efficient solutions and improved exterior visibility at a lower cost [137, 138].

3.5.3 Beyond Binary and Ternary Composites

Advancements in display technology have revealed that the synergy between LCs and other functional entities offers more possibilities for color switching, resulting in better visibility and lower energy consumption as a bonus.

Ongoing research and development aim to extend the boundaries of this interesting era of technology.

Researchers are exploring beyond binary and ternary composites to develop advanced display materials. Quaternary compositions, polymeric blends, multifunctional nanocomposites, responsive hybrid materials, liquid crystal elastomers, self-healing materials, organic-inorganic hybrids, metamaterials, and magnetic nanoparticles offer diverse strategies for enhancing display functionalities. These approaches aim to achieve improved mechanical, optical, and responsive properties, extending the range of possibilities for display technologies.

3.6 Pioneering Advances in Display Technology

Liquid crystal composites, especially when infused with nanoparticles, stand at the front end of elevating display technology. The synergy between liquid crystals and nanostructures empowers precise control over the optical properties of the entire system, resulting in the creation of cutting-edge, high-performance displays. These composites enhance contrast ratios, reduce threshold voltages, and accelerate switching speeds-essential elements driving the progress and efficacy of vibrant displays. The tunability of liquid crystals, coupled with the distinctive attributes of nanostructures, provides a versatile platform for display technology.

Moreover, the integration of LC composites in hologram applications has opened new frontiers in 3D imaging. The capacity of liquid crystals to modulate electromagnetic light, combined with the incorporation of nanoparticles, contributes to the development of dynamic and lifelike holographic displays. This breakthrough holds promise across various domains, spanning entertainment, education, and augmented reality, where immersive three-dimensional visualizations are increasingly in demand. Ongoing research in this field aims to refine and expand the capabilities of liquid crystal composites, paving the way for the next generation of displays and holographic applications.

3.7 Future Perspectives and Emerging Trends

The fusion of liquid crystals with nanostructures represents a significant stride in display technology. Ongoing intensive research is poised to unveil innovative methods for tailoring the properties of liquid crystal composites, ushering in displays with enhanced performance and functionality. Future advancements may center on refining synthesis and manufacturing processes, optimizing the interplay between nanostructures and LCs, and charting new pathways for innovative designs in display applications.

Moreover, the pursuit of sustainable and eco-friendly materials in display technologies presents opportunities to explore alternative nanostructure materials. Researchers may delve into the development of environmentally friendly quantum dots and nanoparticles, aligning the future of liquid crystal composites with the increasing emphasis on green technologies.

The evolution of liquid crystal composites transcends conventional binary and ternary systems. Future exploration may delve into quaternary or more complex compositions, leveraging the combined effects of diverse materials to achieve unprecedented control over optical and electro-optical properties. Such intricacy could open avenues to applications beyond displays, encompassing advanced sensing technologies and adaptable optical devices, all within a concise framework.

In the domain of holographic technology, ongoing efforts strive to push the boundaries of 3D imaging projected onto a 2D space through the utilization of liquid crystal composites. Research and development may focus on enhancing holographic display resolutions, minimizing artifacts, and expanding the range of applications—from medical imaging to interactive educational tools. As technological innovations progress, liquid crystal composites stand poised to play a pivotal role in shaping the future landscape of displays and holography.

3.8 Conclusion

In this chapter, we have offered a comprehensive overview of recent advancements in nanostructured materials doped liquid crystal composites tailored for displays. Bridging the fundamentals of nanostructures and liquid crystals, our aim was to provide insights into fabrication techniques and the intrinsic properties of these innovative materials.

The exploration of various doping materials in binary and ternary composites highlights the versatility of these combinations for display applications. Key findings include the promising applications and versatility of diverse doping materials, such as titania quantum dots, Au and Ag quantum dots, silicon dioxide quantum dots, and ferroelectric and ferromagnetic 0D materials.

Pioneering advances in display technology were showcased, exemplified by the use of photochromic and photo responsive polymers and chiral dopants, indicating the dynamic evolution in this field. To reiterate, these advancements hold significant potential for the broader landscape of nanotechnology, liquid crystal composites, and display technologies.

As we conclude, it is crucial to acknowledge the continuous progression in this field and the potential for future research to explore sustainable materials, alternative nanostructures, and the ongoing integration of liquid crystal composites in holographic applications. This chapter lays the groundwork for understanding the intricate interplay between nanostructures and liquid crystals, fostering continued advancements in the realm of display technologies.

Funding

D. Soubane gratefully acknowledges the support of the MRIF 'Ministère des Relations internationales et de la Francophonie' Quebec – Canada – Morocco Project 'Memristor for Artificial Intelligence Applications' : APQRM 2021-2022 INRS-EMT/UCA-FPS Grant for this work.

References

1. Buzea, C., Pacheco, I.I., Robbie, K.: Nanomaterials and nanoparticles: Sources and toxicity. *Biointerphases*, 2, MR17–MR71, 2007, https://doi.org/10.1116/1.2815690.
2. Kreyling, W.G., Semmler-Behnke, M., Chaudhry, Q., A complementary definition of nanomaterial. *Nano Today*, 5, 165–168, 2010, https://doi.org/10.1016/j.nantod.2010.03.004.
3. Ashby, M.F., Ferreira, P.J., Schodek, D.L., *Nanomaterials, nanotechnologies and design: an introduction for engineers and architects.* Butterworth-Heinemann, Elsevier, Amsterdam Boston, 2009.
4. Reibold, M., Paufler, P., Levin, A.A., Kochmann, W., Pätzke, N., Meyer, D.C., Carbon nanotubes in an ancient Damascus sabre. *Nature*, 444, 286–286, 2006, https://doi.org/10.1038/444286a.
5. Feynman, R.P., There's plenty of room at the bottom [data storage]. *J. Microelectromech. Syst.*, 1, 60–66, 1992, https://doi.org/10.1109/84.128057.
6. Gribbin, J. and Gribbin, M., *Richard Feynman: a life in science.* Plume, New York, 1998.
7. Feynman, R.P. and Sykes, C. (Eds.), No ordinary genius: the illustrated Richard Feynman. Norton, New York, 1994.
8. Ramsden, J., Nanotechnology: an introduction. William Andrew is an imprint of Elsevier, Amsterdam, 2016.
9. Pelaz, B., *et al.*, Diverse Applications of Nanomedicine. *ACS Nano*, 11, 2313–2381, 2017, https://doi.org/10.1021/acsnano.6b06040.
10. Soares, S., Sousa, J., Pais, A., Vitorino, C., Nanomedicine: Principles, Properties, and Regulatory Issues. *Front. Chem.*, 6, 360, 2018, https://doi.org/10.3389/fchem.2018.00360.

11. Shishodia, S., Chouchene, B., Gries, T., Schneider, R., Selected I-III-VI2 Semiconductors: Synthesis, Properties and Applications in Photovoltaic Cells. *Nanomaterials*, 13, 2889, 2023, https://doi.org/10.3390/nano13212889.
12. O'Regan, B. and Grätzel, M., A low-cost, high-efficiency solar cell based on dye-sensitized colloidal TiO2 films. *Nature*, 353, 737–740, 1991, https://doi.org/10.1038/353737a0.
13. Gupta, N., Alapatt, G.F., Podila, R., Singh, R., Poole, K.F., Prospects of Nanostructure-Based Solar Cells for Manufacturing Future Generations of Photovoltaic Modules. *Int. J. Photoenergy.*, 2009, 154059, 2009, https://doi.org/10.1155/2009/154059.
14. Soubane, D., Ihlal, A., Nouet, G., The Role Of Cadmium Oxide Within The Thin Films Of The Buffer Cds Aimed At Solar Cells Based Upon CIGS Films Fabrication. *Rev. Marocaine Matière Condens.*, 9, 2011, https://doi.org/10.34874/PRSM.MJCM-VOL9ISS0.195.
15. Daneshvar, N., Salari, D., Khataee, A.R., Photocatalytic degradation of azo dye acid red 14 in water on ZnO as an alternative catalyst to TiO2. *J. Photochem. Photobiol. Chem.*, 162, 317–322, 2004, https://doi.org/10.1016/S1010-6030(03)00378-2.
16. Ṭāhir, M., Rafique, M., Rafique, M.S. (Eds.), *Nanotechnology and photocatalysis for environmental applications.* Elsevier, Amsterdam, Netherlands, Cambridge, MA, United States, 2020.
17. Luryi, S., Quantum capacitance devices. *Appl. Phys. Lett.*, 52, 501–503, 1988, https://doi.org/10.1063/1.99649.
18. Bisquert, J., Chemical capacitance of nanostructured semiconductors: its origin and significance for nanocomposite solar cells. *Phys. Chem. Chem. Phys.*, 5, 5360, 2003, https://doi.org/10.1039/b310907k.
19. Bueno, P.R., Nanoscale origins of super-capacitance phenomena. *J. Power Sources*, 414, 420–434, 2019, https://doi.org/10.1016/j.jpowsour.2019.01.010.
20. Doerk, G.S., Stein, A., Bae, S., Noack, M.M., Fukuto, M., Yager, K.G., Autonomous discovery of emergent morphologies in directed self-assembly of block copolymer blends. *Sci. Adv.*, 9, eadd3687, 2023, https://doi.org/10.1126/sciadv.add3687.
21. DiVincenzo, D.P. and Loss, D., Quantum information is physical. *Superlattices Microstruct.*, 23, 419–432, 1998, https://doi.org/10.1006/spmi.1997.0520.
22. Kurzweil, R., The singularity is near: when humans transcend biology. Viking, New York, 2005.
23. Moore, G.E., The microprocessor: engine of the technology revolution. *Commun. ACM.*, 40, 112–114, 1997, https://doi.org/10.1145/253671.253746.
24. Taniguchi, N., *On the Basic Concept of "Nano-Technology."* Proc Intl Conf Prod Eng Tokyo Part II 1974, 1974.
25. Parvej, M.S., Khan, M.I., Hossain, M.K., Preparation of nanoparticle-based polymer composites. In *Nanoparticle-Based Polymer Composites*, pp. 55–94, Elsevier, 2022.

26. Kroto, H.W., Heath, J.R., O'Brien, S.C., Curl, R.F., Smalley, R.E., C60: Buckminsterfullerene. *Nature*, 318, 162–163, 1985, https://doi.org/10.1038/318162a0.
27. Soga, T. (Ed.), *Nanostructured materials for solar energy conversion*. Elsevier, Amsterdam, Boston, 2006.
28. Lalwani, G. and Sitharaman, B., Multifunctional fullerene- and metallofullerene-based nanobiomaterials. *Nano Life*, 03, 1342003, 2013, https://doi.org/10.1142/S1793984413420038.
29. Mroz, P., Pawlak, A., Satti, M., Lee, H., Wharton, T., Gali, H., Sarna, T., Hamblin, M.R., Functionalized fullerenes mediate photodynamic killing of cancer cells: Type I versus Type II photochemical mechanism. *Free Radic. Biol. Med.*, 43, 711–719, 2007, https://doi.org/10.1016/j.freeradbiomed.2007.05.005.
30. Soubane, D., Tirbiyine, A., Bellioua, M., Laasri, S., Hajjaji, A., Optical Features of Catalyst-Free Zinc Oxide Nanostructures Confined to One Dimension. *Opt. Spectrosc.*, 127, 522–526, 2019, https://doi.org/10.1134/S0030400X19090261.
31. Mayimele, M.A., Fru, J.N., Nyarige, J.S., Diale, M., Sequential physical vapor deposited methylammonium lead tri-iodide perovskites on FTO and ITO modified zinc oxide nanorods for perovskite solar cells. *Phys. B Condens. Matter.*, 625, 413462, 2022, https://doi.org/10.1016/j.physb.2021.413462.
32. Swathi, S., Babu, E.S., Yuvakkumar, R., Ravi, G., Chinnathambi, A., Alharbi, S.A., Velauthapillai, D., Branched and unbranched ZnO nanorods grown *via* chemical vapor deposition for photoelectrochemical water-splitting applications. *Ceram. Int.*, 47, 9785–9790, 2021, https://doi.org/10.1016/j.ceramint.2020.12.119.
33. Chen, K., Tang, W., Chen, Y., Yuan, R., Lv, Y., Shan, W., Zhang, W.-H., A facile solution processed ZnO@ZnS core–shell nanorods arrays for high-efficiency perovskite solar cells with boosted stability. *J. Energy Chem.*, 61, 553–560, 2021, https://doi.org/10.1016/j.jechem.2021.02.018.
34. Ravi Kiran, A.V.V.V., Kusuma Kumari, G., Krishnamurthy, P.T., Carbon nanotubes in drug delivery: Focus on anticancer therapies. J. *Drug Deliv. Sci. Technol.*, 59, 101892, 2020, https://doi.org/10.1016/j.jddst.2020.101892.
35. De Volder, M.F.L., Tawfick, S.H., Baughman, R.H., Hart, A.J., Carbon Nanotubes: Present and Future Commercial Applications. *Science*, 339, 535–539, 2013, https://doi.org/10.1126/science.1222453.
36. Newman, P., Minett, A., Ellis-Behnke, R., Zreiqat, H., Carbon nanotubes: Their potential and pitfalls for bone tissue regeneration and engineering. *Nanomed. Nanotechnol. Biol. Med.*, 9, 1139–1158, 2013, https://doi.org/10.1016/j.nano.2013.06.001.
37. Zanello, L.P., Zhao, B., Hu, H., Haddon, R.C., Bone Cell Proliferation on Carbon Nanotubes. *Nano Lett.*, 6, 562–567, 2006, https://doi.org/10.1021/nl051861e.
38. Janowska, I., Chizari, K., Ersen, O., Zafeiratos, S., Soubane, D., Costa, V.D., Speisser, V., Boeglin, C., Houllé, M., Bégin, D., Plee, D., Ledoux, M.-J.,

Pham-Huu, C., Microwave synthesis of large few-layer graphene sheets in aqueous solution of ammonia. *Nano Res.,* 3, 126–137, 2010, https://doi.org/10.1007/s12274-010-1017-1.
39. Novoselov, K.S., Geim, A.K., Morozov, S.V., Jiang, D., Zhang, Y., Dubonos, S.V., Grigorieva, I.V., Firsov, A.A., Electric Field Effect in Atomically Thin Carbon Films. *Science,* 306, 666–669, 2004, https://doi.org/10.1126/science.1102896.
40. Hou, C., Tai, G., Liu, Y., Liu, X., Borophene gas sensor. *Nano Res.*, 15, 2537–2544, 2022, https://doi.org/10.1007/s12274-021-3926-6.
41. Polderman, M.C.A., New applications of UVA-1 cold light therapy. s.n.], [S.l., 2006.
42. Norval, M., Lucas, R.M., Cullen, A.P., De Gruijl, F.R., Longstreth, J., Takizawa, Y., Van Der Leun, J.C., The human health effects of ozone depletion and interactions with climate change. *Photochem. Photobiol. Sci.,* 10, 199–225, 2011, https://doi.org/10.1039/c0pp90044c.
43. Smijs, T. and Pavel, Titanium dioxide and zinc oxide nanoparticles in sunscreens: focus on their safety and effectiveness. *Nanotechnol. Sci. Appl.*, 95, 2011, https://doi.org/10.2147/NSA.S19419.
44. Da Costa, L.P., Engineered nanomaterials in the sports industry, in: *Handbook of Nanomaterials for Manufacturing Applications*, pp. 309–320, Elsevier, 2020.
45. Shafique, M. and Luo, X., Nanotechnology in Transportation Vehicles: An Overview of Its Applications, Environmental, Health and Safety Concerns. *Materials*, 12, 2493, 2019, https://doi.org/10.3390/ma12152493.
46. Zhang, R.X., Li, J., Zhang, T., Amini, M.A., He, C., Lu, B., Ahmed, T., Lip, H., Rauth, A.M., Wu, X.Y., Importance of integrating nanotechnology with pharmacology and physiology for innovative drug delivery and therapy – an illustration with firsthand examples. *Acta Pharmacol. Sin.,* 39, 825–844, 2018, https://doi.org/10.1038/aps.2018.33.
47. Gleiter, H., Nanostructured materials: basic concepts and microstructure. *Acta Mater.,* 48, 1–29, 2000, https://doi.org/10.1016/S1359-6454(99)00285-2.
48. Dhand, C., Dwivedi, N., Loh, X.J., Jie Ying, A.N., Verma, N.K., Beuerman, R.W., Lakshminarayanan, R., Ramakrishna, S., Methods and strategies for the synthesis of diverse nanoparticles and their applications: a comprehensive overview. *RSC Adv.*, 5, 105003–105037, 2015, https://doi.org/10.1039/C5RA19388E.
49. Bellah, Md.M., Christensen, S.M., Iqbal, S.M., Nanostructures for Medical Diagnostics. *J. Nanomater.*, 2012, 486301, 2012, https://doi.org/10.1155/2012/486301.
50. Sou, I.K., Lok, S.K., Wang, G., Wang, N., Wong, G.K.L., MBE-Grown II–VI and Related Nanostructures. *J. Electron. Mater.*, 39, 882–892, 2010, https://doi.org/10.1007/s11664-010-1215-3.
51. Cao, G. and Wang, Y., Nanostructures & nanomaterials: synthesis, properties, and applications. World Scientific, New Jersey, 2011.

52. Ishii, H., Usui, S., Douki, K., Kajita, T., Chawanya, H., Shimokawa, T., *Design and lithographic performances of 193-specific photoacid generators*. Presented at the Microlithography 2000, Santa Clara, CA, June 23 2000.
53. Hocevar, M., Immink, G., Verheijen, M., Akopian, N., Zwiller, V., Kouwenhoven, L., Bakkers, E., Growth and optical properties of axial hybrid III–V/silicon nanowires. *Nat. Commun.*, 3, 1266, 2012, https://doi.org/10.1038/ncomms2277.
54. Kumar, R.R., Yuvaraj, D., Rao, K.N., Growth and characterization of germanium nanowires by electron beam evaporation. *Mater. Lett.*, 64, 1766–1768, 2010, https://doi.org/10.1016/j.matlet.2010.05.020.
55. Tang, J., Wang, J., Maurice, J.-L., Chen, W., Foldyna, M., Yu, L., Leshchenko, E.D., Dubrovskii, V.G., Cabarrocas, P.R.I., Tapering-free monocrystalline Ge nanowires synthesized *via* plasma-assisted VLS using In and Sn catalysts. *Nanotechnology*, 33, 405602, 2022, https://doi.org/10.1088/1361-6528/ac57d4.
56. Park, S.J., Chung, S.H., Kim, B.-J., Qi, M., Xu, X., Stach, E.A., Yang, C., Mechanism of vertical Ge nanowire nucleation on Si (111) during subeutectic annealing and growth. *J. Mater. Res.*, 26, 2744–2748, 2011, https://doi.org/10.1557/jmr.2011.313.
57. Soubane, D. and Quitoriano, N.J., Photoluminescence from low thermal budget silicon nano-crystals *in silica*. *Nanotechnology*, 26, 295201, 2015, https://doi.org/10.1088/0957-4484/26/29/295201.
58. Oviroh, P.O., Akbarzadeh, R., Pan, D., Coetzee, R.A.M., Jen, T.-C., New development of atomic layer deposition: processes, methods and applications. *Sci. Technol. Adv. Mater.*, 20, 465–496, 2019, https://doi.org/10.1080/14686996.2019.1599694.
59. Malygin, A.A., Drozd, V.E., Malkov, A.A., Smirnov, V.M., From V. B. Aleskovskii's "Framework" Hypothesis to the Method of Molecular Layering/Atomic Layer Deposition. *Chem. Vap. Depos.*, 21, 216–240, 2015, https://doi.org/10.1002/cvde.201502013.
60. Bigelow, W.C., Pickett, D.L., Zisman, W.A., Oleophobic monolayers. *J. Colloid Sci.*, 1, 513–538, 1946, https://doi.org/10.1016/0095-8522(46)90059-1.
61. Fowkes, F.M. (Ed.), *Contact Angle, Wettability, and Adhesion*. American Chemical Society, Washington, D.C., 1964.
62. Mitov, M., Liquid-Crystal Science from 1888 to 1922: Building a Revolution. *ChemPhysChem.*, 15, 1245–1250, 2014, https://doi.org/10.1002/cphc.201301064.
63. Chandrasekhar, S., *Liquid crystals*. Cambridge University Press, Cambridge [England], New York, NY, USA, 1992.
64. Gennes, P.-G. de and Prost, J., *The physics of liquid crystals*. Clarendon Press, Oxford, 2013.
65. Andrienko, D., Introduction to liquid crystals. *J. Mol. Liq.*, 267, 520–541, 2018, https://doi.org/10.1016/j.molliq.2018.01.175.

66. Stephen, M.J. and Straley, J.P., Physics of liquid crystals. *Rev. Mod. Phys.*, 46, 617–704, 1974, https://doi.org/10.1103/RevModPhys.46.617.
67. Lavrentovich, O.D. and Kleman, M., Cholesteric Liquid Crystals: Defects and Topology, in: *Chirality in Liquid Crystals*, H.-S. Kitzerow and C. Bahr (eds.), pp. 115–158. Springer-Verlag, New York, 2001.
68. Oswald, P. and Pieranski, P., Nematic and Cholesteric Liquid Crystals: Concepts and Physical Properties Illustrated by Experiments. *CRC Press*, 2005.
69. Attard, G. and Williams, G., Novel optical properties of liquid-crystal polymers. *Nature*, 326, 544–545, 1987, https://doi.org/10.1038/326544a0.
70. Alipanah, Z., Zakerhamidi, M.S., Ranjkesh, A., Temperature-dependent optical properties of some mixtures nematic liquid crystal. *Sci. Rep.*, 12, 12676, 2022, https://doi.org/10.1038/s41598-022-16750-x.
71. Neha, Singh, G., Kumar, S., Malik, P., Supreet, Recent trends and insights into carbon dots dispersed liquid crystal composites. *J. Mol. Liq.*, 384, 122225, 2023, https://doi.org/10.1016/j.molliq.2023.122225.
72. Singh, B.P., Sikarwar, S., Pandey, K.K., Manohar, R., Depriester, M., Singh, D.P., Carbon Nanotubes Blended Nematic Liquid Crystal for Display and Electro-Optical Applications. *Electron. Mater.*, 2, 466–481, 2021, https://doi.org/10.3390/electronicmat2040032.
73. Wu, K.-J., Chu, K.-C., Chao, C.-Y., Chen, Y.-F., Lai, C.-W., Kang, C.-C., Chen, C.-Y., Chou, P.-T., CdS Nanorods Imbedded in Liquid Crystal Cells for Smart Optoelectronic Devices. *Nano Lett.*, 7, 1908–1913, 2007, https://doi.org/10.1021/nl070541n.
74. Soubane, D., Ihlal, A., Nouet, G., Cost-Eeffective Fabrication Processes Of The Absorber CuInSe2 (CIS) And The Buffer Cadmium Sulphide. *Rev. Marocaine Matière Condens.*, 9, 28–31, 2011, https://doi.org/10.34874/PRSM.MJCM-VOL9ISS0.194.
75. Garbovskiy, Y.A. and Glushchenko, A.V., Liquid Crystalline Colloids of Nanoparticles, in: *Solid State Physics*, pp. 1–74, Elsevier, 2010.
76. Lagerwall, J.P.F. and Scalia, G. (Eds.), *Liquid crystals with nano and microparticles*. World Scientific, New Jersey, 2017.
77. Prakash, J., Khan, S., Chauhan, S., Biradar, A.M., Metal oxide-nanoparticles and liquid crystal composites: A review of recent progress. *J. Mol. Liq.*, 297, 112052, 2020, https://doi.org/10.1016/j.molliq.2019.112052.
78. Ahmad, F., Luqman, M., Jamil, M., Advances in the metal nanoparticles (MNPs) doped liquid crystals and polymer dispersed liquid crystal (PDLC) composites and their applications - a review. *Mol. Cryst. Liq. Cryst.*, 731, 1–33, 2021, https://doi.org/10.1080/15421406.2021.1954759.
79. Tschierske, C. (Ed.), *Liquid Crystals: Materials Design and Self-assembly*. Springer Berlin Heidelberg, Berlin, Heidelberg, 2012.
80. Draper, M., Saez, I.M., Cowling, S.J., Gai, P., Heinrich, B., Donnio, B., Guillon, D., Goodby, J.W., Self-Assembly and Shape Morphology of Liquid Crystalline Gold Metamaterials. *Adv. Funct. Mater.*, 21, 1260–1278, 2011, https://doi.org/10.1002/adfm.201001606.

81. Qi, H. and Hegmann, T., Impact of nanoscale particles and carbon nanotubes on current and future generations of liquid crystal displays. *J. Mater. Chem.*, 18, 3288, 2008, https://doi.org/10.1039/b718920f.
82. Sharma, A., Urbanski, M., Mori, T., Kitzerow, H.-S., Hegmann, T., Metallic and semiconducting nanoparticles in LCs, in: *Series in Soft Condensed Matter*, pp. 497–535, World Scientific, 2016.
83. Singh, A.K. and Singh, S.P., Revisiting hierarchical arrangement of quantum dots in presence of liquid crystal media. *Mol. Cryst. Liq. Cryst.*, 746, 22–68, 2022, https://doi.org/10.1080/15421406.2022.2078589.
84. Ayeb, H., Derbali, M., Mouhli, A., Soltani, T., Jomni, F., Fresnais, J., Lacaze, E., Viscoelastic and dielectric properties of 5CB nematic liquid crystal doped by magnetic and nonmagnetic nanoparticles. *Phys. Rev. E.*, 102, 052703, 2020, https://doi.org/10.1103/PhysRevE.102.052703.
85. Milette, J., Toader, V., Soulé, E.R., Lennox, R.B., Rey, A.D., Reven, L., A Molecular and Thermodynamic View of the Assembly of Gold Nanoparticles in Nematic Liquid Crystal. *Langmuir.*, 29, 1258–1263, 2013, https://doi.org/10.1021/la304189n.
86. Milette, J., Cowling, S.J., Toader, V., Lavigne, C., Saez, I.M., Bruce Lennox, R., Goodby, J.W., Reven, L., Reversible long range network formation in gold nanoparticle - nematic liquid crystal composites. *Soft Matter.*, 8, 173–179, 2012, https://doi.org/10.1039/C1SM06604H.
87. Orlandi, S., Benini, E., Miglioli, I., Evans, D.R., Reshetnyak, V., Zannoni, C., Doping liquid crystals with nanoparticles. A computer simulation of the effects of nanoparticle shape. *Phys. Chem. Chem. Phys.*, 18, 2428–2441, 2016, https://doi.org/10.1039/C5CP05754J.
88. Gorkunov, M.V. and Osipov, M.A., Mean-field theory of a nematic liquid crystal doped with anisotropic nanoparticles. *Soft Matter.*, 7, 4348, 2011, https://doi.org/10.1039/c0sm01398f.
89. Qi, H. and Hegmann, T., Liquid crystal–gold nanoparticle composites. *Liq. Cryst. Today*, 20, 102–114, 2011, https://doi.org/10.1080/1358314X.2011.610133.
90. Choudhary, A., Singh, G., Biradar, A.M., Advances in gold nanoparticle–liquid crystal composites. *Nanoscale*, 6, 7743–7756, 2014, https://doi.org/10.1039/C4NR01325E.
91. Bisoyi, H.K. and Kumar, S., Liquid-crystal nanoscience: an emerging avenue of soft self-assembly. *Chem. Soc. Rev.*, 40, 306–319, 2011, https://doi.org/10.1039/B901793N.
92. S J, S., Gupta, R.K., Kumar, S., Manjuladevi, V., Enhanced electro-optical response of nematic liquid crystal doped with functionalised silver nanoparticles in twisted nematic configuration. *Liq. Cryst.*, 47, 1678–1690, 2020, https://doi.org/10.1080/02678292.2020.1755901.
93. S. J., S., Gupta, R.K., Kumar, S., V, M., Effect of functionalised silver nanoparticle on the elastic constants and ionic transport of a nematic liquid crystal. *Liq. Cryst.*, 46, 1868–1876, 2019, https://doi.org/10.1080/02678292.2019.1611964.

94. Drzaic, P.S., Liquid Crystal Dispersions. *World Sci.*, 1995.
95. Hsu, C.-J., Lin, L.-J., Huang, M.-K., Huang, C.-Y., Electro-optical Effect of Gold Nanoparticle Dispersed in Nematic Liquid Crystals. *Crystals*, 7, 287, 2017, https://doi.org/10.3390/cryst7100287.
96. Pandey, D.K., Singh, U.B., Dhar, R., Dabrowski, R., Pandey, M.B., Dielectric and electro-optic properties of 6CHBT nematic liquid crystals and silver nanoparticles composites. *Phase Transit.*, 92, 931–938, 2019, https://doi.org/10.1080/01411594.2019.1654107.
97. Singh, U.B., Dhar, R., Dabrowski, R., Pandey, M.B., Influence of low concentration silver nanoparticles on the electrical and electro-optical parameters of nematic liquid crystals. *Liq. Cryst.*, 40, 774–782, 2013, https://doi.org/10.1080/02678292.2013.783136.
98. Huang, S.-Y., Peng, C.-C., Tu, L.-W., Kuo, C.-T., Enhancement of Luminescence of Nematic Liquid Crystals Doped with Silver Nanoparticles. *Mol. Cryst. Liq. Cryst.*, 507, 301–306, 2009, https://doi.org/10.1080/15421400903054071.
99. Chang, C.-H., Lin, R.-J., Tien, C.-L., Yeh, S.-M., Enhanced Photoluminescence in Gold Nanoparticles Doped Homogeneous Planar Nematic Liquid Crystals. *Adv. Condens. Matter Phys.*, 2018, 1–5, 2018, https://doi.org/10.1155/2018/8720169.
100. Chaudhry, M. and Ghildyal, D., Characterization of thermal, electro optical and photoluminescent properties of nematic liquid crystal doped with gold nano particles. *Mater. Today Proc.*, 57, 2061–2066, 2022, https://doi.org/10.1016/j.matpr.2021.11.265.
101. Liu, Y., Zhang, Y., Tang, Y., Zhao, Z., Wang, Q., Jia, G., Plasmonic nanoparticles enhancing blue emission from nematic liquid crystal. *Results Phys.*, 9, 1537–1542, 2018, https://doi.org/10.1016/j.rinp.2018.05.006.
102. Eidenschink, R. and De Jeu, W.H., Static scattering in filled nematic: new liquid crystal display technique. *Electron. Lett.*, 27, 1195–1996, 1991, https://doi.org/10.1049/el:19910744.
103. Bukowczan, A., Hebda, E., Pielichowski, K., The influence of nanoparticles on phase formation and stability of liquid crystals and liquid crystalline polymers. *J. Mol. Liq.*, 321, 114849, 2021, https://doi.org/10.1016/j.molliq.2020.114849.
104. Yadav, S.P., Yadav, K., Lahiri, J., Singh Parmar, A., Ferroelectric liquid crystal nanocomposites: recent development and future perspective. *Liq. Cryst. Rev.*, 6, 143–169, 2018, https://doi.org/10.1080/21680396.2019.1589400.
105. Marinov, Y.G., Hadjichristov, G.B., Petrov, A.G., Prasad, S.K., Thin films of silica nanoparticle doped nematic liquid crystal 7CB for electro-optic modulation. *Photonics Lett. Pol.*, 7, 94–96, 2015, https://doi.org/10.4302/plp.2015.4.03.
106. Marinov, Y.G., Hadjichristov, G.B., Petrov, A.G., Prasad, S.K., Electro-optic modulation by silica-nanostructured nematic system (aerosil/7CB nanocomposite). *Compos. Part B Eng.*, 90, 471–477, 2016, https://doi.org/10.1016/j.compositesb.2016.01.034.

107. Huang, C.-Y., Lai, C.-C., Tseng, Y.-H., Yang, Y.-T., Tien, C.-J., Lo, K.-Y., Silica-nanoparticle-doped nematic display with multistable and dynamic modes. *Appl. Phys. Lett.*, 92, 221908, 2008, https://doi.org/10.1063/1.2938880.
108. Huang, C.-Y., Huang, Y.-J., Tseng, Y.-H., Dual-operation-mode liquid crystal lens. *Opt. Express*, 17, 20860, 2009, https://doi.org/10.1364/OE.17.020860.
109. Shiraishi, Y., Sugihara, K., Okamura, N., Sawai, H., Kobayashi, S., Toshima, N., Fast Electro-Optic Switching of Twisted Nematic LCD Doped with Cyclodextrin Capped Silica Nanoparticles. *Macromol. Symp.*, 317–318, 28–33, 2012, https://doi.org/10.1002/masy.201100066.
110. Meddeb, B., Hbaiebb, S., Guesmi, A., Hamadi, N.B., Soltani, T., Synthesis of β-cyclodextrin Functionalized silica nanoparticles and their effects on the dielectric, optical and electro-optic properties of nematic liquid crystal. *Liq. Cryst.*, 49, 182–191, 2022, https://doi.org/10.1080/02678292.2021.1949053.
111. Liao, S.-W., Hsieh, C.-T., Kuo, C.-C., Huang, C.-Y., Voltage-assisted ion reduction in liquid crystal-silica nanoparticle dispersions. *Appl. Phys. Lett.*, 101, 161906, 2012, https://doi.org/10.1063/1.4760277.
112. Tomylko, S., Yaroshchuk, O., Kovalchuk, O., Maschke, U., Yamaguchi, R., Dielectric and Electro-Optical Properties of Liquid Crystals Doped with Diamond Nanoparticles. *Mol. Cryst. Liq. Cryst.*, 541, 35/[273]–43/[281], 2011, https://doi.org/10.1080/15421406.2011.569658.
113. Mochalin, V.N., Shenderova, O., Ho, D., Gogotsi, Y., The properties and applications of nanodiamonds. *Nat. Nanotechnol.*, 7, 11–23, 2012, https://doi.org/10.1038/nnano.2011.209.
114. Chen, P.-S., Huang, C.-C., Liu, Y.-W., Chao, C.-Y., Effect of insulating-nanoparticles addition on ion current and voltage-holding ratio in nematic liquid crystal cells. *Appl. Phys. Lett.*, 90, 211111, 2007, https://doi.org/10.1063/1.2740581.
115. Chen, P.-S., Huang, C.-C., Liu, Y.-W., Chao, C.-Y., The Effect of Adding Insulating-Nanoparticles to Nematic Liquid Crystal Cells on Ion Current and Voltage-Holding Ratio. *Mol. Cryst. Liq. Cryst.*, 507, 202–208, 2009, https://doi.org/10.1080/15421400903051374.
116. Gaur, D.K., Rastogi, A., Trivedi, H., Parmar, A.S., Manohar, R., Singh, S., Investigation of dielectric and optical properties of pure and diamond nanoparticles dispersed nematic liquid-crystal PCH5. *Liq. Cryst.*, 48, 1257–1267, 2021, https://doi.org/10.1080/02678292.2020.1855679.
117. Barrera, A., Binet, C., Dubois, F., Hébert, P.-A., Supiot, P., Foissac, C., Maschke, U., Dielectric Spectroscopy Analysis of Liquid Crystals Recovered from End-of-Life Liquid Crystal Displays. *Molecules*, 26, 2873, 2021, https://doi.org/10.3390/molecules26102873.
118. Sharma, A., Malik, P., Kumar, P., Electro-Optical and Dielectric Responses of ZnO Nanoparticles Doped Nematic Liquid Crystal in In-Plane Switching (IPS) Mode. *Integr. Ferroelectr.*, 202, 52–66, 2019, https://doi.org/10.1080/10584587.2019.1674824.

119. Eskalen, H., Özğan, Ş., Alver, Ü., Kerlı, S., Electro-Optical Properties of Liquid Crystals Composite with Zinc Oxide Nanoparticles. *Acta Phys. Pol. A.*, 127, 756–760, 2015, https://doi.org/10.12693/APhysPolA.127.756.
120. Sharma, A., Kumar, P., Malik, P., Effect of zinc oxide nanoparticles on dielectric behavior of nematic liquid crystal. *AIP Conf. Proc.*, 1953, 100037, 2018, https://doi.org/10.1063/1.5032973.
121. Jayaprada, P., Rao, M.C., Pardhasaradhi, P., Datta Prasad, P.V., Manepalli, R.K.N.R., Pisipati, V.G.K.M., Optical studies of n-octyloxy-cyanobiphenyl (8ocb) with dispersed ZnO nanoparticles for display device application. *Optik*, 185, 1226–1237, 2019, https://doi.org/10.1016/j.ijleo.2019.04.060.
122. Shcherbakov, A.B., Reukov, V.V., Yakimansky, A.V., Krasnopeeva, E.L., Ivanova, O.S., Popov, A.L., Ivanov, V.K., CeO2 Nanoparticle-Containing Polymers for Biomedical Applications: A Review. *Polymers*, 13, 924, 2021, https://doi.org/10.3390/polym13060924.
123. Mun, H.-Y., Park, H.-G., Jeong, H.-C., Lee, J.H., Oh, B.-Y., Seo, D.-S., Thermal and electro-optical properties of cerium-oxide-doped liquid-crystal devices. *Liq. Cryst.*, 44, 538–543, 2017, https://doi.org/10.1080/02678292.2016.1225838.
124. Reznikov, Y., Buchnev, O., Tereshchenko, O., Reshetnyak, V., Glushchenko, A., West, J., Ferroelectric nematic suspension. *Appl. Phys. Lett.*, 82, 1917–1919, 2003, https://doi.org/10.1063/1.1560871.
125. Ouskova, E., Buchnev, O., Reshetnyak, V., Reznikov, Yu., Kresse, H., Dielectric relaxation spectroscopy of a nematic liquid crystal doped with ferroelectric Sn 2 P 2 S 6 nanoparticles. *Liq. Cryst.*, 30, 1235–1239, 2003, https://doi.org/10.1080/02678290310001601996.
126. Parveen, A., Prakash, J., Singh, G., Impact of strontium titanate nanoparticles on the dielectric, electro-optical and electrical response of a nematic liquid crystal. *J. Mol. Liq.*, 354, 118907, 2022, https://doi.org/10.1016/j.molliq.2022.118907.
127. Basu, R., Soft memory in a ferroelectric nanoparticle-doped liquid crystal. *Phys. Rev. E*, 89, 022508, 2014, https://doi.org/10.1103/PhysRevE.89.022508.
128. Dubey, R., Mishra, A., Singh, K.N., Alapati, P.R., Dhar, R., Electric behaviour of a Schiff's base liquid crystal compound doped with a low concentration of BaTiO3 nanoparticles. *J. Mol. Liq.*, 225, 496–501, 2017, https://doi.org/10.1016/j.molliq.2016.11.077.
129. Lin, Y., Daoudi, A., Dubois, F., Segovia-Mera, A., Legrand, C., Douali, R., Correlation between dielectric properties and phase transitions of 8CB/Sn 2 P 2 S 6 liquid crystal nanocolloids. *J. Mol. Liq.*, 232, 123–129, 2017, https://doi.org/10.1016/j.molliq.2017.02.060.
130. Anu, Varshney, D., Yadav, K., Prakash, J., Meena, H., Singh, G., Tunable dielectric and memory features of ferroelectric layered perovskite Bi4Ti3O12 nanoparticles doped nematic liquid crystal composite. *J. Mol. Liq.*, 369, 120820, 2023, https://doi.org/10.1016/j.molliq.2022.120820.

131. Singh, B.P., Sikarwar, S., Manohar, R., Shah, A., Singh, D.P., Herman, J., Pandey, K.K., Nematic liquid crystals blended ferroelectric nanoparticles (BaTiO3): A perspective way for improving the response time and photoluminescence for electro-optical devices. *J. Appl. Phys.*, 131, 174102, 2022, https://doi.org/10.1063/5.0089449.
132. Lin, Y., Daoudi, A., Dubois, F., Blach, J.-F., Henninot, J.-F., Kurochkin, O., Grabar, A., Segovia-Mera, A., Legrand, C., Douali, R., A comparative study of nematic liquid crystals doped with harvested and non-harvested ferroelectric nanoparticles: phase transitions and dielectric properties. *RSC Adv.*, 7, 35438–35444, 2017, https://doi.org/10.1039/C7RA04154C.
133. Blach, J.-F., Saitzek, S., Legrand, C., Dupont, L., Henninot, J.-F., Warenghem, M., BaTiO 3 ferroelectric nanoparticles dispersed in 5CB nematic liquid crystal: Synthesis and electro-optical characterization. *J. Appl. Phys.*, 107, 074102, 2010, https://doi.org/10.1063/1.3369544.
134. Rasna, M.V., Cmok, L., Evans, D.R., Mertelj, A., Dhara, S., Phase transitions, optical, dielectric and viscoelastic properties of colloidal suspensions of $BaTiO_3$ nanoparticles and cyanobiphenyl liquid crystals. *Liq. Cryst.*, 42, 1059–1067, 2015, https://doi.org/10.1080/02678292.2015.1021719.
135. Mishra, M., Dabrowski, R.S., Dhar, R., Thermodynamical, optical, electrical and electro-optical studies of a room temperature nematic liquid crystal 4-pentyl-4′-cyanobiphenyl dispersed with barium titanate nanoparticles. *J. Mol. Liq.*, 213, 247–254, 2016, https://doi.org/10.1016/j.molliq.2015.11.007.
136. Prodanov, M.F., Buluy, O.G., Popova, E.V., Gamzaeva, S.A., Reznikov, Y.O., Vashchenko, V.V., Magnetic actuation of a thermodynamically stable colloid of ferromagnetic nanoparticles in a liquid crystal. *Soft Matter.*, 12, 6601–6609, 2016, https://doi.org/10.1039/C6SM00906A.
137. Zhang, R., Zhang, Z., Han, J., Yang, L., Li, J., Song, Z., Wang, T., Zhu, J., Advanced liquid crystal-based switchable optical devices for light protection applications: principles and strategies. *Light Sci. Appl.*, 12, 11, 2023, https://doi.org/10.1038/s41377-022-01032-y.
138. Cupelli, D., Pasquale Nicoletta, F., Manfredi, S., Vivacqua, M., Formoso, P., De Filpo, G., Chidichimo, G., Self-adjusting smart windows based on polymer-dispersed liquid crystals. *Sol. Energy Mater. Sol. Cells*, 93, 2008–2012, 2009, https://doi.org/10.1016/j.solmat.2009.08.002.
139. Mesloub, A., Ghosh, A., Kolsi, L., Alshenaifi, M., Polymer-Dispersed Liquid Crystal (PDLC) smart switchable windows for less-energy hungry buildings and visual comfort in hot desert climate. *J. Build. Eng.*, 59, 105101, 2022, https://doi.org/10.1016/j.jobe.2022.105101.
140. Demus, D., Goodby, J., Gray, G.W., Spiess, H. -W., Vill, V., *Handbook of Liquid Crystals Set*. Wiley, 1998.
141. Kim, Y. and Tamaoki, N., Photoresponsive Chiral Dopants: Light-Driven Helicity Manipulation in Cholesteric Liquid Crystals for Optical and Mechanical Functions. *ChemPhotoChem.*, 3, 284–303, 2019, https://doi.org/10.1002/cptc.201900034.

4

Nanocomposites in Aerospace and Defense: Materials for Lightweight Structures, Thermal Management, and Ballistic Protection

Riyadh A. Al-Samarai[1]* and Yarub Al-Douri[2,3,4]

[1]*Electromechanical Engineering Department, Engineering College, University of Samarra, Samarra, Iraq*
[2]*College of Health and Medical Techniques, Al-Bayan University, Baghdad, Iraq*
[3]*Department of Mechanical Engineering, Faculty of Engineering, Piri Reis University, Istanbul, Turkey*
[4]*Nanotechnology and Catalysis Research Centre, University of Malaya, Kuala Lumpur, Malaysia*

Abstract

Advances in nanotechnology have led to the creation of an entirely new class of materials that provide previously unheard-of control over their composition, structure, and defect detection. Traditionally used in low-volume electronics, nanoceramics have recently been used to create stronger, larger materials such as aerospace and aircraft engines and electronic devices. The internal reinforcing fiber, which is usually made of polymer-based ceramics, gives the composite strength and durability and consists of amorphous or nanocrystalline ceramics. One can generate porous nanomaterials or dense homogenous ceramic composites, depending on the ageing and drying conditions. These scientists work on a variety of disciplines, including genetics, biomedical engineering; nanomaterials were examined in fuel cells, anti-corrosion coatings, biotechnology, and photocatalysis.

Keywords: Nanotechnology, nanoceramics, ceramic composites, aerospace materials, polymer-based ceramics, porous nanomaterials, biomedical engineering, photocatalysis

**Corresponding author*: riyadh2824@yahoo.com

Inamuddin, Tariq Altalhi and Mohammad Abu Jafar Mazumder (eds.) Nanostructured Hard Materials: Industrial Applications, (149–174) © 2026 Scrivener Publishing LLC

4.1 Introduction

Ceramic materials are inorganic materials that combine metallic and non-metallic components; the qualities of these materials depend on the specific arrangement of these elements [1–3]. They are good electrical insulators and also have no conduction electrons because they combine to form chemical bonds. The importance of nanomaterials as a by-product of nanotechnology is growing. Ceramics have a complex chemical makeup. However, there are also additional materials including nitrides, carbides, silicides, and borides that can be used to make ceramics. Ceramic alloying is essential for regulating the materials' chemical and physical properties [3–5]. There is a huge demand for advanced materials as a result of developments in aerospace technology. The aerospace industry has given lightweight composite materials much attention as a cutting-edge technology. Using materials with lower density while maintaining comparable or improved performance over traditional materials is the idea behind lightweight materials. Due to their strength-to-weight ratio, low carbon emission, excellent thermal resistance, radiation resistance, fatigue toughness, damage resistance, and superior mechanical performance, these lightweight composite materials have been widely adopted in aerospace industries.

While pottery is simply one facet of the ceramic industry, the name "ceramics" is Greek in origin [6]. Ceramic materials with at least one component between 1 and 100 nm in size are referred to as nanoceramics. They include bulk materials developed for diverse industries and applications, as well as nanosized and nanostructured particles, fibers, tubes, sheets, and rods [7–10]. Ceramics are good thermal and electrical insulators. Brittleness is one of the main physical characteristics of ceramics. Mechanical strength, density, hardness, and other characteristics are other important physical characteristics of ceramics. Nanoceramics are used in several technologies. The durability, biocompatibility, hydrophilicity, and wear resistance of nanoceramics have led to their use in biomedicine, healthcare, and environmental protection. Powder synthesis, mixing, molding, and heat treatment are the steps in the ceramic manufacturing process. The sintering process occurs either due to diffusion in the solid state or due to the creation of a liquid phase between the grains [6]. Ceramic powders can be produced artificially using processes such as sol-gelation, high-temperature autosynthesis (HTS), precipitation, etc., or naturally using raw materials such as rock minerals [11]. Conventional sources of ceramic raw materials require preliminary procedures to remove impurities and dilution techniques to create a powder that can be sintered and molded to create the product. However, high-purity products, mostly synthetically produced, are used as base powders in modern ceramics. Due to their excellent characteristics, ceramics are typically better suited for a wide range of applications. Due to their durability, biocompatibility, hydrophilicity, and resistance to wear, nanoceramics are used in biology, healthcare, and the environment. Due to their insulating and/or conductive qualities, nanoceramics can also be employed to make next-generation chips and other electronic components. Additionally, nanoceramics are utilized in the manufacture of consumer products, vehicles, buildings, and weapons [12].

4.2 Advanced Ceramic Material Preparation

Modern ceramic components are made by heating ceramic powders that have been specially processed to control heterogeneity, chemical composition. The quality of finished ceramic products is significantly influenced by the above-mentioned variables. In theory, finished ceramic products made from natural resources can be distinguished from products made entirely from synthetic materials. Gas-phase processes are used to synthesize ultrapure ultrafine particles. Hydrothermal synthesis and co-precipitation techniques are used in liquid-phase synthesis to create uniform fine particles. Temperature is the main factor controlling the reaction in most synthesis pathways. Recent methods for producing new ceramic nitrides such as -Si_3N_4 or cubic Hf_3N_4 and Zr_3N_4 also use pressure in addition to temperature. Diamond anvil cell and multi-anvil cell methods have also been effectively used for various high-pressure ceramic synthesis studies [13, 14].

The same process is referred to by a number of other names, including gasless combustion, combustion synthesis, self-propagating combustion and self-propagating exothermic processes [15, 16]. This technique, a form of pyrotechnics, was developed from research into the combustion of solid cylindrical compacts such as Ti-B mixtures. A series of events causes combustion recovery to rapidly spread along the combustion wave [17]. In CVD, a low molecular weight and volatile reactant are delivered into a heated reaction zone, where it solidifies and collected as a free powder (the gaseous reactants' thermodynamic instability serves as the driving force behind this process). Diffusion, nucleation and reaction rates all play a role in the total rate of deposition [33, 34]. At higher and lower levels, diffusion and reaction are the determining elements respective to cooler temperatures. Homogeneous nucleation results in the creation of free powders in the reactor's environment. The sol-gel method involves coagulating colloidal gels after peptizing metal oxides or hydrolyzing metal salts to produce ceramics.

The sol particles coagulate when sols are destabilized, for instance, by eliminating the liquid phase or altering the pH level creating a 3D gel [18]. Additionally, these procedures may make it possible to apply in previously inaccessible areas of application. The ceramic solid is created using one of the aforementioned techniques from low molecular weight, elemental organic or inorganic materials. Reduced sintering costs and smaller grains are two key benefits of employing finely ground ceramics made using the sol-gel technique.

4.3 Methods for Creating Nanoceramics

The most popular type of raw material used to make ceramics is chemical powder [19]. To give the ceramic particles a correct shape, additional materials are added to the powder throughout the manufacturing process. Unfortunately, the production of these powders remains difficult, despite the fact that the excellent mechanical, electrical and thermal properties of the resulting materials distinguish them from conventional homologs [20, 21]. After this, various methods are used including two-photon lithography and sintering. Sintering uses pressure and heat to create a solid. In this chapter, we will exclusively discuss this approach for creating nanoceramics, mixing, molding and heat treatment are all steps in the manufacture of ceramic parts, and during sintering, either solid-state diffusion or the development of an intergranular liquid phase occurs [23–25].

Ceramic powders can be produced synthetically through processes such as sol-gel, high-temperature self-propagating synthesis (HTS), precipitation, or naturally by processing natural raw materials (rock minerals). In general, traditional ceramics use ceramic powders made from minerals, whereas modern ceramics more often use particle systems produced by synthesis methods. In the aerospace sector, the use of lightweight composites has improved manufacturability, reduced industrial waste, and reduced energy consumption during production. These benefits have prompted the scientific community to propose many avenues for research into improving the performance of aircraft and aerospace materials. On another hand, in modern ceramics, high-purity materials obtained mainly synthetically are used as starting powders. Depending on the composition of their constituent parts, high-tech ceramics can be divided into oxide and non-oxide. There are many different types (Figure 4.1).

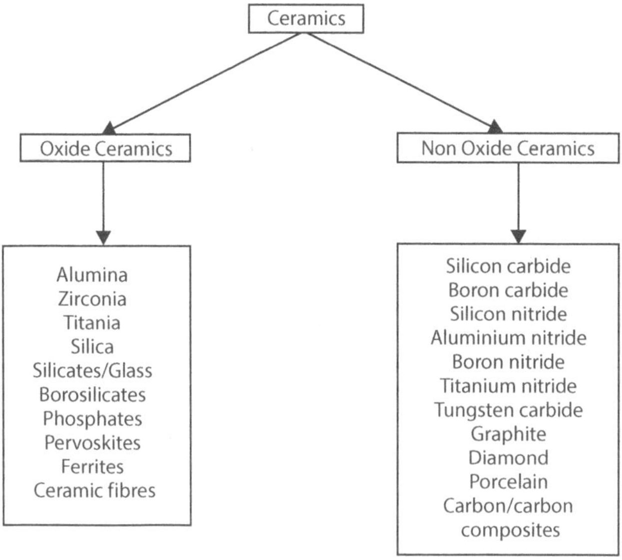

Figure 4.1 Ceramic classification [26].

4.3.1 Sol-Gel Chemistry for Nanoceramic

Ceramic materials with special electrical properties find application in portable gas sensors. These systems may be customized by adjusting structural and compositional factors, and they are cheap and straightforward. There is a fantastic review on this topic [1]. H_2 separation systems are another field connected to electronic ceramic materials [2]. These systems make use of membranes built of dense proton-conducting ceramic [25–28].

The term "sol" refers to a colloidal solution consisting of solid particles (several hundred nm in diameter) suspended in a liquid phase. Sol-gel technology is a well-known colloidal chemistry technique that has great potential to produce various materials with predefined properties through a simple procedure at fairly low process costs. During water-induced inorganic polymerization processes, the precursor solution is converted into an inorganic solid. The first substance or precursor is either an inorganic metal salt (like chloride, nitrate or sulfate) or an organic metal complex such as an alkoxide. Basically, this process consists of three stages [29]. Polymerization and cross-linking proceeds until the sol-gel transition point, where it increases rapidly and gelation occurs.

To create durable products, these procedures are accompanied by heat treatment or sintering. Al, Ti, Si, and Zr alkoxides arc examples of inexpensive commercially available alkoxides, while Fe, Mn, Cu, Ni, Nb, Y, Co and Ta alkoxides are expensive and difficult to prepare [30]. Research over the past few decades has focused on nanoceramics, a unique type of material. These particular nanoparticles are composed of ceramics, which are classified as inorganic, non-metallic, heat-resistant solids consisting of both metallic and non-metallic substances. It is very important to remember that ceramics are hard and brittle on a macro level, meaning that they can break if hit hard against other hard objects, for example. At the nanoscale, researchers have shown that nanoceramics serve a wider range of purposes than commonly believed [31, 32]. Due to the fact that sol-gel method allows one to obtain pure and homogeneous nanoceramic materials over a long period of time, products are homogeneous under mild conditions. In addition, by combining sols of different chemicals, sol-gel technology offers many opportunities to create a variety of compounds with controlled stoichiometry. Ceramics have already shown that of useful qualities, and heat resistance, as well as increased wear resistance.

However, most of them are fragile, making them difficult to use on a large scale industrially. Typical ceramic nanocomposite structures (Figure 4.2) consist of a micro- or nano-sized matrix and mixed phases of nanoparticle components. The interfacial area of ceramic nanocomposites increases sharply which is the main characteristic of their properties [6]. Improvements porosity and hydrophilicity have been discovered, two functional qualities that have new applications in environmental treatment and remediation.

The powder in the mold does not shrink during this operation and no parts are removed. Instead, the setting agent either traps voids or molecules within the particle system or hardens through polymerization, binders are removed and sintering occurs [32–35]. In addition, gel casting is significantly different from PIM and shares some features with it. (1) Removing binder from gel casting is less difficult because the binder content is 5 to 10 times lower than PIM. (2) Gel casting does not require high-temperature injection and only involves pouring a ceramic slurry into the non-porous mold of the tools and equipment used in PIM. There are two key things to consider when using gel casting to make ceramic components. The ideal gelation rate must first be established, and it must be high enough to prevent the foam from breaking down. Gels with high fluidity are necessary, especially for the fabrication of extremely complex structures, since the rheology of the gel must be able to fill all the subtleties of the mold [36–39]. Without reacting with the monomer or polymer solution, it forms a gel that a similar gel would form in the absence of the ceramic, suspending the powder and fixing the particles in place. Ceramic precursors are an important component of the gelation process in sol-gel processing (by polycondensation, hydrolysis, etc.). The gel network is based on metal hydroxides and alkoxides, which are subsequently processed into ceramics [40].

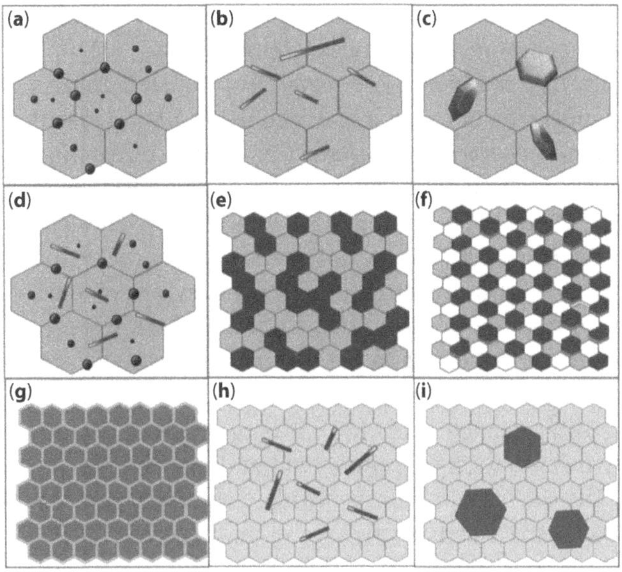

Figure 4.2 Ceramic nanocomposites fall under the category of micro/nanocomposites, which contain rounded nanoparticles. (b) Micro/nanocomposite with a micron-sized matrix and elongated nano-reinforcing components. (c) Nano-reinforcement plates inserted into a micron-sized matrix in a micro/nanocomposite. (d) Micro/nanocomposite with a micron-sized matrix and embedded round and elongated nano-reinforcement components. (e) Two-phase composite consisting of two ultrafine immiscible phases. (f) Multiphase composite consisting of three immiscible nanophases (or more). (g) Composite nano/nanolayer. (h) thin matrix contains nano- or microfibers. A thin matrix containing significant second-phase precipitates. Reproduced from source [26].

4.4 Recent Developments in General and Advanced Nanoceramics Fields

Since the dawn of civilization, ceramics have been the earliest type of plastic material used by humans. The earliest ceramic compounds are always created using natural resources such as clay minerals. Kaolinite, pyrophyllite, halloysite and montmorillonite are some examples. Glass, ceramics and enamel are also made from feldspar and various ceramic substances. Asbestos, talc, sillimanite and wollastonite are other natural elements used in the production of ceramic products [60–63].

In fact, large materials such as bricks, tiles, toilets (white ware) and ceramics are used to create traditional pottery. Traditional ceramics are also characterized by minimal technology, although advanced production techniques have been used, including sophisticated tooling, advanced equipment and computerized process control [41–45].

Essentially, they are crystalline materials produced under strict control from highly purified and/or characterized basic components that are given precisely defined characteristics [66]. Carries out research on biomaterials, including biomarkers, biodiagnostics, and biosensors; nanoelectronics and photonic materials, including MEMS and NEMS; and related nanomaterials for use in solar materials, polymers, textiles, fuel cell coatings, and composites. Application research on BN nanocrystals involves their usage as coatings for precision cutting tools to change their properties according to the desired wear resistance, superhydrophobicity, abrasiveness, and heat resistance. Furthermore, nanoBN has prospective applications in plastics and other cutting-related fields [68]. It is particularly useful as a thermal and oxidative shield for fiber-reinforced building materials, such as carbon fiber materials coated with boron nitride. Additional uses could occur in optoelectronics, space constructions, and sensor technology.

Table 4.1 displays applications for ceramic nanocomposites in the treatment of water and waste [46–50]. The two kinds of nanoceramic materials that are now available are dry powder and liquid dispersions. Two other

Table 4.1 Examples of the use of nanocomposites and waste treatment [26].

Applications	Examples	Some novel properties
Adsorption	CNTs/metal oxide and nanofibers	Minimal intra-particle diffusion distance, high specific surface area, easily accessible adsorption sites, more selective adsorption sites, adjustable surface chemistry, and ease of reuse
Disinfection	Ag/TiO2, ZnO and CNTs	High chemical stability, minimal cost and toxicity, potent antibacterial activity, and simplicity of use
Photocatalysis	TiO2/CNTs	Low cost, and activity under sun radiation
Membranes	Ag/TiO2/Magnetite/ Zeolites with CNTs	High mechanical and chemical stability, high permeability and selectivity, hydrophilicity, minimal toxicity to humans, strong antibacterial action, and photocatalytic activity

cutting-edge nanoceramic materials that are frequently utilized today are silicon carbide and tungsten carbide [51]. They are utilized in abrasive situations, where other ceramic materials cannot be used, like wear plates in crushing machinery in the mining industry current ceramics. The materials are also utilized in the electrical, aerospace, and medical sectors [52–55].

4.4.1 Nanoceramics Based on Silicate

According to the literature, the use of nano-powders to create fine-grained ceramics leads to the production of materials that are stronger than their bulk or coarse-grained counterparts [56–58]. Nanoscale silica (SiO_2) materials are now widely used as fillers in engineering composites as a result of recent advances in nanotechnology [59, 67]. Crystals are smaller than those used in the two-stage sol-gel sintering technique, 100 nm, to produce nanocrystalline corundum abrasive. Grain growth is reduced in this technique by regulating the sintering temperature [31]. The increasing pace and intensity of development have led to significant advances in technology and society throughout modern history. However, the physical techniques used to improve device- and die-level capabilities have also created very challenging conditions for keeping electronic devices cool—a critical component in determining their speed, efficiency, and reliability. More than ever, there are needs and opportunities for new materials that will help solve some of these pressing thermal management challenges due to advances in nanoelectronics and the emergence of new applications including 3D chip array topologies and flexible electronics.

To generate multilayer coatings for carbon composites, refractory oxides like silica, alumina, and aluminosilicate nanoceramics [32] were used. In this study, ceramic silica, alumina, and aluminosilicate sols are electrophoretically deposited onto carbon composites to create a multilayer ceramic coating. The findings demonstrated that the modified composites integrating nanoceramic material have a significantly lower rate of oxidation loss than the unmodified infiltrated carbon composite [64].

4.4.2 Nanoceramics Based on Zirconium Dioxide

Optoelectronics are widely used in nanoscale ceramic materials. A recent review discusses new results from modern experiments with solid-state lasers and theoretical advances in the field of luminescence quenching kinetics. The study of nanoscale luminous particles often uses and benefits greatly from luminescence quenching kinetics. This is critical in the creation of new lasers and the production of laser materials. Multiple emitting particles may participate in complex quenching kinetics. Distinguishing between different types of releasing chemicals usually requires lifetime experimentation [12, 69–70]. When a crack can develop at the fiber/matrix interface, CMC can reach its maximum strength. This prevents cracks from entering the fiber and promotes fiber pullout, a process that wastes energy [65]. Direct atomistic models can be used to analyze the strengthening of nanosized particles in CMC [75–78]. Using these models, researchers can study how cracks in the matrix respond to changes in interfacial connectivity and subsequent slip [71–79]. The technology for producing electricity from fuel is called a fuel cell. Besides efficiency exceeding 70% compared to traditional power generation (33 to 38%) [39, 41]. In addition, photocatalysts and electrochemical electrodes are used [80–87].

4.4.3 Nanoceramics Based on Titania

The peptized sol–gel method is often used to produce nanoscale titanium ceramic materials [45, 46]. Knowledge of improved $BaTiO_3$-based electrical materials opens up a unique characteristic. According to published studies, Al_2O_3-TiO_2 powder is mixed with various concentrations of impurities and the particles are compacted to form aluminum titanate [90, 92–93]. Ferroelectric ceramic materials can have larger grains [94]. Ceramics made of aluminum titanate (Al-TiO) exhibit strong thermal conductivity, low coefficient of thermal expansion (*1106 K), and good heat resistance [88, 89, 91]. Due to these qualities, these materials are more suitable for automotive applications such as thermal barriers, engine parts, turbochargers, and the metallurgical industry [95]. Due to conveying high-temperature, steel liquid act as a heat shield and catalyst carrier for the purification of automobile exhaust, among other uses [96]. This ceramic's susceptibility to fracture at temperatures between 750 and 1300°C is a drawback [97]. Elements that can be added to the matrix of barium titanate It has been found that doping, which significantly increases temperature stability, is the cause of ceramic [124, 125].

4.4.4 Nanoceramics Based on Borate

The processing, production, microstructure evolution, and characteristics of bulk ceramic nanomaterials were reviewed. The relationship between reduced microstructural complexity and increased mechanical strength of bulk engineering ceramics, as well as the potential benefits, is also reviewed [98]. Glass ceramics are produced using the conventional processes of melt quenching and subsequent heat treatment as 20MO55Bi2O325BiO3

(M=Ba,Sr) systems [99]. Boron oxide enhances the aesthetics of ceramic materials, acts as a flux, and raises their chemical and mechanical resistance. The thermal expansion coefficient of ceramic materials is unaffected by this, though. A crucial part of electro-optics is transparent glass-ceramics [100], notably in laser systems with transparent laser cladding and other parts. The largest borate reserves are those present in boron oxide (B2O3) [101]. The primary minerals of sedimentary borate ores are borax decahydrate (Na2O2B2O310H2O), colemanite (2CaO3B2O35H2O), and ulexite (Na2O2CaO5B2O316H2O) [102]. Glass ceramics are produced using the conventional processes of melt quenching and subsequent heat treatment as 20MO55Bi2O325BiO3 systems (M = Ba, Sr) systems [103]. Boron oxide enhances the aesthetics of ceramic materials, serves as a flux, and raises their chemical and mechanical resistance. However, this does not affect the coefficient of thermal expansion of ceramic materials. Transparent glass-ceramics are a critical component of electro-optics [104], especially in laser systems with laser claddings and other transparent components used to create low-viscosity frits for enameling steel, aluminum, and cast-iron metals.

4.4.5 Nanoceramics Based on Silicon Carbide

Pressurized water reactor (PWR) claddings can be made from complex, three-layer silicon carbide-coated inner tubes. Chemical vapor deposition techniques enable the production of these materials. One of the most important processing characteristics that must be controlled is the surface roughness of substrates such as graphite. The effect on surface roughness was investigated using different types of graphite. The availability of different graphite materials will determine the types of substrates that can be used in combination with certain ingredients [105–107]. In addition, it indicates potential directions for future research and development in the field of boron carbide matrix nanocomposites with improved physical and mechanical properties including microhardness and fracture toughness. We find that for all the microhardness and fracture toughness values given here, it is 10 g/15 s due to its strong dependence on the applied load when measuring these parameters (the numbers are given below). It is calculated as the average of about ten measurements with a relative error of about five percent (calculated as the mean absolute deviation from the mean of the measurements) [108, 109].

4.4.6 Nanoceramics Based on Boron Carbide

Boron carbide, with the approximate chemical formula B4C, is one of the most popular and successful ceramic materials used today. The special combination of mechanical and physical properties of boron carbide, such as its low density, good wear resistance in harsh environments, high hardness, high ballistic resistance, and high elastic modulus, provides a wide range of applications. Notably, boron carbide has the highest hardness-to-density ratio of any hard material used in practical applications [89–92]. However, some engines require materials that combine high strength, strong thermal conductivity, high hardness and adequate dimensional stability under thermal shock. Pure boron carbide products and products cannot be used as parts in these engines. This is due to the relatively low fracture toughness and brittleness of boron carbide. The performance parameters of boron carbide, such as electrical conductivity, flexural strength and sinterability, need to be improved to meet the requirements of modern technology [110].

4.5 Nanoceramics Applications

The characteristics of these nanoscale ceramic materials have recently attracted a lot of attention. The creation of ceramic materials has been the focus of research over the past several decades. Biomedical and storage systems, technology, evaporation, anti-fog, anti-freeze and anti-corrosion, wastewater and water treatment and manufacturing technologies are just a few of these applications [110, 111].

4.5.1 Electronics

By reducing insertion loss and power dissipation, low-loss ceramic components increase overall circuit efficiency, suppressing electrical noise in oscillators [111, 112]. In addition, the radiation efficiency of a dielectric resonator antenna (DRA) can be improved by using a low-loss ceramic material since DRAs do not produce surface waves and have low conductor and dielectric losses [113]. Electronics, industrial raw materials, refractory materials, and other industries are now investigating the potential applications of nanoceramics [98].

Because there are so many industrial uses for electroceramics, we won't cover them all in this chapter. We'll instead concentrate on a couple of them. BaTiO3, a ferroelectric ceramic that is environmentally friendly and free of lead is one of them. This has a lot of potential applications. Barium titanate is the primary ferroelectric in perovskite structures. multilayer electrical components, for example, ceramic capacitors (MLCCs), positive temperature coefficient (PTC) thermistors, electro-optical devices, memory devices, sensors, actuators, and others, to name a few [14, 104].

By utilizing the complementary effects of holmium (Ho) and dysprosium (Dy) modified rare earth elements in BaTiO3 core-shell dielectric ceramics, multilayer ceramic capacitors (MLCCs) are produced [105–107].

Since the annealing temperature used determines the degree of crystallinity, nanostructuring of BN remains difficult. By increasing the degree of crystallinity, increasing temperature will directly affect the nanostructure of the material. In addition, porosity can be added to the material to give it a nanostructure. During the replication process, the nanostructured template can provide porosity. More specifically, while nanostructure at the microscopic level is based on the formation of microporosity in the walls surrounding the mesopores, the described BN nanostructure is expected to be observed at the microscopic level with controlled porosity organization. At the cyanoethylcellulose/barium titanate/antimony films and tin oxide nanocomposite, the improvement in mechanical properties and dielectric constant reached 147.30 at 100 Hz, 30 and 28%, respectively [107]. High dielectric constant polymer crystalline ceramics are frequently employed for a range of applications in the electronics industry [108, 109].

4.5.2 Applications of Biomedical

The need for the creation of new medical materials has increased in recent decades as a result of an increase in life expectancy. The application of materials in healthcare is one of the most fascinating and gratifying fields of materials research. Over the past 50 years, bio-ceramics have undergone significant advancement [114]. Bio-ceramics are ceramics used in biologically related fields. For the production of diagnostic tools, glasses, controlled drug delivery, implants, radiation therapy, artificial organs, medical devices, and fiber optics for endoscopy, among other things, the healthcare industry specifically uses ceramic materials and bioactive glasses [115, 116]. In fact, all materials have recently found use in orthopedic procedures. In addition, ceramic components are used in biomedical devices such as implants, hip prostheses, biomedical sensors, and microsystems [117–119].

For example, femoral heads made of $ZrOn_2$ (in tetragonal form) is stronger and more durable than Al_2O_3, which has better thermal conductivity. Al_2O_3-ZrO_2 ceramic composites have recently been considered for the creation of joint prostheses. Connecting ceramic femoral heads to femoral stems in hip replacement is one application of purely mechanical technology [120]. ZTA or Al_2O_3 are the two materials most commonly used to make ceramics for these purposes. Since the mid-1970s, heads to pins have been connected using a Morse taper, as opposed to earlier technologies using solder, screws and glue [121–123].

4.5.3 Applications of Environmental

One of the most important environmental problems facing the world today is waste disposal. As population and human activity grow, waste production will inevitably increase dramatically. In addition, ceramic adsorbents have a large surface area with moderate pressure drop and good liquid penetration. Regarding wastewater treatment, ceramic microbeads produced by the ionotropic gelation method of sodium alginate is used to create porous membranes from various inorganic colloids. It has excellent potential for further functionalization for water and wastewater treatment due to their surface area, variable pore diameter, and programmable shape [126].

Removal of microorganisms, organics, ions, dyes and other contaminants have been reported to be achieved using functionalized reactive ceramic ultrafiltration membranes [127–129]. Additionally, titanium dioxide thin films and nanoparticles are typically created through a chemical process known as sol-gel. Au/TiO_2 thin nanocrystalline films [139]. For splitting azo dyes [130]. The authors found that the increase in the photocatalytic efficiency of titanium dioxide films caused by the deposition of gold ions exceeds 100%. This improvement is attributed to the attraction of Au particles and holes by photoelectrons in the conduction band [138]. Sulfur-doped TiO_2 compounds are used to protect the environment [140]. For the purpose of this study, the scientists used sol-gel nanoscale films of pure and S-doped TiO_2 to accelerate the degradation of organochlorines in aqueous solutions. The results showed that sulfur-doped TiO_2 films consistently outperformed undoped films in terms of chlorobenzene removal efficiency (97% yield), regardless of the number of layers, irradiation duration, or sulfur doping concentration. The technique of developing hybrid alginate-ceramic membranes also addressed cadmium (II) absorption from wastewater. The findings suggested that wastewater treatment might make use of alginate-modified ceramic membranes with good metal retention efficiency [141].

4.5.4 Applications in Space and Defense

Space shuttle project tiles, rocket nose cones, jet engine blades, aerospace components, etc. are examples of ceramic materials [140–143]. To achieve efficient system integration on space-constrained platforms, engineers focus on minimizing the size of components. The result is more options for more functions and payloads, as well as increased mobility and maneuverability [144]. The mass of an aircraft system is its weight. Weight becomes an important factor when selecting a motion control solution for A&D applications as they often operate in challenging environments. Lighter components help improve portability and ease of transport, while reduced weight is essential for improved mobility, fuel efficiency, and payload capacity. Power refers to the energy requirements of a system, with special emphasis placed on maximizing power requirements to ensure reliable operation. Optimizing energy efficiency results in missions lasting longer and requiring fewer logistics [144].

4.5.5 Strategic Applications

Ceramic materials, for instance, are essential for the production, distribution, and storage of energy as well as for usage in the nuclear sector as waste immobilization and fuel ceramics [145, 146]. Particularly important parts of motors, filters, and communication equipment include ceramic magnets like hexaferrites [147–149]. Nanomaterials, especially intermetallic nanomaterials and nanoceramics, can be used as coating materials for components that are designed to withstand high temperatures, thermal stress, and corrosion. Nanomaterials can be synthesized using a variety of methods, and any approach can be used, depending on the material being created and the limitations of the method. In addition to improving mechanical or chemical properties, nanomaterial coatings can save on the expensive materials and alloys needed to make such components. Table 4.1 lists some popular forming methods in ceramic processing. Due to their exposure to high temperatures and moisture, intermetallic nanomaterials like Fe3Al, which have good oxidation and corrosion resistance qualities, can be employed in steam turbine blades, turbine casing, and bolts. ZrO_2, a nanoceramic material with a melting point of roughly 2700°C, is another example. Suitable for high-temperature applications, industrial furnaces, rocket engines and gas jet turbines.

4.6 Conclusion and Prospective Future of Nanoceramics

Promising materials for the production of ceramic parts include oxides of aluminum, boron, titanium, silicon, zirconium, and their nitride derivatives. Several synthetic processes for producing ceramic components have been identified. The most popular method for producing ceramic powders in large quantities and at low cost is through solid-state processes. The development of advanced technologies is inextricably linked with the development of modern nano- and biotechnologies materials for various applications based on nanostructured metal oxides. Given the size of today's high-tech market, this is a critical responsibility. Based on scientific assessment from the literature, we can conclude that biomaterials science is a newly created scientific discipline whose main goal is problem-solving. The creation of medical biomaterials based on oxide nanocomposites. scientific method to solve this problem creates new methods based on which biomaterials with the required chemical and physical properties will be developed in the future for the production of powder precursors.

In addition, ceramic precursors of high purity and required qualities have been prepared using sol-gel, CVD, and polymer pyrolysis processes. However, 3Y-TZP has problems with uneven grain growth, uneven grain distribution, and heat aging, as evidenced by ZTA powders produced by combining multiphase suspension with the sol-gel process. The crystalline phase transition, densification and porosity of the ceramic are primarily stimulated by the sintering temperature used during production. These parameters control the hardness, corrosion resistance, excellent conductivity and microwave absorption of the ceramic. Ceramics of the desired shape, polish, strength and size are produced mechanically. However, the cost of typical machining ranges from 60% to 80% of the cost of making ceramics. This article examines the use of thermal molding or free-form manufacturing as an alternative to producing defect-free final ceramics. In particular, 3D printing is receiving greater interest as it has significant potential for widespread commercialization.

Ceramics are a desirable material due to their many applications in electrical engineering, environmental engineering and biomedicine. Because ceramics have higher breakdown voltages and lower switching losses than other materials used in power electronics, they have interesting potential for use in high-temperature cycling. They additionally provide an isolating effect between the circuit and the heatsinks, a thermal path to the heatsink, and a conductive path with comparatively low resistance. Ceramic materials would be the best and most attractive options from this perspective. It should be emphasized that fifty years after their invention, silicon nitride (Si3N4) ceramics have not achieved their full potential. The other novel nanoceramic materials will enhance their functionality and make them more appropriate for a range of applications that are unlikely to be accomplished with alternative materials. The improvement of conventional techniques for processing ceramic materials must be quickened, albeit right now the focus should be on creating novel ceramic materials for fresh uses.

References

1. Sharma, S.D., Sowntharya, L., Kar, K.K., Polymer-based composite structures: Processing and applications, in: *Composite materials: processing, applications, characterizations*, pp. 1–36, Springer, Berlin Heidelberg, 2016.
2. Liu, Y., Ma, H., Hsiao, B.S., Chu, B., Tsou, A.H., Improvement of meltdown temperature of lithium-ion battery separator using electrospun polyethersulfone membranes. *Polymer (Guildf)*, 107, 163–169, 2016.
3. Joshi, M. and Chatterjee, U., Polymer nanocomposite: an advanced material for aerospace applications, in: *Advanced composite materials for aerospace engineering*, pp. 241–264, Woodhead Publishing, 2016.
4. Ma, L., Hendrickson, K.E., Wei, S., Archer, L.A., Nanomaterials: Science and applications in the lithium-sulfur battery. *Nano Today*, 10, 315–338, 2015.
5. Šutka, A. and Gross, K.A., Spinel ferrite oxide semiconductor gas sensors. *Sens. Actuators, B*, 222, 95–105, 2016.
6. Tao, Z., Yan, L., Qiao, J., Wang, B., Zhang, L., Zhang, J., A review of advanced proton-conducting materials for hydrogen separation. *Prog. Mater Sci.*, 74, 1–50, 2015.
7. Srivastava, V., Gusain, D., Sharma, Y.C., Critical Review on the Toxicity of Some Widely Used Engineered Nanoparticles. *Ind. Eng. Chem. Res.*, 54, 6209–6233, 2015.
8. Agrawal, V. and Kapoor, S., Effect of fiber reinforcement on microleakage of class II cavities restored with a novel G-aenial posterior composite, silorane composite, and nanohybrid composite: an in vitro comparative study. *J. Investig. Clin. Dent.*, 8, 2017.
9. Ghoreishi, S.M., Karamali, E., Khoobi, A., Enhessari, M., Preparation of a manganese titanate nanosensor: Application in electrochemical studies of captopril in the presence of para-aminobenzoic acid. *Anal. Biochem.*, 487, 49–58, 2015.
10. Bhat, K.A., Leo Prakash, P., Manoharan, N., Lakshmibai, A., Sangeetha, D., Fabrication of polymethyl methacrylate/polysulfone/nanoceramic composites for orthopedic applications. *J. Appl. Polym. Sci.*, 127, 2764–2775, 2013.
11. He, J., Li, X., Su, D., Ji, H., Wang, X.J., Ultra-low thermal conductivity and high strength of aerogels/fibrous ceramic composites. *J. Eur. Ceram. Soc.*, 36, 1487–1493, 2016.
12. Kumar, N. and Dixit, A., *Nanotechnology for defence applications*, pp. 155–203, Springer International Publishing, Cham, 2019.
13. Dudzevičiūtė, G., Bekesiene, S., Meidute-Kavaliauskiene, I., Ševčenko-Kozlovska, G., An assessment of the relationship between defence expenditure and sustainable development in the Baltic countries. *Sustainability*, 13, 12, 6916, 2021.
14. Brow, R.K. and Schmitt, M.L., A survey of energy and environmental applications of glass. *J. Eur. Ceram. Soc.*, 29, 1193–1201, 2009.

15. Bamberger, L., International institute for strategic studies: the military balance 2017. *2017 SIRIUS-Zeitschrift Strategische Analysen*, 1, 3, 309–310, 2017.
16. Saiz, A., Statistical model for intensities in the rotational spectra of linear and non-linear molecules. *Wuli Huaxue Xuebao/ Acta Phys. - Chim. Sin.*, 27, 1053–1060, 2011.
17. Fontanel, J., New global and national military expenditure series developed by SIPRI, Economie de la sécurité internationale, 2019, October.
18. Bhaduri, S. and Bhaduri, S.B., Recent developments in ceramic nanocomposites. *Jom*, 50, 44–51, 1998.
19. Ahmad, I., Yazdani, B., Zhu, Y., Recent advances on carbon nanotubes and graphene reinforced ceramics nanocomposites. *Nanomaterials*, 5, 90–114, 2014.
20. Peer review statement. *IOP Conf. Ser. Mater. Sci. Eng.*, 20, 011002–01102, 2011.
21. Yang, Y. and Wu, Y., Tape-casted transparent alumina ceramic wafers. *J. Mater. Res.*, 29, 2312–2317, 2014.
22. Jingxian, Z., Dongliang, J., Weisensel, L., Greil, P., Binary solvent mixture for tape casting of TiO2 sheets. *J. Eur. Ceram. Soc.*, 24, 147–155, 2004.
23. Madou, M.J., *Fundamentals of microfabrication: the science of Sminiaturization*, CRC Press, 2018.
24. Ruys, A., Processing, structure, and properties of alumina ceramics, in: *Alumina Ceram.*, pp. 71–121, 2019.
25. Kuvarega, A.T., Shivani, B., Mamba, B.B., Hybrid ceramic materials for environmental applications, in: *Sol-gel Based Nanoceramic Mater. Prep. Prop. Appl.*, pp. 275–297, 2016.
26. Wang, C. and Zeng, X., Study of laser carving three-dimensional structures on ceramics: Quality controlling and mechanisms. *Opt. Laser Technol.*, 39, 1400–1405, 2007.
27. Kelly, J.R. and Denry, I., Stabilized zirconia as a structural ceramic: An overview. *Dent. Mater.*, 24, 289–298, 2008.
28. Fontanel, J., Is the SIPRI estimate of military expenditure a reliable indicator of the power of states? Defence economics, 2022, February.
29. Wederni, A., Daza, J., Ben Mbarek, W., Saurina, J., Escoda, L., Suñol, J.J., Crystal structure and properties of Heusler alloys: a comprehensive review. *Metals*, 14, 6, 688, 2024.
30. Mæland, D., Suciu, C., Wærnhus, I., Hoffmann, A.C., Sintering of 4YSZ (ZrO2 + 4 mol% Y2O3) nanoceramics for solid oxide fuel cells (SOFCs), their structure and ionic conductivity. *J. Eur. Ceram. Soc.*, 29, 2537–2547, 2009.
31. Hanaor, D.A.H. and Sorrell, C.C., Review of the anatase to rutile phase transformation. *J. Mater. Sci.*, 46, 855–874, 2011.
32. Heo, U. and Ye, M., Defense spending and economic growth around the globe: the direct and indirect link. *Int. Interact.*, 42, 5, 774–796, 2016.

33. Bhaduri, S.B., Bhaduri, S., Huang, J.-G., Auto Ignition Synthesis and Microwave Sintering of ZrO2-CaO, in: *23nd Annu. Conf. Compos. Adv. Ceram. Mater. Struct. B Ceram. Eng. Sci. Proceedings, Vol. 20 Issue 4*, pp. 227–234, 2008.
34. Siqueira, J.R.R., Simões, A.Z., Stojanovic, B.D., Paiva-Santos, C.O., Santos, L.P.S., Longo, E., Varela, J.A., Influence of milling time on mechanically assisted synthesis of Pb0.91Ca0.1TiO3 powders. *Ceram. Int.*, 33, 937–941, 2007.
35. Yesilyurt, F. and Yesilyurt, M.E., Meta-analysis, military expenditures and growth. *J. Peace Res.*, 56, 3, 352–363, 2019.
36. Eom, J.H., Kim, Y.W., Raju, S., Processing and properties of macroporous silicon carbide ceramics: A review. *J. Asian Ceram. Soc.*, 1, 220–242, 2013.
37. Rooney, B., Johnson, G., Priebe, M., How does defense spending affect economic growth? Rand Corp, Santa Monica CA, 2021.
38. Liu, S., Tao, W., Li, J., Yang, Z., Liu, F., Study on the formation process of Al2O3-TiO 2 composite powders. *Powder Technol.*, 155, 187–192, 2005.
39. Buscaglia, V., Caracciolo, F., Leoni, M., Nanni, P., Viviani, M., Lemaitre, J., Synthesis, sintering and expansion of Al0.8Mg0.6Ti1.6O5: A low-thermal-expansion material resistant to thermal decomposition. *J. Mater. Sci.*, 32, 6525–6531, 1997.
40. Buscaglia, V., Alvazzi Delfrate, M., Leoni, M., Bottino, C., Nanni, P., The effect of MgAl2O4 on the formation kinetics of Al2TiO5 from Al2O3 and TiO2 fine powders. *J. Mater. Sci.*, 31, 1715–1724, 1996.
41. Lashkari, S. and Ebadzadeh, T., Microwave sintering of Al2(1-x)MgxTi (1+x) O5 ceramics obtained from mixture of nano-sized oxide powders. *Ceram. Int.*, 40, 12669–12674, 2014.
42. Mudinepalli, V.R., Song, S., Li, J., Murty, B.S., Effect of grain size on the electrical properties of high dense BPT nanocrystalline ferroelectric ceramics. *Ceram. Int.*, 40, 1781–1788, 2014.
43. Mudinepalli, V.R., Song, S., Li, J., Murty, B.S., A comparative study of structural and electrical properties of Ba 0.8Pb0.2TiO3 nanocrystalline ceramics prepared by microwave and spark plasma sintering. *Mater. Chem. Phys.*, 142, 686–691, 2013.
44. Barg, S., de Moraes, E.G., Koch, D., Grathwohl, G., New cellular ceramics from high alkane phase emulsified suspensions (HAPES). *J. Eur. Ceram. Soc.*, 29, 2439–2446, 2009.
45. Akartuna, I., Studart, A.R., Tervoort, E., Gauckler, L.J., Macroporous ceramics from particle-stabilized emulsions. *Adv. Mater.*, 20, 4714–4718, 2008.
46. Kroll, S., Soltmann, C., Koch, D., Kegler, P., Kunzmann, A., Rezwan, K., Colored ceramic foams with tailored pore size and surface functionalization used as spawning plates for fish breeding. *Ceram. Int.*, 40, 15763–15773, 2014.
47. Du, Y., Hedayat, N., Panthi, D., Ilkhani, H., Emley, B.J., Woodson, T., Freeze-casting for the fabrication of solid oxide fuel cells: A review. *Materialia*, 1, 198–210, 2018.

48. Shinoda, Y., Nagano, T., Gu, H., Wakai, F., Superplasticity of silicon carbide. *J. Am. Ceram. Soc.*, 82, 2916–2918, 1999.
49. Sun, G., Chen, D., Zhu, G., Li, Q., Lightweight hybrid materials and structures for energy absorption: a state-of-the-art review and outlook. *Thin Walled Struct.*, 172, 108760, 2022.
50. Khalifa, H.E., Deck, C.P., Gutierrez, O., Jacobsen, G.M., Back, C.A., Fabrication and characterization of joined silicon carbide cylindrical components for nuclear applications. *J. Nucl. Mater.*, 457, 227–240, 2015.
51. Lorrette, C., Réau, A., Briottet, L., Mechanical properties of nanostructured silicon carbide consolidated by spark plasma sintering. *J. Eur. Ceram. Soc.*, 33, 147–156, 2013.
52. Fleischer, J., Teti, R., Lanza, G., Mativenga, P., Moehring, H.C., Caggiano, A., Composite materials parts manufacturing. *CIRP Ann.*, 67, 2, 603–626, 2018.
53. Kumar, R., Singh, K., Chakravarty, D., Chowdhury, A., Attaining near-theoretical densification in nanograined pyrochlore La2Zr2O7 (LZ) ceramic at 1150 °C by spark plasma sintering. *Scr. Mater.*, 117, 37–40, 2016.
54. Suyama, S., Kameda, T., Itoh, Y., Development of high-strength reaction-sintered silicon carbide. *Diamond Relat. Mater.*, 12, 1201–1204, 2003.
55. Rubio, E.M., Blanco, D., Marín, M.M., Carou, D., Analysis of the latest trends in hybrid components of lightweight materials for structural uses. *Procedia Manuf.*, 41, 1047–1054, 2019.
56. Antou, G., Pradeilles, N., Gendre, M., Maître, A., New approach of the evolution of densification mechanisms during Spark Plasma Sintering: Application to zirconium (oxy-)carbide ceramics. *Scr. Mater.*, 101, 103–106, 2015.
57. Zhang, W., Lu, T., Ma, B., Wei, N., Lu, Z., Li, F., Guan, Y., Chen, X., Liu, W., Qi, L., Improvement of optical properties of Nd:YAG transparent ceramics by post-annealing and post hot isostatic pressing. *Opt. Mater. (Amst)*, 35, 2405–2410, 2013.
58. Sun, Y., Chu, Y., Wu, W., Xiao, H., Nanocellulose-based lightweight porous materials: a review. *Carbohydr. Polym.*, 255, 117489, 2021.
59. Chen, F., Zhang, F., Wang, J., Zhang, H., Tian, R., Zhang, J., Zhang, Z., Sun, F., Wang, S., Microstructure and optical properties of transparent aluminum oxynitride ceramics by hot isostatic pressing. *Scr. Mater.*, 81, 20–23, 2014.
60. Ganesan, K., Dennstedt, A., Barowski, A., Ratke, L., Design of aerogels, cryogels and xerogels of cellulose with hierarchical porous structures. *Mater. Des.*, 92, 345–355, 2016.
61. Roh, M.H., Kim, Y.W., Kim, W., Nishimura, T., Seo, W.S., Ko, S.I., Lee, S.J., Effect of aluminum nitride-scandia content on the microstructural and mechanical properties of sintered silicon carbide ceramics. *Met. Mater. Int.*, 15, 937–941, 2009.
62. Neuman, E.W., Hilmas, G.E., Fahrenholtz, W.G., Mechanical behavior of zirconium diboride-silicon carbide ceramics at elevated temperature in air. *J. Eur. Ceram. Soc.*, 33, 2889–2899, 2013.

63. Neuman, E.W., Hilmas, G.E., Fahrenholtz, W.G., Mechanical behavior of zirconium diboride-silicon carbide-boron carbide ceramics up to 2200°C. *J. Eur. Ceram. Soc.*, 35, 463–476, 2015.
64. Zimmermann, J.W., Hilmas, G.E., Fahrenholtz, W.G., Dinwiddie, R.B., Porter, W.D., Wang, H., Thermophysical properties of ZrB2 and ZrB2-SiC ceramics. *J. Am. Ceram. Soc.*, 91, 1405–1411, 2008.
65. Antadze, M., Chedia, R., Tsagareishvili, O., Mikeladze, A., Gacheciladze, A., Margiev, B., Gabunia, D., Tsuladze, T., Khantadze, D., Metal-ceramics based on nanostructured boron carbide. *Solid State Sci.*, 14, 1725–1728, 2012.
66. Ebrahimi, S., Heydari, M.S., Baharvandi, H.R., Ehsani, N., Effect of iron on the wetting, sintering ability, and the physical and mechanical properties of boron carbide composites: A review. *Int. J. Refract. Met. Hard Mater*, 57, 78–92, 2016.
67. Baharvandi, H.R. and Hadian, A.M., Investigation on addition of kaolinite on sintering behavior and mechanical properties of B 4C. *J. Mater. Eng. Perform.*, 18, 433–437, 2009.
68. Gosset, D. and Provot, B., Boron carbide as a potential inert matrix: an evaluation. *Prog. Nucl. Energy*, 38, 263–266, 2001.
69. Costa, C.M., Costa, P., Lanceros-Mendez, S., Overview on lightweight, multifunctional materials, in: *Advanced lightweight multifunctional materials*, pp. 1–24, Woodhead Publishing, 2021.
70. Bharadwaj, S.R., Varma, S., Wani, B.N., Electroceramics for Fuel Cells, Batteries and Sensors, in: *Funct. Mater. Prep. Process. Appl.*, pp. 639–674, 2011.
71. Taub, A.I. and Luo, A.A., Advanced lightweight materials and manufacturing processes for automotive applications. *MRS Bull.*, 40, 12, 1045–1054, 2015.
72. Hou, X. (Ed.), *Design, fabrication, properties and applications of smart and advanced materials*, CRC Press, 2016.
73. Zheng, T. and Wu, J., Quenched bismuth ferrite-barium titanate lead-free piezoelectric ceramics. *J. Alloys Compd.*, 676, 505–512, 2016.
74. Sharma, P., Kumar, P., Kundu, R.S., Juneja, J.K., Ahlawat, N., Punia, R., Structural and dielectric properties of substituted barium titanate ceramics for capacitor applications. *Ceram. Int.*, 41, 13425–13432, 2015.
75. Kambale, K.R., Kulkarni, A.R., Venkataramani, N., Grain growth kinetics of barium titanate synthesized using conventional solid state reaction route. *Ceram. Int.*, 40, 667–673, 2014.
76. Wei, B., Wang, P., Yang, T.H., Dai, G.L., Jiang, L.Z., Wen, Y., Effects of friction variability on isolation performance of rolling-spring systems. *J. Cent. South Univ.*, 23, 233–239, 2016.
77. Jia, C., Shao, Z., Fan, H., Feng, R., Wang, F., Wang, W., Wang, J., Zhang, D., Lv, Y., Barium titanate as a filler for improving the dielectric property of cyanoethyl cellulose/antimony tin oxide nanocomposite films. *Compos. Part A Appl. Sci. Manuf.*, 86, 1–8, 2016.
78. Kumar Patel, P. and Yadav, K.L., Study of barrier layer capacitance effect in lead free Ba 0.95Sr0.05(Fe0.5Nb0.5)O 3-BaZr0.1Ti0.9O3 ceramics. *Phys. B Condens. Matter*, 452, 136–141, 2014.

79. Scott, R.I., Thomas, M., Hampson, C., Development of low cost, high performance Ba(Zn1/3Nb2/3O3) based materials for microwave resonator applications. *J. Eur. Ceram. Soc.*, 23, 2467–2471, 2003.
80. Puli, V.S., Pradhan, D.K., Riggs, B.C., Chrisey, D.B., Katiyar, R.S., Investigations on structure, ferroelectric, piezoelectric and energy storage properties of barium calcium titanate (BCT) ceramics. *J. Alloys Compd.*, 584, 369–373, 2014.
81. Liu, G., Fan, H., Dong, G., Shi, J., Chang, Q., Enhanced energy storage and dielectric properties of Bi0.487Na0.427K0.06Ba0.026TiO3-xCeO2 antiferroelectric ceramics. *J. Alloys Compd.*, 664, 632–638, 2016.
82. Li, L., Yu, X., Cai, H., Liao, Q., Han, Y., Gao, Z., Preparation and dielectric properties of BaCu(B2O 5)-doped SrTiO3-based ceramics for energy storage. *Mater. Sci. Eng., B*, 178, 1509–1514, 2013.
83. Ivanov, M., Kopylov, Y., Kravchenko, V., Li, J., Medvedev, A., Pan, Y., Highly transparent ytterbium doped yttrium lanthanum oxide ceramics. *J. Rare Earths*, 32, 254–258, 2014.
84. Sujith, S.S., Arun Kumar, S.L., Mangalaraja, R.V., Peer Mohamed, A., Ananthakumar, S., Porous to dense LaPO4 sintered ceramics for advanced refractories. *Ceram. Int.*, 40, 15121–15129, 2014.
85. Shimizu, T., Matsuura, K., Furue, H., Matsuzak, K., Thermal conductivity of high porosity alumina refractory bricks made by a slurry gelation and foaming method. *J. Eur. Ceram. Soc.*, 33, 3429–3435, 2013.
86. Lu, T.J. and Fleck, N.A., The thermal shock resistance of solids. *Acta Mater.*, 46, 4755–4768, 1998.
87. Wang, B.L., Mai, Y.W., Zhang, X.H., Thermal shock resistance of functionally graded materials. *Acta Mater.*, 52, 4961–4972, 2004.
88. Amrane, B., Ouedraogo, E., Mamen, B., Djaknoun, S., Mesrati, N., Experimental study of the thermo-mechanical behaviour of alumina-silicate refractory materials based on a mixture of Algerian kaolinitic clays. *Ceram. Int.*, 37, 3217–3227, 2011.
89. Sadik, C., El Amrani, I.E., Albizane, A., Processing and characterization of alumina-mullite ceramics. *J. Asian Ceram. Soc.*, 2, 310–316, 2014.
90. Pieta, A., Bućko, M.M., Januś, M., Lis, J., Jonas, S., Calcium hexaaluminate synthesis and its influence on the properties of CA2-Al2O3-based refractories. *J. Eur. Ceram. Soc.*, 35, 4567–4571, 2015.
91. Soltan, A.M., Pöllmann, H., Kaden, R., König, A., Abd EL-Raoof, F., Eltaher, M., Serry, M., Degradation of aluminosilicate refractories: An integrated approach. *J. Eur. Ceram. Soc.*, 35, 4573–4592, 2015.
92. Zhi, J., Cui, H., Chen, A., Xie, Y., Huang, F., Efficient highly flexible dye sensitized solar cells of three dimensional graphene decorated titanium dioxide nanoparticles on plastic substrate. *J. Power Sources*, 281, 404–410, 2015.
93. Zhu, B., Zhu, Y., Li, X., Zhao, F., Effect of ceramic bonding phases on the thermo-mechanical properties of Al2O3-C refractories. *Ceram. Int.*, 39, 6069–6076, 2013.

94. Howatt, G.N., Breckenridge, R.G., Brownlow, J.M., Fabrication of Thin Ceramic Sheets for Capacitors. *J. Am. Ceram. Soc.*, 30, 237–242, 1947.
95. Ma, C., Tang, F., Zhu, J., Du, M., Yuan, X., Yu, Y., Wang, K., Wen, Z., Zhang, J., Long, J., Guo, W., Cao, Y., Spectral and Laser Properties of Yb:LuAG Transparent Ceramics Fabricated by Tape Casting Method. *J. Am. Ceram. Soc.*, 99, 3267–3272, 2016.
96. Kumar, A., Chaudhary, S., Chandra, S., Effect of point defects and lattice distortions on the structural, electronic, and magnetic properties of Co2MnAl Heusler alloy. *Phys. Rev. Mater.*, 8, 034405, 2024.
97. Friák, M., Zelený, M., Mazalová, M., Miháliková, I., Turek, I., Kaštil, J., *et al.*, The impact of disorder on the 4O-martensite of Ni-Mn-Sn Heusler alloy. *Intermetallics*, 151, 107708, 2022.
98. Kirat, G. and Ali Aksan, M., Investigation of martensitic transformation and magnetoresistance properties of Cu-substituted Ni-Mn-Sn-B melt spun ribbons. *J. Magn. Magn. Mater.*, 529, 167858, 2021.
99. Ruys, A., Processing, structure, and properties of alumina ceramics, in: *Alumina Ceram*, pp. 71–121, 2019.
100. Monash, P., Pugazhenthi, G., Saravanan, P., Various fabrication methods of porous ceramic supports for membrane applications. *Rev. Chem. Eng.*, 29, 357–383, 2013.
101. Asemi, M. and Ghanaatshoar, M., Conductivity improvement of CuCrO2 nanoparticles by Zn doping and their application in solid-state dye-sensitized solar cells. *Ceram. Int.*, 42, 6664–6672, 2016.
102. Liang, M.S., Khaw, C.C., Liu, C.C., Chin, S.P., Wang, J., Li, H., Synthesis and characterisation of thin-film TiO2 dye-sensitised solar cell. *Ceram. Int.*, 39, 1519–1523, 2013.
103. Sujinnapram, S. and Moungsrijun, S., Additive SnO2-ZnO Composite Photoanode for Improvement of Power Conversion Efficiency in Dye-sensitized Solar Cell. *Procedia Manuf.*, 2, 108–112, 2015.
104. Zhang, W. and Xu, J., Advanced lightweight materials for Automobiles: a review. *Mater. Des.*, 221, 110994, 2022.
105. Kawakami, S., Myojin, T., Cho, H.S., Hatamachi, T., Gokon, N., Kodama, T., Thermochemical two-step water splitting cycle using Ni-ferrite and CeO 2 coated ceramic foam devices by concentrated Xe-light radiation. *Energy Procedia*, 49, 1980–1989, 2014.
106. Raso, R.A., Stepuk, A., Mohn, D., Paunescu, D., Koehler, F.M., Stark, W.J., Regenerable cerium oxide based odor adsorber for indoor air purification from acidic volatile organic compounds. *Appl. Catal. B Environ.*, 147, 965–972, 2014.
107. Faruk, O., Tjong, J., Sain, M. (Eds.), *Lightweight and sustainable materials for automotive applications*, CRC Press, Boca Raton, 2017.
108. Amin, M.T., Alazba, A.A., Manzoor, U., A review of removal of pollutants from water/wastewater using different types of nanomaterials. *Adv. Mater. Sci. Eng.*, 2014, 1–24, 2014.

109. Ohira, T. and Yamamoto, O., Correlation between antibacterial activity and crystallite size on ceramics. *Chem. Eng. Sci.*, 68, 355–361, 2012.
110. Raveendra, R.S., Prashanth, P.A., Hari Krishna, R., Bhagya, N.P., Nagabhushana, B.M., Raja Naika, H., Lingaraju, K., Nagabhushana, H., Daruka Prasad, B., Synthesis, structural characterization of nano ZnTiO 3 ceramic: An effective azo dye adsorbent and antibacterial agent. *J. Asian Ceram. Soc.*, 2, 357–365, 2014.
111. Sabbani, S., Gallego-Perez, D., Nagy, A., James Waldman, W., Hansford, D., Dutta, P.K., Synthesis of silver-zeolite films on micropatterned porous alumina and its application as an antimicrobial substrate. *Microporous Mesoporous Mater.*, 135, 131–136, 2010.
112. Miedaner, M.M., Weerasooriya, R., Tobschall, H.J., Chapter 17 1-pK modeling strategies for the adsorption of some trace elements onto gibbsite, in: *Interface Sci. Technol.*, vol. 11, pp. 469–490, 2006.
113. Wijnja, H. and Schulthess, C.P., Vibrational spectroscopy study of selenate and sulfate adsorption mechanisms on Fe and Al (Hydr)oxide surfaces. *J. Colloid Interface Sci.*, 229, 286–297, 2000.
114. Kumar, E., Bhatnagar, A., Hogland, W., Marques, M., Sillanpää, M., Interaction of anionic pollutants with Al-based adsorbents in aqueous media - A review. *Chem. Eng. J.*, 241, 443–456, 2014.
115. Li, Y.H., Wang, S., Cao, A., Zhao, D., Zhang, X., Xu, C., Luan, Z., Ruan, D., Liang, J., Wu, D., Wei, B., Adsorption of fluoride from water by amorphous alumina supported on carbon nanotubes. *Chem. Phys. Lett.*, 350, 412–416, 2001.
116. Das, S. and Jayaraman, V., Erratum: "SnO2: A comprehensive review on structures and gas sensors" (Progress in Materials Science (2014) 66 (112-255)). *Prog. Mater. Sci.*, 67, 161, 2015.
117. Feng, Q., Li, X., Wang, J., Gaskov, A.M., Reduced graphene oxide (rGO) encapsulated Co3O4 composite nanofibers for highly selective ammonia sensors. *Sens. Actuators, B*, 222, 864–870, 2016.
118. Li, X., Zhao, Y., Wang, X., Wang, J., Gaskov, A.M., Akbar, S.A., Reduced graphene oxide (rGO) decorated TiO2 microspheres for selective room-temperature gas sensors. *Sens. Actuators, B*, 230, 330–336, 2016.
119. Galstyan, V., Comini, E., Baratto, C., Faglia, G., Sberveglieri, G., Nanostructured ZnO chemical gas sensors. *Ceram. Int.*, 41, 14239–14244, 2015.
120. Cernat, A., Tertiş, M., Săndulescu, R., Bedioui, F., Cristea, A., Cristea, C., Electrochemical sensors based on carbon nanomaterials for acetaminophen detection: A review. *Anal. Chim. Acta*, 886, 16–28, 2015.
121. Arvand, M. and Gholizadeh, T.M., Simultaneous voltammetric determination of tyrosine and paracetamol using a carbon nanotube-graphene nanosheet nanocomposite modified electrode in human blood serum and pharmaceuticals. *Colloids Surf., B*, 103, 84–93, 2013.

122. Banerjee, S., Dionysiou, D.D., Pillai, S.C., Self-cleaning applications of TiO2 by photo-induced hydrophilicity and photocatalysis. *Appl. Catal. B Environ.*, 176–177, 396–428, 2015.
123. Houmard, M., Berthomé, G., Joud, J.C., Langlet, M., Enhanced cleanability of super-hydrophilic TiO2-SiO2 composite surfaces prepared via a sol-gel route. *Surf. Sci.*, 605, 456–462, 2011.
124. Zhu, W., Feng, X., Feng, L., Jiang, L., UV-Manipulated wettability between superhydrophobicity and superhydrophilicity on a transparent and conductive SnO2 nanorod film. *Chem. Commun.*, 26, 2753–2755, 2006.
125. André, R., Natalio, F., Tahir, M.N., Berger, R., Tremel, W., Self-cleaning antimicrobial surfaces by bio-enabled growth of SnO 2 coatings on glass. *Nanoscale*, 5, 3447–3456, 2013.
126. Sun, R.D., Nakajima, A., Fujishima, A., Watanabe, T., Hashimoto, K., Photoinduced surface wettability conversion of ZnO and TiO2 Thin Films. *J. Phys. Chem. B*, 105, 1984–1990, 2001.
127. Wang, P., Xu, N., Shi, J., A pilot study of the treatment of waste rolling emulsion using zirconia microfiltration membranes. *J. Membr. Sci.*, 173, 159–166, 2000.
128. Kwan, S.M. and Yeung, K.L., Zeolite micro fuel cell. *Chem. Commun.*, 31, 3631–3633, 2008.
129. Du, L., Liu, W., Hu, S., Wang, Y., Yang, J., Preparation and photocatalytic properties of macroporous honeycomb alumina ceramics used for water purification. *J. Eur. Ceram. Soc.*, 34, 731–738, 2014.
130. Kocakuşakoğlu, A., Dağlar, M., Konyar, M., Yatmaz, H.C., Öztürk, K., Photocatalytic activity of reticulated ZnO porous ceramics in degradation of azo dye molecules. *J. Eur. Ceram. Soc.*, 35, 2845–2853, 2015.
131. Athanasekou, C.P., Morales-Torres, S., Likodimos, V., Romanos, G.E., Pastrana-Martinez, L.M., Falaras, P., Faria, J.L., Figueiredo, J.L., Silva, A.M.T., Prototype composite membranes of partially reduced graphene oxide/TiO2 for photocatalytic ultrafiltration water treatment under visible light. *Appl. Catal. B Environ.*, 158–159, 361–372, 2014.
132. Makwana, N.M., Hazael, R., McMillan, P.F., Darr, J.A., Photocatalytic water disinfection by simple and low-cost monolithic and heterojunction ceramic wafers. *Photochem. Photobiol. Sci.*, 14, 1190–1196, 2015.
133. Kujawa, J., Cerneaux, S., Kujawski, W., Removal of hazardous volatile organic compounds from water by vacuum pervaporation with hydrophobic ceramic membranes. *J. Membr. Sci.*, 474, 11–19, 2015.
134. Çam, G. and İpekoğlu, G., Recent developments in joining of aluminum alloys. *Int. J. Adv. Manuf. Technol.*, 91, 5, 1851–1866, 2017.
135. Mallick, P.K. (Ed.), *Materials, design and manufacturing for lightweight vehicles*, Woodhead Publishing, 2020.
136. Carter, C.B. and Norton, M.G., *Ceramic materials: Science and engineering*, vol. 716, pp. 1–766, Springer, 2007.
137. Rahaman, M.N., *Ceramic processing*, pp. 1–475, CRC Press, 2017.

138. Garcia-Valles, M., Avila, G., Martinez, S., Terradas, R., Nogués, J.M., Acoustic barriers obtained from industrial wastes. *Chemosphere*, 72, 1098–1102, 2008.
139. Luo, A.A., Sachdev, A.K., Apelian, D., Alloy development and process innovations for light metals casting. *J. Mater. Process. Technol.*, 306, 15–16, 117606, 2022.
140. Andreola, F., Barbieri, L., Lancellotti, I., Leonelli, C., Manfredini, T., Recycling of industrial wastes in ceramic manufacturing: State of art and glass case studies. *Ceram. Int.*, 42, 13333–13338, 2016.
141. Shang, R., Goulas, A., Tang, C.Y., de Frias Serra, X., Rietveld, L.C., Heijman, S.G.J., Atmospheric pressure atomic layer deposition for tight ceramic nanofiltration membranes: Synthesis and application in water purification. *J. Membr. Sci.*, 528, 163–170, 2017.
142. Leonardi, S.A., Zanuttini, M.A., Miró, E.E., Milt, V.G., Catalytic paper made from ceramic fibres and natural ulexite. Application to diesel particulate removal. *Chem. Eng. J.*, 317, 394–403, 2017.
143. Yang, H., Liu, P., Yan, F., Lin, Y., Wang, T., A novel lead-free ceramic with layered structure for high energy storage applications. *J. Alloys Compd.*, 773, 244–249, 2019.
144. Sung, P.J., Chang, C.Y., Chen, L.Y., Kao, K.H., Su, C.J., Liao, T.H., Fang, C.C., Wang, C.J., Hong, T.C., Jao, C.Y., Hsu, H.S., Luo, S.X., Wang, Y.S., Huang, H.F., Li, J.H., Huang, Y.C., Hsueh, F.K., Wu, C.T., Huang, Y.M., Hou, F.J., Luo, G.L., Huang, Y.C., Shen, Y.L., Ma, W.C.Y., Huang, K.P., Lin, K.L., Samukawa, S., Li, Y., Huang, G.W., Lee, Y.J., Li, J.Y., Wu, W.F., Shieh, J.M., Chao, T.S., Yeh, W.K., Wang, Y.H., Voltage Transfer Characteristic Matching by Different Nanosheet Layer Numbers of Vertically Stacked Junctionless CMOS Inverter for SoP/3D-ICs applications. *Tech. Dig. - Int. Electron Devices Meet. IEDM*, 2018-December, pp. 21.4.1–21.4.4, 2018.
145. Xing, C.F., Bao, J., Sun, Y.F., Sun, J.J., Wu, H.T., Ba2BiSbO6: A novel microwave dielectric ceramic with monoclinic structure. *J. Alloys Compd.*, 782, 754–760, 2019.
146. Cao, D., Xu, X., Jiang, S., Corrigendum to "Ultrasound-electrochemistry enhanced flotation and desulphurization for fine coal" [Sep. Purif. Technol. 258 (2021) 117968] (Separation and Purification Technology (2021) 258(P2), (S1383586620324412), (10.1016/j.seppur.2020.117968)). *Sep. Purif. Technol.*, 278, 119585, 2022.
147. Qu, Z., Zhang, Q., He, R., Pei, Y., Fang, D., Characterization of oxidation film on SiC ceramic substrate based on indentation method. *Ceram. Int.*, 43, 4399–4404, 2017.
148. Dai, R.Y., Zhao, X.F., Li, J.J., Chen, R., Luo, Z.L., Yu, L.X., Chen, S.K., Zhang, C.Y., Duan, C.Y., Liu, Y.P., Feng, C.H., Xia, X.M., Li, H., Fu, J., Wang, H.Y., Implication of transcriptional repression in compound C-induced apoptosis in cancer cells. *Cell Death Dis.*, 4, e883–e883, 2013.
149. Holthaus, M.G., Treccani, L., Rezwan, K., Comparison of micropatterning methods for ceramic surfaces. *J. Eur. Ceram. Soc.*, 31, 2809–2817, 2011.

5

Cermet: Cutting Tool Materials for High Speed and Advancements in Nanocoating Technology

Riyadh A. Al-Samarai[1]* and Yarub Al-Douri[2,3,4]

[1]*Electromechanical Engineering Department, Engineering College, University of Samarra, Samarra, Iraq*
[2]*College of Health and Medical Techniques, Al-Bayan University, Baghdad, Iraq*
[3]*Department of Mechanical Engineering, Faculty of Engineering, Piri Reis University, Istanbul, Turkey*
[4]*Nanotechnology and Catalysis Research Centre, University of Malaya, Kuala Lumpur, Malaysia*

Abstract
High-Speed Machining (HSM) is a new development in machining technology. HSM is a metal-cutting method that prioritizes high feed rates and speeds to increase productivity and improve surface quality. HSM is widely used in the production of dies and molds, precision components, automotive, optical, micromechanical, and household appliances, among other industries. The main purpose of this chapter is to introduce high-speed machining technology. An introduction to HSM covers its definitions, historical evolution, and differences from traditional processing. What follows is an examination of the instrumentation and management system in HSM. The chapter concludes with a summary and future scope.

***Keywords*:** High-speed machining (HSM), metal-cutting, feed rates, surface quality, precision components, dies and molds, instrumentation, manufacturing technology

**Corresponding author*: riyadh2824@yahoo.com

Inamuddin, Tariq Altalhi and Mohammad Abu Jafar Mazumder (eds.) Nanostructured Hard Materials: Industrial Applications, (175–206) © 2026 Scrivener Publishing LLC

5.1 Introduction

A sophisticated and newly developed processing method called high-speed machining (HSM) is widely used to produce complex parts with increased productivity, better quality, environmental friendliness, and cost-effectiveness. HSM was first created for the processing of aluminum and rocket and aircraft components based on its alloys. With continuous advances in HSM [1], newly introduced machine tools can machine hard metals, difficult-to-cut materials, complex 3D geometries, and micro/nano features with great precision and accuracy [1]. In addition, the modern machine used today can cut burr-free edges and eliminate built-up edges (BUE), providing an exceptional surface finish. HSM is a metal cutting technology that uses cutting speeds ten to fifteen times faster than traditional machining. Researchers have used some parameters that characterize HSM and classify it into many areas, including high-throughput, high-speed machining, high-feed machining, and high-speed spindle machining. Many factors, including work material, tool material, cutting conditions, and machine tools, influence the cutting speed range in HSM. On the other hand, cutting speed is highly dependent on the material being processed. For example, aluminum has a significantly higher cutting speed than titanium alloy (see Figure 5.1) cutting circumstances, as well as equipment. On the other hand, cutting speed is highly dependent on the material being processed. For example, compared to titanium alloy, aluminum has a significantly higher cutting speed (Figure 5.1). As a result, HSM varies depending on the type of material. The range of cutting speeds for various processed materials is shown in Figure 5.1. The graph shows that different materials have different

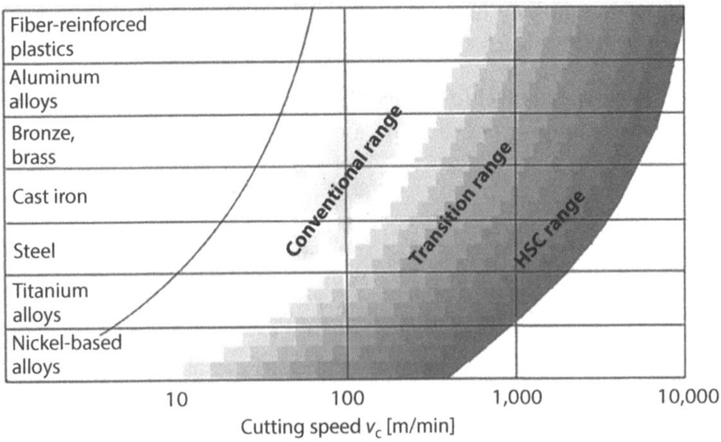

Figure 5.1 The cutting speeds of various materials.

cutting speed ranges for conventional and high-speed cutting (HSC). The term "transitional range" refers to the area that lies between the normal and HSC zones. This chapter discusses the history of HSM, its benefits, features, and scientific differences from traditional processing. It also covers the tool system, machine tools, currently available software, and machine tools.

5.2 Applications of High-Speed Machining

The effectiveness of high-speed machining is determined by the following machine subsystem components: spindles, axes, motor drives, tool holders, and cutting tools. Since the spindle is arguably the most critical component [1], runout must be kept as low as possible to maintain optimal accuracy and performance. With a cutter, the shorter the interval between changing inserts, the smaller the runout. The runout size is important for accuracy, especially in HSM applications [1, 5]. At the cutting edge, the total indicator reading (TIR) should not exceed 10 microns [2]. An additional 10 microns of runout results in a 50% reduction in tool life. There can be multiple sources of imbalance, even in cases where the tool, tool holder, and spindle are perfectly balanced and unpredictable. The contact between the tool holder and the spindle contact is one of the causes of instability. With this grip, there is usually a noticeable gap. In addition, chips or dirt may form inside the cone. There is usually no need to balance a 20,000 rpm end mill to anything greater than 20 g/mm, and 5 g/mm is suitable for much higher speeds.

5.3 Machine Tools for HSM – Requirements

Machine tool designers and engineers have only recently begun to build machines for HSM, although high-speed alternatives to spindles for conventional machining centers have been available for some time. As mentioned earlier, HSM requires the configuration of many factors related to the machines. Some general requirements for the machine and data transfer to the HSM (ISO/BT40 or similar size, 3 axes) are listed below.

5.3.1 Hard Material Cutting

Removal of solid material; rigid milling; hard turning of parts; ard turn; hard cutting; hard drilling; heavy machining; hard machining. hard cutting is the process of cutting hard materials such as hardened steel with hardness values greater than 47 HRC. This manufacturing method includes all cutting operations such as turning, drilling, milling, broaching, and other processes with geometrically defined cutting edges (as opposed to abrasive processes). In the past, hard materials were only processed using abrasive methods and worked into their final shape unless they needed repair. In the 1980s and 1990s, suitable cutting tools, machines, and processing methods were developed [13–15].

Today, most hard materials can be cut, and since the 1990s, cutting hard materials has become an industry standard. In particular, instrumental materials. The main technological advances have been high hardness and sufficient strength at elevated temperatures, mainly polycrystalline cubic boron nitride (PCBN), as well as ultra-fine-grain carbide and mixed oxide ceramics (Al2O3/TiCN) with improved toughness. Alumina ceramics with a very negative effective rake angle have been used to successfully machine steels with hardnesses greater than 60 HRC as early as the mid-1960s [31]; however, it was not until the development of CBN/TiCN composites in the 1980s that satisfactory and consistent levels of tool performance were achieved. PVD coatings have been successfully applied to PCBN tools over the last ten years or so, allowing for increased cutting speeds and, in some cases, more than doubling tool life. Hardened steels account for a significant portion of the hard materials cut in industry, and case-hardened steels account for the majority in the automotive sector . The main technological advances have been high hardness and sufficient strength at elevated temperatures, mainly in polycrystalline cubic boron nitride (PCBN), as well as ultra-fine-grain carbide and mixed oxide ceramics (Al2O3/TiCN) with improved toughness. Alumina ceramics with very negative effective rake angles have been used to successfully machine steels with hardness greater than 60 HRC. however, it was not until the mid-1960s that satisfactory and consistent levels of tool performance were successfully applied to PCBN tools following the development of CBN/TiCN composites; over the last ten years or so, allowing cutting speeds to increase and, in some cases, tool life to more than double [16, 17]. Hardened steels make up a significant portion of the hard materials cut in industry, with the majority being case-hardened steels in the automotive sector. As can be seen from the graph in Figure 5.2, the hardness of the material being cut and the cutting speed have a significant impact on the tool's life. Another important factor is the composition of the steel, namely the amount of carbides. Low alloy steels are not only susceptible to thermochemical wear mechanisms but also experience more aggressive abrasive wear due to higher concentrations of carbide-forming components. The size of the carbide grains also matters; for example, the larger carbides in conventional steels cause significantly higher wear rates than the smaller carbides found in steels produced from PM [31].

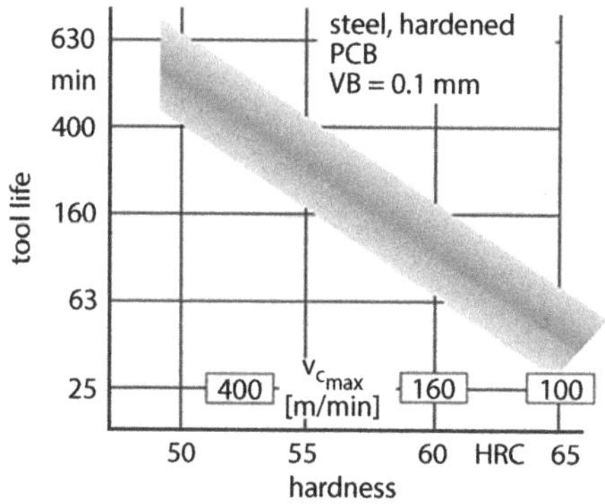

Figure 5.2 Tool life and hardness with recommended cutting speeds.

Polycrystalline cubic boron nitride (PCBN), mixed oxide ceramics (SC), and fine-grained tungsten carbides (TC) are the best tool materials; however, the latter should only be used in special circumstances. PCBN is the preferred material. Ceramics is used for low-volume operations with few breaks in the workpiece, while heavy roughing and finishing turning processes (H05–H30) are used for higher volumes.

Cemented carbides are primarily used in milling applications, where more geometrically complex tools can be used with them; however, PCBN is now used in very high-volume applications such as ball milling with constant velocity joints. Defined edge tools are often used to cut hardened steels, hot and cold work steels, sintered steels, high chromium, white cast irons, and even carbide. A Taylor diagram illustrating the relationship between PCBN tool life and cutting speed is shown in Figure 5.3. Unlike straight lines when cutting mild steel, unusual curves show that there is usually a narrow ideal speed range for hard turning. Tests revealed differences in the hardness and composition of the processed steels, which were partially reflected in the hardness of the workpiece. The material used to make cutting tools must meet these two requirements: it must be more resistant to fracture and plastic deformation. In addition, the material used to make cutting tools must be heat-resistant, chemically stable, and have an ideal strength-hardness ratio [20]. Many manufactured cutting tool materials have been found to differ in two specific properties: toughness and hardness. These two qualities must be present in an ideal ratio in the material of the cutting tool. The evolution of cutting tool materials has revealed numerous trends and changes. The initial and sustained transition occurs from high-speed steel to carbide. The second change concerns the tool material: from carbide to polycrystalline diamond (PCD).

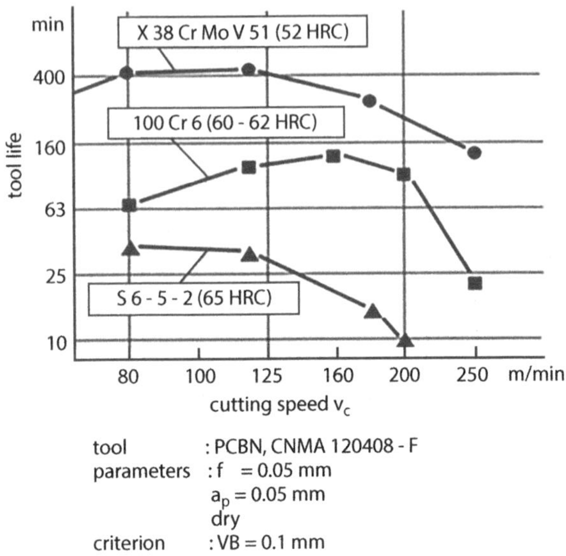

Figure 5.3 Cutting hard material, PCBN finishing, Taylor curves.

Figure 5.4 shows the solid end mills used in HSM. There are tools with different coatings, helix angles, number of grooves, and structures available for a wide variety of applications. Before selecting cutting tools for HSM, it is important to consider the following factors: contact conditions, cutting tool material, cutting tool geometry, tool design, and associated processing methods [19]. Hard turning is a common finishing method that replaces grinding processes. Figure 5.5 shows a real-life example where a sharp turn is preferable. The friction disc, made of bearing steel DIN100Cr6 (SAE 52100), has three surfaces machined: the outer cylinder (A), the end (B), and the shorter inner cylinder (C). The benefits of turning grinding largely depend on component tolerances, surface integrity criteria, removal allowances, and geometric complexity of components. The economic and environmental benefits shown in the figure above should be interpreted as typical only and do not apply to every component or process. When working with case-hardening steels, the drill must be able to reliably cut through both soft and hard areas of the part. As shown in Figure 5.6, there is a limited processing window for hard drilling of hardened materials, with the two main characteristics being low feed rates and relatively low cutting speeds (Spintig 1995). Typically, short tungsten carbide drill bits are used. The ideal cutting material is ultra-fine-grained tungsten carbide. Boring tools with replaceable tips, mainly titanium nitride (TiN) coated,

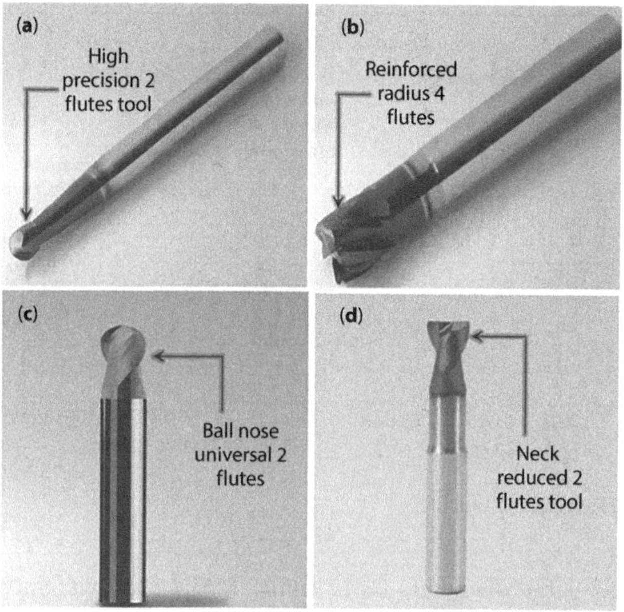

Figure 5.4 Shows the following tools: (a) solid end mill with hard surface for high-precision operations; (b) hard coated toroidal end mill to ensure high precision; (c) hard coated short flute tool; and (d) a CVD-coated two-flute tool. (Image taken from Seco Tools website) [23].

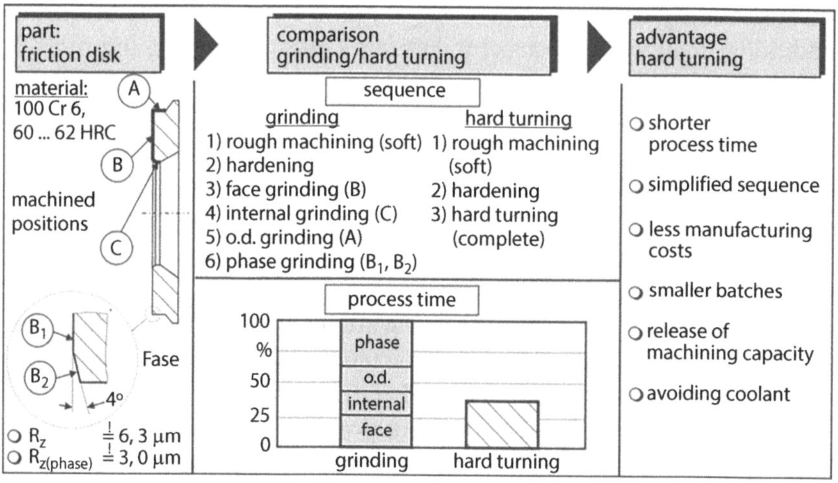

Figure 5.5 Cutting hard materials, Cost-effective production due to strict rotation.

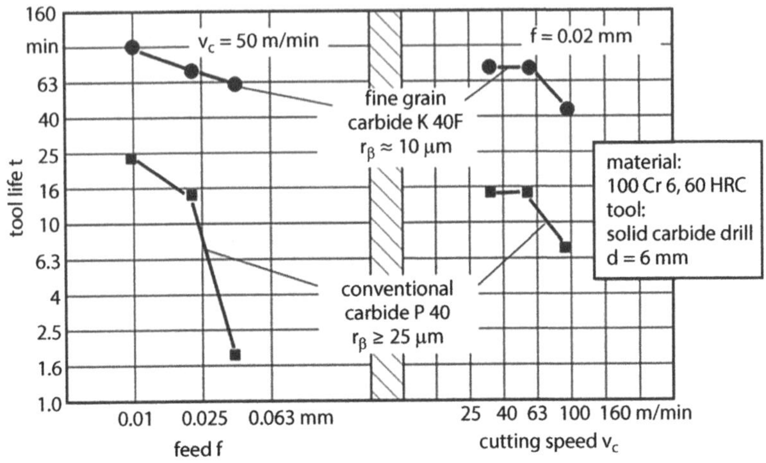

Figure 5.6 Hard drilling tool life.

are used for large diameters (> 12 mm), TiAlN coatings. Such coatings reduce friction between the tool and the hole surface, as well as between the tool and the chips, and also increase wear resistance. In finishing conditions, solid carbide drills can provide diameter quality from IT 7 to IT 9, as well as surface roughness Rz ¼ 2–4 mm and, under certain circumstances, Rz ¼ 1 mm.

The production chain of cemented parts can be significantly shortened, as shown in the example in Figure 5.6. It is important to remember that the carburization process does not always require separate heat treatment or additional coating. Microstructural changes near the surface and residual stresses can have the same effect. The microstructure, passive force, and residual stresses of hardened steel DIN16MnCr5 (corresponding to SAE 5115) after turning are shown in Figure 5.7. The distinct freshly quenched layer, which appears white under an optical microscope, is the result of increased thermo-mechanical stress on the workpiece surface caused by increased tool wear [31]. as a result of a thermomechanical load of the processed material.

Low alloy hardened steels have been widely used in manufacturing for decades due to their outstanding characteristics. Tool wear has an impact on the surface's physical state (Figure 5.8). These materials are increasingly being used, requiring high-throughput processing methods, resulting in ongoing challenges. Numerous studies have been carried out to improve the machining characteristics of low-alloy steels [1–4]. Hard machining is the machining process of low-alloy hardened steels (>45 HRC), especially

up to now	part	in future
1. stress free annealing	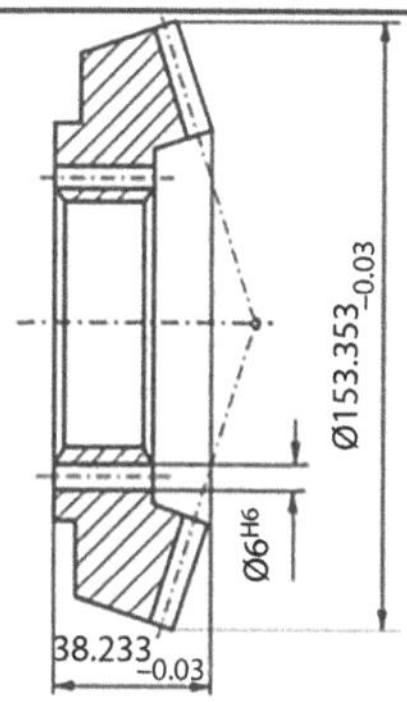	1. stress free annealing
2. turning both sides		2. turning both sides
3. drilling holes		3. gear cutting
4. intermediate		4. case hardening
5. inspection		5. drilling and reaming or tapping
6. set cylinder pins in holes		
7. gear cutting		
8. remove pins	material : 15 CrNi 6	
9. reaming of fit holes or tapping	number of teeth : z_n = 36 real module : m = 2.5	

Figure 5.7 Rigid drilling reduces the number of work steps.

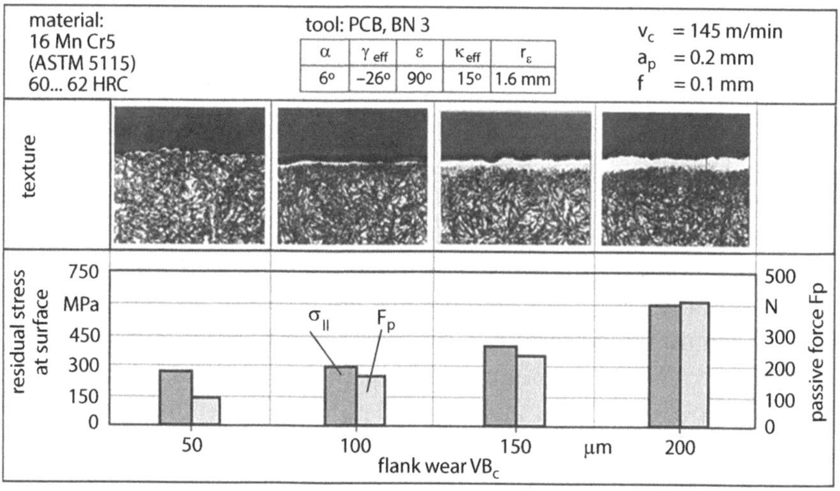

Figure 5.8 Tool wear affects the physical condition of the surface.

when using single-point turned bearings [5]. Bearing steel, defined as high-chromium, low-alloy carbon steel AISI 52100, is characterized by high strength, abrasion resistance, and dimensional stability [5, 6]. Instead of grinding, bearing steel can be hard machined, providing a cost-effective alternative with high material removal rates. However, due to its extreme strength, AISI 52100 bearing steel is a difficult material to machine. High temperatures between the workpiece, tool, and chips also reduce tool life and result in poor surface finish. Coolant is a useful tool because it controls the temperature generated during machining, reducing surface roughness, cutting force, and tool wear. However, flood cooling methods can be hazardous to human health and the environment. They may also require additional steps, such as post-processing cleaning, and increase production costs [7]. In addition, green cutting technology is a new idea that adapts well to strict environmental laws [8]. The industry can reduce carbon emissions and help create a cleaner, healthier environment by using green-cutting technology. Adopting green cutting technology can also improve the quality of life of industrial workers by reducing their exposure to hazardous pollutants and improving workplace safety. All things considered, green-cutting technology represents a viable option for businesses looking to maintain sustainable practices and meet environmental standards. Another machining method that has gained popularity is dry machining, which has numerous environmental and economic benefits [9]. By reducing machine downtime associated with fluid changes and cleaning, dry machining can also improve productivity and efficiency. While dry machining produces chips that are dry and easy to dispose of, cutting fluids often leads to chip removal problems and can increase the frequency of machine tool maintenance. Another lubrication method is MQL, which effectively supplies coolant to the cutting zone [10]. But with MQL, the amount of coolant is dramatically reduced - typically to 30-100 ml/h - and it is delivered directly to the cutting-edge using precision nozzles.

The main obstacles facing metal cutting today are the creation and implementation of creative solutions designed to firstly reduce cycle times and costs to improve profitability and productivity, as well as increase productivity and, most importantly, reduce emissions and energy. and resources used. The use of hard machining modes (HMC and HPC) is often limited by machine performance or tool wear and workpiece machinability, such as when cutting lightweight materials or high-temperature alloys. Generally speaking, using high speed can result in reduced cutting forces, power, and energy, which is beneficial for machine and tool design.

CIRP's main activities are in the field of High-Speed Machining (HSM) or High. STC "C" High-Performance Cutting (HPC) and High-Speed Cutting (HPC) subjects. Figure 5.18 shows the range of various characteristics of the tool material, such as temperature hardness and toughness, which are related to cutting speed and toughness, respectively. Cutting tool materials are the main limitation for high cutting speeds and high loads. Cutting tool material evolution is shown in Figure 5.9. High Performance Computing and HSM.

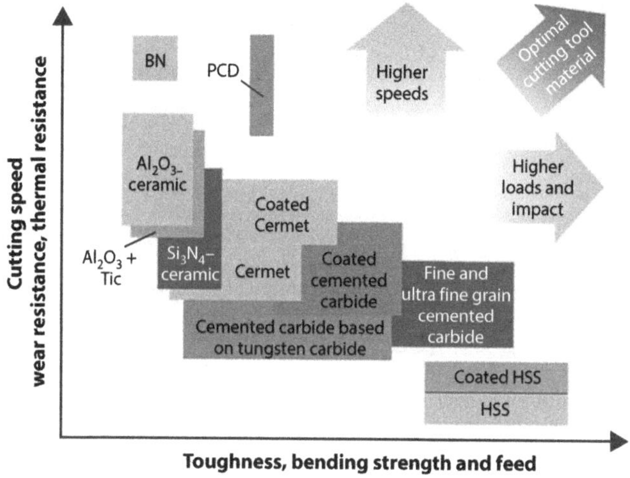

Figure 5.9 Evolution of cutting tool materials HSM and HPC (Klocke 2004).

Typically, after earlier precision machining procedures such as grinding, honing, and superfinishing (sometimes called short stroke honing), abrasive machining operations using bonded abrasive tools are used. However, superfinishing also uses fine-grit abrasive tools. The overall goal of honing and superfinishing is to improve the tribological properties of the workpiece, surface quality, dimensional accuracy, and shape. The latter is usually achieved by increasing the surface bearing ratio—the ratio of the bearing contact area to the total area of the workpiece—and the surface roughness parameter Rvk, or reduced cavity depth, which affects the lubricity of the surface. In addition, subsurface damage to the workpiece caused by previous manufacturing operations, such as grinding, can be eliminated through cleaning and enhancement.

Bonded honing and superfinishing tools typically use the same identification system as grinding tools, which can be manufactured using super abrasives or conventional abrasive grains. Before initial use, these tools must be prepared, that is, sharpened and profiled. Standard tool preparation procedures for bonded honing and superfinishing include machining using diamond grinding and grinding wheels (Figure 5.10), silicon carbide bonded grinding wheels or corundum grinding wheels, and hardened steel or abrasive coating alignment sleeves. Like grinding, honing involves fixing abrasive grains (or grits as they are most commonly known) in a bonded tool. The main purpose of honing is to give the cylinder bore a smooth surface. The honing stones opposite the hole are subjected to radial outward pressure. Two things differentiate honing from grinding.

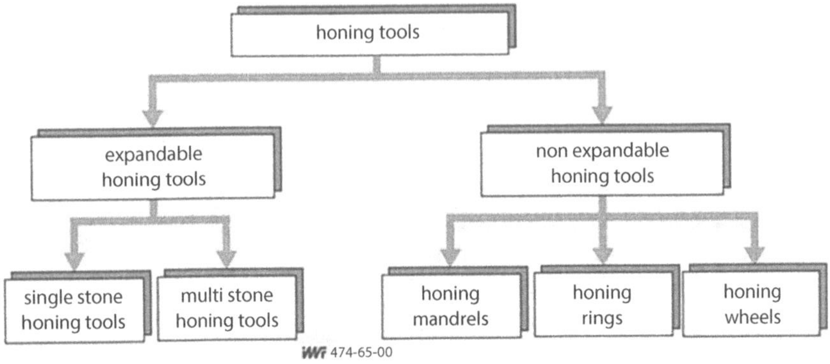

Figure 5.10 A summary of instruments for honing.

First, the abrasive tool moves slowly relative to the workpiece during honing. The surface speed is usually between 0.2 and 2 m/s. The vibrating and rotating actions of the tool work together to provide average material removal across the entire surface of the workpiece and create the characteristic "hatch" pattern that is preferred for retaining oil in engine cylinder bores. Another difference between grinding and honing is that the tool is freely positioned with the surface of the workpiece. This indicates that the hole eccentricity relative to the outer diameter cannot be adjusted.

Abrasive machining techniques called honing (long stroke honing) and superfinishing (short stroke honing) are used to improve the dimensional and shape accuracy, surface quality, and tribological performance of the workpiece. By increasing the surface bearing coefficient and reducing the cavity depth Rvk, which affects the lubrication properties of the surface, the tribological properties can be improved. In addition, honing and superfinishing can be used to eliminate subsurface damage to the workpiece, such as grinding burns caused by previous manufacturing operations. Thus, honing can be used to maximize the surface integrity of the workpiece, including microhardness, residual stresses, and lattice structure. Honing is divided into sub-processes such as external and internal honing, surface honing, gear honing (internal and external), and form honing, as well as the distinction between long-stroke and short-stroke honing (superfinishing). Workpiece geometry and process kinematics. Some techniques have also been developed to improve honing efficiency, such as high-speed honing, spiral slide honing, and BOREHONE.

Honing is a crucial step in the production of highly functional surfaces, especially for hydraulic components and the automotive industry. The shape of the cylinder block holes, dimensional accuracy, and surface quality (Figure 5.11) have a significant impact on engine characteristics such as fuel consumption. By eliminating surface protrusions and reducing surface cores, the modern multi-stage flat honing process is often used to improve the surface finish of bores in cylinder blocks. In this particular case, the cylinder head block and the thermal load of the engine lead to static and thermal deformation of the cylinder head. the working surface of the cylinder (Figure 5.12). Consequently, crankcase breakthrough and oil consumption increase.

The growing need for faster and more efficient supply networks has led to the development of robotics and automation as critical supply chain management technologies. By reducing long-haul costs, increasing work and utilization efficiency, reducing errors, reducing repetitive inventory checks, updating orchestration, taking care of time, and gathering information in problem and dangerous regions, robotics and automation are improving supply chain management. Robotics helps in creativity, design, etc. Through the use of computer software, other technologies, and autonomous physical devices, automation helps perform jobs that are often done by humans. Although artificial intelligence and machine learning are widely recognized as decision support tools, supply chain management (SCM) has rarely ever used them. This article covers several major subfields of artificial intelligence and machine learning. They are ideally suited to solve practical problems associated with SCM to maximize the potential benefits of artificial intelligence and machine learning for SCM. In doing so, this article examines the history of successful implementations

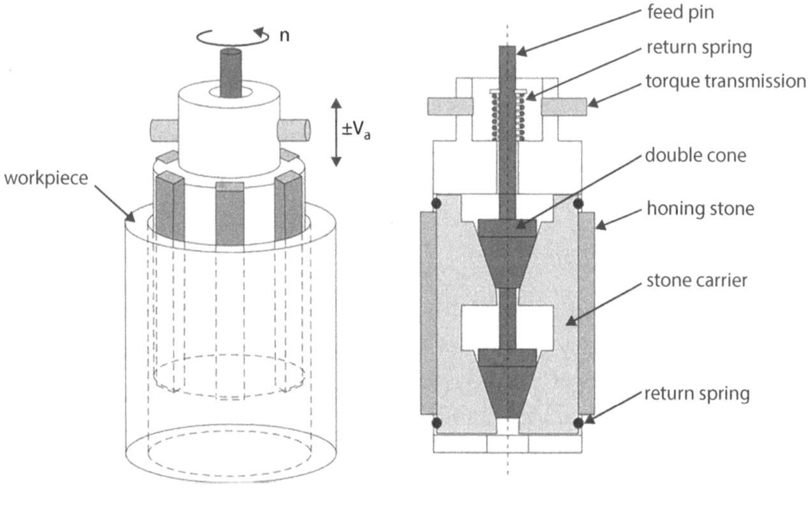

Figure 5.11 Basic design of a multi-stone sharpening tool (according to WBK).

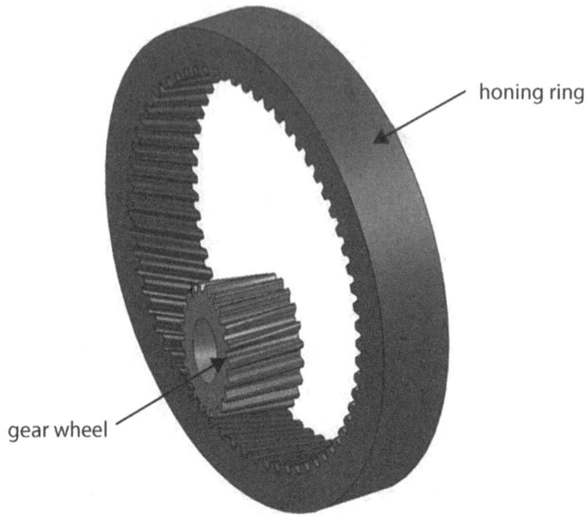

Figure 5.12 Honing ring for gears.

of artificial intelligence and machine learning in supply chain management and identifies the most productive areas of SCM for the application of these technologies. Also, we learn how automation and robotics help in warehouse management. This article provides a comprehensive analysis of recent advances in robotics and automation in supply chain management. We first provide a general overview of the challenges facing supply chain management and then discuss the many applications of robots and automation at different stages of the supply chain. We also discuss the benefits of automation and robotics in supply chain management, such as increased safety, reduced costs, accuracy, and efficiency. Finally, some current challenges and limitations associated with automation and robotics in supply chain management and suggest some potential research directions. Figure 5.13: Projected future hybrid automation expenses (right) and robotic system cost savings in relation to labor costs (left).

5.3.1.1 Hybrid Cutting

Innovative hybrid cutting model is for predicting workpiece temperature distribution during dry milling of compact graphite cast iron (CGI). The hybrid model consists of a finite element analysis (FEA) based on a thermal model and an analytical force model based on a mechanistic approach. The advection heat distribution model is used to calculate the amount of heat generated during the milling process and transferred to the workpiece. The temperature field's development during laser cutting is shown in Figure 5.14.

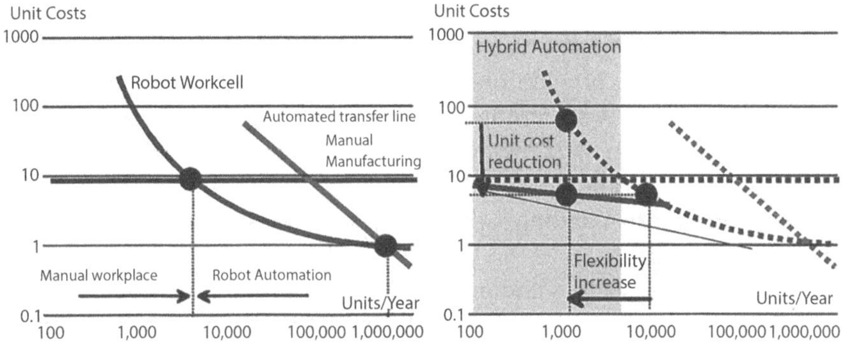

Figure 5.13 Estimated potential costs of future hybrid automation (right) and cost savings for robotic systems relative to labor costs (left).

The cutting forces were used to calculate the temperature distribution of the workpiece, which was obtained using thermal loads and used as boundary conditions in FEA. The constructed power and thermal models were tested experimentally, and good agreement between the calculated and measured results was obtained. Energy and active force calculations show that by doubling the feed during CGI milling, nearly doubling the cutting forces is achieved, although only about 10% of the energy is used [45]. However, thermal cutting tools such as arc plasma torches inevitably

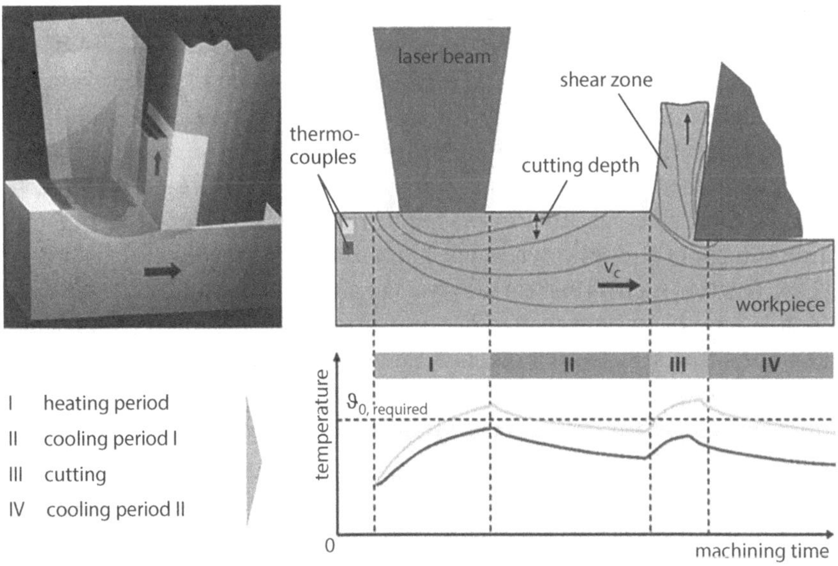

Figure 5.14 Development of the temperature field during laser cutting.

produce radioactive aerosols and gaseous vapors when cutting radioactively contaminated metal structures, increasing the risk of radiation exposure to workers [7]. In addition, molten materials can interfere with the cutting process [6]. Also, strong arc currents of 200–900 A when cutting thick metal [6, 8] are detrimental to the service life of the plasma torch, increasing the cost and duration of dismantling. Thus, using only an arc plasma torch for dismantling radioactively contaminated metal structures is ineffective [1–5].

Milling cutters, such as end mills, emit very little aerosols compared to arc torches and other mechanical tools, except for easily removed chips, making them suitable for safely cutting metal structures contaminated with radioactivity [3]. Additionally, since the final cutting pass splits the inner surface of the workpiece [3], these mechanical cutting tools can be useful for segmenting metal structures (such as pipes and tanks) whose internal layers are radioactively contaminated. Despite these clear advantages, cutters have not found widespread use in nuclear power plant dismantling (except for a few specialized applications such as pre-preparation) due to the high time and tooling costs, especially when using tools without cutting fluid [9].

5.3.1.2 Chemical Vapor Deposition

Chemical vapor deposition reactors must be built with the ability to control several aspects of the film, including composition, surface morphology, thickness, and crystal structure. The direction of gas flow to the substrate holder (susceptor) can be used to distinguish between horizontal and vertical systems. Reactors designed for air and low-pressure deposition have been specially designed. Using radio frequency, which is conductive, the substrate holder is heated by radiation. Cold or hot wall reactors can heat up from the inside out. Rotating substrate holders are designed to improve film uniformity and flow dynamics. Large-scale CVD production is possible, so reactors such as barrel reactors, pancake reactors, and spinning disk reactors have been built that allow simultaneous deposition onto multiple substrates. The selected deposit is created on the surface of the substrate as a result of the reaction or destruction of one or more volatile precursors when the substrate is exposed to them by chemical vapor deposition (Figure 5.15).

In addition, volatile by-products are often formed and are ejected from the reaction chamber by the gas stream [23]. System conditions play an important role in determining the success of a process. During this procedure, the temperature in the area varies from 800 to 2000°C. Thanks to

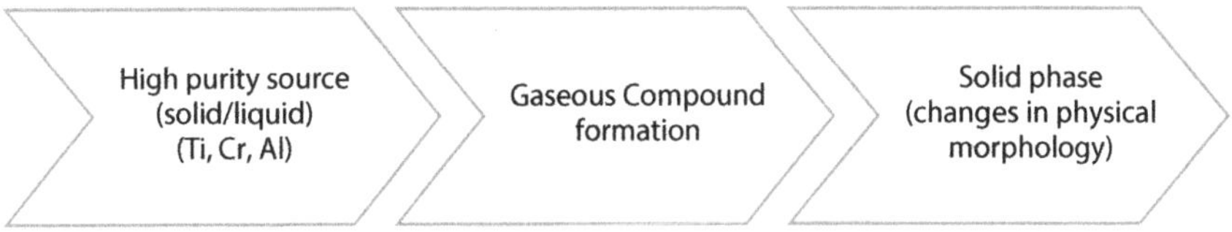

Figure 5.15 Physical vapour deposition flow chart.

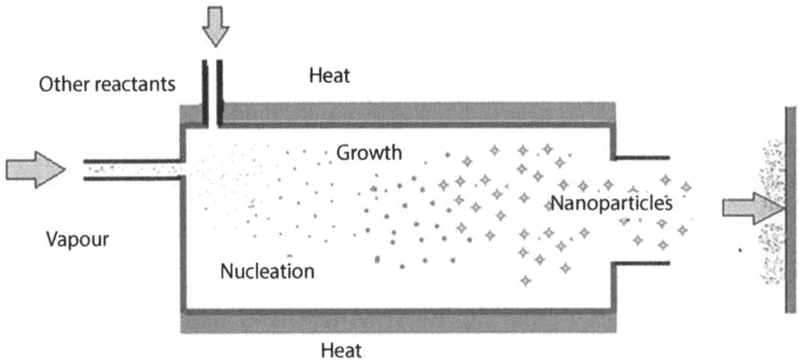

Figure 5.16 Chemical vapour deposition schematic diagram for nanocoatings.

this method, excellent adhesion and coating thickness of up to 20 μm are achieved [24, 25]. Characterization techniques such as SEM and XRD, concluded that the pore surface characteristics changed as a result of CVD nanocoating [26], (Figure 5.16).

5.3.1.3 *Plasma Arc*

The electrically conductive material is heated, melted, and cut into unique patterns and designs using a plasma torch. This manufacturing technology is known as plasma cutting, sometimes known as fusion plasma cutting or plasma art cutting.

Plasma cutting can cut a wide range of metal materials such as copper, aluminum, alloy steel, and structural steel, which can cut materials from 0.5mm to 180mm thick. Rubber, plastic, and other non-conductive materials are not suitable for plasma cutting. There are different types of plasma cutting, such as water injection plasma arc cutting, secondary plasma arc cutting, and standard (or traditional) plasma arc cutting. Alternatives to plasma cutting include oxy-fuel, waterjet, and laser cutting.

The technique, known as plasma cutting, uses a jet of hot plasma that is accelerated and cuts electrically conductive materials. When an electrically conductive gas, such as oxygen, nitrogen, argon, or a combination of gases, is ionized and forms a plasma arc, it binds to the material being cut, transferring energy and melting it. This is exactly how plasma cutting works. When a gas is heated to such a high temperature that electrons are stripped from the atoms, creating a mixture of free electrons and ions (atoms that have gained or lost electrons), a unique state of matter known as plasma is formed. Ionization is the process by which atoms lose their electrons.

Three states of matter are most often mentioned: solid, liquid, and gasoline. But plasma, which is found both inside and outside the galaxy (for example, in the centers of stars and the form of lightning), is sometimes called the fourth state of matter. From a technical point of view, plasma is the most common state of matter, both by volume and mass. When more energy is given to a gas, the molecules can move faster and collide with greater force, resulting in plasma.

Molecules can be separated into their constituent atoms by collisions, and these individual atoms can be separated from their outer shell electrons to form ions. The gas reaches a critical point where the number of atoms gaining free electrons and the number of atoms losing electrons balance out as more atoms lose their electrons [28]. Figure 5.17 shows a schematic illustration of the plasma arc process. Sylvain Coulomb and Cao Qin demonstrated how to create organic layer-coated metal nanoparticles. Comparing substrates with low Critical Process Spray Parameter (CPSP) and high CPSP coatings, the latter was found to exhibit the least wear resistance [30], (Figure 5.18).

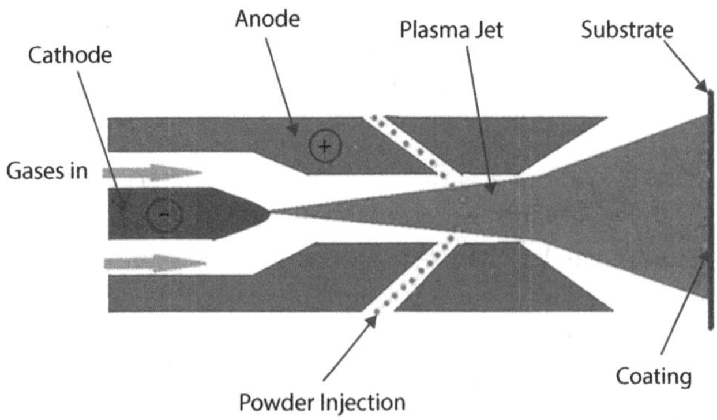

Figure 5.17 Plasma arc process schematic diagram.

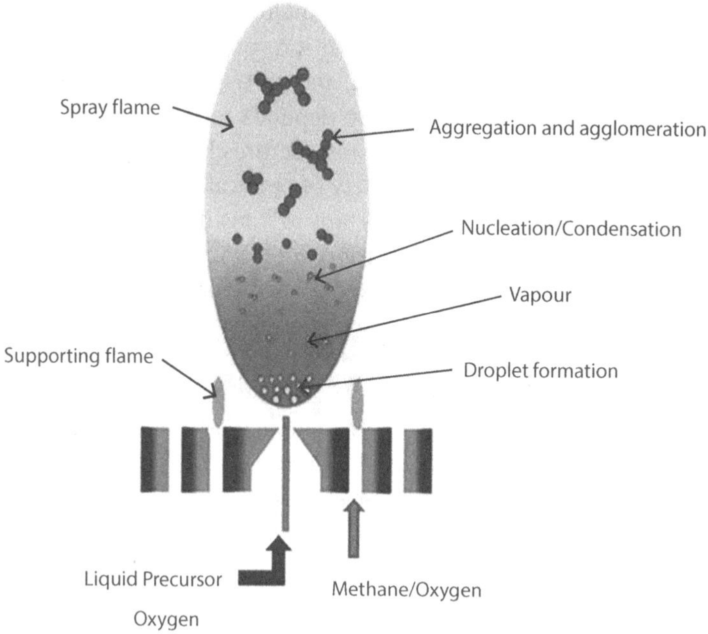

Figure 5.18 Diagrammatic schematic of the flame synthesis process.

5.3.1.4 *Flame Synthesis*

An industrially scalable method for creating nanomaterials is called flame spray pyrolysis (FSP), which can produce a wide range of metal or metal oxide nanoparticles with specific physicochemical properties [1]. The complex process architecture, including precursor atomization, combustion, and nanoparticle production, is the basis of FSP technology [2]. This enables rapid, one-step synthesis and eliminates the need for post-production modifications that are often necessary with other approaches.

Recently, the importance of FSP has increased in the creation of nanodevices, in which nanoparticles can be inserted *in situ* into complex functional arrays [3]. Consequently, FSP is not only transforming the manufacturing of nanomaterials but with recent advances, it is also opening the door to the development of next-generation nanodevices [4]. In this review paper, we offer a revised summary of the current state of the art in nanodevice development, new materials synthesis, and innovative reactor and process projects in FSP technology [12].

In the field of nanotechnology, the synthesis of nanostructures is an important area of research that includes a wide range of techniques in addition to FSP. These various methods, including electrospinning, sol-gel processing, and chemical vapor deposition, provide specific qualities in terms of shape, chemical composition, and particle size control. The choice of an appropriate synthesis technique depends on several factors directly related to the intended use of the nanostructures [18, 21]. When assessing the suitability of a method for a variety of applications, from solar devices to drug delivery systems, it is important to consider factors such as temperature range, synthesis time, environmental impact, and material adaptability. Several synthesis methods used to create nanomaterials (including FSP). and describes the specific standards applicable to their use. This careful consideration ensures that the synthesis procedure meets the functional specifications of the end application, thereby optimizing the efficiency and practicality of the generated nanostructures. An alternative to FSP is atomic layer deposition (ALD) [5, 6], especially for thin film technology. ALD provides atomic-level precision and high levels of consistency, resulting in extremely stable and flawless films. Disadvantages include the need for expensive controlled environment equipment, slower deposition rates, and possibly significant precursor prices [5, 6, 32].

5.3.1.5 Laser Ablation

One method for creating several types of nanoparticles is laser ablation. These include core-shell nanoparticles, semiconductor quantum dots, carbon nanotubes, and nanowires. Using this method, particles subjected to laser evaporation in a background gas are nucleated and multiplied to form nanoparticles. It is beneficial to produce high-purity nanoparticles in the quantum size range (<10 nm) by extremely rapid steam cooling. Schematic representation of the process of laser ablation of nanocoatings is seen in Figure 5.19.

This provides an overview of the process of creating nanoparticles using laser ablation. Next, we review recent developments in nanoparticle size control and the challenges of synthesizing functional nanoparticles using modern laser ablation technologies.

5.3.2 Method of Liquid Phase

Another method for creating phosphorus compounds based on silicates is liquid-phase synthesis. This procedure can produce approximately 90% more transformation than the high-pressure process, which produces only 20%. Solvents are widely used in liquid-phase synthesis as reaction media [22]. The response speed and system can vary significantly depending on

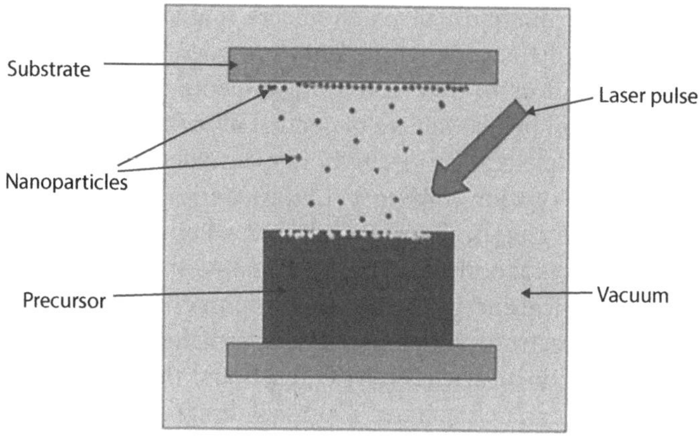

Figure 5.19 Diagrammatic representation of the laser ablation procedure for nanocoatings.

the range of solvents. The effect of solvation is the term for this process, which can be analyzed quantitatively using the free energy of solvation. Understanding the factors influencing appearance during liquid-phase synthesis has received much less attention than gas-phase synthesis methods. Using the method of liquid-phase precursors, a green-emitting phosphor (Ba1−x, Sr1−x)SiO4:2xEu2+, where x = 0.03, 0.05, 0.1 and 0.15, was created [77]. When phosphors are produced using a liquid-phase precursor, their emission intensity is reported to be significantly higher than other synthesis methods. The most widely used film is SiO2, which is easily hydrolyzed by metal, boric acid, or aluminum. In addition to basic solutions, extremely acidic solutions are also formed when creating a thin film. A block diagram of the liquid-phase method is shown in Figure 5.9. The types of liquid phase methods used to produce nanocoatings are shown in Figure 5.10.

5.3.2.1 Electrochemical

Electrochemistry is the study of chemical processes involving the separation of charges, usually in liquid media such as solutions. Charge transfer, which can occur heterogeneously on the surface of electrodes or homogeneously in solution between different chemical compounds, is often associated with charge separation. Thus, it is immediately obvious that its application is very wide and will be discussed later.

At least two charge transfer half-reactions: oxidation (loss of electrons or increase in oxidation state) and reduction (gain of electrons or decrease in oxidation state) occur simultaneously and in opposite directions to ensure electrical neutrality [27].

Oxidation and reduction half-reactions are spatially separated at electrode surfaces and often occur at different electrodes immersed in solution within the same cell or in different compartments of the cell. The electrodes are connected by conductive paths that allow the transfer of charge and completion of an electrical circuit both externally (via electrical wires, etc.) and internally (via ion transport) [29]. Batteries and fuel cells can be used to capture the electrical energy released when the total number of Gibbs energy changes at both electrodes is negative. If the difference is positive, external electrical energy can be used to counteract the positive Gibbs energy and drive electrode reactions that convert chemicals (electrolysis). Therefore, as in industrial electrolytic reactions, the products of two-electrode reactions can be separated if the cell design allows it [43].

5.3.2.2 *Precipitation*

The most popular method for extracting dissolved (ionic) metals from solutions such as hazardous metal-containing process wastewater is chemical precipitation. A chemical reaction between the soluble metal compounds and the precipitating reagent converts the ionic metals into an insoluble form (particle). By settling and/or filtration, the particles resulting from this process are removed from the solution. In this method, the typically required unit steps are neutralization, sedimentation, coagulation/flocculation, separation of particles from liquids, and dewatering. The efficiency of the chemical precipitation process depends on several variables, such as the type and amount of ionic metals in the solution, the precipitant used, the reaction conditions (especially the pH of the solution), and the presence of additional substances that may interfere with precipitation. reaction. The most widely used chemical precipitation method is hydroxide precipitation, also known as pH precipitation, which uses precipitants such as sodium hydroxide (caustic) or calcium hydroxide (lime) to create metal hydroxides. The ideal pH for the precipitation of hydroxides of each dissolved metal ranges from 7.5 for chromium to 11.0 for cadmium. Because metal hydroxides are amphoteric, they become more soluble at both high and low pH levels. Thus, the metal may dissolve or begin to reenter the solution depending on the ideal pH for that metal to precipitate. Precipitation of these various metals as hydroxides can be a complex procedure since most process wastewaters contain mixed metals. Liquid phase method flow chart (Figure 5.20) cyanide. Lead(II) iodide precipitate is formed when potassium iodide combines with lead(II) nitrate. Nucleation is a crucial step in the deposition process. The precipitation approach requires the creation of an interface for the formation of solid particles,

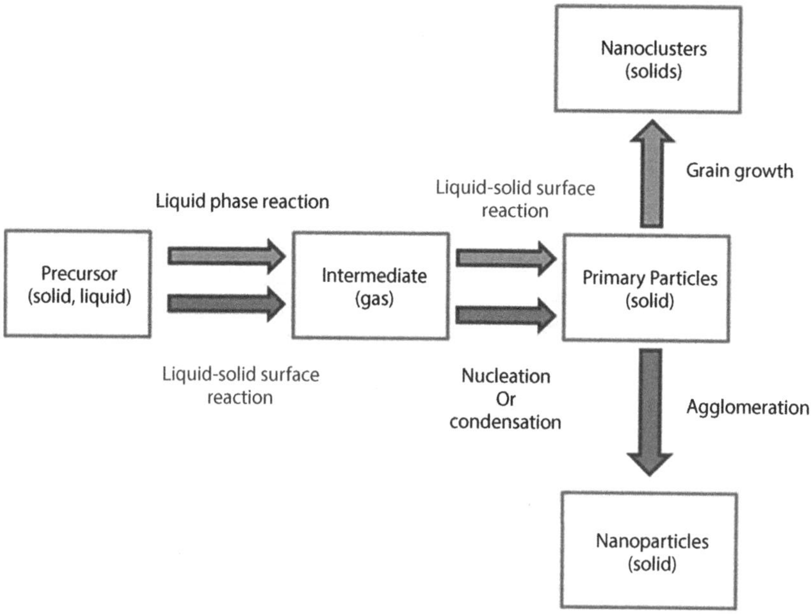

Figure 5.20 Flow chart of liquid phase method.

and this requires energy based on the relative surface energies of the solid and solution. Many liquid phase methods for producing nanocoatings are shown in Figure 5.21. Super saturation occurs when this energy is absent and there is no nucleation platform. Figure 5.22 lists the various sol-gel techniques in summary [46].

5.3.2.3 Sol–Gel

One of the common wet chemical methods for applying nanocomposite films is the sol-gel procedure. Material sources that were first prepared in this process included metal oxide and alkoxide precursors, metallic nanoparticles, tetraethoxysilane for the silica matrix, stabilizers, catalysts, and other porosity additives. After this, the sol undergoes processes of polycondensation and hydrolysis, which gradually lead to the development of a network resembling a gel. The fundamental morphology or structure of the solid phase can take several forms, such as continuous polymer networks or discrete colloidal particles.

The sol-gel precursor is converted into a nanocomposite film by spin-coating or dip-coating onto a substrate, followed by pyrolysis and decomposition of compounds, and extraction of organics and water from

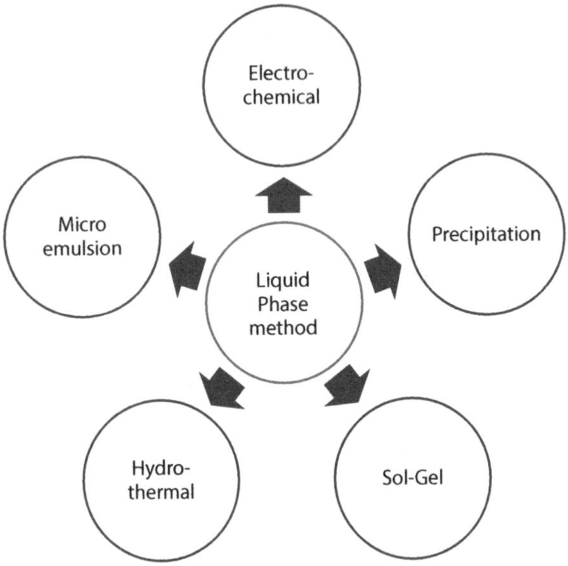

Figure 5.21 Varieties of liquid phase techniques used to create nanocoatings.

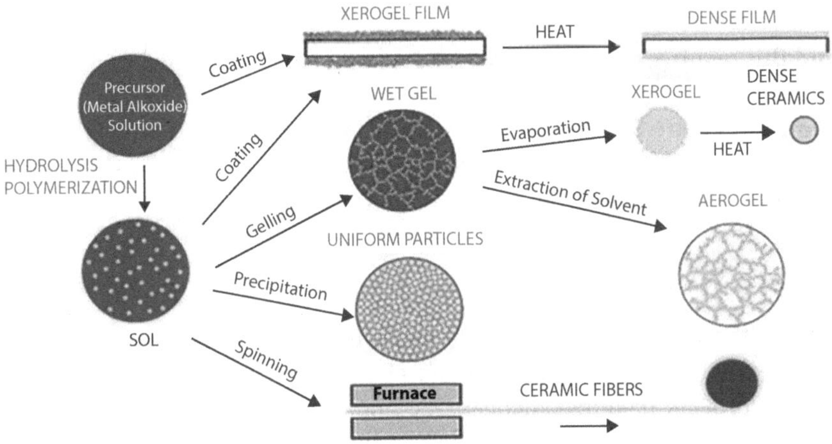

Figure 5.22 An overview of the several methods of sol-gel [46].

the resulting network, after which crystallites nucleate and develop. The crystallite size and film porosity are significantly influenced by the thermal decomposition behavior of the gel precursor [46]. In contrast, in ash, the dispersed phase is a solid and the continuous phase is a liquid. The fact that solid particles are scattered throughout the liquid in the ash distinguishes it from a non-colloidal liquid.

Scattered particles in a liquid can be precipitated by centrifuging the sol. A non-colloidal liquid solution, such as salt dissolved in water, does not experience this [47]. An overview of the various sol-gel methods is shown in Figure 5.22. (A) Sol-gel creation. The chemical processes of sol-gel formation are depicted in Figure 5.12 and include the stages of hydrolysis, condensation, gelation, aging, drying, and compaction. Then the addition of an acid (e.g., hydrochloric acid) or a basic substance (e.g., potassium hydroxide) activates the precursor. Subsequently, the activated precursors combine to form a network. With time and temperature, the network expands and ages until it reaches the size of a container. From this point on, the viscosity of the liquid increases exponentially until gelation occurs [48]. The structure of the gels obtained in two different modes is shown in Figure 5.23. Several variables, including pH, solvent, temperature, time, catalyst, and agitation, affect sol-gel chemistry.

5.3.2.4 *Hydrothermal*

Hydroxyapatite (HAp) is a naturally occurring form of calcium apatite that is an inorganic component of bones. About 80% of it is minerals, 22% is protein and 8% is water. Hexagonal Crystal Structure: HAp is a form of calcium phosphate with the chemical formula Ca10(PO4)6(OH)2. Hydroxyapatite, a needle-like material 60 nm long and 5–20 nm wide, imparts rigidity and strength to bones and accounts for about 65% of bone mass [1–2].

The wide range of possible applications of nanostructured HAp has attracted the interest of investors, industry, and researchers. The osteoconductivity and biocompatibility qualities of hydroxyapatite attract attention. The substance is widely and effectively used in drug delivery systems and

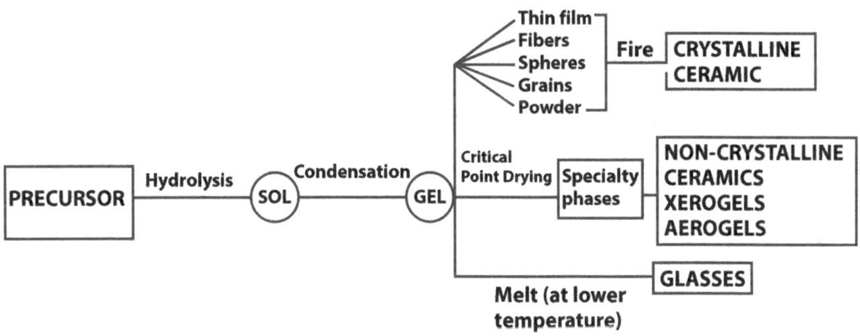

Figure 5.23 Sol-gel method chemical processing [60].

regenerative medicine [3, 4]. Hydroxyapatite particles with a nanostructure can be used in caries therapy as building blocks for damaged enamel [5]. Numerous studies show that nanosized HAp can improve its mechanical properties due to its high surface energy, which allows for faster surface renewal of implants [6]. Hydroxyapatite materials are used as sunscreen, anti-wrinkle agents, and bone replacement. They help speed up wound healing and are used in toothpaste formulations. Polish scientists have created patches and ointments that use the special properties of HAp [7]. There is also great interest in hydroxyapatite to restore bone loss in animals.

It has been demonstrated that HAp nanoparticles (nano-HAp) rather than micrometer-sized hydroxyapatite (micro-HAp) are more suitable to function as a bone apatite substitute in biomedical applications [8]. The effects of differently shaped nano-HAp particles on extremely aggressive melanoma cells were investigated. The results obtained showed that the appearance of a nanoscale effect, in contrast to the morphology of HAP particles, more effectively suppresses the development of such malignant cells [9]. A further study examined the effect of different sizes of nano-HAp, from 20 to 80 nm, on the growth of bone-associated cells, specifically bone marrow mesenchymal stem cells and osteosarcoma cells [10]. A cell culture experiment confirmed that the hydrophilicity of the nanophase mineral is increased by using nano-HAp compared to traditional HAp. In addition, nano-HAP was found to enhance stem cell viability and proliferation, especially for the smallest particles of 20 nm. Similarly, the smallest 20 nm particles were found to be the most effective inhibitor of osteosarcoma cell development, while each nano-HAp slowed down the growth rate. The anticancer activity of nano-HAp was also found in another study [11].

5.3.2.5 Microemulsion

Nanotechnology is a branch of research that originated in the physical, chemical, biological, and engineering sciences. These domains are known for their innovative methods of manipulating individual atoms and molecules. In the context of nanotechnology, a nanoparticle is a small object with transport characteristics and possible applications that behaves as a whole. Figure 5.24 shows a simplified representation of the different shapes that can be produced using gel processing techniques. The image illustrates the different steps that can be followed to produce ceramic or composite materials from gels; the final shape depends on the sintering, drying, and processing conditions.

Simple or compound nanoparticles are useful building blocks for creating innovative nanodevices with a wide range of physical, biological,

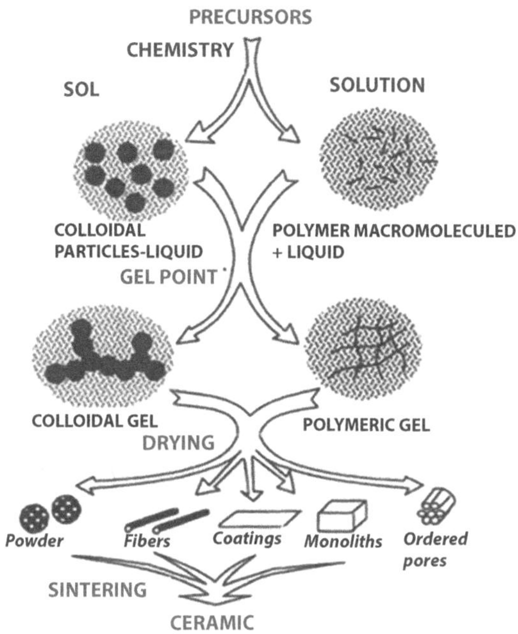

Figure 5.24 Simplified diagram showing the many gel processing structures [48].

biomedical, and pharmacological applications. For many years, scientists have been fascinated by nanoparticles and nanostructured materials due to their applications in Raman scattering, catalysis, photonics, electronics, labeling, imaging, and sensing. They involve intriguing modifications in their optical, magnetic, catalytic, electrical, and drug delivery properties, coupled with improved physical properties such as mechanical hardness, thermal stability, or chemical passivity, and are one of the most challenging areas of modern scientific and technological research.

When comparing the optical characteristics of single crystals of the same chemical composition with nanoparticles, significant differences are often found. The properties of nanomaterials differ from their bulk phases due to the influence of a significant number of surface atoms and the effect of quantum confinement of electronic states on nanosized crystallites. Compared to bulk materials and individual atoms or molecules, the properties of matter at this scale are very different. These nanoparticles, both organic and inorganic, have been prepared using a range of physical and chemical methods such as thermal and irradiation methods. One of the many preparation methods that can control many particle properties, including size, geometry, morphology, uniformity, and surface area, is the microemulsion method, which covers various aspects of microemulsions. Holmes *et al.*, Colloidal metals, colloidal AgCl, nanocrystalline Fe2O3, TiO2, Al2O3, and high-temperature oxide YBa2Cu3O7 were synthesized by microemulsion method. In addition to summarizing the most recent studies conducted on the synthesis of both organic and inorganic nanoparticles using this technology, this review aims to provide a clear overview of the use of microemulsions for the production of nanoparticles and the control of grain size and morphology.

5.4 Advancements in Nanocoating Technology

In every field, including mechanical, electrical, and electronic devices, and computer applications, which are mainly used to improve performance, prevent corrosion, prevent weathering, and prevent water damage, nanotechnology is the fastest-growing field. Thus, it is used everywhere nowadays. In addition, as they become more accessible, we are trying to use nanotechnology in turning and other machining operations. In this case, the nanoparticles will be coated using traditional methods, and the hard, difficult-to-cut material will be turned. Before going deep into this project, we will first learn about nanotechnology, the materials that we will use, the place of our work, and the types of tools that will be used, as we can see one by one. nanotechnology one billionth of a meter (109 m) is called a nanometer; it is approximately 100,000 times smaller than the diameter of a human hair, 1,000 times smaller than a red blood cell, and half the size of a DNA molecule. Nanotechnology is defined as the creation and application of structures, devices, and systems that, due to their small size, have unique properties and functions; the ability to control or manipulate matter on an atomic scale; and research and development of technologies at the atomic, molecular, or macromolecular levels using a length scale of approximately 1 to 100 nm in any dimension. Engineering at the molecular level (groups of atoms) is known as nanotechnology. It is a group as a whole word used to describe a variety of methods, instruments, and procedures involving the manipulation of matter at the smallest scale (1 to 100 nm2). On such an extremely small scale, the classical rules of physics and chemistry are not easily applied for two reasons. First, compared to their larger counterparts, very small particles can have completely different electrical characteristics. Second, the material's characteristics change because the surface atoms are often the most reactive, resulting in a much larger surface area-to-volume ratio [51].

5.5 Application of Nanocoatings

Typically, single- or multi-phase solid structures called nanocoatings, about 100 nm thick or less, are deposited onto an immediate surface to impart specific properties or purposes to the substrate material. Different materials have different characteristics at the nanoscale. The particle size of a material has a significant impact on overall properties such as hydrophobicity, electrical conductivity, magnetic permeability, melting point, and reactivity. Intermolecular forces such as hydrogen bonds, van der Waals interactions, and electrostatic forces control material properties at the nanoscale. The length scales over which these intermolecular forces act vary from a few angstroms to several nanometers.

When the dimensions of a material reach the nanoscale, or the same order of magnitude as the intermolecular force distances, the properties of the material change significantly. Thus, many new applications can be explored by improving the physical and chemical properties of nanocoatings made from nanoparticles or multiple nanoscale layers. The main industrial applications of nanocoatings are wear-resistant coatings, self-cleaning (non-stick) surfaces, corrosion prevention, and thermal protection. These applications are particularly important to the aviation sector as they allow aircraft engine, cabin, and frame components to be modified to improve performance, reduce operating costs, and reduce fuel efficiency [57]. It is reported that the solar absorber used in the collector panel determines how well solar radiation is converted into thermal energy. A solar absorber with low emissivity and high absorption coefficient is ideal. It has been found that electrochemical deposition of selective nickel-based coatings can provide both cost-effective and good optical performance required for solar energy absorbers [58]. However, scientists are moving toward physical vapor deposition technology to be used to coat medium- and high-temperature solar photovoltaic panels due to the poor thermal stability of the chemical vapor deposition method and environmental concerns. Sputtering is one of the PVD processes that has been successfully used to coat receiver tubes and flat plate solar collectors [59].

Acknowledgements

Riyadh A. Al-Samarai and Y. A. would like to acknowledge the Abdus-Salam International Center for Theoretical Physics (ICTP, Trieste, Italy) for their support.

References

1. Tan, D.-W., Zhu, L.-L., Wei, W.-X., Yu, J.-J., Zhou, Y.-Z., Guo, W.-M., Lin, H.-T., Performance improvement of Si3N4 ceramic cutting tools by tailoring of phase composition and microstructure. *Ceram. Int.*, 46, 26182–26189, 2020.
2. Souza, J.V.C., Nono, M.C.A., Ribeiro, M.V., Machado, J.P.B., Silva, O.M.M., Cutting forces in turning of gray cast iron using silicon nitride based cutting tool. *Mater. Des.*, 30, 2715–2720, 2009.
3. Lü, Z., Deng, L., Tian, Q., Zhao, X., Cutting performance of Si3N4/TiC micro-nanocomposite ceramic tool in dry machining of hardened steel. *Int. J. Adv. Manuf. Technol.*, 95, 3301–3307, 2017.
4. Kwon, W.T. and Kim, Y.W., Cutting performance of Si3N4 based SiC ceramic cutting tools. *KSME Int. J.*, 18, 388–394, 2004.
5. Zhao, J., Yuan, X., Zhou, Y., Cutting performance and failure mechanisms of an Al2O3/WC/TiC micro- nano-composite ceramic tool. *Int. J. Refract. Met. Hard Mater.*, 28, 330–337, 2010.
6. van Luttervelt, C.A., Childs, T.H.C., Jawahir, I.S., Klocke, F., Venuvinod, P.K., Altintas, Y., Armarego, E., Dornfeld, D., Grabec, I., Leopold, J., *et al.*, Present Situation and Future Trends in Modelling of Machining Operations Progress Report of the CIRP Working Group 'Modelling of Machining Operations'. *CIRP Ann.*, 47, 587–626, 1998.
7. Tu, L., Tian, S., Xu, F., Wang, X., Xu, C., He, B., Zuo, D., Zhang, W., Cutting performance of cubic boron nitride-coated tools in dry turning of hardened ductile iron. *J. Manuf. Process.*, 56, 158–168, 2020.
8. Liu, W., Chu, Q., He, R., Huang, M., Wu, H., Jiang, Q., Chen, J., Deng, X., Wu, S., Preparation and properties of TiAlN coatings on silicon nitride ceramic cutting tools. *Ceram. Int.*, 44, 2209–2215, 2018.
9. Şap, S., Usca, U., Uzun, M., Kuntoğlu, M., Salur, E., Performance evaluation of AlTiN coated carbide tools during machining of ceramic reinforced Cu-based hybrid composites under cryogenic, pure-minimum quantity lubrication and dry regimes. *J. Compos. Mater.*, 56, 3401–3421, 2022.
10. Xing, Y., Deng, J., Zhang, K., Wang, X., Lian, Y., Zhou, Y., Fabrication and dry cutting performance of Si3N4/TiC ceramic tools reinforced with the PVD WS2/Zr soft-coatings. *Ceram. Int.*, 41, 10261–10271, 2015.
11. Mikolajczyk, T., Paczkowski, T., Kuntoglu, M., Patange, A., Binali, R., Research on Using an Unconventional Tool for Increasing Tool Life by Selective Exchange of Worn Cutting Edge. *Appl. Sci.*, 13, 460, 2023.
12. Wang, L., Qi, Q., Zhang, H., Yang, X., Liu, Z., Ge, S., Peng, X., Liu, X., Huang, Z., High tough W-added silicon nitride ceramics. *Ceram. Int.*, 45, 19055–19059, 2019.
13. Lucchini, E., Lo Casto, S., Sbaizero, O., The performance of molybdenum toughened alumina cutting tools in turning a particulate metal matrix composite. *Mater. Sci. Eng., A*, 357, 369–375, 2003.

14. Hu, Z., Zhao, Z., Deng, X., Lu, Z., Liu, J., Qu, Z., Microstructure and mechanical behavior of TiCN reinforced AlSi10Mg composite fabricated by selective laser melting. *Mater. Chem. Phys.*, 283, 125996, 2022.
15. Yang, Y., Dang, W., Liu, J., Zhang, H., Gu, S., Lei, C., Chen, Y., Effect of the Addition of Ultrafine Powders on the Microstructure and Mechanical Properties of TiCN-Based Cermets. *J. Ceram. Sci. Technol.*, 13, 15–22, 2022.
16. Meng, J., Lu, J., Wang, J., Yang, S., Tribological behavior of TiCN-based cermets at elevated temperatures. *Mater. Sci. Eng., A*, 418, 68–76, 2006.
17. Wang, B. and Liu, Z., Cutting performance of solid ceramic end milling tools in machining hardened AISI H13 steel. *Int. J. Refract. Met. Hard Mater.*, 55, 24–32, 2016.
18. Xiong, H., Wen, Y., Gan, X., Li, Z., Chai, L., Influence of coarse TiCN content on the morphology and mechanical properties of ultrafine TiCN-based cermets. *Mater. Sci. Eng., A*, 682, 648–655, 2017.
19. Lee, B.T., Yoon, Y.J., Lee, K.H., Microstructural characterization of electroconductive Si3N4-TiN composites. *Mater. Lett.*, 47, 71–76, 2001.
20. Kgoete, F.M., Popoola, A.P.I., Fayomi, O.S.I., Adebiyi, I.D., Spark plasma sintered Ti-6Al-4V-Si3N4-TiN ternary composites: Effect of combined microsized Si3N4 and TiN addition on microstructure and mechanical properties for aerospace application. *J. Alloys Compd.*, 769, 817–823, 2018.
21. Huang, J.L., Chang, Y.L., Lu, H.H., Fabrication of multilaminated Si3N4-Si3N4/TiN composites and its anisotropic fracture behavior. *J. Mater. Res.*, 12, 2337–2344, 1997.
22. Miyazaki, H., Hyuga, H., Yoshizawa, Y.-I., Hirao, K., Ohji, T., Correlation of wear behavior and indentation fracture resistance in silicon nitride ceramics hot-pressed with alumina and yttria. *J. Eur. Ceram. Soc.*, 29, 1535–1542, 2009.
23. Tsuge, A., Nishida, K., Komatsu, M., Effect of Crystallizing the Grain-Boundary Glass Phase on the High-Temperature Strength of Hot-Pressed Si3N4 Containing Y2O3. *J. Am. Ceram. Soc.*, 58, 323–326, 1975.
24. Hu, J., Zhang, B., Li, C., Wang, L., Wang, S., Shi, Z., Yang, J., Fabrication of Si3N4 ceramics with high thermal conductivity and flexural strength via novel two-step gas-pressure sintering. *J. Eur. Ceram. Soc.*, 42, 4846–4854, 2022.
25. Hampshire, S., Silicon nitride ceramics—Review of structure, processing and properties. *J. Achiev. Mater. Manuf. Eng.*, 24, 43–50, 2007.
26. Liao, S., Zhou, L., Jiang, C., Wang, J., Zhuang, Y., Li, S., Thermal conductivity and mechanical properties of Si3N4 ceramics with binary fluoride sintering additives. *J. Eur. Ceram. Soc.*, 41, 6971–6982, 2021.
27. Cinibulk, M.K., Thomas, G., Johnson, S.M., Fabrication and Secondary-Phase Crystallization of Rare-Earth Disilicate-Silicon Nitride Ceramics. *J. Am. Ceram. Soc.*, 75, 2037–2043, 1992.
28. Shen, Z., Zhao, Z., Peng, H., Nygren, M., Formation of tough interlocking microstructures in silicon nitride ceramics by dynamic ripening. *Nature*, 417, 266–269, 2002.

29. Xiong, J., Guo, Z., Wen, B., Li, C., Shen, B., Microstructure and properties of ultra-fine TiC0.7N0.3 cermet. *Mater. Sci. Eng., A*, 416, 51–58, 2006.
30. Gao, J., Song, J., Lv, M., Cao, L., Xie, J., Microstructure and mechanical properties of TiC0.7N0.3-HfC cermet tool materials. *Ceram. Int.*, 44, 17895–17904, 2018.
31. Toenshoff, H.K., Hard material cutting, in: CIRP Encyclopedia of Production Engineering. pp. 619–627, Springer, Berlin, Heidelberg, 2014.

6

Cermet: Composite Materials for Cutting Tools, Electrical Contacts, and Thermal Management

Riyadh A. Al-Samarai[1]* and Yarub Al-Douri[2,3,4]

[1]*Electromechanical Engineering Department, Engineering College, University of Samarra, Samarra, Iraq*
[2]*College of Health and Medical Techniques, Al-Bayan University, Baghdad, Iraq*
[3]*Department of Mechanical Engineering, Faculty of Engineering, Piri Reis University, EflatunSk, Tuzla, Istanbul, Turkey*
[4]*Nanotechnology and Catalysis Research Centre, University of Malaya, Kuala Lumpur, Malaysia*

Abstract
Cermet, sintered titanium carbonitride carbides, and titanium nitride are covered in this chapter. A unique class of cermets known as TiN and TiCN compounds offers an excellent combination of high-temperature resistance and hardness when exposed to hostile, oxidizing, or abrasive conditions. TiN and TiCN coatings have a variety of uses, but their mechanical and tribological properties are dependent on the substrates they are put to. Their work focuses mostly on the performance of TiCN and TiN coatings on machined tool steel. The enhancement of high stiffness at low hardness is the only new development. Finding patterns and advancements in cutting tool technology requires evaluating gains in hardness and toughness. Furthermore, it is anticipated that to overcome, the extremely hard, binder-free TiCN phase will be needed the drawbacks of utilizing TiCN at high temperatures as a cutting tool.

Keywords: Cermet, titanium carbonitride (TiCN), titanium nitride (TiN), high-temperature resistance, hardness, tribological properties, tool steel coatings, cutting tool technology

**Corresponding author*: riyadh2824@yahoo.com

Inamuddin, Tariq Altalhi and Mohammad Abu Jafar Mazumder (eds.) Nanostructured Hard Materials: Industrial Applications, (207–236) © 2026 Scrivener Publishing LLC

6.1 Introduction

For machining, drilling, and mining applications, today's manufacturing industries are continuously developing their wear-resistant coatings and tools to improve equipment performance and durability while lowering maintenance costs. Producers of technology can be applied to improve goods, maintain competitiveness, and eventually boost revenue. Some essential properties, such as superior wear and corrosion resistance, good toughness and hardness, environmental friendliness, high performance at high temperatures, and reasonably cheap production costs, are ideal for materials to meet the aforementioned aims [1]. To endure prolonged severe wear, the component needs to have a high level of wear resistance and corrosion resistance. If the part can be successfully used in these situations regularly, then it is not. It lowers maintenance and replacement costs while raising safety [2].

Equally crucial as having a material with excellent corrosion resistance is having a material with robust mechanical properties. A wide range of metal-ceramic systems are now the focus of extensive study. Among them are titanium carbide (TiC) and titanium carbonitride. Ti(C, N) ceramic phases. Conventional "carbide" compounds, which are made of tungsten carbide (WC) embedded in a cobalt link, serve the same purpose as cermets; however, most applications cannot employ them because of their high density and relatively low resistance to oxidation and corrosion. applications involving high temperatures in structures. Their primary applications are in cutting and machining due to their excellent wear resistance, which is provided by their high strength, toughness, and hardness combination [3–8]. Historically, metal carbides and nitrides have been made with a range of techniques, the most cost-effective and efficient being solid-phase reaction techniques [9]. Depending on Ti(C, N), the hardness of the material and cermet can be changed. Because the hardness and toughness grow with increasing TiC content and TiN ratio, respectively, there is a significant amount of TiC and TiN. The common metal binder materials utilized in these ceramics include nickel, cobalt, stainless steel, and nickel aluminide (Ni3Al). Ni3Al is particularly interesting because of its great heat resistance and outstanding strength retention at high temperatures [11–13]. However, TiB2 low self-diffusion coefficient and poor fracture toughness make it difficult to compact, limiting its application [4]. Among other benefits, the combination of TiB2 and AlN provides increased impact resistance due to unique strengthening mechanisms. When it comes to hardening ceramic materials, these mechanisms are critical. A class of extremely hard, refractory, wear-resistant alloys produced by powder metallurgy are known as carbides and cermets. These materials are widely used as high-pressure components, cutting tools, wear parts in mining and industrial applications, and drill inserts. Properties including toughness, microstructure, hardness, strength, corrosion, and wear resistance have a significant impact on component life. While the binder (or matrix) is the dominant phase in MMC, the hard carbide phase predominates in carbide alloys and cermets. The long-term plan has two parts. First, the relationship between properties, microstructure, composition, production conditions, carbides, and cermets may perform better. Second, to gain new insights and expand potential markets for these materials, new carbides, ceramics, and hard coatings are being developed using primary resources from South Africa [1]. Co3O4 and WO3 [2–4], two wear products of WC-Co, are registered under REACH, the European Union's system of chemical registration, evaluation, authorization, and restriction [5]. A significant barrier to thermal and mechanical applications of ceramic materials is their low fracture toughness, as shown by KIC tests [2]. Among these applications, objects of high hardness, such as those made of hardened steel, require the use of ceramics exclusively as cutting tools.

6.2 Structural Formation

6.2.1 Nitride Titanium (TiN)

Reducing the amount of titanium dioxide in the ore that is of the "rutile" form will yield TiN. It has been found that large homogeneous mononitride and nitride forms of different stoichiometries, such as sodium nitride (Na3N), exist. For instance, an oxygen or carbon atom in the interstitial region can readily replace the nitrogen atom in titanium nitride in the TiN0.1 and TiN0.5 phases. As seen in Figures 6.1 and 6.2, this leads to the production of individual atoms (C, O, and N). TiN is exceptionally stable at temperatures below 1300°C. Product materials now have better mechanical qualities, particularly increased hardness. Older or traditional machining processes like drilling, turning, and milling involve cutting tools with harder materials than the material's hardness when cutting it. The short service life of cutting tools used in different machining processes is caused by progressively diminishing hardness, which raises the cost of machining [4–7].

For example, when the temperature increased to 500°C, the curve took a parabolic shape [20]. Refractory carbides and nitrides like titanium carbide and titanium nitride are typically not found in normal conditions. They frequently take a long time and cost a lot of money to install [14].

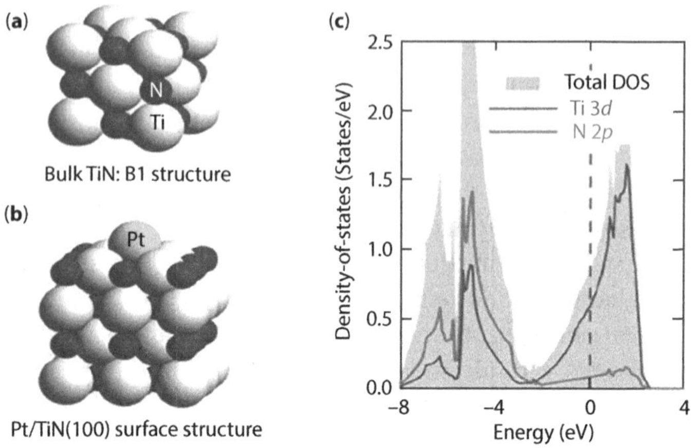

Figure 6.1 Titanium nitride's (TiN) crystal structure (VESTA software are used to calculate the crystal structure of titanium nitride) [17].

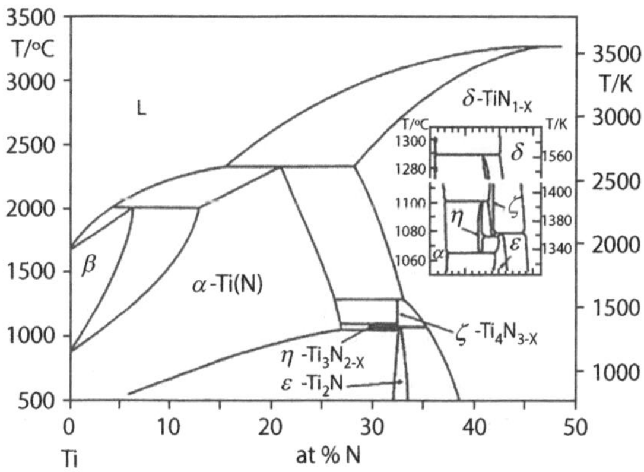

Figure 6.2 Titanium Nitride (TiN) phase diagram, reproduced with consent [17].

These materials feature robust covalent bonding due to their excellent heat resistance and inertness. Therefore, it cannot be acquired through conventional means. This rescue prevented it from being widely used before now [21].

6.2.2 Carbonitride of Titanium (TiCN)

Powders, coatings, fibers, and filaments are typical forms of refractory carbides and nitrides. Nevertheless, powdered TiC, TiN, and Ti (C, N) materials that are currently made by various methods will be taken into consideration later in the study [15, 16]. TiCN is well known as an intriguing material for cutting tools because of its wear resistance and capacity to be utilized in the production of intricate ceramic composites or refractory materials [22–24]. The inclusion of carbon or nitrogen atoms in the interstices may be the reason for the titanium carbonitride ceramic particles' broad dispersion. The octahedral lattice location has randomly provided nitrogen and carbon atoms. Cubic structural categorization TiCN are still face-centered cubic, in contrast to TiN and TiC [24–26]. The elements can be directly carbonized or nitrided to produce TiC, TiN, and Ti(C, N). Because it requires energy, this process consumes a lot of energy. The time ranges from 5 to 20 hours, with temperatures ranging between 1900°C and 2900°C [27]. Finally, Finarhyd is a very clumpy powder that needs to be ground more finely. To remove contaminants, additional chemical cleaning is required due to the small particle size, constructed from milling machines.

As a result, the procedure is very expensive due to the cost of pure titanium and because it uses a lot of energy [18], which leads to an ordered arrangement of titanium atoms and a bold formation in the presence of N and C atoms.

6.3 Coating of TiCN and TiN

Researchers are currently looking into the production of titanium nitride and its many uses in engineering services. Solid coatings remain at the forefront of research due to their versatility across numerous fields of study [1–3]. There has been a lot of interest in the mechanical compatibility of coating materials such as titanium nitride (TiN) and titanium carbonitride (TiCN) with different metal substrates. Producing these ceramics on metal substrates is challenging because of differences in mechanical characteristics, especially tribological and mechanical aspects [1–6]. Consequently, a great deal of research has been conducted to improve these coatings' properties and increase their range of applications. Cutting tool development plays a crucial role in improving mechanical properties, especially during long-term high-speed machining. However, there are a number of issues with how cutting tools are made. Prior to being used, the material must fulfill the standards for the required mechanical qualities. Second, the cost of the finished product is a reflection of the limits of traditional sintering techniques, which include the quantity of materials used and the time needed to finish the sintering cycle. This presents significant challenges for the manufacture of these instruments [8]. Furthermore, the materials used to make cutting tools need to be able to withstand abrasive cutting environments, like high temperatures and friction between the cutting tool's surface and the workpiece. Cutting tool materials need to have several characteristics to achieve these criteria [9].

TiCN coatings provide strong mechanical and wear-resistant qualities as well as great thermal stability. They also demonstrated minimal internal stresses and great resistance to corrosion [4, 6, 7]. Additionally, they are still utilized today to safeguard mechanical components and the materials used in cutting tools [28–31] due to their resistance to damage during transportation [5, 8–10].

6.4 Challenges and Restrictions of TiN and TiCN

Despite this, a variety of failure modes, such as high load, temperature, stability, and time, have an impact on the service life of H13 tools due to their demanding operating conditions [32, 33]. The main consequences of such working conditions are adhesive and abrasive tool surface deterioration and thermal fatigue [35, 36]. Because of this, working instruments need to be changed after a certain time, which raises the overall cost of production processes [12–13]. The melting point of these composites is only determined by the stoichiometric arrangement of these constituents [38–41]. These defining parameters are reportedly used in the manufacturing of molten metal crucibles, cutting tools, and refractory linings for furnaces and hypersonic vehicles.

Although TiN and TiCN offer valuable features, their applications are often limited by drawbacks such as low fracture toughness and poor oxidation resistance. In addition, important elements such as a long sintering period can limit grain formation and increase particle size. As a result, the performance and overall efficiency of the sintered body may also be reduced indirectly [42–47]. In any case, it is critical to improve the synthesis of TiN and TiCN composites, and properties such as tensile strength, elastic modulus, fracture toughness, and oxidation resistance require urgent research.

6.5 Treatment Techniques

Microstructural integrity or material and component morphology. Collectively, they affect how composites are processed and performed [34]. The processing techniques and sintering are the same because they are both extraordinarily high-temperature ceramics [48, 49]. Pseudothermal pressure (PHIP), quasi-pneumatic pressure (QIP), spark plasma (SPS), and response spark pressure. TiCN-based cermet is a promising material for various modern applications, such as semi-finished and final manufacture [50, 51]. Its limited wear resistance, however, prevents it from being utilized in other industries, such as high-speed cutting. The surface layer containing a significant quantity is called wear because of the increased hardness and lack of a ductile metal binder phase that contrasts with the entire cermet. This can result in better cutting operations when the carbide phase is layers of repetition generated in the nitride environment [52–55].

Carbide-containing cermets are often manufactured mostly from titanium carbides, tungsten carbides, and possibly cermets, or solid materials based on titanium and tungsten carbides, in small quantities [55, 56]. Therefore, the harder the material, which reduces the wear rate and increases wear resistance, the greater the number of binder phases [56].

6.6 Factors that Improve Sintering Kinetics

- Diffusion kinetics are mostly improved by spark plasma sintering when compared to other traditional sintering methods. It is commonly recognized that any composite specimen's compaction will increase when pressure is used to create plastic [57]. The spark plasma sintering is produced by the high pulsed energy source [58, 59]. Research indicates that to achieve sufficient densification of any sintered product in IPS, a continuous pulsed current application is necessary [60].
- When creating nanocrystalline Al2O3 composites from the resulting powders [62–66], high heating rates are required to achieve the greatest densification, according to a recent study by Chandran *et al.*, [61].
- Furthermore, research has shown that DC pulsing raises the degree of condensation in titanium carbonitride.

6.7 Binderless

In particular, binder-free WC shows remarkable resilience against wear when compared to titanium carbonitride and certain other materials. Because of their remarkable wear resistance, binderless carbide systems are advantageous and significant in numerous manufacturing industries, including high-speed cutting tools, etc. [72, 75]. Because the binder phase deteriorates in some severe situations, bondless carbides and cermets are rapidly replacing typical cemented carbides in industry. The most productive and well-suited for combination qualities are bondless WC-MoxC carbides [79].

In addition, fracture toughness is reduced compared to previous sintering methods. Figure 6.3 grain distribution of metal and ceramic materials used at high temperatures, either with or without binders [78]. It has two intertwined fcc lattices, one of which has a comparable crystal structure and is composed of carbon and metal. Two-dimensional (2D) nanomaterials have garnered a lot of interest as a novel kind of nanomaterial. There are ultrathin 2D nanomaterials that have a special planar structure, in which the side measurements are larger than sub-microns. Furthermore, its lengthwise dimensions is restricted to atomic or nanoscale levels. Frameworks, typically, a bond is created by combining d orbitals with 2s and 2p orbitals.

The structure of sp orbitals changes with increasing concentration of electron clouds. The bulk modulus and lattice constants of the carbides are theoretically analyzed and correlated using the Murnaghan equation. Based

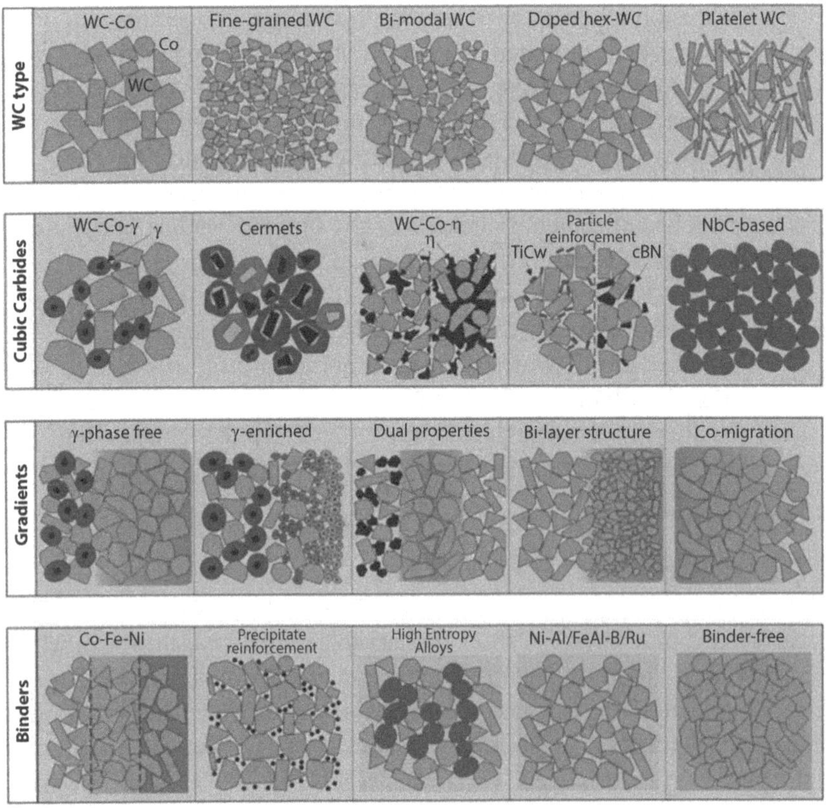

Figure 6.3 High-temperature granular dispersion of ceramic, with or without binders [70].

on individual two-dimensional transition metal atoms, explore 66 transition metal carbides due to their high content, more than 800 of them have different computer characteristics of MAX phases.

Zhou *et al.* [55], have conducted analysis of effect of toughness at different metal of Ti(C,N). It has been demonstrated that when using metal bonds ranging from 4–5 and 2–3 μm, respectively, the fracture toughness of TiCN approaches 20% and 25%. Consequently, as the proportion of binder metal increases, these values begin to decrease.

6.8 Prior Studies on Titanium Carbonate Nitride and Titanium Nitride without a Binder Used Sintering Techniques

In place of conventional carbides, unbonded carbides have been designed to augment the conventional carbide cermet composition [79]. It is interesting to note that it is very difficult to fuse any materials, including tungsten carbides with titanium carbides, using antiquated standard sintering procedures [68, 69]. Owing to its remarkable properties, including thermal stability and refractoriness, Rosias, Cardinal, and Young [78].

Superior tolerance control is possible with titanium carbonitride cermet enhanced surface quality, feed rate, and any complex machine part [72–73]. Hard coatings made of Ti(C,N), TiN and TiC are applied to the surface of hard alloys. These coating materials have recently undergone modifications to better suit particular applications in sectors like [64]. The various structural distributions of hard alloys are shown in Figure 6.4. The majority of materials used in friction and wire drawing applications develop these microstructures. Different grain is produced by various reinforcing chemicals, buildings, as depicted in Figure 6.5. However, it has been demonstrated that TiC inclusions can enhance the processing capabilities of high-speed cutting tools [75].

The mechanism of tungsten carbide is created during the initial alloying with 0.5% Cr3C2-0.3%VC. According to Kim *et al.* [73], binder-free ($TiC_{0.5}N_{0.5}$) titanium carbonate nitride, which are ball at different times and then sintered at 1170°C using high-frequency induction heating methods, showed good results. The compaction process has been significantly improved by developing the crystalline structure of the ball milled powders. Shredding is reported to be decreasing. After 10 hours of grinding, the powder crystallite size decreased from 241 nm to 9 nm as illustrated in Figure 6.6.

After sintering, the rapid sintering procedure preserved the fine grain microstructure. The relationship between variable crystallites and relative

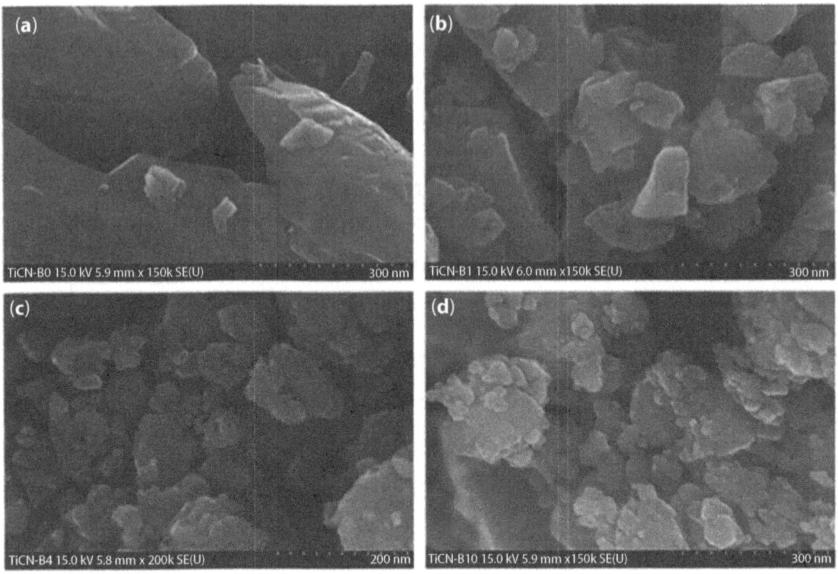

Figure 6.4 FESEM microstructures with different grinding times on TiCN without binder, including (a) as received, (b) 1 hour, (c) 4 hours and (d) 10 hours [70].

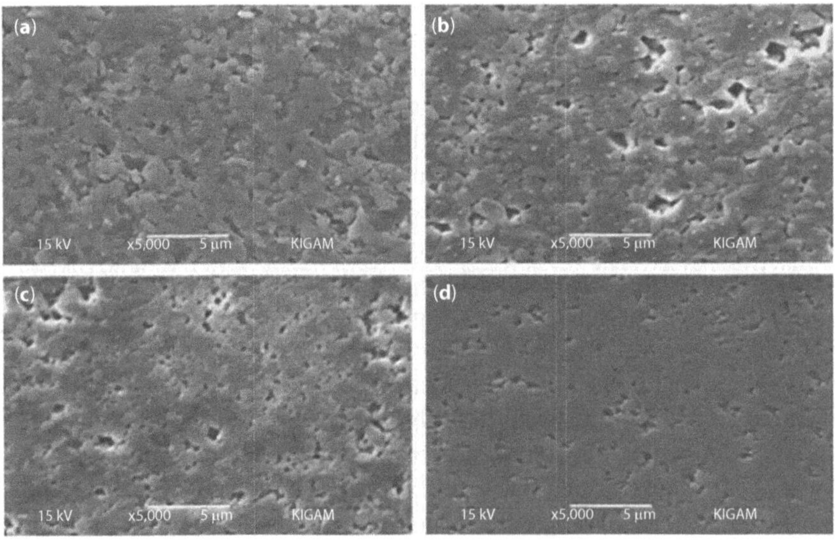

Figure 6.5 FESEM microstructures on sintered TiCN without binder after milling for (a) 1 hour, (b) 1 hour, (c) 4 hours, and (d) 10 hours [70].

density and fracture toughness are insignificant. It is estimated that the crystallite size decreased by 231-9 nm after treatment for 10 hours. HFIHS, sometimes called fast combined high frequency induction thermal sintering [74, 75]. After sintering, the process retained a fine microstructure with acceptable mechanical properties, as shown in Figure 6.7 graphite. According to the results, 99% compaction are achieved after 4 minutes of exposure.

Relative density and microhardness are in direct inverse relationship, since the grinding time reached 40 hours. Using HEBM high-energy ball milling technology, the microstructure of binder-free titanium nitride particles has been thoroughly improved, as shown in Figure 6.8. The X-ray patterns show after 40 hours of grinding, broadening due to deformation

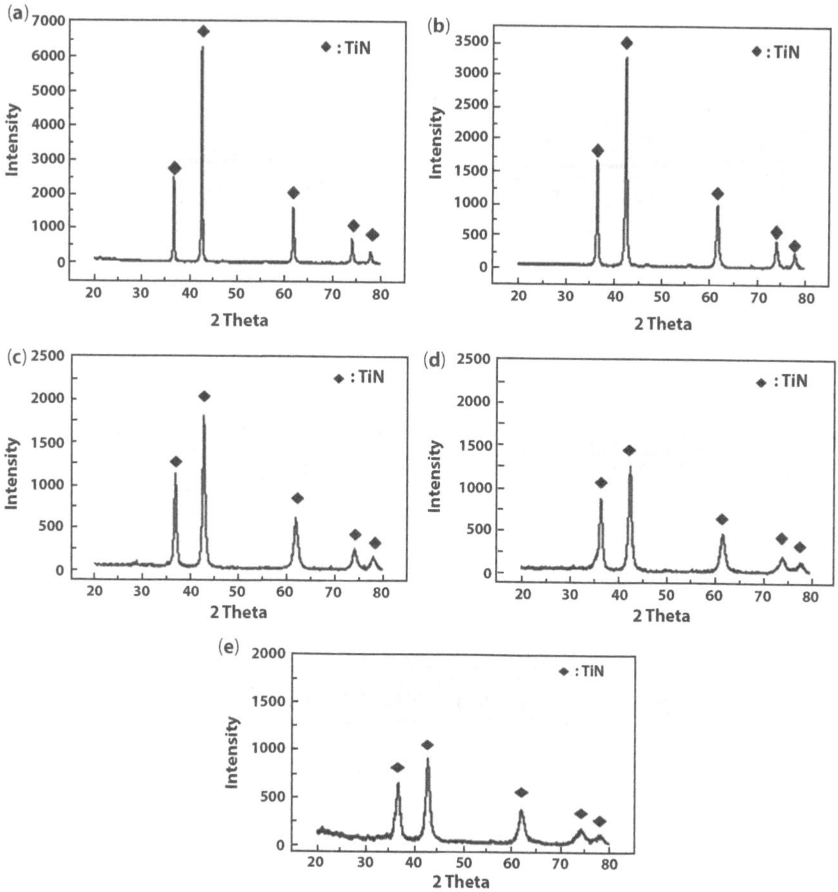

Figure 6.6 The XRD patterns below illustrate titanium nitride to milling durations without the use of binders: Cp-TiN, TiN-1 h, TiN-4 h, TiN-10 h, TiN-40 h, and TiN-1 h [70]).

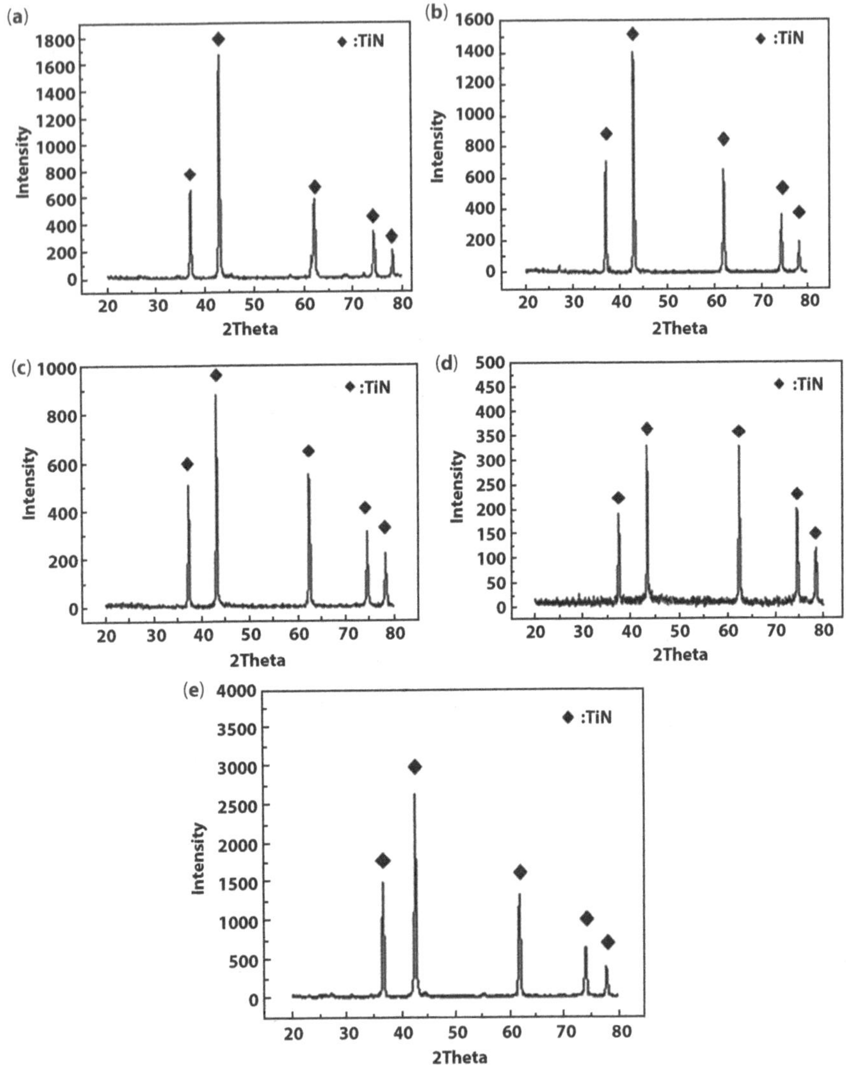

Figure 6.7 Demonstrates the results through the X-ray diffraction patterns of titanium nitride to milling periods (without binders): 40 h, 40 h, 10 h, 1 h, and Cp- TiN [70].

of the crystallites and a decrease in crystallite size, noticeable after 1 hour of grinding. The ball mills used have Due to the constant contact of the granules with the powder in the grinding bottle, there are a tendency for unwanted particles or contaminants to enter. In Figure 6.9, it clearly shows the X-ray of grinding. The peak broadening of titanium nitride decreases as a result of grain development during the sintering process.

Figure 6.8 Titanium nitride powder SEM pictures for (a) Cp-TiN, (b) 1 h, (c) 4 h, (d) 10 h, and (e) 40 h, respectively, were obtained [70].

As grinding time progresses, as shown in Figure 6.8, the formula developed determines the final size of each particle. After long milling segments of 1 hour, 4 hours, 10 hours, and 40 hours, Norton determined them to be 56 nm, 21 nm, 15 nm, and 10 nm. When TiN is added, the concentration of Mo in the bond phase increases, which greatly increases the corrosion resistance of cermet [71]. The grain size and edge thickness of solid particles generally decrease with increasing nitrogen content in cermet. In this regard, TiCN-based cermet shows interesting properties, including low metal sliding friction, excellent chemical stability, high hardness at high temperatures, and good wear resistance. In addition to better performance in terms of wear resistance and good surface quality of mechanical

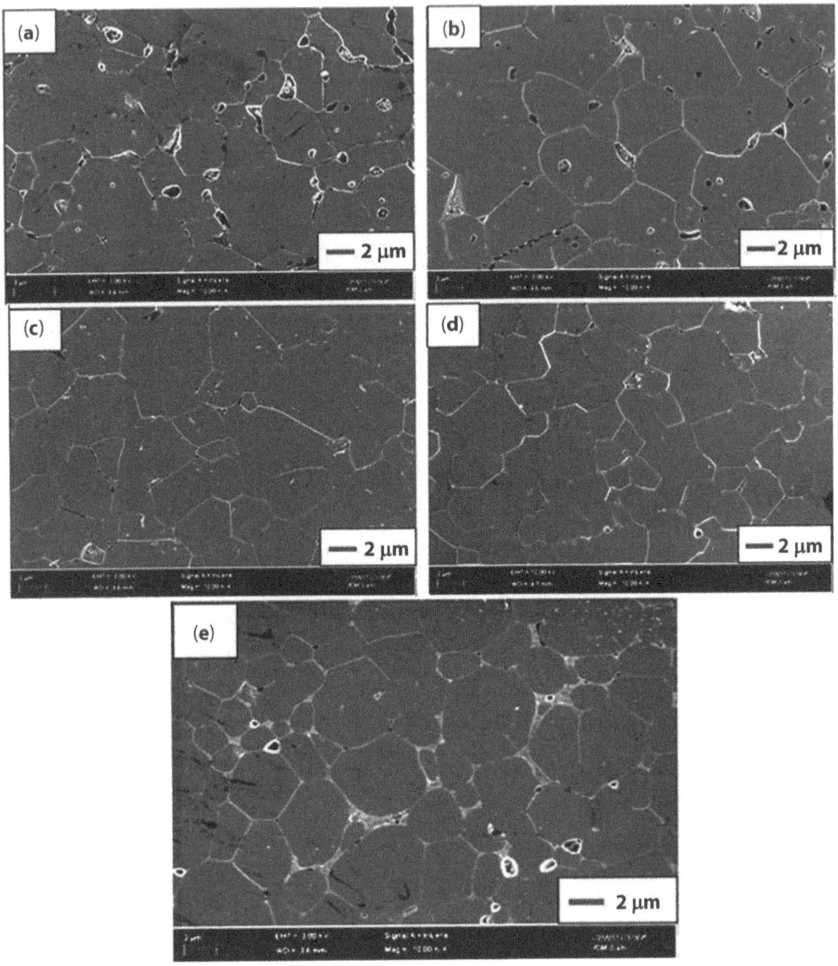

Figure 6.9 A scanning microscopy (SEM) of titanium nitride taken at milling times for the following Cp-TiN, (a) 1 h, (b) 4 h, (d) 10 h, and (e) 40 h [70] are the times.

workpieces, it can replace traditional carbide alloys. The effect of adding nanometric or micrometric TiN on the mechanical properties and microstructure of TiC-based cermet has been the subject of recent research [6]. These results demonstrated that the addition of Nano TiN could stop the melting of TiC grains and that the mechanical performance are improved when TiN are added at a weight percentage of less than 6 wt%. It would be interesting to investigate how the TiN particles generated by SHS affect the final mechanical properties of the bulk cermet in this case and compare its frictional behavior with that of other solid materials. Nitrogen addition to

TiC-Ni-Mo ceramic alloys has been the subject of numerous studies aimed at optimizing the alloys' microstructure, particularly the solids' grain size [76]. These days, titanium carbide (TiC)-based modified cermets are frequently utilized as cutting tools due to their superior hardness and wear resistance compared to TiC–Mo_2C–Ni cermets [3–5].

As the TiC content areas, so does the hardness. To ensure that the metal binder adequately wets the ceramic phase and forms a mixed carbide (Ti, Mo, C) ring around the TiC grains, molybdenum is often supplied as Mo_2C. Harder surfaces are attributed to nickel. The parent combination forms a complex microstructure after sintering, with the development of the distinctive fundamental edge structure of the grains (Ti, Mo) (C, N). The edge is enriched with the heaviest element Mo and shares the same crystal structure as the nucleus, whereas the center is virtually unresolved TiC or TiCN.

6.9 Niobium Carbide

Recently, there has been a renewed focus on the search for novel materials with remarkable properties such as high incompressibility, hardness, chemical inertness, and thermal stability. Numerous materials, such as trimetallic carbides, diamonds, nitrides, and boron derivatives, have been studied in this context. The primary objectives of these activities are the investigation, production, and enhancement of materials for strength-related applications. In the past, the primary techniques for characterizing these materials were indentation tests like the Brinell and Rockwell tests [18–22]. For each of these applications, it is essential to guarantee good performance and long component life by optimizing different qualities such as hardness, heat resistance, fracture toughness, strength, and wear resistance [23]. Although in theory other factors such as tool constants and use are first disregarded, most entire material units that are taken from the radioactive substance are the only basis for theoretical hypotheses. These materials are the only ones that use conventional glass materials like Almaz and C-BN; however, even these materials have certain uses and needs [23].

6.10 NbC Cermets

Because NbC is brittle in the absence of a binder, metal matrix composites, and cermets combine the advantages of, say, the hardness of the binder with the fracture toughness of the ceramic. Several parameter studies, including those based on cobalt, nickel, and iron binders, have been conducted to examine various metal-bonded NbC ceramics. The binder content is expressed as either a percentage by weight (wt%) or a percentage by volume (vol%).

6.11 NbC-Co Cermets

Cho *et al.* [30], completed the first fundamental studies on the development of NbC grains in a Co matrix during sintering in 1968, 2001 and 2004, respectively. Due to the fact that a cobalt binder, namely WC-6Co, is used for the production and use of cemented WC carbides, Co-bonded NbC cermets with a binder content of 8 or 12 wt.% Co is created [8].

6.12 NbC-Ni Cermets

Since the usage of Co is linked to health issues and switching from WC to NbC results in rather poor binder wettability, Co is being replaced by a Ni binder [17, 37]. NbC-Ni cermets are harder but stronger than NbC-Co cermets [4]. Cemented carbides, also known as metal matrix composites (MMCs), are composed of a hard carbide phase, usually a tungsten carbide (WC) monolith, surrounded by a ductile metal binder [77].

The hard carbide phase predominates in hard alloys and ceramics, while the binder (or matrix) is the main phase in MMC. The word "cermet", a combination of the "metal" and "ceramic", refers to a subgroup of cemented carbides where TiC or NbC, rather than WC, are the predominant hard carbide phase and are embedded in a carbide matrix. Co, Ni or Fe bonds. Until now, 70% of cutting tools produced are made from tungsten carbide (WC), the material of choice for cemented carbide.

6.13 Secondary Carbides' Impact

Corrosion is a major issue that often arises in the hydropower, chemical, oil, and petroleum refining industries [4–6]. Pumps and valves employed in the transportation of slurries or marine oil are among the components of flow that are most frequently prone to failure. In such extreme settings, ceramic-metal composites offer an excellent combination of high strength and high hardness to successfully combat aggressive media. The corrosion behavior of metal and ceramic composites has been the subject of numerous fundamental investigations, with some significant findings already achieved [38–39].

6.14 Mechanical Properties

Because of their porosity and susceptibility to degradation regardless of the joining technique, the materials exhibit low mechanical strength and toughness. The study [42], states that carbides only become ductile in the presence of significant stress close to the crack source. They claim that preventing fractures is the sole method to make carbides stronger. Grain boundary phases, microstructure grain size, and surface porosity all affect hardness [43]. NbC ceramics are developed and studied with great attention. Between 2000 and 2014, NbC-Co metal ceramics were the subject of extensive research. NbC-Co has a harder consistency than WC-cemented carbide alloys, although it is not as resilient. By substituting NbC1.0, NbC0.88-12Co has a higher hardness. Lately, NBC cements with iron bonds have been examined. In contrast to NbC-Ni cermets, the hardness-to-fracture toughness ratio is often worse. The durability and endurance of cutting tools, particularly those used for heavy lifting and heavy-duty applications, are now subject to stronger limitations due to recent technological advancements [1]. Enhancing the mechanical and physical characteristics of diamond composites, which are utilized to create cutting tools for a variety of technical applications, is crucial in this regard. High physical, mechanical, and performance characteristics, including wear resistance, heat and friction resistance, hardness, elastic modulus, fracture toughness, thermal conductivity, ductility, and high diamond holding capacity, must all be present in these materials [2–4]. It is commonly known that the wear resistance of diamond composites is influenced by the elemental composition and matrix properties, whether they are metallic or carbide [5–7].

In this regard, a large number of studies have focused on nickel-bonded NbC ceramics with varying concentrations of secondary carbides. Metal ceramics are produced using numerous technological methods. The main focus are metal powder production including mechanical grinding, colloidal powder, hot pressing and sintering, SPS, LPS and further compression using GPS or HIP. Casting Fe3Al-NbC metal matrix composites has also been successful in some cases. Table 6.1 describes a few ceramic nanocomposites' properties [79]. The study of wear, lubrication, and friction are known as tribology. Materials can become worn out either by contact between two surfaces or through contact between an exposed surface and abrasive particles, liquids, chemicals, or heat [67].

All in all, there are two basic categories of wear mechanisms: mechanical and all other cases that include an additional degree of active chemistry, like oxidation or corrosion. Four categories can be used to separate the different types of mechanical wear. Categories include surface fatigue, adhesion,

Table 6.1 Some ceramic nanocomposites' characteristics [79].

Ceramic composite	Properties
SiN/SiC	increased toughness and fracture strength
Al_2O_3/SiC	durability, strength at high temperatures
Al_2O_3/Co	creepmagnetic, electrical, and optical qualities
Al_2O_3/Ni	thermal conductivity, electrical, and electronic qualities
Fe_2O_3/CNT	enhanced electrical conductivity, microhardness, friction/wear coefficients, and fracture toughness
Al_2O_3/CNT	superior strength and toughness to fracture over monolithic SiO_2
SiO_2/CNT	enhanced fracture toughness, optical qualities, and tensile strength
SiC/CNT	enhanced storage modulus, reduced flammability, and increased
PE/TiO_2	tensile Young's modulus and robust matrix-silica layer interaction
Nylon-6/layered-silicates	enhanced stiffness, Young's modulus, strain at break, tensile stress, and tensile strength
Nylon-6/montmorillonite	improved thermal stability and glass transition
PP/montmorillonite	Improved heat stability
PVA/Ag	bending strength, break, bending modulus, and storage modulus
PPX/Pd or Sn NPs	
PMMA/montmorillonite	Increased modulus and tensile strength
Polylactide/layered-silicates	Strength and Young's modulus improvements
Epoxy/layered-silicates	enhanced strain to failure, and strength a decrease in ultimate
Epoxy/CNTs	strain and tensile strength and an increase in Young's modulus

(Continued)

Table 6.1 Some ceramic nanocomposites' characteristics [79]. (*Continued*)

Ceramic composite	Properties
Epoxy/CNTs	increased ductility and strain energy density
Epoxy/DWCNT	heightened yielding and breaking energy
Polyimide/MWCNT	increases in strength and stiffness
PE/PAN/MWCNT	Growth in electrical conductivity, tensile strength, and Young's modulusSignificant increase in hardness
Silicone/elastomer SWCNT	strengthened microhardness Higher hardness and Young's modulus
PVC/MWCNT	
a-Fe/$Fe_{23}C6/Fe_3B$	
Al/SiC	
Cu/Nb	

abrasion, and erosion. Additionally, there is also a fretting mechanism operating under similar wear circumstances, which combines adhesive, abrasive, and occasionally corrosive wear. From an industrial perspective, all of the aforementioned modes result in loss of dimensional tolerances, higher operating costs, and supplementary costs like part replacement [67, 68]. In addition, according to some studies, the main wear mechanisms include brittle pullout or separation of carbon nitride grains, as well as extrusion and spreading of the binder phase. From a mechanical point of view, "water" and "dry" corrosion have certain similarities. However, in solid and liquid corrosive environments, the fluid effect and more complex particle trajectories lead to more complex material failure mechanisms [12].

6.15 Technical Developments

The use of ceramic metal (cermet) cutting tools for hard machining of steel's brief history of hard machining technology is given, as well as an analysis of the importance of cutting tools made of hard alloys (cermet and carbide) for hard machining. In addition, research on increasing the intrinsic factor of cermets is presented. Several machinability parameters are evaluated that control the performance of the cermet. Ultimately, the performance of cermets as a cutting tool for hard machining is restored. To maximize productivity and efficiency, hardened ferrous metals have been processed since the 1980s using hard machining technologies. Grinding was a machining method used to machine hardened black steel before hard finishing was introduced. Due to the need for grinding equipment, longer setup times, and lower metal removal rates than hard machining, the efficiency and productivity of grinding operations were lower. This ultimately affected the cost and duration of production. Despite the long history of ceramic and metal cutting tools, tool manufacturers continue to improve them. According to Wilshire, one such innovation at Kyocera uses a combination of techniques used in the insert molding and sintering processes to create a gradient of cermet material. Because they have a harder outer layer and greater strength in the center, we call them "hybrid cermets." Hardness and strength are also combined in Kyocera ceramic inserts. The company can create a honeycomb-like grain structure, with the hexagonal-shaped honeycomb consisting of a stronger ceramic material than the surrounding, harder central core, Wilshire said. If a tear appears in the middle part, it will not extend beyond the body and will not allow insertion. Carbide is used in the shank of the MC275 Walter Prototyp ceramic burr (Figure 6.10).

The solid carbide shank of the Walter Prototyp MC275/MC075 ceramic cutters is soldered to the ceramic head. In addition, the tool manufacturer offers a design with replaceable ceramic heads that screw into the shank. The purpose of the application is the milling of high-temperature alloys. While the MC075 has a geometry for high feeds when machining difficult materials, the MC275 offers a general geometry for a variety of applications. Both models offer cutting speeds of up to 3,300 square feet per minute. Walter also produces sintered inserts for turning. However, nickel-bonded cermet cutting

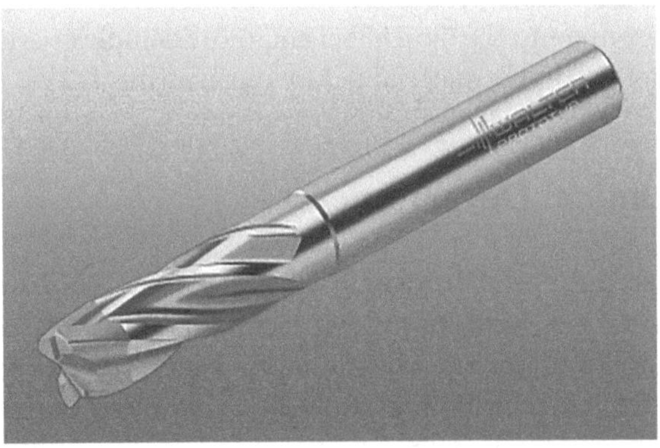

Figure 6.10 The MC275 Walter Prototyp ceramic burr uses carbide in its shank.

tools are not suitable for machining nickel-based superalloys since superalloys often contain nickel. “You get a lot of build-ups on the cutting edge, the cobalt leaches out of the cutting edge, the carbide grains become loose, the edge chips, and the tool doesn’t last very long,” Benkochi explained. “If there is cobalt in the cutting grade and there is cobalt in the workpiece.” According to Joachim Beck, marketing manager at CeramTec GmbH in Plochingen, Germany, cermet cutting tools offer noticeably high cutting-edge stability and moderate toughness when used correctly. (Lawrence, South Carolina, is home to CeramTec North America Corp.) These characteristics make ceramics ideal for precision machining of a range of workpiece components, such as case-hardened steels, free-cut steels, and ductile irons (Figure 6.11). Kyocera alters the grain structure in some of its most recent SiAlON ceramic turning grades, such KS6040, so that some particles resemble needles and function as fracture-resistant whiskers. A literature review showed that coated cemented carbide cutting tools are more commonly used for hard machining than coated cermet [1, 2]. Yang *et al.* [20], reported the performance of TiCN-based cermet with TiAlN and TiAlN/CrAlN coating among the available journal publications on the application of coated cermet for hard machining. Figure 6.12 shows TiAlN and TiAlN/CrAlN coating layers deposited on a cermet substrate. The figure shows that the TiAlN coating layer is firmly adhered to the surface of the metal-ceramic substrate. As can be seen in Figure 6.12(b), on the contrary, a crack is observed at the interface between the TiAlN/CrAlN coating layer and the cermet substrate. Ti(CN) based cermet was found to be suitable for hard turning of steel (62 HRC) at cutting speeds of 60 and 80 m/min, feed of 0.1 mm/rev, and depth of cut

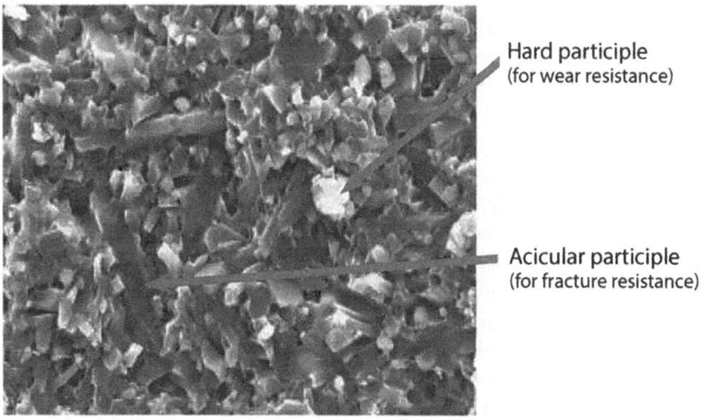

Figure 6.11 In some of its newer SiAlON ceramic turning grades, such as KS6040, Kyocera modifies the grain structure so that some particles take on the shape of needle-like needles that act as fracture-resistant whiskers.

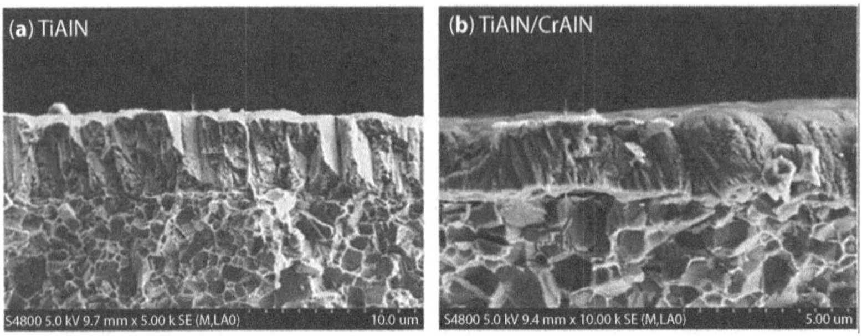

Figure 6.12 Ceramic cutting tools coated with TiAlN (a) and TiAlN/CrAlN (b) [20].

of 0.2 and 0.5 mm. provide longer tool life than coated cemented carbide (TiCN/Al2O3/TiN) (see Figure 6.13 for graphs). However, when cutting at a speed of 60 m/min and a depth of cut of 0.2 mm, cermet showed a higher Ra value than coated cemented carbide in the case of surface roughness (see Figure 6.14 for graphs). In this study, due to crater and lateral wear, the service life of the cermet was completed.

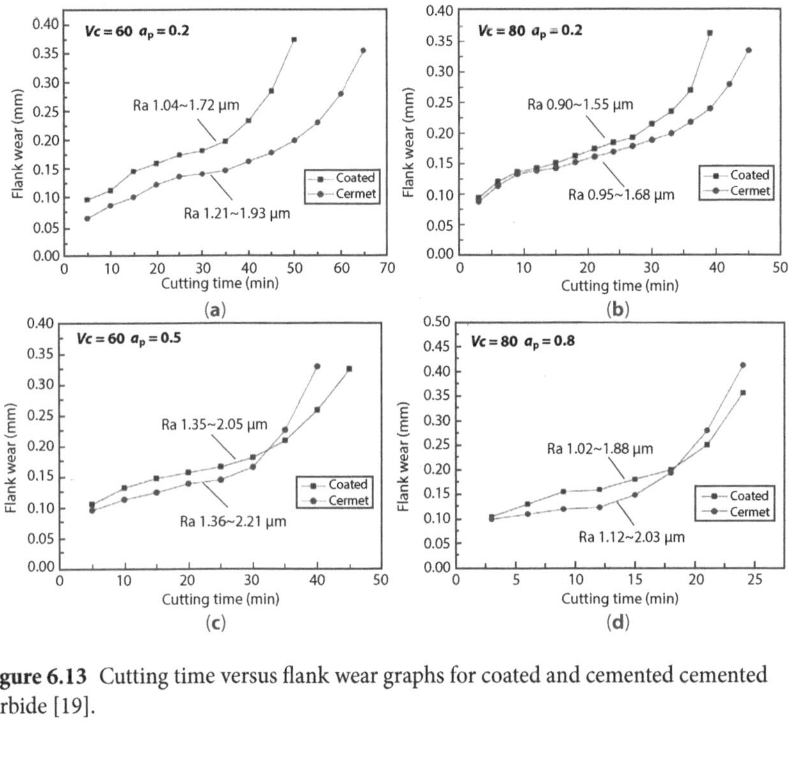

Figure 6.13 Cutting time versus flank wear graphs for coated and cemented cemented carbide [19].

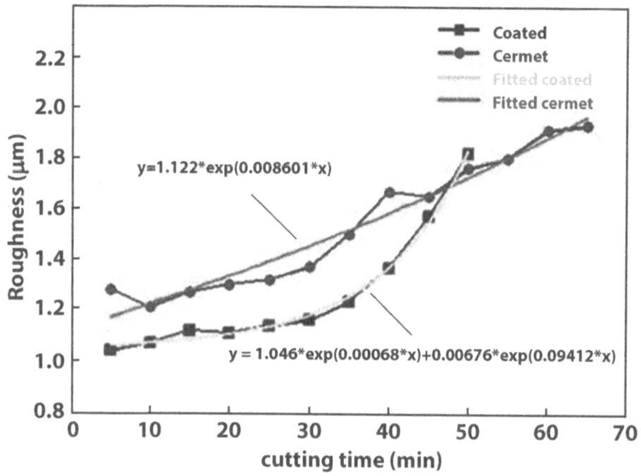

Figure 6.14 Cutting time vs. roughness plots for coated cemented carbide and cement [19].

6.16 Concluded Remarks

In many applications of modern industrial manufacturing, UHTC materials close a gap. Because TiCN/TiN-based cermet is resistant to oxidation, it can be accomplished with ultra-hard ceramic materials that can tolerate the high temperatures of high-speed cutting tools, molten metal crucibles, etc. The primary drawbacks, however, are inadequate mechanical stability of characteristics and poor sintering. TiCN/TiN is brittle while having high-temperature characteristics, which may restrict its use in various applications, particularly in the aerospace sector and in high-speed cutting tools. The primary determinant of TiCN/TiN composites' application and standardization is their manufacturing procedures. Titanium nitride, one of several compounds used as an additional reinforcement to improve the properties of TiCN, can impart special properties to TiCN and TiN that make them very attractive as advanced technologies.

When many carbon allotropes are combined, TiCN performs better. Enhancing TiCN's qualities, such as its oxidation resistance and fracture toughness, is also crucial. The production of binder-free titanium carbonitride is examined in further detail in this review to clarify the impacts of several additions, such as WC, WC-Cr2C3, and TiC. It demands a great deal of attention in research labs and academic institutions.

Additionally, the computed wear factors were found to be incredibly low—on the scale of 10-8. For both coatings, abrasive wear is the primary source of wear, though. Furthermore, some plastic deformation was found by the wear trace profiles. EDS analyses and SEM investigations show that coating integrity is maintained during lubrication conditions. Secondary carbides possess a notable influence on the hardness and wear resistance of the coating, contributing to enhanced tool performance under severe cutting conditions.

Methods for producing binder-free TiCN by powder metallurgy are not addressed in the literature, nor are starting powders, powder impurity preparation, dispersion techniques, equipment requirements (mixing, grinding), sintering method, sintering parameters, or best methods for characterizing composite materials. Last, but not least, it is very important to characterize the mechanical behavior (varicosity and hardness) of the obtained titanium carbonitride cermet without binder.

References

1. Belgacem, L., Ouinas, D., Viña Olay, J.A., Amado, A.A., Experimental Investigation of Notch Effect and Ply Number on Mechanical Behavior of Interply Hybrid Laminates (Glass/Carbon/Epoxy). *Compos. Part B Eng.*, *145*, 189–196, 2018.
2. Yu, J., Pan, Z., Ye, W., Li, Q., Wu, Z., Dynamic Temperature Field and Drilling Damage Mechanism of Plain Woven Carbon/Glass Hybrid Composites. *Compos. Struct.*, *304*, 116375, 2023.
3. Gao, M., Li, S., Guan, W., Xie, H., Wang, X., Liu, J., Wang, H., Excellent thermal shock resistance of NiCrAlY coatings on copper substrate *via* laser cladding. *J. Mater. Sci. Technol.*, 130, 93–102, 2022.
4. Geier, N., Davim, J.P., Szalay, T., Advanced Cutting Tools and Technologies for Drilling Carbon Fibre Reinforced Polymer (CFRP) Composites: A Review. *Compos. Part A Appl. Sci. Manuf.*, *125*, 105552, 2019.
5. Nagy, S.R., Pop, A.B., Titu, A.M., Determination of processing precision of hole in industrial 2 plastic materials. *Polymers*, *15*, 347, 2023.
6. Sorrentino, L., Turchetta, S., Bellini, C., A new method to reduce delaminations during drilling of FRP laminates by feed rate control. *Compos. Struct.*, *186*, 154–164, 2018.
7. Zeng, W., Gan, X., Li, Z., Zhou, K., Effect of TiC addition on the microstructure and mechanical properties of TiN-based cermets. *Ceram. Int.*, 43, 1, 1092–1097, 2017.
8. Imasato, S., Tokumoto, K., Kitada, T., Sakaguchi, S., Properties of ultra-fine grain binderless cemented carbide 'RCCFN'. *Int. J. Refract. Met. Hard Mater.*, 13, 5, 305–312, 1995.
9. Kim, H., Kim, J., Kwon, Y., Mechanical properties of binderless tungsten carbide by spark plasma sintering. *Proceedings. The 9th Russian-Korean International Symposium on Science and Technology, KORUS 2005*, IEEE, pp. 458–461, 2005.

10. Al-wandi, S., Ding, S., Mo, J., An approach to evaluate delamination factor when drilling carbon fiber-reinforced plastics using different drill geometries: Experiment and finite element study. *Int. J. Adv. Manuf. Technol.*, 93, 4043–4061, 2017.
11. Demirskyi, D., Agrawal, D., Ragulya, A., Compararеons of grain size-density trajectory during microwave and conventional sintering of titanium nitride. *J. Alloys Compd.*, 581, 498–501, 2013.
12. Karimi, N.Z., Heidary, H., Minak, G., Critical thrust and feed prediction models in drilling of composite laminates. *Compos. Struct.*, *148*, 19–26, 2016.
13. Fernández-Pérez, J., Cantero, J.L., Díaz-Álvarez, J., Miguélez, M.H., Hybrid Composite-Metal Stack Drilling with Different Minimum Quantity Lubrication Levels. *Materials*, 12, 448, 2019.
14. Gaugel, S., Sripathy, P., Haeger, A., Meinhard, D., Bernthaler, T., Lissek, F., Kaufeld, M., Knoblauch, V., Schneider, G., A comparative study on tool wear and laminate damage in drilling of carbon-fiber reinforced polymers (CFRP). *Compos. Struct.*, 155, 173–183, 2016.
15. Qiu, X., Li, P., Niu, Q., Chen, A., Ouyang, P., Li, C., Ko, T.J., Influence of machining parameters and tool structure on cutting force and hole wall damage in drilling CFRP with stepped drills. *Int. J. Adv. Manuf. Technol.*, 97, 857–865, 2018.
16. Xu, J., Li, C., Mi, S., An, Q., Chen, M., Study of drilling-induced defects for CFRP composites using new criteria. *Compos. Struct.*, 201, 1076–1087, 2018.
17. Che, D., Saxena, I., Han, P., Guo, P., Ehmann, K.F., Machining of carbon fiber reinforced plastics/polymers: A literature review. *J. Manuf. Sci. Eng.*, 136, 034001, 2014.
18. Su, F., Zheng, L., Sun, F., Wang, Z., Deng, Z., Qiu, X., Novel Drill Bit Based on the Step-Control Scheme for Reducing the CFRP Delamination. *J. Mater. Process. Technol.*, 262, 157–167, 2018.
19. Hocheng, H. and Tsao, C.C., Effects of Special Drill Bits on Drilling-Induced Delamination of Composite Materials. *Int. J. Mach. Tools Manuf.*, 46, 1403–1416, 2006.
20. Feito, N., Muñoz-Sánchez, A., Díaz-Álvarez, A., Miguelez, M.H., Multi-Objective Optimization Analysis of Cutting Parameters When Drilling Composite Materials with Special Geometry Drills. *Compos. Struct.*, 225, 111187, 2019.
21. Bilgin, S., Güler, O., Alver, U., Erdemir, F., Aslan, M., Çanakçı, A., Effect of TiN, TiAlCN, AlCrN, and AlTiN ceramic coatings on corrosion behavior of tungsten carbide tool. *J. Australas. Ceram. Soc.*, 57, 263–273, 2021.
22. Dabees, S., Mirzaei, S., Kaspar, P., Holcman, V., Sobola, D., Characterization and Evaluation of Engineered Coating Techniques for Different Cutting Tools—Review. *Materials*, 15, 5633, 2022.
23. Taheri, R., Jalali, M., Al-Yaseri, A., Yabesh, G., Improving TC drill bit's efficiency and resistance to wear by graphene coating. *Pet. Res.*, 7, 430–466, 2022.

24. Aamir, M., Tolouei-Rad, M., Giasin, K., Nosrati, A., Recent advances in drilling of carbon fiber–reinforced polymers for aerospace applications: A review. *Int. J. Adv. Manuf. Technol.*, 105, 2289–2308, 2019.
25. Wang, X., Shen, X., Zeng, C., Sun, F., Combined influences of tool shape and as-deposited diamond film on cutting performance of drills for CFRP machining. *Surf. Coat. Technol.*, 347, 390–397, 2018.
26. Iliescu, D., Gehin, D., Gutierrez, M.E., Girot, F., Modeling and tool wear in drilling of CFRP. *Int. J. Mach. Tools Manuf.*, 50, 204–213, 2010.
27. Hrechuk, A., Bushlya, V., M'Saoubi, R., Ståhl, J.E., Experimental investigations into tool wear of drilling CFRP. *Procedia Manuf.*, 25, 294–301, 2018.
28. Karpat, Y., Deger, B., Bahtiyar, O., Drilling thick fabric woven CFRP laminates with double point angle drills. *J. Mater. Process. Technol.*, 212, 2117–2127, 2012.
29. Cavatorta, M.P., A comparative study of the fatigue and post-fatigue behavior of carbon-glass/epoxy hybrid RTM and hand lay-up composites. *J. Mater. Sci.*, 42, 8636–8644, 2007.
30. Nagaraja, K.C., Rajanna, S., Prakash, G.S., Koppad, P.G., Alipour, M., Studying the effect of different carbon and glass fabric stacking sequence on mechanical properties of epoxy hybrid composite laminates. *Compos. Commun.*, 21, 100425, 2020.
31. Dong, C. and Davies, I.J., Flexural and tensile strengths of unidirectional hybrid epoxy composites reinforced by S-2 glass and T700S carbon fibres. *Mater. Des.*, 54, 955–966, 2014.
32. Tan, C.L., Azmi, A.I., Mohamad, N., Performance Evaluations of carbon/glass hybrid Polymer composites. *Adv. Mater. Res.*, 980, 8–12, 2014.
33. Selmy, A., Abd El-baky, M.A., Hegazy, D.A., Mechanical properties of inter-ply hybrid composites reinforced with glass and polyamide fibers. *J. Thermoplast. Compos. Mater.*, 32, 267–293, 2019.
34. Attia, M.A., Abd El-baky, M.A., Alshorbagy, A.E., Mechanical performance of intraply and inter-intraply hybrid composites based on e-glass and polypropylene unidirectional fibers. *J. Compos. Mater.*, 51, 381–394, 2017.
35. Abd El-baky, M.A., Evaluation of mechanical properties of jute/glass/carbon fibers reinforced hybrid composites. *Fibers Polym.*, 18, 2417–2432, 2017.
36. Franz, G., Vantomme, P., Hassan, M.H., A Review on Drilling of Multilayer Fiber-Reinforced Polymer Composites and Aluminum Stacks: Optimization of Strategies for Improving the Drilling Performance of Aerospace Assemblies. *Fibers*, 10, 78, 2022.
37. Gemi, L., Köklü, U., Yazman, S., Morkavuk, S., The Effects of Stacking Sequence on Drilling Machinability of Filament Wound Hybrid Composite Pipes: Part-1 Mechanical Characterization and Drilling Tests. *Compos. Part B Eng.*, 186, 107787, 2020.

38. Sonkar, V., Abhishek, K., Datta, S., Mahapatra, S.S., Multi-Objective Optimization in Drilling of GFRP Composites: A Degree of Similarity Approach. *Procedia Mater. Sci.*, 6, 538–543, 2014.
39. Tan, C.L., Azmi, A.I., Muhammad, N., Delamination and Surface Roughness Analyses in Drilling Hybrid Carbon/Glass Composite. *Mater. Manuf. Process.*, 31, 1366–1376, 2016.
40. Krishnaraj, V., Prabukarthi, A., Ramanathan, A., Elanghovan, N., Senthil Kumar, M., Zitoune, R., Davim, J.P., Optimization of Machining Parameters at High Speed Drilling of Carbon Fiber Reinforced Plastic (CFRP) Laminates. *Compos. Part B Eng.*, 43, 1791–1799, 2012.
41. Gaitonde, V.N., Karnik, S.R., Rubio, J.C., Correia, A.E., Abrão, A.M., Davim, J.P., Analysis of Parametric Influence on Delamination in High-Speed Drilling of Carbon Fiber Reinforced Plastic Composites. *J. Mater. Process. Technol.*, 203, 431–438, 2008.
42. Campos Rubio, J., Abrao, A.M., Faria, P.E., Correia, A.E., Davim, J.P., Effects of High Speed in the Drilling of Glass Fibre Reinforced Plastic: Evaluation of the Delamination Factor. *Int. J. Mach. Tools Manuf.*, 48, 715–720, 2008.
43. Rawat, S. and Attia, H., Wear Mechanisms and Tool Life Management of WC–Co Drills during Dry High Speed Drilling of Woven Carbon Fibre Composites. *Wear*, 267, 1022–1030, 2009.
44. Navarro-Mas, M.D., Meseguer, M.D., Lluch-Cerezo, J., García-Manrique, J.A., Comparison of Different Parameters to Evaluate Delamination in Edge Trimming of Basalt Fiber Reinforced Plastics (Bfrp). *Materials*, 13, 5326, 2020.
45. Mkaddem, A., Soussia, A.B., El Mansori, M., Wear resistance of CVD and PVD multilayer coatings when dry cutting fiber reinforced polymers (FRP). *Wear*, 302, 946–954, 2013.
46. Meinhard, D., Haeger, A., Knoblauch, V., Drilling induced defects on carbon fiber-reinforced thermoplastic polyamide and their effect on mechanical properties. *Compos. Struct.*, 256, 173–183, 2021.
47. Heisel, U. and Pfeifroth, T., Influence of point angle on drill hole quality and machining forces when drilling CFRP. *Proc. CIRP*, 1, 471–476, Jan. 2012.
48. Poulachon, G., Outeiro, J., Ramirez, C., André, V., Abrivard, G., Hole surface topography and tool wear in CFRP drilling. *Proc. CIRP*, 45, 35–38, 2016.
49. Findenig, G., Buchegger, C., Lengauer, W., Veitsch, C., Demoly, A., Investigation of the main influencing parameters on the degassing behavior of titanium carboni-trides using mass spectrometry. *Int. J. Refract. Met. Hard Mater.*, 63, 38–46, 2017.
50. Suib, S.L., A review of nanoceramic materials for use in ceramic matrix composites, in: *Sol-gel Based Nanoceramic Materials: Preparation, Properties and Applications*, pp. 185–230, 2017.
51. Nguyen, T.K., Fundamental Tool Wear Study in Turning of Ti-6Al-4V Alloy (Ti64) and Nano-Enhanced Minimum Quantity Lubrication (MQL) Milling, Michigan State University, East Lansing, MI, USA, 2015.

52. Lagarde, Q., Wagner, V., Dessein, G., Harzallah, M., Effect of Temperature on Tool Wear During Milling of Ti64. *J. Manuf. Sci. Eng.*, *143*, 071007, 2021.
53. Veiga, C., Davim, J.P., Loureiro, A., Properties and applications of titanium alloys: A brief review. *Rev. Adv. Mater. Sci.*, *32*, 14–34, 2012.
54. Pramanik, A., Islam, M.N., Basak, A., Littlefair, G., Machining and Tool Wear Mechanisms during Machining Titanium Alloys, in: *Engineering Materials and Application*, pp. 338–343, Trans Tech Publications Ltd., Bäch, Switzerland, 2013.
55. Ulutan, D. and Ozel, T., Machining induced surface integrity in titanium and nickel alloys: A review. *Int. J. Mach. Tools Manuf.*, *51*, 250–280, 2011.
56. Amin, A.K.M.N., Ismail, A.F., Khairusshima, M.K.N., Effectiveness of uncoated WC–Co and PCD inserts in end milling of titanium alloy—Ti–6Al–4V. *J. Mater. Process. Technol.*, *192–193*, 147–158, 2007.
57. Sahoo, P., Singh, V.K., Patra, K., Influences of tribological and mechanical properties of (Ti/Al)N and diamond like carbon coated tungsten carbide tool on machining and dynamic stability in micro milling of Ti6Al4V. *J. Manuf. Process.*, *85*, 915–934, 2023.
58. Rahman, M., Wang, Z., Wong, Y., A Review on High-Speed Machining of Titanium Alloys. *JSME Int. J. Ser. C Mech. Syst. Mach. Elem. Manuf.*, *49*, 11–20, 2006.
59. Jianxin, D., Yousheng, L., Wenlong, S., Diffusion wear in dry cutting of Ti–6Al–4V with WC/Co carbide tools. *Wear*, *265*, 1776–1783, 2008.
60. Hosseini, A., Kishawy, H.A., Hussein, H.M., Machinability of Titanium and Its Alloys, in: *Machinability of Advanced Materials*, pp. 95–118, John Wiley & Sons, Inc., Hoboken, NJ, USA, 2014.
61. Umbrello, D., Finite element simulation of conventional and high speed machining of Ti6Al4V alloy. *J. Mater. Process. Technol.*, *196*, 79–87, 2008.
62. Kumar, K.M., Mathew, N.T., Baburaj, M., Sustainable milling of Ti-6Al-4V super alloy using AlCrN and TiAlN coated tools. *Mater. Today Proc.*, *50*, 1732–1738, 2022.
63. Hosseini, A. and Kishawy, H.A., Cutting Tool Materials and Tool Wear, in: *Machining of Titanium Alloys*, J.P. Davim, (Ed.), pp. 31–56, Springer, Berlin/Heidelberg, Germany, 2014.
64. Nabhani, F., Machining of aerospace titanium alloys. *Robot Comput. Integr. Manuf.*, *17*, 99–106, 2001.
65. Soroka, O.B., Evaluation of residual stresses in PVD-coatings. Part 1. Review. *Strength Mater.*, *42*, 287–296, 2010.
66. Abadias, G., Chason, E., Keckes, J., Sebastiani, M., Thompson, G.B., Barthel, E., Doll, G.L., Murray, C.E., Stoessel, C.H., Martinu, L., Review Article: Stress in thin films and coatings: Current status, challenges, and prospects. *J. Vac. Sci. Technol., A*, *36*, 20801, 2018.
67. Patro, S.S., Roy, S., Sahoo, A.K., Investigation of influence of design on residual stress at free edge of multilayer Ti/AlN coatings by PVD using FEM. *Int. J. Interact. Des. Manuf. JIDeM*, *17*, 2987–2998, 2023.

68. Liu, J., Wang, Y., Liu, G., Hua, J., Deng, X., Properties and Performance of TiAlSiN and AlCrN Monolayer and Multilayer Coatings for Turning Ti-6Al-4V. *Coatings*, *13*, 1229, 2023.
69. Mishra, S.K., Ghosh, S., Aravindan, S., Characterization and machining performance of laser-textured chevron shaped tools coated with AlTiN and AlCrN coatings. *Surf. Coat. Technol.*, *334*, 344–356, 2018.
70. Reiter, A.E., Derflinger, V.H., Hanselmann, B., Bachmann, T., Sartory, B., Investigation of the properties of Al1−xCrxN coatings prepared by cathodic arc evaporation. *Surf. Coat. Technol.*, *200*, 2114–2122, 2005.
71. Mo, J.L., Zhu, M.H., Leyland, A., Matthews, Impact wear and abrasion resistance of CrN, AlCrN and AlTiN PVD coatings. *Surf. Coat. Technol.*, *215*, 170–177, 2013, [Google Scholar] [CrossRef].
72. Feng, Y., Zhang, L., Ke, R., Wan, Q., Wang, Z., Lu, Z., Thermal stability and oxidation behavior of AlTiN, AlCrN and AlCrSiWN coatings. *Int. J. Refract. Met. Hard Mater.*, *43*, 241–249, 2014.
73. Zhang, J., Ling, L., Luo, D., Deng, C., Huang, X., Tao, G., Cao, H., Cutting performance and surface quality of Ti-6Al-4V by longitudinal ultrasonic vibration-assisted high-speed dry milling with coated carbide tools. *Int. J. Adv. Manuf. Technol.*, *126*, 5583–5596, 2023.
74. Kim, D., Swan, S.R., He, B., Khominich, V., Bell, E., Lee, S.-W., Kim, T.-G., A study on the machinability of advanced arc PVD AlCrN-coated tungsten carbide tools in drilling of CFRP/titanium alloy stacks. *Carbon Lett.*, *31*, 497–507, 2021.
75. ISO 19252:2008, Plastics—Determination of Scratch Properties, Technical Committee ISO/TC 61, Plastics, Subcommittee SC 2, Mechanical Properties, ISO, Geneva, Switzerland, 2008.
76. ISO 8688-2:1989, Tool Life Testing in Milling—Part 2: End Milling, ISO, Geneva, Switzerland, 1989.
77. Panjan, P. and Cekada, M., Growth defects in PVD hard coatings Growth defects in PVD hard coatings. *Vaccum*, *84*, 209–214, 2009.
78. Polcar, T. and Cavaleiro, A., High temperature properties of CrAlN, CrAlSiN and AlCrSiN coatings—Structure and oxidation. *Mater. Chem. Phys.*, *129*, 195–201, 2011.
79. Kumar, T.S., Prabu, S.B., Manivasagam, G., Padmanabhan, K.A., Comparison of TiAlN, AlCrN, and AlCrN/TiAlN coatings for cutting-tool applications. *Int. J. Miner. Metall. Mater.*, *21*, 796–805, 2014.

7

Advancements in Nanocomposite Technology for Environmental Remediation: Innovations in Water Purification, Air Filtration, and Pollutant Capture

Ashish Mogra and Ranvir Singh Panwar*

School of Technology Management & Engineering, SVKM's Narsee Monjee Institute of Management Studies (NMIMS) Deemed-to-University, Chandigarh, India

Abstract

One of the most urgent issues facing the modern world is environmental degradation. A potential approach that offers notable improvements in environmental remediation—more especially in air filtration, water purification and pollution capture—is nanocomposite technology. The most recent advancements in nanocomposite materials and their uses in environmental cleaning are examined in this study. Nanocomposites have demonstrated remarkable capabilities in the field of water filtration, including the elimination of organic contaminants, heavy metals, and microbes. Traditional filters' adsorption capacity and selectivity are increased by adding nanoscale materials like metallic nanoparticles, graphene oxide, and carbon nanotubes. Furthermore, nanocomposites play a key role in the photocatalytic destruction of contaminants, which improves the effectiveness and sustainability of water purification.

***Keywords*:** Nanocomposite technology, environmental remediation, water purification, air filtration, pollutant capture

**Corresponding author*: ranvirccs@gmail.com

Inamuddin, Tariq Altalhi and Mohammad Abu Jafar Mazumder (eds.) Nanostructured Hard Materials: Industrial Applications, (237–268) © 2026 Scrivener Publishing LLC

7.1 Introduction

Nanocomposites are advanced materials that combine nanoparticles with other materials, such as polymers, ceramics, or metals, to create new materials with enhanced properties. In the context of environmental remediation, nanocomposites are particularly valuable due to their small size, high surface area, and unique chemical and physical properties, which make them highly effective in removing pollutants from the environment [1–3]. In a broader sense, environmental remediation includes the elimination of pollutants from industry and agriculture, as well as the purification of airborne contaminants like nitrogen dioxide (NO_2) and volatile organic compounds (VOCs), as well as difficult-to-decompose organic compounds, bacteria, and fungus found in wastewater [4]. The development of nanocomposites for environmental remediation is a rapidly growing field, driven by the need for more efficient and cost-effective ways to reduce pollution and protect the environment. However, it's also important to consider the life cycle and potential environmental impacts of these nanomaterials themselves to ensure that they do not introduce new environmental challenges.

Nanocomposites are characterized by their composition, which typically involves a matrix material and a nanoscale filler or reinforcement. The diversity in the composition of nanocomposites allows for the tailoring of materials to specific needs and applications, making them highly versatile and valuable across various industries. Figure 7.1 shows the classifications of nanocomposite based on matrix material with their applications [5–7].

Nanoparticles, due to their nanoscale dimensions, exhibit unique physical, chemical, and biological properties that differentiate them from their bulk material counterparts. These properties are a result of their small size, high surface-to-volume ratio, and quantum effects.

These unique properties of nanoparticles open up a vast range of applications, from advanced materials and electronics to medical therapies and environmental technologies. However, these same properties also necessitate careful consideration of potential health and environmental impacts, as they may behave differently in biological and ecological systems compared to larger particles. These properties make nanocomposites highly versatile, finding applications in a wide range of fields from aerospace and automotive to electronics, biomedical devices, and environmental remediation. The choice of matrix and filler, and how they are combined, dictate the specific properties and applications of the resulting nanocomposite [8–10].

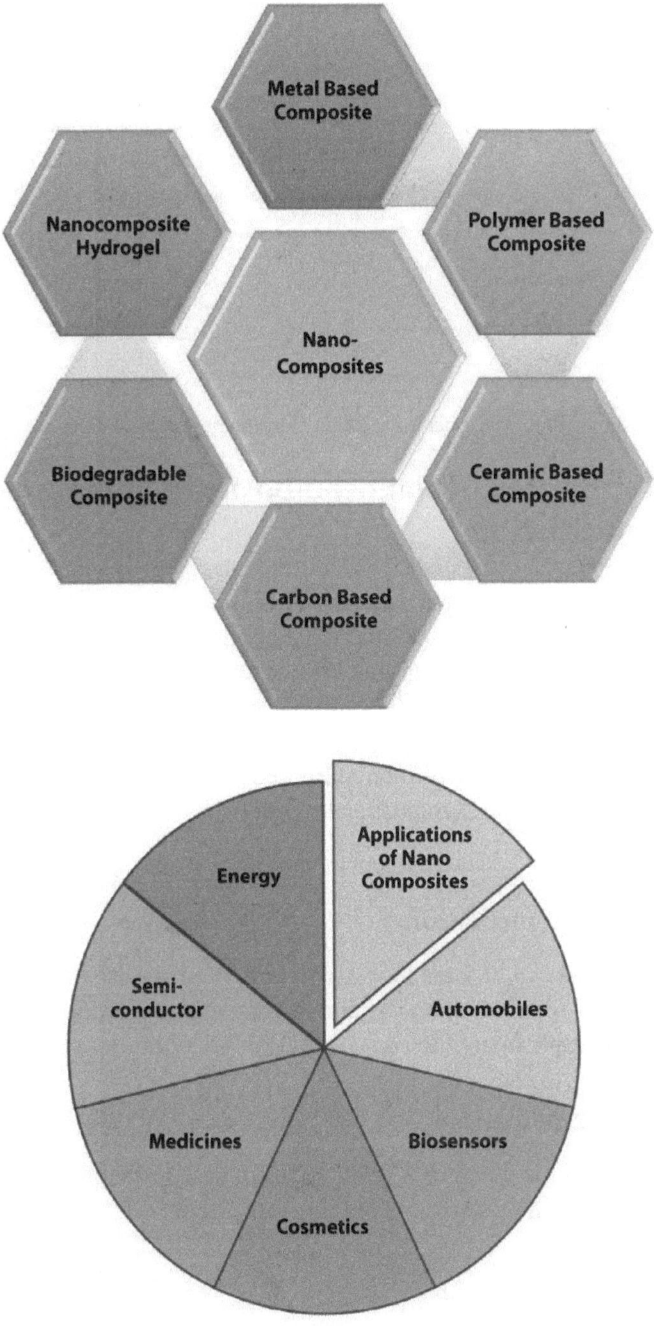

Figure 7.1 Classifications and applications of nanocomposite.

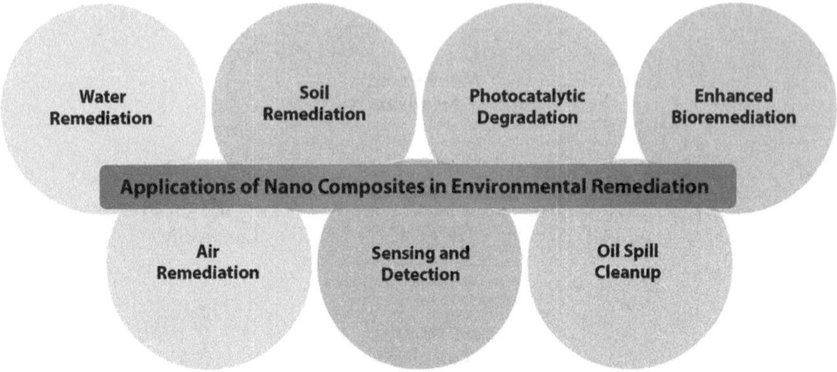

Figure 7.2 Applications of nanocomposite in environmental remediation.

7.2 Application of Nanocomposite in Environmental Remediation

Figure 7.2: A conceptual diagram showing the different ways that nanocomposites are used in environmental remediation. Water Remediation, Soil Remediation, Air Remediation, Photocatalytic Degradation, Enhanced Bioremediation, Sensing and Detection, and Oil Spill Cleanup are among the domains that branch out from a central node titled "Applications of Nano Composites in Environmental Remediation".

Here are some key applications of nanocomposites in environmental remediation [11–13]:

7.2.1 Water Remediation

Nanocomposites can be used in water treatment processes to remove toxic heavy metals, organic pollutants, and pathogens. For example, magnetic nanocomposites facilitate the easy separation of contaminants from water.

7.2.2 Air Remediation

They are used in air filters to remove harmful gases, volatile organic compounds (VOCs), and particulate matter from the air.

7.2.3 Soil Remediation

Nanocomposites can help in the removal of pollutants like heavy metals and organic contaminants from soil, either by immobilizing them or breaking them down into less harmful substances.

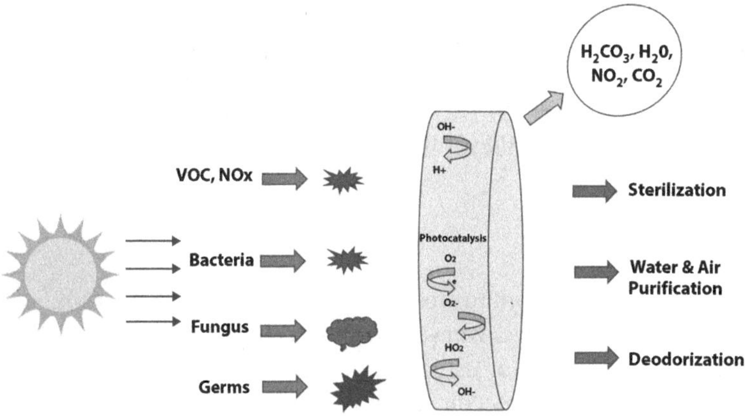

Figure 7.3 Environmental remediation using photocatalysis approach.

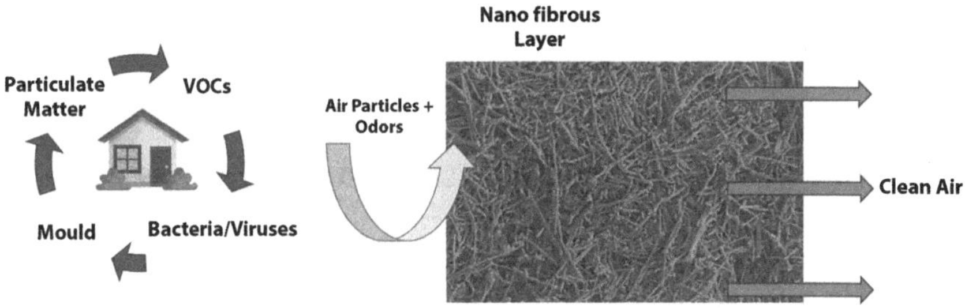

Figure 7.4 Schematic showing air filtration through the nano fibrous layer.

7.2.4 Photocatalytic Degradation

Certain nanocomposites have photocatalytic properties, meaning they can use light energy to catalyze the degradation of pollutants. This is especially useful in breaking down persistent organic pollutants in water and air.

7.2.5 Sensing and Detection

Nanocomposites can be used to develop sensors for the detection of environmental pollutants at very low concentrations, which is crucial for early warning and prevention of environmental contamination.

7.2.6 Oil Spill Cleanup

Some nanocomposites are hydrophobic and oleophilic, meaning they repel water but absorb oils, making them ideal for cleaning up oil spills in aquatic environments.

7.2.7 Enhanced Bioremediation

By combining nanomaterials with certain microbes, nanocomposites can enhance the efficiency of bioremediation processes, where biological agents are used to clean up contaminated sites.

7.3 Objectives

The chapter on "Nanocomposites for Environmental Remediation" aims to provide a comprehensive overview of the technologies, applications, and real-world implications of using nanocomposites in addressing environmental challenges. The specific objectives and coverage of the chapter are outlined as follows:

1. Understanding Nanocomposite Technologies
2. Exploring Applications in Environmental Remediation
3. Addressing Challenges and Future Directions

7.4 Nanocomposites in Water Remediation

Nanocomposites have emerged as a powerful tool in water remediation, addressing a range of water pollutants through innovative approaches. Nanocomposites in water remediation refer to materials composed of nanoparticles combined with other substances to enhance their properties for cleaning and purifying water. These materials are at the forefront of innovative solutions for tackling water pollution. Their application in water treatment leverages their unique properties, such as high surface area, reactivity, and the ability to be engineered for specific purposes. Common nanoparticles used include metal oxides (like titanium dioxide), carbon-based materials (like graphene), and metal-based particles (like silver or iron) [14–16]. Figure 7.3: An illustration of the photocatalytic remediation procedure. The diagram illustrates how photocatalytic action breaks down pollutants like VOCs, NOx, bacteria, and fungi. Reactive species involved in the degradation of contaminants under light activation are identified by labels such as H+, OH^-, O_2^-, and HO_2.

Here's how nanocomposites are utilized in water remediation:

7.4.1 Removal of Heavy Metals

The removal of heavy metals from water sources is a critical aspect of environmental remediation due to the toxic and non-biodegradable nature of these metals. Various methods have been developed to address this challenge, each with its own mechanisms and efficiencies. Here's an overview of the key techniques used for the removal of heavy metals:

7.4.1.1 Adsorption

Adsorption is a process where molecules, atoms, or ions from a substance (like a gas, liquid, or dissolved solid) adhere to the surface of an adsorbent. It's a critical principle in many environmental, chemical, and biochemical applications, including water purification and air filtration. Unlike absorption, which involves the whole volume of the material, adsorption is a surface phenomenon. The adsorbate (the substance being adsorbed) accumulates only at the surface and not in the bulk of the solid or liquid. The use of nanomaterials as adsorbents, like graphene oxide or metal-organic frameworks (MOFs), has shown great promise due to their high surface areas, tuneable properties and specific affinity for metal ions can effectively adsorb heavy metals from water. Materials like graphene oxide, carbon nanotubes, and metal oxide nanoparticles are often used. Nanocomposites can remove pollutants from water by adsorbing them onto their surface. Due to their large surface area, they are highly effective at trapping contaminants. Adsorption is a versatile and essential process used extensively in environmental remediation, particularly in water purification. Its effectiveness depends on the properties of both the adsorbate and the adsorbent, and ongoing research is focused on enhancing efficiency, selectivity, and sustainability in adsorption technologies [17–19].

7.4.1.2 Catalysis

Catalysis is a process in which the rate of a chemical reaction is increased by a substance called a catalyst. Unlike other reactants, a catalyst is not consumed in the reaction and can continue to act repeatedly. This process is fundamental in various industrial, environmental, and biological systems. Catalysts work by providing an alternative reaction pathway with a lower activation energy compared to the uncatalyzed reaction. This allows more molecules to have enough energy to react at a given temperature. Catalysts are used to remove pollutants from air and water. For example, catalytic converters in vehicles reduce harmful emissions. The development of nanoscale catalysts has opened new possibilities, offering high surface area and the ability to tailor properties at the molecular level. Some nanocomposites act as catalysts to break down pollutants into less harmful substances. For example, photocatalytic titanium dioxide can degrade organic pollutants under UV light. Catalysis is a key principle in accelerating chemical reactions, making processes more efficient, selective, and environmentally friendly. Its applications span from industrial manufacturing to environmental protection and energy production, underpinning many modern technologies and industrial processes. The ongoing development of new catalysts and catalytic methods continues to be a vital area of research in chemistry and materials science [20–22].

7.4.1.3 *Magnetic Nanocomposites*

Magnetic nanocomposites in water remediation are an innovative class of materials that combine the properties of nanoscale particles with magnetic features, offering a unique and efficient approach to purifying water. These composites typically consist of magnetic nanoparticles embedded in various matrix materials. Commonly used magnetic nanoparticles include iron oxides like magnetite (Fe3O4) and maghemite (γ-Fe2O3). These nanoparticles exhibit superparamagnetic, meaning they become magnetized in the presence of a magnetic field and lose their magnetism when the field is removed, minimizing particle aggregation. The magnetic particles are often embedded in other materials, such as carbon-based substances (like activated carbon or graphene), polymers, silica, or metal-organic frameworks. These matrices provide additional functionalities like adsorption, catalytic activity, or biocompatibility. These are particularly useful as they can be easily separated from water after treatment using a magnetic field [23–25]. Magnetic nanocomposites represent a versatile and effective tool in the field of water remediation. Their unique combination of adsorption capabilities, catalytic potential, and ease of separation under magnetic fields positions them as a promising solution for addressing the global challenge of water pollution. Continuous research and development in this area are aimed at enhancing their efficiency, specificity, and sustainability [26]. The removal of heavy metals from water is a multifaceted challenge that requires a combination of techniques, each suited to specific scenarios and types of metal contamination. Ongoing research and development are focused on improving the efficiency, selectivity, and sustainability of these methods, ensuring the safety and cleanliness of water resources [27].

7.4.2 Degradation of Organic Pollutants

The degradation of organic pollutants in water is a critical environmental process aimed at reducing the presence of harmful organic compounds in water sources. These organic pollutants include a wide range of substances such as pesticides, pharmaceuticals, petrochemicals, and industrial solvents. The degradation processes are designed to either completely mineralize these pollutants into harmless substances like water and carbon dioxide or transform them into less harmful forms [28]. Here's a detailed look at the various methods employed:

7.4.2.1 Photocatalytic Degradation

Photocatalytic degradation in water remediation is a process that utilizes light, typically ultraviolet (UV), and a photocatalyst to break down pollutants in water into less harmful substances. This technology is gaining attention for its efficiency and eco-friendliness in treating water contaminants. The most widely used photocatalyst is titanium dioxide (TiO2) due to its strong oxidative power, stability, and non-toxicity. Other materials like zinc oxide (ZnO) and various metal-doped or composite photocatalysts are also explored. Nanocomposites that exhibit photocatalytic properties, such as TiO2-based nanocomposites, can degrade organic pollutants under light irradiation. This process breaks down complex organic molecules into simpler, less harmful substances. It can be used in both small-scale and large-scale water treatment facilities to purify drinking water and treat wastewater [29–31]. Photocatalytic degradation is a promising technology in water remediation, harnessing the power of light and catalysts to degrade pollutants. Its eco-friendly nature and potential for wide application make it a key area of research and development in environmental technology, especially in the pursuit of sustainable water treatment solutions [32].

7.4.2.2 Adsorption and Breakdown (Degradation)

Adsorption and breakdown are two fundamental processes used in water remediation to remove and neutralize contaminants. Both play critical roles in purifying water, but they operate through different mechanisms and are effective against various types of pollutants. Adsorption involves contaminants (adsorbates) being captured on the surface of a material (adsorbent). It doesn't involve the entire volume of the adsorbent. On the other hand, breakdown involves chemical reactions that convert harmful pollutants into less harmful or non-toxic substances. In many water remediation systems, adsorption and breakdown are used together for a more comprehensive treatment approach. Certain nanocomposites can adsorb organic contaminants and then catalyze their breakdown, often using reactive oxygen species generated on the nanoparticle surface [33, 34]. Adsorption and breakdown are complementary processes in water remediation, each addressing different types of pollutants and stages of purification. Adsorption is excellent for capturing and concentrating contaminants, while breakdown processes are key to neutralizing and transforming these pollutants into less harmful forms. The integration of both methods offers a more robust and effective approach to purifying water, especially in complex and varied contamination scenarios [35].

The degradation of organic pollutants in water encompasses a range of techniques, each with its own strengths and limitations. The choice of method often depends on the nature of the pollutants, the volume of water to be treated, and the desired purity levels. Continuous advancements in these technologies aim to enhance their efficiency, reduce costs, and minimize environmental impacts.

7.4.3 Pathogen Elimination

Eliminating pathogens from water is a crucial aspect of water treatment to prevent waterborne diseases and ensure safe drinking water. Pathogens in water include bacteria, viruses, protozoa, and parasites. Various methods are employed to remove or inactivate these harmful microorganisms, each with its own mechanisms and effectiveness. Here's an overview:

7.4.3.1 Antibacterial Properties

The use of antibacterial properties in water remediation is an essential aspect of ensuring water safety and quality. Antibacterial agents and technologies play a crucial role in eliminating or inhibiting the growth of bacteria in water sources, preventing waterborne diseases.

Nanocomposites containing silver or copper nanoparticles can effectively kill bacteria and other pathogens in water, making them useful in disinfection and purification processes. Silver ions released from nanoparticles exhibit strong antibacterial properties, interfering with bacterial cell membranes and DNA.

Incorporating antibacterial properties into water remediation systems is essential for providing safe and clean water. The choice of antibacterial strategy depends on various factors, including the type of bacteria, water source, and treatment scale. Ongoing research is focused on enhancing the efficacy, sustainability, and safety of these methods, as well as understanding and mitigating any potential impacts on the environment and public health [36].

7.4.3.2 Virus Removal

Virus removal in water remediation is a critical aspect of ensuring water safety, as viruses can be highly resistant to many traditional water treatment methods and are capable of causing serious health issues even in small quantities. Several technologies and strategies are employed to effectively remove or inactivate viruses from water sources. Physical removal of viruses based on size exclusion. Membrane filtration technologies like ultrafiltration, nanofiltration, and reverse osmosis are particularly effective.

Specialized nanocomposites have been developed to capture and neutralize viruses in water, an area of growing importance in public health. The removal of viruses from water is a complex challenge that often requires a combination of physical, chemical, and sometimes biological methods. The selection of appropriate virus removal techniques depends on the types of viruses present, water quality, the scale of treatment, and resource availability. Ongoing advancements in water treatment technologies aim to enhance the efficiency and accessibility of virus removal methods while ensuring the safety of treated water for various uses [37].

Pathogen elimination from water requires a multifaceted approach, often involving a combination of physical, chemical, and biological methods. The choice of treatment depends on the type of pathogens present, water quality, scale of treatment required, and resource availability. Ongoing research and development in this field focus on enhancing efficiency, reducing costs, and ensuring the safety and accessibility of clean water.

7.4.4 Desalination and Ion Removal

7.4.4.1 Selective Ion Exchange

Selective ion exchange in water remediation using nanocomposites is an advanced and highly efficient method for purifying water. Nanocomposites in this context refer to materials that combine ion exchange resins with nanoparticles to enhance their properties and performance. The use of nanocomposites in ion exchange offers several advantages over traditional ion exchange methods This method is particularly effective in treating water for various applications, such as drinking water purification, wastewater treatment, and industrial processes. Nanocomposites can be designed to selectively remove specific ions from water, useful in softening hard water or in desalination processes [38].

While promising, the use of nanocomposites in ion exchange also presents challenges. The production and disposal of these materials must be managed carefully to avoid potential environmental impacts. Moreover, the cost and complexity of manufacturing these advanced materials can be higher than traditional ion exchange resins.

The integration of nanocomposites into selective ion exchange processes for water remediation represents a significant advancement in water treatment technology. It offers improved efficiency, selectivity, and stability, making it an attractive option for addressing complex water contamination issues. However, careful consideration of environmental and economic factors is essential in the development and application of these advanced materials [39].

7.4.4.2 Membrane Technology

Membrane technology utilizing nanocomposites in water remediation is a cutting-edge approach that integrates the efficiency of membrane filtration with the enhanced properties provided by nanomaterials. This technology is highly effective in removing a wide range of contaminants from water, including pollutants, pathogens, and heavy metals. These are advanced membranes incorporated with nanoparticles or nanomaterials. The integration of nanomaterials into the membrane matrix can significantly improve its properties, such as selectivity, permeability, and resistance to fouling. Common nanomaterials used in these membranes include carbon nanotubes, metal oxides (like titanium dioxide), silver nanoparticles, and graphene oxide. Each of these materials contributes unique characteristics like antimicrobial properties, enhanced mechanical strength, or improved filtration efficiency.

Nanocomposite membranes can operate through various filtration mechanisms, including microfiltration, ultrafiltration, nanofiltration, and reverse osmosis. The choice of mechanism depends on the size of the contaminants to be removed and the specific application.

Membrane technology enhanced by nanocomposites represents a significant advancement in water remediation, offering superior performance in terms of contaminant removal, durability, and efficiency. This technology holds great promise for addressing global water quality challenges, although it is essential to balance the technological benefits with considerations of cost, environmental impact, and scalability. Incorporating nanoparticles into membranes can enhance their efficiency and selectivity in desalination and water purification processes [40–42].

7.4.5 Sensing and Monitoring

Sensing and monitoring using nanocomposite materials in water remediation is a progressive field that focuses on the detection, analysis, and real-time monitoring of various contaminants in water. Nanocomposites, due to their unique properties, have become increasingly important in developing advanced sensors for water quality assessment. Here's an overview of how this technology works and its applications:

7.4.5.1 Nanocomposite-Based Sensors

These sensors integrate nanomaterials with conventional sensing materials to enhance sensitivity, selectivity, and response time. Nanocomposites can be engineered to react with specific contaminants, making them highly effective for detecting a wide range of substances, including heavy metals, organic pollutants, and biological agents. Commonly used nanomaterials in these sensors include graphene, carbon nanotubes, metal oxides, and quantum dots. Each of these materials offers unique properties, such as high surface area, electrical conductivity, and chemical reactivity, which are beneficial for sensing applications. Nanocomposite-based sensing and monitoring in water remediation represent a significant step forward in environmental technology. These sensors offer enhanced capabilities for detecting and monitoring a wide range of water contaminants, contributing greatly to water safety and sustainability efforts. However, their development and implementation must consider factors such as cost, stability, and data management [43].

7.4.5.2 Detection of Pollutants

The detection of pollutants in water using nanocomposite-based technologies is a rapidly evolving area in the field of environmental monitoring and remediation. Nanocomposites, due to their unique physicochemical properties, are highly effective in sensing and identifying various contaminants in water, including organic pollutants, heavy metals, and biological agents. Nanocomposites used for pollutant detection typically consist of a matrix material (like a polymer or silica) combined with nanoparticles (such as metal oxides, carbon nanotubes, graphene, or quantum dots). These nanoparticles enhance the sensitivity and selectivity of the composite material towards specific pollutants [44].

Some nanocomposites are used to develop sensors that can detect trace amounts of pollutants in water, aiding in monitoring and managing water quality. Nanocomposites can be designed to detect various heavy metals like lead, mercury, arsenic, and cadmium, which are hazardous even at low concentrations.

In summary, nanocomposites offer a multifaceted approach to water remediation, capable of addressing a wide range of pollutants through adsorption, catalysis, and antibacterial action. Their effectiveness in both removing contaminants and purifying water makes them a valuable asset in the ongoing efforts to ensure clean and safe water supplies globally. However, it is essential to continue researching the long-term stability and potential environmental impacts of these nanomaterials to ensure sustainable and safe water treatment solutions.

7.5 Nanocomposites in Air Purification

Nanocomposites have become increasingly important in the field of air purification, offering advanced solutions for removing various air pollutants, including particulate matter, volatile organic compounds (VOCs), and harmful gases. Their unique properties, such as high surface area, reactivity, and ability to be tailored for specific applications, make them highly effective in air purification processes. Figure 7.4: An illustration of a nanofibrous layer-based air filtration system. VOCs, mold, bacteria, viruses, air particles, and smells enter from one side, go through the nanofibrous filter, and exit as clean air on the other. Icons and arrows indicate the elimination of microbiological and particulate matter pollutants.

7.5.1 Filtering of Particulate Matter

Filtering particulate matter in air remediation involves using various filtration technologies to capture and remove airborne particles from the environment. These particles can range from large dust particles to fine particulates like PM2.5 (particulate matter with a diameter of 2.5 micrometer or smaller), which are particularly harmful to human health. This technology leverages the unique properties of nanoscale materials to enhance the efficiency and effectiveness of filtration processes. Nanofiber filters are being developed to improve the efficiency of capturing ultrafine particles [45]. Here's an overview of how particulate matter is filtered in air remediation:

7.5.1.1 Nanocomposite Filters

These filters are made by incorporating nanoparticles into traditional filter materials, like activated carbon, cellulose, or polymers. The nanoparticles used can vary, including carbon nanotubes, graphene, metal oxides (like titanium dioxide or iron oxide), and silica nanoparticles. Nanocomposite filters in air remediation represent a significant advancement in air purification technology. These filters integrate nanoscale materials into traditional filtration matrices, enhancing their ability to capture and remove a wide range of airborne contaminants, including ultrafine particulate matter, pollutants, and pathogens.

Nanocomposite filters in air remediation offer a highly effective and advanced solution for improving air quality. Their ability to efficiently remove a wide spectrum of air pollutants, combined with potential antimicrobial properties, makes them a promising technology in the field of environmental engineering. However, it's important to address the associated economic, environmental, and health challenges to ensure their sustainable and safe application in air purification systems [46, 47, 52].

7.5.1.2 *Enhanced Filters*

Enhanced filters in air remediation are advanced filtration systems designed to improve air quality by effectively removing a wide range of airborne contaminants, including particulate matter, pollutants, and biological agents. These filters incorporate innovative materials and technologies to enhance their filtration efficiency and effectiveness. The nanocomposite filters are capable of physically trapping a wide array of contaminants, including microorganisms, particulate matter, and microplastics, due to their enhanced pore structures.

Enhanced filters in air remediation offer a sophisticated and highly effective solution for improving air quality. They are capable of addressing a broad spectrum of air quality issues, from removing particulate matter to neutralizing harmful gases and biological contaminants. While they represent a significant advancement in air purification technology, considerations around cost, maintenance, and proper usage are essential for their effective implementation [48].

7.5.1.3 *Electrostatic Capture*

Electrostatic capture in air remediation is a technology that uses electrical forces to remove particulate matter and certain types of pollutants from the air. This method is particularly effective for capturing very fine particles that traditional mechanical filters might miss. Electrostatic capture is a valuable technology in air remediation, particularly for removing fine particulate matter. Its energy efficiency and reusable nature make it an attractive option for various applications. However, considerations regarding ozone production and maintenance requirements are important for ensuring the safe and effective use of this technology.

Nanocomposites can be incorporated into air filters to improve their efficiency in trapping particulate matter, including ultrafine particles. Filtering particulate matter in air remediation is crucial for maintaining healthy indoor and outdoor air quality. Various technologies, each with their strengths and limitations, are employed to capture different sizes of particles. For example, filters with nanofibers or carbon nanotubes offer high filtration efficiency while maintaining good airflow. Ongoing research is focused on identifying and utilizing novel nanomaterials that offer greater effectiveness and environmental safety. Continuous innovation and improvement in filtration technologies are essential to effectively address the challenges posed by air pollution [49].

7.5.2 VOCs and Hazardous Gas Adsorption

Volatile Organic Compounds (VOCs) and hazardous gas adsorption in air remediation involve the use of materials and technologies to capture and remove these harmful substances from the air. VOCs are a wide range of organic chemicals that can evaporate at room temperature and include substances like formaldehyde, benzene, and toluene. Hazardous gases may include sulfur dioxide, nitrogen oxides, ammonia, and other toxic or corrosive gases. Effective adsorption of these compounds is crucial for maintaining air quality and protecting human health and the environment. Here's an overview of the methods and materials used for VOC and hazardous gas adsorption:

7.5.2.1 Adsorption Capability

Adsorption capability in air remediation refers to the ability of certain materials or systems to capture and hold pollutants, such as volatile organic compounds (VOCs), particulate matter, and hazardous gases, from the air. This capability is crucial for improving air quality, especially in industrial, urban, and indoor environments. The effectiveness of adsorption depends on various factors, including the type of adsorbent material, the nature of the pollutants, and environmental conditions. Nanocomposites with high surface area and specific surface functionalities are capable of adsorbing VOCs and other hazardous gases, removing them from the air. The adsorption capability in air remediation is a critical factor in determining the effectiveness of different materials and systems in purifying air. Advances in material science and technology continue to enhance the efficiency and application range of adsorption-based air remediation solutions [50].

7.5.2.2 Selective Adsorption

Selective adsorption in air remediation refers to the process where certain materials selectively target and capture specific types of pollutants from the air, rather than indiscriminately adsorbing various particles and gases. This specificity is particularly important in scenarios where certain contaminants are more harmful or prevalent than others. Selective adsorption is achieved through the careful design and choice of adsorbent materials, each with unique properties that make them suitable for capturing specific pollutants.

Selective adsorption in air remediation represents a targeted approach to air purification, focusing on specific pollutants of concern. The ongoing development of new materials and technologies continues to enhance the effectiveness and applicability of this approach in various settings. Certain nanocomposites can be designed to target specific gases, making them useful in environments where selective removal of certain contaminants is required [51].

The adsorption of VOCs and hazardous gases is a critical component of air remediation strategies. A variety of materials and technologies are available, each with its strengths and suitable applications. Ongoing research and development continue to improve the efficiency, selectivity, and sustainability of these adsorption methods.

7.5.3 Photocatalytic Applications

7.5.3.1 Photocatalytic Degradation

Photocatalytic degradation in air remediation is a process that uses light-activated catalysts to decompose organic and some inorganic pollutants in the air into less harmful substances. This technology harnesses the power of photocatalysis, a phenomenon where a substance (the photocatalyst) becomes reactive upon exposure to light, typically UV light. Nanocomposites, particularly those containing materials like titanium dioxide (TiO_2), can act as photocatalysts to degrade organic pollutants in the air. Materials like zinc oxide (ZnO) and modified TiO_2 (doped with other elements to enhance performance) are also used. Under light irradiation, these nanocomposites generate reactive oxygen species that break down harmful organic compounds into less harmful substances.

Photocatalytic degradation represents a promising and environmentally friendly approach to air remediation, particularly effective against organic pollutants and pathogens. As research progresses, advancements in photocatalyst efficiency and the integration of this technology into various applications are expected to broaden its scope and effectiveness in air quality management [52].

7.5.3.2 Self-Cleaning Surfaces

Self-cleaning surfaces in air remediation refer to materials or coatings that can break down pollutants and contaminants upon contact, effectively reducing the concentration of these harmful substances in the air. This technology typically leverages the principles of photocatalysis, using light-activated catalysts to initiate chemical reactions that degrade pollutants. Titanium dioxide (TiO_2) is commonly used as a photocatalyst in self-cleaning surfaces due to its effectiveness, stability, and non-toxicity. When exposed to UV light (from sunlight or artificial sources), TiO_2 generates reactive oxygen species (ROS) like free radicals [53].

Self-cleaning surfaces in air remediation offer an innovative and environmentally friendly approach to reducing air pollution and maintaining clean surfaces. While their effectiveness is influenced by factors like light availability and material durability, ongoing advancements in photocatalytic technology are expanding their potential applications and efficacy in improving air quality.

7.5.4 NOx and SOx Removal

The removal of nitrogen oxides (NOx) and sulfur oxides (SOx) in air remediation is crucial due to the harmful effects these gases have on both human health and the environment. NOx and SOx are primarily produced by industrial processes and the burning of fossil fuels, contributing to problems like acid rain, smog, and respiratory issues. Various technologies and methods have been developed to effectively reduce the concentrations of these gases in the atmosphere. Specific nanocomposites are effective in capturing and breaking down NOx and SOx from industrial emissions and vehicle exhausts, which are significant air pollutants.

The removal of NOx and SOx is a critical aspect of air remediation efforts, especially given the global push for cleaner air and reduced environmental impact. Continued advancements in technology and more stringent environmental regulations are driving improvements in the effectiveness and efficiency of these removal methods [54].

7.5.5 Sensing and Monitoring of Air Quality

Sensing and monitoring of air quality in air remediation are essential processes for assessing and managing the levels of various pollutants in the atmosphere. These activities involve the use of sophisticated sensors and monitoring systems to continuously track air quality parameters, including the concentrations of specific pollutants like particulate matter (PM2.5 and PM10), nitrogen oxides (NOx), sulfur oxides (SOx), carbon monoxide (CO), ozone (O3), volatile organic compounds (VOCs), and others [55].

Nanocomposites are also used in developing sensors for monitoring air quality. These sensors can detect trace levels of various air pollutants, aiding in environmental monitoring and control systems. Effective sensing and monitoring of air quality are pivotal for successful air remediation efforts. They provide the necessary data to understand pollution dynamics, evaluate the effectiveness of remediation strategies, and make informed decisions to protect public health and the environment.

7.6 Nanocomposites in Soil Remediation

Nanocomposites have emerged as a promising solution for soil remediation, offering efficient and innovative approaches to address various soil contaminants, including heavy metals, organic pollutants, and even radioactive materials. The unique properties of nanocomposites, such as high surface area, reactivity, and the ability to be engineered for specific targets, make them highly effective for soil decontamination [56].

7.6.1 Heavy Metal Immobilization

Heavy metal immobilization in soil remediation is a technique used to reduce the mobility and bioavailability of heavy metals in contaminated soils. This process is crucial because heavy metals like lead, cadmium, mercury, and arsenic can pose significant health risks to humans and ecosystems. Immobilization prevents these metals from leaching into groundwater or being taken up by plants, thus mitigating environmental and health hazards. Here's an overview of the methods and principles involved in heavy metal immobilization:

7.6.1.1 Adsorption and Chelation

Adsorption and chelation are two key mechanisms used in heavy metal immobilization in soil remediation, especially when employing nanocomposite materials. These approaches are particularly effective due to their high efficiency and specificity in capturing and stabilizing heavy metals from contaminated soils. Nanocomposites used for adsorption typically consist of a matrix (like clay, carbon-based materials, or polymers) integrated with nanoparticles (such as metal oxides, graphene, or biochar).

In chelation process, organic compounds can form stable, water-soluble complexes with heavy metal ions. In nanocomposites, chelating agents can be either incorporated into the matrix or attached to the surface of nanoparticles. These are organic compounds that can form stable, water-soluble complexes with heavy metal ions. In nanocomposites, chelating agents can be either incorporated into the matrix or attached to the surface of nanoparticles. Chelating nanocomposites are particularly useful in treating soils contaminated with multiple heavy metals, as they can be designed to target specific ions.

Nanocomposites can adsorb heavy metals from soil, reducing their mobility and bioavailability. Some nanocomposites are designed to chemically bind (chelate) heavy metals, transforming them into less toxic forms. Ongoing research focuses on developing new types of nanoparticles and chelating agents that are more efficient, cost-effective, and environmentally friendly. Combining nanocomposite-based adsorption and chelation with other remediation techniques for more comprehensive soil treatment.

The use of nanocomposites for adsorption and chelation in heavy metal soil remediation offers a highly effective and targeted approach to mitigating heavy metal pollution. These methods capitalize on the unique properties of nanomaterials, providing enhanced capacity and selectivity for heavy metal immobilization. However, careful consideration of their cost, environmental impact, and regulatory compliance is essential for their sustainable application in soil remediation efforts [57–60].

7.6.1.2 Phytoremediation Enhancement

Enhancing phytoremediation for heavy metal immobilization in soil through the use of nanocomposites is a promising area of environmental biotechnology. Phytoremediation is a green and cost-effective method that uses plants to extract, stabilize, or degrade pollutants, including heavy metals, from soil. The integration of nanocomposites can significantly improve the efficiency and effectiveness of phytoremediation processes. Nanocomposites used in phytoremediation may include nanoparticles such as nano-zerovalent iron, titanium dioxide, or modified biochar. These materials can enhance the soil's properties or directly interact with plants to improve their metal uptake and tolerance. Nanocomposites can aid in immobilizing heavy metals in soil, making them less bioavailable, or they can help in mobilizing them for easier uptake by plants. Nanocomposites can convert heavy metals into forms that are more easily absorbed by plants, thereby enhancing phytoextraction.

Phytoremediation enhanced with nanocomposites presents a novel approach to addressing heavy metal contamination in soils. By leveraging the synergistic effects of plants and nanotechnology, it offers a more efficient and sustainable solution for soil remediation. However, the responsible deployment of this technology requires thorough understanding and management of potential risks and environmental impacts [61].

Heavy metal immobilization is a critical component of soil remediation strategies, especially in areas where soil excavation and removal are impractical or too costly. Its success depends on a thorough understanding of soil chemistry, pollutant characteristics, and the careful selection and application of immobilization techniques.

7.6.2 Organic Contaminant Breakdown

The breakdown of organic contaminants in soil remediation using nanocomposites is a sophisticated approach that leverages the unique properties of nanoscale materials to enhance the degradation or removal of pollutants. Organic contaminants, such as petroleum hydrocarbons, pesticides, and industrial chemicals, pose significant environmental and health risks, and their effective remediation is crucial. Nanocomposites can be particularly effective in addressing these challenges due to their high reactivity, large surface area, and customizable properties. Here's an overview of how nanocomposites work in the breakdown of organic contaminants in soil:

7.6.2.1 Degradation of Pollutants

The degradation of pollutants and organic contaminant breakdown in soil remediation using nanocomposites involves advanced techniques that harness the unique properties of nanoscale materials. These methods are particularly effective for decomposing complex organic pollutants, such as hydrocarbons, pesticides, and industrial solvents, which are often challenging to treat using conventional remediation methods. Nanocomposites, due to their high surface area, reactivity, and customizable structure, can significantly enhance the degradation of these contaminants.

Nanocomposites, especially those with photocatalytic properties, can degrade organic pollutants in soil. For instance, TiO_2-based nanocomposites can break down pesticides, herbicides, and other organic contaminants under light exposure. The use of nanocomposites for the degradation of organic pollutants in soil remediation represents a forward-thinking approach that offers efficiency and specificity in breaking down complex contaminants. This method has the potential to address some of the most challenging pollution issues in soil environments. However, it's essential to balance the benefits with considerations regarding the cost, stability, and environmental impact of these advanced materials.

7.6.2.2 *Sorption and Destruction*

Sorption and destruction are two key processes involved in the breakdown of organic contaminants in soil remediation using nanocomposites. This approach combines the ability of nanocomposites to adsorb (sorb) pollutants onto their surfaces (sorption) and then degrade or destroy these contaminants (destruction). They typically involve carbon-based nanomaterials like activated carbon, biochar, graphene, or carbon nanotubes, as well as metal oxides or bimetallic nanoparticles. Many nanocomposites possess catalytic properties that can stimulate chemical reactions to degrade adsorbed contaminants. For example, photocatalytic nanoparticles like TiO_2 can degrade organic compounds under UV light [62].

The combination of sorption and destruction mechanisms in nanocomposite-based soil remediation provides a comprehensive approach to dealing with organic contaminants. This method enhances the efficiency of remediation by ensuring that pollutants are not just adsorbed but also effectively broken down, offering a promising solution for complex soil contamination issues. However, careful consideration of the environmental impact, cost, and practical application challenges is essential for sustainable and effective use of this technology.

Nanocomposites offer a promising solution for the breakdown of organic contaminants in soil remediation. Their ability to effectively adsorb, degrade, or transform these pollutants into less harmful forms can significantly aid in environmental restoration efforts. However, their use must be carefully managed to balance remediation goals with potential environmental impacts and economic considerations.

7.6.3 Stabilization of Contaminated Soil

Stabilization of contaminated soil using nanocomposite materials is an innovative approach in soil remediation. This method involves the use of nanoscale materials to stabilize and reduce the mobility of contaminants, particularly heavy metals and organic pollutants, in the soil. The goal is to prevent the spread of contamination to the surrounding environment and reduce the risk to human health and ecosystems.

Nanocomposites can be used to stabilize contaminated soil, preventing the spread of pollutants to surrounding areas and groundwater. This is particularly important in managing sites with loose or erodible soils. Materials like nano-zerovalent iron (nZVI), titanium dioxide (TiO_2), and zinc oxide (ZnO) are commonly used. Combinations of polymers with nanoparticles can enhance the mechanical properties and barrier effects of the composites, reducing the leachability of contaminants. Nanocomposites can adsorb contaminants onto their surfaces or chemically bind with them, reducing their mobility and bioavailability in the soil. Nanocomposites can create physical barriers that encapsulate contaminants, preventing their migration [63].

Nanocomposite-based stabilization offers a promising approach for remediating and managing contaminated soils. Its ability to immobilize and transform contaminants effectively can significantly mitigate environmental and health risks associated with soil pollution. However, the application of this technology requires careful consideration of its long-term stability, environmental impact, and cost-effectiveness.

7.6.4 Radioactive Waste Treatment

7.6.4.1 Radioactive Contaminant Removal

Radioactive waste treatment in soil remediation using nanocomposite materials is a specialized and highly technical area of environmental engineering. This approach aims to safely manage and mitigate the risks posed by radioactive contaminants in soil, which can arise from nuclear accidents, industrial processes, or improper disposal of radioactive materials. Nanocomposites, due to their unique properties, can offer effective solutions to addressing these challenges. Nano-zerovalent iron (nZVI), titanium dioxide (TiO_2), zeolites, graphene, carbon nanotubes, fullerenes, and clay nanoparticles can be used for the reduction, adsorption, stabilization, and immobilization of radioactive elements.

The use of nanocomposites in the treatment of radioactive waste in soil remediation presents a promising but complex approach. It offers the potential for more efficient and targeted remediation of radioactive contaminants. However, the application of this technology requires careful consideration of safety, regulatory, and environmental impact factors. Continued research and development are crucial to optimize these methods and ensure their safe and effective implementation [64].

7.6.5 Sensing and Monitoring of Soil Quality

Sensing and monitoring of soil quality in the context of soil remediation using nanocomposite materials is a critical aspect of environmental management and restoration efforts. This process involves using advanced sensors and monitoring systems, often enhanced with nanocomposites, to assess soil health, track the presence and concentration of contaminants, and evaluate the effectiveness of remediation strategies. Carbon nanotubes, graphene, metal oxides, and quantum dots are commonly employed in sensor fabrication for their unique electrical, optical, and chemical properties. Nanocomposite sensors can detect trace levels of heavy metals like lead, mercury, arsenic, and cadmium in the soil. These sensors are also capable of identifying organic contaminants, including hydrocarbons, pesticides, and industrial chemicals. Nanocomposite-enhanced sensors can provide continuous, real-time data on soil quality and contaminant levels. The development of wireless networks of these nanocomposite sensors enables extensive and efficient soil quality monitoring across large areas.

The use of nanocomposite-enhanced sensors in sensing and monitoring soil quality represents a significant advancement in environmental remediation. These technologies offer enhanced sensitivity, specificity, and real-time monitoring capabilities, providing valuable insights into soil health and the effectiveness of remediation efforts. Ongoing development in this field focuses on improving affordability, scalability, and environmental safety, ensuring a sustainable approach to soil quality monitoring and management [65].

7.7 Challenges and Future Perspectives

Nanocomposites, while offering innovative solutions for environmental remediation, also present several challenges. Addressing these challenges and looking toward future developments is essential for the sustainable and responsible use of nanocomposites in environmental applications. While nanocomposites hold great promise for environmental remediation, there are challenges, such as the potential release of nanoparticles into the environment and the need for thorough life-cycle assessments to understand their overall environmental impact.

7.7.1 Challenges

7.7.1.1 Environmental Impact

Understanding the long-term environmental impact of nanocomposites is crucial. This includes assessing the potential toxicity, persistence, and bioaccumulation of nanoparticles released into the environment.

7.7.1.2 Health Risks

The small size of nanoparticles raises concerns about their interaction with living organisms, including potential inhalation risks and cellular damage.

7.7.1.3 Regulatory Framework

There is a need for comprehensive regulations and guidelines specific to the use and disposal of nanocomposites, ensuring safe handling and minimizing environmental risks.

7.7.1.4 Economic Viability

The cost-effectiveness of nanocomposite-based remediation technologies is a significant factor, especially for large-scale applications and in developing countries.

7.7.1.5 Technological Challenges

Issues such as scalability, stability, and efficiency of nanocomposites under various environmental conditions need to be addressed.

7.7.1.6 Public Perception

Public awareness and acceptance of nanotechnology-based solutions can be a barrier, necessitating transparent communication about their benefits and risks.

7.7.2 Future Perspectives

7.7.2.1 Sustainable Development

Research is focusing on developing more sustainable and eco-friendly nanocomposites, including biodegradable and green synthesis methods.

7.7.2.2 *Advanced Functionalities*

Future nanocomposites are expected to have multifunctional capabilities, such as simultaneous detection and removal of pollutants.

7.7.2.3 *Smart Nanocomposites*

The development of smart nanocomposites that can respond to environmental stimuli (like pH, temperature, or light) for targeted remediation actions.

7.7.2.4 *Integrated Systems*

Combining nanocomposites with other remediation technologies, like bioremediation or traditional filtration methods, to enhance overall effectiveness.

7.7.2.5 *Nanocomposite Recycling and Reuse*

Developing methods for the recovery, recycling, and reuse of nanocomposites to enhance their economic viability and reduce environmental impact.

7.7.2.6 *Advanced Sensing Technologies*

Utilizing nanocomposites in advanced sensors for real-time monitoring of environmental pollution and remediation progress.

7.7.2.7 *Global Collaboration and Regulation*

Strengthening international collaboration for the development of standardized regulations and guidelines for nanocomposite use in environmental applications.

7.8 Conclusion

In conclusion, nanocomposites hold great promise for environmental remediation, with ongoing advancements expected to address current challenges and open new avenues for their application. The focus remains on developing sustainable, safe, and effective nanocomposite-based solutions, coupled with a robust regulatory framework, to harness their full potential in combating environmental pollution. Nanocomposites represent a powerful and versatile tool in the fight against environmental pollution. Their ability to be engineered for specific tasks, combined with their efficiency in removing various types of pollutants, positions them at the forefront of innovative solutions for environmental remediation. However, ongoing research and careful consideration of their environmental impact are crucial to ensure their sustainable and safe application.

References

1. Al-Jawhari, I.F.H., Nanocellulose for sustainable future applications, in: *Handbook of Nanomaterials and Nanocomposites for Energy and Environmental Applications*, pp. 1–12, 2020.
2. Almasi, H., Mehryar, L., Ghadertaj, A., Photocatalytic activity and water purification performance of *in situ* and *ex situ* synthesized bacterial cellulose-CuO nanohybrids. *Water Environ. Res.*, *92*, 9, 1334–1349, 2020.
3. Alves, A.A., Silva, W.E., Belian, M.F., Lins, L.S.G., Galembeck, A., Bacterial cellulose membranes for environmental water remediation and industrial wastewater treatment. *Int. J. Environ. Sci. Technol.*, *17*, 3997–4008, 2020.
4. Liu, X., He, S., Yang, Y., Yao, B., Tang, Y., Luo, L., Zhou, Y., A review on percarbonate-based advanced oxidation processes for remediation of organic compounds in water. *Environ. Res.*, 200, 111371, 2021.
5. Zhou, H. and Smith, D.W., Advanced technologies in water and wastewater treatment. *Can. J. Civ. Eng.*, *28*, S1, 49–66, 2001.
6. Nagtode, V.S., Cardoza, C., Yasin, H.K.A., Mali, S.N., Tambe, S.M., Roy, P., Singh, K., Goel, A., Amin, P.D., Thorat, B.R., Cruz, J.N., Pratap, A.P., Green Surfactants (Biosurfactants): A Petroleum-Free Substitute for Sustainability—Comparison, Applications, Market, and Future Prospects. *ACS Omega*, 8, 11674–11699, 2023.
7. Ul-Islam, M., Wajid Ullah, M., Khan, S., Kamal, T., Ul-Islam, S., Shah, N., Kon Park, J., Recent advancement in cellulose based nanocomposite for addressing environmental challenges. *Recent Pat. Nanotechnol.*, *10*, 3, 169–180, 2016.
8. Lu, N., Zhang, M., Jing, X., Zhang, P., Zhu, Y., Zhang, Z., Electrospun Semiconductor-Based Nano-Heterostructures for Photocatalytic Energy Conversion and Environmental Remediation: Opportunities and Challenges. *Energy Environ. Mater.*, *6*, 2, e12338, 2023.
9. Ul-Islam, M., Ullah, M.W., Khan, S., Manan, S., Khattak, W.A., Ahmad, W., Park, J.K., Current advancements of magnetic nanoparticles in adsorption and degradation of organic pollutants. *Environ. Sci. Pollut. Res.*, *24*, 12713–12722, 2017.

10. Asahi, R., Morikawa, T., Irie, H., Ohwaki, T., Nitrogen-doped titanium dioxide as visible-light-sensitive photocatalyst: designs, developments, and prospects. *Chem. Rev.*, *114*, 19, 9824–9852, 2014.
11. Baig, N., Kammakakam, I., Falath, W., Nanomaterials: A review of synthesis methods, properties, recent progress, and challenges. *Mater. Adv.*, *2*, 6, 1821–1871, 2021.
12. Bhat, S.A., Sher, F., Hameed, M., Bashir, O., Kumar, R., Vo, D.V.N., Lima, E.C., Sustainable nanotechnology based wastewater treatment strategies: Achievements, challenges and future perspectives. *Chemosphere*, *288*, 132606, 2022.
13. Ahmed, S.N. and Haider, W., Heterogeneous photocatalysis and its potential applications in water and wastewater treatment: a review. *Nanotechnology*, *29*, 34, 342001, 2018.
14. Cai, Z., Zhao, X., Duan, J., Zhao, D., Dang, Z., Lin, Z., Remediation of soil and groundwater contaminated with organic chemicals using stabilized nanoparticles: Lessons from the past two decades. *Front. Environ. Sci. Eng.*, *14*, 1–20, 2020.
15. Eyvazi, B., Jamshidi-Zanjani, A., Darban, A.K., Immobilization of hexavalent chromium in contaminated soil using nano-magnetic MnFe2O4. *J. Hazard. Mater.*, *365*, 813–819, 2019.
16. Fan, G., Cang, L., Qin, W., Zhou, C., Gomes, H.I., Zhou, D., Surfactants-enhanced electrokinetic transport of xanthan gum stabilized nanoPd/Fe for the remediation of PCBs contaminated soils. *Sep. Purif. Technol.*, *114*, 64–72, 2013.
17. Gong, Y., Zhao, D., Wang, Q., An overview of field-scale studies on remediation of soil contaminated with heavy metals and metalloids: Technical progress over the last decade. *Water Res.*, *147*, 440–460, 2018.
18. He, Y., Peng, G., Jiang, Y., Zhao, M., Wang, X., Chen, M., Lin, S., Nano-Photocatalysts: Environmental Hazard Potential of Nano-Photocatalysts Determined by Nano-Bio Interactions and Exposure Conditions (Small 21/2020). *Small*, *16*, 21, 2070118, 2020.
19. Huang, B., Lei, C., Wei, C., Zeng, G., Chlorinated volatile organic compounds (Cl-VOCs) in environment—sources, potential human health impacts, and current remediation technologies. *Environ. Int.*, *71*, 118–138, 2014.
20. Ken, D.S. and Sinha, A., Recent developments in surface modification of nano zero-valent iron (nZVI): Remediation, toxicity and environmental impacts. *Environ. Nanotechnol. Monit. Manage.*, *14*, 100344, 2020.
21. Liu, W., Tian, S., Zhao, X., Xie, W., Gong, Y., Zhao, D., Application of stabilized nanoparticles for *in situ* remediation of metal-contaminated soil and groundwater: a critical review. *Curr. Pollut. Rep.*, *1*, 280–291, 2015.
22. Mondal, K. and Sharma, A., Recent advances in the synthesis and application of photocatalytic metal–metal oxide core–shell nanoparticles for environmental remediation and their recycling process. *RSC Adv.*, 6, 83589–83612, 2016.

23. Mu, Y., Jia, F., Ai, Z., Zhang, L., Iron oxide shell mediated environmental remediation properties of nano zero-valent iron. *Environ. Sci.: Nano*, 4, 27–45, 2017.
24. Arabyarmohammadi, H., Darban, A.K., Abdollahy, M., Yong, R., Ayati, B., Zirakjou, A., van der Zee, S.E., Utilization of a novel chitosan/clay/biochar nanobiocomposite for immobilization of heavy metals in acid soil environment. *J. Polym. Environ.*, 26, 2107–2119, 2018.
25. Salehi, E., Khajavian, M., Sahebjamee, N., Mahmoudi, M., Drioli, E., Matsuura, T., Advances in nanocomposite and nanostructured chitosan membrane adsorbents for environmental remediation: a review. *Desalination*, 527, 115565, 2022.
26. Pathak, D., Sharma, A., Sharma, D.P., Kumar, V., A review on electrospun nanofibers for photocatalysis: Upcoming technology for energy and environmental remediation applications. *Appl. Surf. Sci. Adv.*, *18*, 100471, 2023.
27. Kayanja, O., Abdel-Aty, A.A., Hassan, M.A., Hassanin, A., Ohashi, H., Khalil, A.S., A Review on TMDCs Nanomaterials and Their Surface Engineered Polymeric Membrane Nanocomposites for Water Remediation and Wastewater Treatment. *Surf. Interfaces*, 38, 103578, 2023.
28. Abdulhamid, M.A. and Muzamil, K., Recent progress on electrospun nanofibrous polymer membranes for water and air purification: A review. *Chemosphere*, *310*, 136886, 2023.
29. Namasivayam, S.K.R., Prakash, P., Babu, V., Paul, E.J., Bharani, R.A., Kumar, J.A., Moovendhan, M., Aquatic biomass cellulose fabrication into cellulose nanocomposite and its application in water purification. *J. Clean. Prod.*, 421, 136386, 2023.
30. Zhang, Y., Wang, B., Wang, R., Functionally decorated metal–organic frameworks in environmental remediation. *Chem. Eng. J.*, 455, 140741, 2023.
31. Rasouli, K., Rasouli, J., Mohtaram, M.S., Sabbaghi, S., Kamyab, H., Moradi, H., Chelliapan, S., Biomass-derived activated carbon nanocomposites for cleaner production: A review on aspects of photocatalytic pollutant degradation. *J. Clean. Prod.*, 471, 138181, 2023.
32. Issahaku, I., Tetteh, I.K., Tetteh, A.Y., Chitosan and chitosan derivatives: Recent advancements in production and applications in environmental remediation. *Environ. Adv.*, 13, 100351, 2023.
33. Shahzad, A., Ullah, M.W., Ali, J., Aziz, K., Javed, M.A., Shi, Z., Yang, G., The versatility of nanocellulose, modification strategies, and its current progress in wastewater treatment and environmental remediation. *Sci. Total Environ.*, 858, 159937, 2023.
34. Ahmad, M.A., Adeel, M., Shakoor, N., Javed, R., Ishfaq, M., Peng, Y., Deng, X., Modifying engineered nanomaterials to produce next generation agents for environmental remediation. *Sci. Total Environ.*, 891, 164861, 2023.
35. Arsenov, D., Beljin, J., Jović, D., Maletić, S., Borišev, M., Borišev, I., Nanomaterials as endorsed environmental remediation tools for the next

generation: Eco-safety and sustainability. *J. Geochem. Explor.*, 253, 107283, 2023.

36. Cui, H., Gu, X., Lu, S., Fu, X., Zhang, X., Fu, G.Y., Qiu, Z., Sui, Q., Degradation of ethylbenzene in aqueous solution by sodium percarbonate activated with EDDS–Fe(III) complex. *Chem. Eng. J.*, 309, 80–88, 2017.
37. Ilyas, R.A., Aisyah, H.A., Nordin, A.H., Ngadi, N., Zuhri, M.Y.M., Asyraf, M.R.M., Sapuan, S.M., Zainudin, E.S., Sharma, S., Abral, H., others, Natural-fiberreinforced chitosan, chitosan blends and their nanocomposites for various advanced applications. *Polymers (Basel)*, 14, 874, 2022.
38. Li, H., Wang, Z., Zhang, H., Pan, Z., Nanoporous PLA/(chitosan nanoparticle) composite fibrous membranes with excellent air filtration and antibacterial performance. *Polymers (Basel)*, 10, 1085, 2018.
39. Peter, S., Lyczko, N., Gopakumar, D., Maria, H.J., Nzihou, A., Thomas, S., Chitin and chitosan based composites for energy and environmental applications: A review. *Waste Biomass Valorization*, 12, 4777–4804, 2021.
40. Gao, Y., Hu, M., Mi, B., Membrane surface modification with TiO2–graphene oxide for enhanced photocatalytic performance. *J. Membr. Sci.*, 455, 349–356, 2014.
41. Shaumbwa, V.R., Liu, D., Archer, B., Li, J., Su, F., Preparation and application of magnetic chitosan in environmental remediation and other fields: a review. *J. Appl. Polym. Sci.*, 138, 51241, 2021.
42. Baruah, M., Ezung, S.L., Sharma, S., Sinha, U.B., Sinha, D., Synthesis and characterization of Ni-doped TiO2 activated carbon nanocomposite for the photocatalytic degradation of anthracene. *Inorg. Chem. Commun.*, 144, 109905, 2022.
43. Apopei, P., Orha, C., Popescu, M.I., Lazau, C., Manea, F., Catrinescu, C., Teodosiu, C., Diclofenac removal from water by photocatalysis- assisted filtration using activated carbon modified with N-doped TiO2. *Process Saf. Environ. Prot.*, 138, 324–336, 2020.
44. Chauhan, A., Sudhaik, A., Raizada, P., Khan, A.A.P., Singh, A., Le, Q.V., Nguyen, V.-H., Ahamad, T., Thakur, S., Singh, P., Asiri, A.M., Enhancement strategies for ZnSe based photocatalysts: application to environmental remediation and energy conversion. *Process Saf. Environ. Prot.*, 170, 415–435, 2023.
45. Müller, A.-K., Xu, Z.-K., Greiner, A., Preparation and performance assessment of low-pressure affinity membranes based on functionalized, electrospun polyacrylates for gold nanoparticle filtration. *ACS Appl. Mater. Interfaces*, 13, 13, 15659–15667, 2021.
46. Bortolassi, A.C.C., Guerra, V.G., Aguiar, M.L., Soussan, L., Cornu, D., Miele, P., Bechelany, M., Composites based on nanoparticle and pan electrospun nanofiber membranes for air filtration and bacterial removal. *Nanomaterials*, 9, 1740, 2019.

47. Celebioglu, A., Sen, H.S., Durgun, E., Uyar, T., Molecular entrapment of volatile organic compounds (VOCs) by electrospun cyclodextrin nanofibers. *Chemosphere*, 144, 736–744, 2016.
48. Lu, T., Liang, H., Cao, W., Deng, Y., Qu, Q., Ma, W., Xiong, R., Huang, C., Blow spun nanofibrous composite Self-cleaning membrane for enhanced purification of oily wastewater. *J. Colloid Interface Sci.*, 608, 2860–2869, 2022.
49. Wang, L., Kang, Y., Xing, C.Y., Guo, K., Zhang, X.Q., Ding, L.S., Zhang, S., Li, B.J., B-Cyclodextrin based air filter for high-efficiency filtration of pollution sources. *J. Hazard. Mater.*, 373, 197–203, 2019.
50. Kim, B., Lee, Y.R., Kim, H.Y., Ahn, W.S., Adsorption of volatile organic compounds over MIL-125-NH2. *Polyhedron*, 154, 343–349, 2018.
51. Mahmoodi, N.M., Taghizadeh, A., Taghizadeh, M., Abdi, J., *In situ* deposition of Ag/ AgCl on the surface of magnetic metal-organic framework nanocomposite and its application for the visible-light photocatalytic degradation of Rhodamine dye. *J. Hazard. Mater.*, 378, 120741, 2019.
52. Liu, H., Cao, C., Huang, J., Chen, Z., Chen, G., Lai, Y., Progress on particulate matter filtration technology: basic concepts, advanced materials, and performances. *Nanoscale*, 12, 2, 437–453, 2020.
53. Zhang, K., Huo, Q., Zhou, Y.Y., Wang, H.H., Li, G.P., Wang, Y.W., Wang, Y.Y., Textiles/Metal- organic frameworks composites as flexible air filters for efficient particulate matter removal. *ACS Appl. Mater. Interfaces*, 11, 17368–17374, 2019.
54. Yang, F., Wu, J., Zhu, X., Ge, T., Wang, R., Enhanced stability and hydrophobicity of LiX@ZIF-8 composite synthesized environmental friendly for CO2 capture in highly humid flue gas. *Chem. Eng. J.*, 410, 128322, 2021.
55. Chen, C., Wang, H., Chen, Y., Wei, X., Zou, W., Wan, H., Dong, L., Guan, G., Layer-by layer self-assembly of hierarchical flower-like HKUST-1-based composite over amino-tethered SBA-15 with synergistic enhancement for CO2 capture. *Chem. Eng. J.*, 413, 127396, 2021.
56. Vo, T.K., Nguyen, V.C., Quang, D.T., Park, B.J., Kim, J., Formation of structural defects within UiO-66(Zr)-(OH)2 framework for enhanced CO2 adsorption using a microwave-assisted continuous-flow tubular reactor. *Microporous Mesoporous Mater.*, 312, 110746, 2021.
57. Hoseinzadeh, H., Hayati, B., Shahmoradi Ghaheh, F., Seifpanahi-Shabani, K., Mahmoodi, N.M., Development of room temperature synthesized and functionalized metal-organic framework/graphene oxide composite and pollutant adsorption ability. *Mater. Res. Bull.*, 142, 111408, 2021.
58. Sarfraz, M.H., Zubair, M., Aslam, B., Ashraf, A., Siddique, M.H., Hayat, S., Cruz, J.N., Muzammil, S., Khurshid, M., Sarfraz, M.F., Hashem, A., Dawoud, T.M., Avila-Quezada, G.D., Abd_Allah, E.F., Comparative analysis of phyto-fabricated chitosan, copper oxide, and chitosan-based CuO nanoparticles: antibacterial potential against Acinetobacter baumannii isolates and anticancer activity against HepG2 cell lines. *Front. Microbiol.*, 14, 1188743, 2023.

59. Karami, K., Beram, S.M., Bayat, P., Siadatnasab, F., Ramezanpour, A., A novel nanohybrid based on metal–organic framework MIL101– Cr/PANI/Ag for the adsorption of cationic methylene blue dye from aqueous solution. *J. Mol. Struct.*, 1247, 131352, 2022.
60. Kumar, G. and Masram, D.T., Sustainable synthesis of MOF-5@GO nanocomposites for efficient removal of rhodamine B from water. *ACS Omega*, 6, 9587–9599, 2021.
61. Mahmoodi, N.M., Oveisi, M., Asadi, E., Synthesis of NENU metal-organic framework-graphene oxide nanocomposites and their pollutant removal ability from water using ultrasound. *J. Cleaner Prod.*, 211, 198–212, 2019.
62. Eltaweil, A.S., Mamdouh, I.M., Abd El-Monaem, E.M., El-Subruiti, G.M., Highly Efficient removal for methylene blue and Cu2+ onto UiO-66 metal-organic framework/carboxylated graphene oxide-incorporated sodium alginate beads. *ACS Omega*, 6, 23528–23541, 2021.
63. Kasula, M., Le, T., Thomsen, A., Rabbani Esfahani, M., Silver metal organic frameworks and copper metal organic frameworks immobilized on graphene oxide for enhanced adsorption in water treatment. *Chem. Eng. J.*, 446, 135542, 2022.
64. Khajeh, M., Oveisi, A.R., Barkhordar, A., Sorinezami, Z., Co-Fe-layered double hydroxide decorated amino-functionalized zirconium terephthalate metal-organic framework for removal of organic dyes from water samples. *Spectrochim. Acta, Part A*, 234, 118270, 2020.
65. Huang, G., Huang, C., Tao, Y., Li, H., Localized heating driven selective growth of metal-organic frameworks (MOFs) in wood: A novel synthetic strategy for significantly enhancing MOF loadings in wood. *Appl. Surf. Sci.*, 564, 150325, 2021.

8

Nanoscale Coatings for Tribology: Applications, Challenges and Future Directives

Ranvir Singh Panwar and Ashish Mogra*

School of Technology Management & Engineering, SVKM's Narsee Monjee Institute of Management Studies (NMIMS) Deemed-to-University, Chandigarh, India

Abstract

Nanoscale coatings and thin films have gained significant attention in recent years due to their remarkable properties and versatile applications. Tribology at the nanoscale, particularly concerning nanoscale coatings, is a highly specialized area that focuses on understanding and manipulating the interactions between surfaces at extremely small scales to optimize friction, wear, and lubrication properties. This book chapter provides an overview of nanoscale coatings and their diverse tribology applications, challenges and future directives. It discusses the fundamental principles, techniques, and the latest developments in the field, highlighting the potential for future innovations.

***Keywords*:** Nanoscale coating, tribology, surface phenomenon, nanoparticle

8.1 Introduction

Nanoscale coatings in tribology refer to very thin layers of material, typically on the scale of nanometres (one nanometre is one-billionth of a meter), applied to surfaces to modify their tribological properties. These can include metals, ceramics, polymers, and composite materials. Each material offers unique properties like hardness, chemical resistance, or specific frictional characteristics. The development and application of

**Corresponding author*: ashish.mogra@nmims.edu

Inamuddin, Tariq Altalhi and Mohammad Abu Jafar Mazumder (eds.) Nanostructured Hard Materials: Industrial Applications, (269–282) © 2026 Scrivener Publishing LLC

nanoscale coatings represent a significant advancement in nanotechnology and materials science. A lot of study has been done on nanoscale coatings for enhancement of tribological performance. Numerous studies have demonstrated how applying such coatings can significantly increase the performance of cutting tools. Researchers found various advantages of nanoscale coatings for tribology, some have shown in Figure 8.1.

Tribology is the science and engineering of interacting surfaces in relative motion, encompassing the study of friction, lubrication, and wear. The Greek term "tribo," which means "to rub," and the suffix "logia," which means "study or learning," are the sources of the English word "tribology" [1]. It involves the principles of physics, chemistry, materials science, biology and mechanical engineering to understand and control friction, wear, and lubrication as shown in Figure 8.2.

This is the science of understanding how surfaces interact in motion, including aspects of friction (resistance to motion), wear (degradation of surfaces), and lubrication (use of substances to reduce friction and wear). One of the primary reasons for using nanoscale coatings in tribological applications is to reduce friction and wear between surfaces. This can lead to improved efficiency and longer lifespan of mechanical components. Wear and friction are dependent upon usage conditions and experimental factors rather than being inherent qualities. Because different materials have varied physical, chemical, mechanical, and adhesive properties, they exhibit different wear behaviours [2].

Friction, wear, and lubrication as shown in Figure 8.3 are the three branches of tribology science, which deals essentially with interacting solid

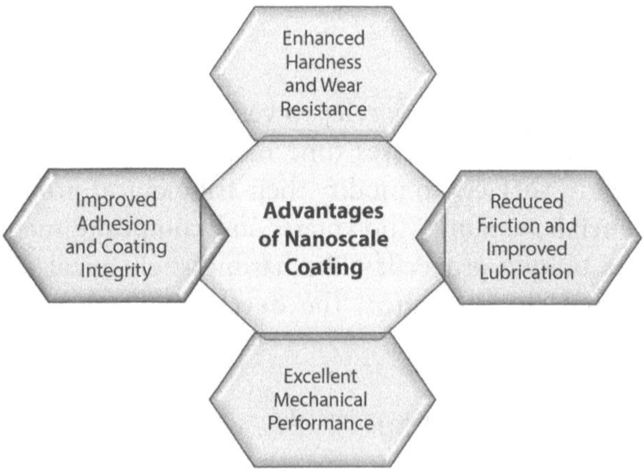

Figure 8.1 Advantages of nanoscale coating.

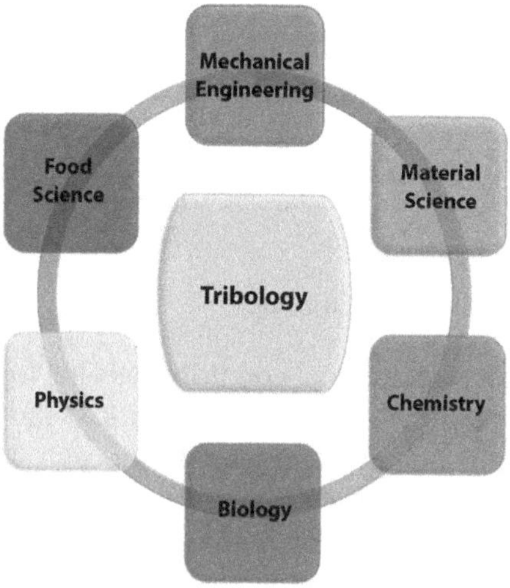

Figure 8.2 Tribology in various domains of science.

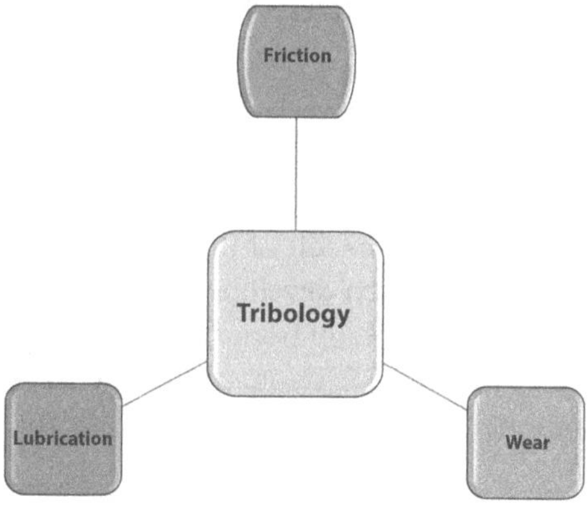

Figure 8.3 Branches of tribology.

surfaces in relative motion. At the nanoscale, friction can be influenced by adhesion, deformation, and various surface forces that differ significantly from macro-scale friction. Nanoscale roughness and surface chemistry play crucial roles in how two surfaces interact, affecting adhesion and therefore friction. The thickness of the coating can significantly influence frictional behaviour. Ultra-thin coatings may behave differently compared to thicker layers. Nanocoating are often designed to be more wear-resistant than their bulk counterparts. This is due to the unique properties of materials at the nanoscale, like increased hardness or self-healing capabilities. The interaction between asperities (tiny contact points) on surfaces can be critical in determining wear rates. The chemical and structural stability

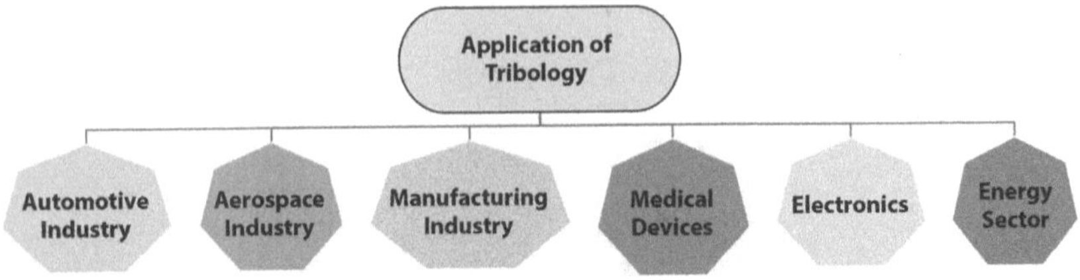

Figure 8.4 Industrial applications of tribology.

of nanocoating under operational conditions is vital for maintaining their wear-resistant properties. A diagram (Figure 8.4) illustrating the industrial applications of tribology, represented by a central rounded rectangle labeled 'Application of Tribology' connected to six octagons representing different industries: energy sector, automotive industry, aerospace industry, manufacturing industry, medical devices, and electronics.

8.2 Overview of the Role of Tribology in Various Industries

Over time, the industrial revolution has spurred scientists to create scientific knowledge addressing problems with friction and wear-related energy loss. The study of tribology is a result of this issue with energy loss. Many materials and methods have been developed in response to the significant rise in demand for environmentally friendly technology and fuel-efficient machinery. These aim to reduce friction, preserve energy, prevent damage to machinery components, and ultimately decrease the amount of trash generated [3, 4].

Actuality, we deal with many tribological applications on a daily basis. Examples include writing with a pencil on paper, brushing down a surface, using a nail file, applying oil to noisy door hinges, and more. Its application can be found in almost every aspect of modern technology, manufacturing, transportation, healthcare, space exploration, machinery, equipment and more [5, 6]. Here is an overview of the role of tribology in different industries:

8.2.1 Automotive Industry: Engine Component Coatings

Nanoscale coatings in engine components play a crucial role in enhancing tribological performance. These coatings are designed to reduce friction, wear, and tear, improve fuel efficiency, and extend the lifespan of engine parts.

Types of Nanoscale Coatings Used in Engine Components:
Titanium Nitride (TiN): Titanium Nitride (TiN) is a widely used nanoscale coating in engine components for enhancing tribological performance. This ceramic material is favoured for its hardness, wear resistance, and low coefficient of friction. TiN is extremely hard, which helps in reducing wear and extending the life of engine parts. It has a low coefficient of friction, which minimises resistance between moving parts. It offers good stability at high temperatures, which is crucial in engine environments. Resistant to oxidation and corrosion, thereby protecting the underlying material. Ensures smoother valve operation, reduces the risk of valve seizure, and prolongs valve life. TiN coatings can protect fuel injectors from wear and corrosion. It applied to engine bearings to reduce friction and wear, particularly in areas with high contact stress. It increases bearing life and reliability, which is crucial for engine durability. It extends the life of the crankshaft and maintains engine balance and efficiency [7, 8].

Diamond-Like Carbon (DLC): These coatings are increasingly used in engine components to enhance tribological performance. DLC coatings are unique due to their high hardness, low friction coefficient, and good wear resistance, characteristics that are similar to diamond. These coatings are among the hardest materials available, providing excellent wear resistance. They have a low coefficient of friction, which reduces the energy loss due to friction in engine components. They are chemically inert, which means they don't react with engine oils or additives and maintains its properties over a wide range of temperatures. DLC coatings improve engine efficiency by reducing friction between piston rings and cylinder walls [9].

Aluminium Oxide (Al_2O_3):- It is commonly known as alumina, and Alumina-Titanium Nitride (AlTiN). Their coatings are utilised in engine components to enhance tribological performance. These materials are chosen for their exceptional hardness, thermal stability, and resistance to wear and corrosion. They maintain their structural integrity at high temperatures, a critical feature for engine components. Coating pistons with these materials can protect against high temperatures and abrasive wear. They are resistant to oxidation and corrosion, which is essential for longevity in harsh engine environments. These coatings are used on cylinder liners to reduce wear and friction with piston rings. Al2O3 coatings in engine components significantly contribute to improved wear resistance, thermal stability, and overall engine efficiency. Their application showcases the advancements in materials engineering, directly impacting the performance and longevity of automotive engines, developing composite coatings that combine the benefits of alumina and titanium nitride with other materials for improved performance [10].

Chromium Nitride (CrN):- It is a valuable material for nanoscale coating which is used in various engine components to enhance tribological performance. CrN is favoured for its high hardness, excellent corrosion resistance, and good wear resistance. These properties make it an ideal coating for improving the longevity and efficiency of engine parts. Applying CrN on piston rings and cylinder surfaces reduces friction and wear. It improves efficiency, decreases oil consumption, and extends the lifespan of these components. CrN coatings on bearings and gears reduce friction and protect against wear. It prolongs the life of these critical components, leading to reduced maintenance and improved reliability. Chromium Nitride coatings provide significant benefits in reducing friction and wear in engine components, leading to more efficient, reliable, and durable engines. Their application in the automotive industry is a clear testament to the advancements in materials science and its direct impact on improving engine technology, combining CrN with other materials to create coatings with enhanced properties [11, 12].

8.2.2 Aerospace and Energy Industry: Steam and Wind Turbine Blade Coatings

Turbine blade coatings at the nanoscale play a crucial role in enhancing the tribological performance of turbines, especially in high-stress environments like those found in aerospace and power generation industries. These coatings are designed to protect the blades from wear, corrosion, and high-temperature degradation. Turbines operate at extremely high temperatures, so coatings must maintain their integrity and functionality under thermal stress. Blades are exposed to oxidizing and corrosive environments, requiring coatings that can protect against such conditions. Coatings need to withstand the mechanical wear caused by high-speed operation and particulate matter in the air or fuel. Reducing friction is essential for efficient turbine operation and minimizing energy losses. Coatings protect against the high-temperature gas stream and improve overall engine efficiency. TBCs are crucial for allowing higher operating temperatures, which translates to more efficient fuel consumption. Coatings must withstand the stress of rapid temperature changes during operation, ensuring strong adhesion to the blade material and durability over extended periods. Considering the ecological implications of the materials and processes used in coating, it is essential to focus on developing more advanced thermal barrier coatings with greater thermal insulation and durability, while utilizing nanotechnology to improve coating properties such as toughness and resistance to thermal shock. Nanoscale coatings on turbine blades significantly contribute to the tribological performance, enhancing the efficiency, reliability, and longevity of turbines in various industries. The continuous advancement in coating technologies and materials science plays a pivotal role in addressing the challenging operating conditions of turbines. Nanoscale coatings on wind turbine blades are crucial for enhancing their tribological performance, significantly impacting the efficiency and durability of wind turbines. These coatings are designed to withstand harsh environmental conditions and mechanical stresses experienced by the blades; essential for protecting blades from wear caused by rain, hail, dust, and airborne particles. Nanoscale coatings on wind turbine blades are essential for protecting them against erosion, corrosion, and environmental factors, directly impacting the efficiency and operational lifespan of wind turbines. Continuous advancements in material science and nanotechnology are pivotal in developing more effective and sustainable coating solutions for the renewable energy sector [13–17].

8.2.3 Manufacturing Industry: Cutting Tool Coatings

Composite coatings at the nanoscale may improve the performance of cutting tools. Such coatings can be engineered to offer more wear protection and enhanced heat dissipation, hence increasing tool durability and cutting efficiency. Additionally, friction can be decreased using the nanoscale composite coatings, increasing productivity and requiring fewer tools. Additionally, the coatings contain particular additives that improve the coating's resistance to corrosion, ability to retain lubricant, and increased thermal stability. By enhancing the tribological (friction, lubrication, and wear) performance of components, nano-scale composite coatings can lessen wear and tear on components during usage. Nanoscale coatings for cutting tools are a pivotal advancement in the field of material science and tribology, significantly enhancing the performance, durability, and efficiency of these tools. These coatings are designed to withstand the extreme conditions of cutting, such as high temperature, friction, and wear. Nanoscale coatings on cutting tools represent a significant advancement in enhancing the efficiency, lifespan, and performance of these tools. The continuous evolution in coating technologies is a testament to the ongoing innovation in the field of tribology and materials science [18–20].

8.2.4 Medical Devices: Orthopaedic Implant Coatings

Nanoscale coatings for orthopaedic implants are a significant advancement in biomedical engineering and tribology, aimed at enhancing the performance, biocompatibility, and longevity of these implants. These coatings are crucial in orthopaedic applications where implants interact with human bone and tissue, requiring optimised surface properties to ensure successful integration and function. These coating have the ability to support bone growth and integration with the implant, high resistance to wear is critical to prevent implant degradation and particulate debris. These coating have the properties which are essential to mimic the natural movement of joints and reduce wear. Nanoscale coatings on orthopaedic implants represent a critical area of innovation, addressing key challenges in implant longevity, biocompatibility, and functionality. The ongoing advancements in materials science and nanotechnology continue to significantly impact the field of orthopaedic surgery and patient care [21, 22].

8.2.5 Electronics: Hard Disk Drive Coatings

Nanoscale coatings in hard disk drives (HDDs) play a vital role in enhancing their tribological performance, which is crucial for the reliability, efficiency, and longevity of data storage devices. These coatings are engineered to address the challenges posed by the high-speed, high-precision environment of HDDs. They are also withstanding the continuous contact and friction between the read/write head and the disk surface. These coatings maintain performance under the varying temperatures experienced during HDD operation. They provide additional protection against wear and corrosion, and reduces stiction [23].

8.3 Challenges and Future Directives

Tribology of nanoscale coatings, while offering numerous advantages, also faces several challenges and problems. These issues are often complex and multifaceted, stemming from the inherent difficulties of working at the nanoscale and the high expectations for performance in varied and demanding applications [24]. Here are some of the key problems in this field:

8.3.1 Durability and Longevity

Adhesion Issues: Ensuring that nanoscale coatings remain firmly attached to their substrates under various operating conditions is challenging. Poor adhesion can lead to rapid coating failure.

Wear Resistance: Although nanocoating are designed to be wear-resistant, in some cases, they may wear out faster than expected, especially under extreme conditions of load, speed, or temperature.

8.3.2 Uniformity and Consistency

Coating Uniformity: Achieving uniform thickness and consistent properties across the entire coated surface can be difficult, especially for complex geometries.

Quality Control: Ensuring consistent quality in mass production of nanocoating is challenging, which can lead to variations in performance.

8.3.3 Environmental, Sustainability and Health Concerns

Toxicity: Some materials used in nanocoating can be toxic or harmful to the environment, raising concerns about their safe production, usage, and disposal.

Nano-particle Release: There's a risk of nanoparticles being released into the environment during wear, which might pose health and environmental risks.

8.3.4 Technical and Fabrication Challenges

Scaling Up: Translating lab-scale successes to industrial-scale production without loss of performance or significant cost increase is a major challenge.

Complex Fabrication Processes: Some nanocoating processes are complex and require precise control, making them expensive or difficult to implement in standard manufacturing settings.

8.3.5 Performance Under Extreme Conditions

High-Temperature Stability: Maintaining coating integrity and performance at high temperatures can be problematic for certain nanocoating materials.

Corrosive Environments: In chemically aggressive environments, nanocoating might degrade or fail more quickly than expected.

8.3.6 Compatibility with Substrates

Material Interactions: The interactions between the coating material and the substrate can sometimes lead to unexpected issues, like interfacial reactions or stress-induced failures. Thermal Expansion Mismatch: Differences in thermal expansion coefficients between the coating and the substrate can lead to cracking or delamination.

8.3.7 Characterisation and Testing

Measuring Properties: Accurately measuring the tribological properties of nanoscale coatings can be difficult due to their small scale and the complexity of interactions at the nanoscale.

Standardisation: There is a lack of standardised testing methods specific to nanoscale tribological coatings, making comparison and validation of results challenging.

8.3.8 Cost and Economic Viability

Production Costs: The production of certain nanoscale coatings can be costly, limiting their application to high-value or critical-use cases.

Return on Investment: Balancing the cost of implementing nanoscale coatings against the performance benefits and lifespan improvements is a key economic challenge.

A Table 8.1 summarizing key properties and applications of Titanium Nitride (TiN), including its composition (titanium and nitrogen), hardness

Table 8.1 Summary of Titanium Nitride (TiN).

Property	Description	Application
Composition	Composed of titanium and nitrogen	Cutting tools, surgical instruments, automotive
Hardness	Hardness up to 30 GPa	Drills and milling machines to enhance durability
Friction coefficient	Typically, around 0.4 to 0.6	Bearings and gears in high-load environments
Thickness	Generally, 1-5 micrometres	Decorative finishes for wear and corrosion resistance
Wear resistance	High wear resistance	Protective coatings in maritime environments

(up to 30 GPa), friction coefficient (0.4 to 0.6), coating thickness (1–5 micrometres), and high wear resistance. Applications include cutting tools, surgical instruments, and automotive parts. A Table 8.2 detailing the properties and uses of Diamond-Like Carbon (DLC) coatings. It covers composition (amorphous carbon with sp3 and sp2 bonds), very high hardness (up to 80 GPa), low friction coefficient (0.1 to 0.2), variable thickness, and excellent wear resistance. Applications include automotive parts, hard drives, medical implants, and aerospace components.

Table 8.2 Summary of Diamond-Like Carbon (DLC).

Property	Description	Application
Composition	Consists of amorphous carbon with a mix of sp3 and sp2 bonds.	Automotive components, hard drives, medical implants.
Hardness	Very high, up to 80 GPa.	Engine components to reduce wear and friction
Friction coefficient	Very low, typically between 0.1 to 0.2.	Used in pumps and compressors.
Thickness	Ranges from a few nano meters to several micro meters.	Precision instruments where thickness control is crucial.
Wear resistance	Exceptional wear resistance	Aerospace bearings and gears

Table 8.3 Summary of Aluminium Oxide (Al_2O_3).

Property	Description	Application
Composition	Consists primarily of aluminium oxide.	Textile machinery, food processing equipment.
Hardness	Very hard, up to 25 GPa.	Orthopaedic implants to prevent wear.
Friction coefficient	Moderate, typically around 0.5.	Thermal barrier coatings in engines.
Thickness	Usually between 2-10 micrometres.	Electrical insulation applications.
Wear resistance	High resistance to wear and thermal shock.	Gas turbines and other high-temperature applications.

Table 8.4 Summary of Chromium Nitride (CrN).

Property	Description	Application
Composition	Composed of chromium and nitrogen	Automotive and aerospace components, tooling industry
Hardness	Typically around 20-25 GPa	Used for cutting tools, molds, and dies to increase lifespan and performance.
Friction coefficient	Relatively low, around 0.3 to 0.6	Ideal for components requiring low friction, such as gears and bearings.
Thickness	Typically ranges from 2 to 5 micrometres	Suitable for both thin films in precision components and thicker coatings for harsh environments.
Wear resistance	High wear resistance due to its hardness and toughness	Protective coatings in machining, especially for wear-prone areas.

A Table 8.3 outlining characteristics and applications of Aluminium Oxide (Al_2O_3) coatings. Key points include high hardness (up to 25 GPa), moderate friction coefficient (around 0.5), thickness (2–10 micrometres), and high thermal shock resistance. Common applications include orthopaedic implants, engines, and electrical insulation. A Table 8.4 presenting the properties of Chromium Nitride (CrN) coatings. Features include composition (chromium and nitrogen), hardness (20–25 GPa), friction coefficient (0.3 to 0.6), coating thickness (2–5 micrometres), and high wear resistance. Applications involve automotive and aerospace components, gears, bearings, and tooling.

8.4 Conclusion

Nanoscale coatings in tribology represent a frontier in engineering and materials science, offering significant potential for enhancing the performance and lifespan of a wide range of mechanical systems while also opening avenues for new technologies and applications. As we look ahead, the field of nanoscale coatings is poised for continued growth, offering new and exciting opportunities for innovation and problem solving. This chapter has provided an overview of the applications, challenges and future directives in the field of tribology.

References

1. Sundus, F., Fazal, M.A., Masjuki, H.H., Tribology with biodiesel: A study on enhancing biodiesel stability and its fuel properties. *Renew. Sustain. Energy Rev.*, 70, 399–412, 2017.
2. Czichos, H., *Tribology: a systems approach to the science and technology of friction, lubrication, and wear*, Elsevier, 1978.
3. Ren, Y., Zhang, L., Xie, G., Li, Z., Chen, H., Gong, H., Xu, W., Guo, D., Luo, J., A review on tribology of polymer composite coatings. *Friction*, 9, 429–470, 2021.
4. Tichy, J.A. and Meyer, D.M., Review of solid mechanics in tribology. *Int. J. Solids Struct.*, 37, 391–400, 2000.
5. Jojibabu, P., Jagannatham, M., Haridoss, P., Janaki Ram, G.D., Deshpande, A.P., Bakshi, S.R., Effect of different carbon nano-fillers on rheological properties and lap shear strength of epoxy adhesive joints. *Compos. Part A Appl. Sci. Manuf.*, 82, 53–64, 2016.
6. Schwartz, C.J. and Bahadur, S., Studies on the tribological behavior and transfer film-counterface bond strength for polyphenylene sulfide filled with nanoscale alumina particles. *Wear*, 237, 261–273, 2000.
7. Chang, L. and Friedrich, K., Enhancement effect of nanoparticles on the sliding wear of short fiber-reinforced polymer composites: A critical discussion of wear mechanisms. *Tribol. Int.*, 43, 2355–2364, 2010.
8. Briscoe, B.J. and Sinha, S.K., Tribology of Polymeric Solids and Their Composites, in: *Wear - Mater. Mech. Pract.*, pp. 223–267, 2005.
9. Friedrich, K. and Schlarb, A.K., Foreword, in: *Tribol. Interface Eng. Ser.*, vol. 55, pp. ix–xi, 2008.
10. Cho, D.H. and Bhushan, B., Friction and wear of various polymer pairs used for label and wiper in labeling machine. *Tribol. Int.*, 98, 10–19, 2016.
11. Laux, K.A., Jean-Fulcrand, A., Sue, H.J., Bremner, T., Wong, J.S.S., The influence of surface properties on sliding contact temperature and friction for polyetheretherketone (PEEK). *Polymer (Guildf)*, 103, 397–404, 2016.
12. Araújo Junior, A.S., Sales, W.F., da Silva, R.B., Costa, E.S., Rocha Machado, A., Lubri-cooling and tribological behavior of vegetable oils during milling of AISI 1045 steel focusing on sustainable manufacturing. *J. Clean. Prod.*, 156, 635–647, 2017.
13. Singh, R., Dureja, J.S., Dogra, M., Performance evaluation of textured carbide tools under environment-friendly minimum quantity lubrication turning strategies. *J. Braz. Soc. Mech. Sci. Eng.*, 87, 41, 2019.

14. Şirin, S. and Kıvak, T., Effects of hybrid nanofluids on machining performance in MQL-milling of Inconel X-750 superalloy. *J. Manuf. Process.*, 70, 163–176, 2021.
15. Singh, J., Gill, S.S., Dogra, M., Singh, R., A review on cutting fluids used in machining processes. *Eng. Res. Express*, 3, 12002, 2021.
16. Rajaganapathy, C., Vasudevan, D., Murugapoopathi, S., Tribological and rheological properties of palm and brassica oil with inclusion of CuO and TiO2 additives. *Mater. Today Proc.*, 37, 207–213, 2020.
17. Wang, Y., Zou, B., Wang, J., Wu, Y., Huang, C., Effect of the progressive tool wear on surface topography and chip formation in micro-milling of Ti–6Al–4V using Ti(C7N3)-based cermet micro-mill. *Tribol. Int.*, 141, 105900, 2020.
18. Wu, X., Du, M., Shen, J., Jiang, F., Li, Y., Liu, L., Experimental research on the top burr formation in micro milling. *Int. J. Adv. Manuf. Technol.*, 117, 3477–3486, 2021.
19. Pawar, R.V., Hulwan, D.B., Mandale, M.B., Recent advancements in synthesis, rheological characterization, and tribological performance of vegetable oil-based lubricants enhanced with nanoparticles for sustainable lubrication. *J. Clean. Prod.*, 378, 134454, 2022.
20. Rawat, S.S. and Harsha, A.P., Recent Progress in Vegetable Oil-Based Lubricants for Tribological Applications, in: *Tribol. Sustain.*, pp. 115–142, 2021.
21. Wang, Y., Zou, B., Huang, C., Tool wear mechanisms and micro-channels quality in micro-machining of Ti-6Al-4V alloy using the Ti(C7N3)-based cermet micro-mills. *Tribol. Int.*, 134, 60–76, 2019.
22. Roushan, A., Rao, U.S., Patra, K., Sahoo, P., Performance evaluation of tool coatings and nanofluid MQL on the micro-machinability of Ti-6Al-4V. *J. Manuf. Process.*, 73, 595–610, 2022.
23. Singh, J., Gill, S.S., Dogra, M., Singh, R., Singh, M., Sharma, S., Singh, G., Li, C., Rajkumar, S., State of the art review on the sustainable dry machining of advanced materials for multifaceted engineering applications: progressive advancements and directions for future prospects. *Mater. Res. Express*, 9, 64003, 2022.
24. Liang, J., Gao, H., Li, D., Lei, Y., Li, S., Guo, L., Chen, L., Leng, Z., Sun, Y., Li, C., Study on milling tool wear morphology and mechanism during machining superalloy GH4169 with PVD-TiAlN coated carbide tool. *Tribol. Int.*, 182, 108298, 2023.

9

Metal Borides: High-Temperature Materials for Aerospace, Automotive, Nuclear, and Electromagnetic Industries

Arun K. Chattopadhyay[1], Tuncay Simsek[2*], Alican Yakın[3] and Telem Simsek[4]

[1]Uniformity Labs, Fremont, CA, USA
[2]Kirikkale University, Department of Mechanical and Metal Technologies, Kirikkale, Turkey
[3]Gazi University, Departments of Industrial Desing Engineering, Ankara, Turkey
[4]Hacettepe University, Department of Nanotechnol. and Nanomed., Ankara, Turkey

Abstract
Metal diborides are favored for extremely high-temperature environments because of their high melting temperatures and exceptional mechanical features. Challenges for instance, poor fracture toughness have led to the development of composites. Research aims to enhance fracture resistance and other properties through refining processing techniques and microstructural development. High-entropy ceramics, particularly diborides, show promise in hypersonic vehicle components and solar power concentrators, but addressing engineering challenges remains crucial. Significant efforts have tackled processing challenges to improve fracture properties, oxidation resistance, and thermal shock resilience. Metal borides are of considerable importance in both aerospace and automotive industries, offering high strength, thermal stability, lightweight nature, and resistance to extreme conditions. They find applications in turbine blades, rocket nozzles, heat shields, and corrosion-resistant coatings. Metal borides are also used in nuclear industries as effective radiation shielding materials. They exhibit versatile properties, with research focusing on understanding their magnetic properties, particularly in transition metal borides, which show variations from weak paramagnetism to robust ferromagnetism.

***Keywords*:** Metal borides, ceramics, high temperature materials, nanostructures

**Corresponding author:* tuncaysimsek@kku.edu.tr

Inamuddin, Tariq Altalhi and Mohammad Abu Jafar Mazumder (eds.) Nanostructured Hard Materials: Industrial Applications, (283–316) © 2026 Scrivener Publishing LLC

9.1 Introduction and Overview

The use of high-temperature materials is crucial for advancing technologies in aerospace, jet propulsion, defense, nuclear, thermal, and chemical industries. Efforts are underway to enhance existing materials and explore new compositions, including advanced ceramics, ultra-high-temperature ceramics (UHTC), intermetallic compounds, and ceramic-metal composites [1]. Understanding and controlling the microstructures and properties of these materials are essential, considering the potential of processing methods to introduce unique configurations and properties. Among the materials for ultra-high-temperature applications, metal boride-based composites have gained recognition and interest for new material development.

NASA conducted pioneering experiments in 2000 to evaluate the real-world performance of Ultra-High-Temperature Ceramic (UHTC) materials in space shuttle reentry conditions. The experiments used composites based on diborides, particularly ZrB_2 or HfB_2 [2]. The $ZrB_2/SiC/C$ composite survived reentry without failure, while others experienced issues due to large grain sizes and cracks along grain boundaries. This prompted refinements in UHTC production techniques and further fundamental research. UHTC development traces back to the 1960s, driven by the need for advanced structural components in nuclear reactors, hypersonic vehicles and turbine engines. Traditional materials fell short in meeting critical property requirements at high temperatures, prompting research into transition metal diborides such as TaB2 ZrB_2, NbB_2, TiB_2, and HfB_2, [3, 4].

Metal diborides exhibit high melting points and Refractory oxide scales, thereby conferring upon them the capacity to endure elevated temperatures up to 1900–2500°C. The addition of silicon carbide (SiC) enhances oxidation resilience and reduces diboride grain enlargement. Metal borides show promise for airframes on reentry vehicles but face different challenges in hypersonic systems, requiring more rigorous testing conditions. Various compositions, including ZrB_2 with SiC and SiC fiber-reinforced ZrB_2, have been examined under propulsion environment conditions, highlighting the need for comprehensive understanding of diboride properties under different stresses [5]. It is important that a comprehensive understanding of the properties of diborides under various thermal and mechanical stress conditions to unlock the full potential of diborides as high-temperature substance. The earlier studies on the alloying aspects of the composite materials and their properties were highly beneficial to make informed decisions based on a thorough comprehension of their properties, which laid the foundation for future UHTC developments. Investigating the chemical, physical, and thermodynamic properties of metal borides in high-temperature oxidizing environments further contributed to understanding their behavior. Among the category of UHTC materials hitherto identified, a distinctive subset comprising metal diborides emerges as the most potent ones that encompasses materials with melting points exceeding 3000°C [6–16]. The unique attribute of UHTCs lies in their anticipated utility at temperatures surpassing approximately 2000°C in ambient air and under challenging aerodynamic circumstances. In the class of di-borides belonging to IVB and VB transition metals, the metal diborides are identified based on their elevated melting points and other pertinent mechanical properties. The mechanical traits of boride based UHTCs, crucial for their performance, are outlined in Table 9.1.

Despite their excellent mechanical properties with respect to their hardness and flexural strength, fracture toughness of the metal diborides is still a concern. Composites, incorporating materials like SiC, enhance toughness and other mechanical characteristics. Further efforts are ongoing to improve fracture toughness and flexural strength, critical for structural applications. Processing challenges include achieving high sintering temperatures and minimizing voids and flaws to enhance material properties. Extensive research focuses on enhancing fracture properties and optimizing compositions and microstructures to ensure compatibility with extreme conditions. UHTCs typically have a hardness exceeding 20 GPa because of their stable covalent bonds. However, variations in hardness can arise from different processing methods. Notably, ZrB_2 and HfB_2 exhibit flexural strengths exceeding 200 MPa at 1,800°C, with finer-grained particles showing advanced strengths. Incorporating silicon carbide (SiC) into diboride ceramics enhances material toughness by densifying the structure and reducing grain size during processing. This synergistic effect improves mechanical properties, making composites a strategic choice for optimizing UHTC performance.

In applications requiring robust structural integrity, focusing solely on hardness is inadequate. Therefore, a comprehensive approach must address essential mechanical features for instance flexural strength and fracture toughness. Di-borides and covalently bonded ceramics, despite their high hardness and moderate fracture strength, often exhibit inferior fracture toughness compared to alternative materials. This inherent limitation limits the widespread use of di-borides in applications involving shock loading or impact unless specific measures, such as composite development, are employed to enhance fracture toughness [14, 15, 17–19]. Overcoming this challenge involves minimizing voids, flaws, and porosities while also improving surface finish. The traits of ceramics in bulk polycrystalline structure heavily depend on microstructural development and composition that are influenced by the chosen processing method. Compared to

Table 9.1 Physical properties of some metal boride compounds.

Metal diborides	Melt. point °C	Hardness, GPa	Bulk modulus, GPa	Young's mo, GPa	Flexural strength, MPa	Fracture toughness MPa/m2	Ref.
TiB_2	3230	25-33	240-250	565-581	370 (23 °C)	700-1000	[2, 12]
VB_2	2980	27	239-282	577	-		[20]
CrB_2	2200	22.1	230-295		-		[20]
ZrB_2	3246	16.7-19.7	221-236	471-487	380 (23°C), 200 (1800 °C)	300-400	[2, 21, 22]
ZrB_2–20%SiC				540 (23°C)	270 (1800 °C)		[2]
NbB_2	2900	20-25	539		-	414	[2, 12]
MoB_2	2180	16-22	299-334	463-522	-		[23, 24]
HfB_2	3250	21.2-28.2	530		480 (23 °C), 280 (1800 °C)	350-450	[2, 12]
HfB_2-20%SiC					280 (1800 °C)		[2]
TaB_2	3200	25.5	341-360	551	555	555	[2, 12]
WB_2	2670	20	293-326		-		[20]
ReB_2	2400	26.6	350-360	712	-		[20]

traditional ceramic materials, UHTCs display an exceptional array of features, among them notably high melting points, exceptional hardness, a robust elastic modulus, commendable wear resilience, proficient thermal and electrical conductivity, substantial creep resilience, and impeccable chemical durability. This diverse range of characteristics positions boride-based UHTCs as formidable contenders for applications requiring resilience and high performance in extreme conditions. While metal diborides hold promise for extreme environments, their application requires addressing engineering challenges and understanding material-environment interactions. Experimental tests and computational simulations are essential for evaluating material behavior under different conditions and optimizing performance for practical applications [11–16].

Research efforts aim to overcome processing challenges associated with high sintering temperatures and improve fracture properties of UHTCs. Strategies include incorporating additives and reinforcements and optimizing processing parameters [7–15]. Composition and microstructure tailoring are crucial for achieving desired material properties compatible with extreme conditions.

This chapter presents a summary of the production techniques utilized for metal borides and explores their applications in automotive, aviation, aerospace, and nuclear reactors. Additionally, it emphasizes the unique crystal structures of borides and their significance in determining the electromagnetic properties of these materials.

9.2 Production Methods of Metal Borides

Metal borides are compounds composed of boron and one or more metallic elements. These compounds typically exhibit metallic properties and are characterized by stable the covalent connection established between boron atoms and metal atoms. Metal borides can exhibit diverse structures and bonding behaviors due to the unique electronic configuration of boron atoms, which deviates from the typical valence concept observed in other metallic compounds. These forms range from structures with low boron content, such as M_5B (M:Metal), to those with high boron content, like MB_{66}-MB_{99} structures. The most common ones include tetraborides (MB4), diborides (MB_2), monoborides (MB), hexaborides (MB_6), icosahedron dodecaborides (MB_{12}), and super-icosahedron borides (MB25, MB50, MB66 and MB80). The diverse forms exhibited by metal borides significantly impact their crystal structures, which in turn affect a broad spectrum of characteristics. Table 9.2 presents the most common crystal

Table 9.2 Crystallographic characteristics of some metal boride forms [12, 25–28].

Formula	Compound metals	Crystal system and structure type	Configuration of boron atoms
M_4B	Pd, Pt	Cubic, Pt4B type	Isolated atoms
M_2B	Co, Ta, Cr, N Mo, Fe, W	Tetragonal, CuAl2 type	Isolated atoms
M_5B_3	Cr	Tetragonal, Cr5B3 type	
M_3B_2	Nb, V, Mo, Ta,	Tetragonal, U3Si2 type	Isolated pairs
MB	V, Ta, Nb, Mo, Cr, W, Ni	Rhombohedral, CrB type	Single chain
MB	W, Mo	Tetragonal, MoB type	
MB	Ti, Hf, Fe, Co	Rhombohedral, FeB type	
M_3B_4	V, Cr, Ta, Nb,	Rhombohedral, Ta3B4 type	Double chain
MB_2	Sc, Zr, Y, Ti, Hf, Nb, V, Ta, U, Cr, Mo, W, Al, Pu,	Hexagonal, AlB2 type	Two-dimensional hexagonal lattice
M_2B_5	W, Mo	Hexagonal,Mo2B5 type	
MB_4	Y, Actinides, Lanthanides	Tetragonal, UB4 type	Three-dimensional lattice
MB_6	Y, Actinides, Lanthanides	Cubic, CaB6 type	Octahedral B6 linkage
MB_{12}	Sc, Mo, Y, Zr, W, actinides Lanthanides,	Cubic, UB12 type	Central metallic atom encircled by 24 interconnected boron atoms
MB_{66}	Pu,Y, Lanthanides	YB66 type	Three-dimensional B12 network

structures observed in metal borides. The arrangements of boron atoms within these structures influences their physical, chemical, thermal, magnetic, corrosion, and mechanical characteristics. Metal borides form structures in one, two, and three dimensions depending on the ratio of the included metal (M) and boron (B). This ratio results in various structural motifs such as binary B atom-containing (M_3B_2), zigzag chain-like (MB), branched chain-like ($M_{11}B_8$), double-chain (M_3B_4), planar network (MB_2) structures, and even three-dimensional networks with cross-linked boron atoms (MB_4). Other structures include interconnected hexagonal arrangements containing B_6 (MB_6), cubic hexagonal structures containing B_{12} (MB_{12}), and triangular-faced structures containing B_{12} (MB_{66}). An escalation in the M:B ratio results in a corresponding augmentation in B-B bonds, impacting all properties of metal borides [12, 25–28].

Metal borides are capable of being manufactured across a broad spectrum of purities, both on laboratory and industrial scales, employing various manufacturing methods. These methods include, for example, chemical reactions conducted at temperatures ranging between 1000-1800°C, mechanical alloying, chemical vapor deposition (CVD), self-propagating high-temperature synthesis (SHS), and several others. Each method carries its unique advantages and disadvantages concerning economic factors, impurity levels, grain size distribution, production quantity and time. These methods can be outlined as follows [12, 25–28].

- Self-propagating high-temperature synthesis (SHS) starting from the elements:
  It is an eco-friendly method that reduces reliance on toxic solvents and has a lower environmental impact than traditional methods. However, it's unsuitable for nanoparticle production due to particle sintering from high temperatures. Despite challenges in energy dissipation and reactor vessel requirements, some systems harness excess heat for other processes.

$$13Al + 6B_2O_3 \rightarrow AlB_{12} + 6Al_2O_3, \text{ or } Ti + B_2 \rightarrow TiB_2 \text{ etc.}$$

- Borothermal reduction of metal oxides:
  Borothermal reductions offer benefits such as accessible starting materials and simplified reaction systems, in contrast to SHS, which presents contamination risks. Metal boride phases resulting from borothermal reduction are

obtained by carefully controlling the ratios of metallic oxides and boron. Generally, borothermal reduction processes occur at temperatures, > 1000°C. e.g, $3ZrO_2 + 10B \rightarrow 3ZrB_2 + 2B_2O_3$

- Aluminothermic, silicothermic, or magnesiothermic reductions by mechanochemical methods involve the reduction of blends comprising metal oxides and boron oxides. These methods effectively synthesize high-melting metals and ceramics. For instance, TiB2 can be produced by ball milling TiO_2, $B2O_3$, and metallic magnesium, aluminum or silicon for 10-15 hours in an inert atmosphere, resulting in pure TiB2 powder after acid leaching to remove MgO.

$$TiO_2 + B_2O_3 + 5Mg \rightarrow TiB_2 + 5MgO$$

- Carbothermal reduction on blends of metal oxides and boron oxides:
  Large-scale production of UHTC metal borides can be achieved through carbothermic reduction of a blend of metal oxides and B2O3 in electric furnaces. For example,

$$TiO_2 + B_2O_3 + 5C \rightarrow TiB_2 + 5CO_2$$

  Alternatively, another method, boro-carbothermal reduction method, for large-scale production of UHTC metal borides, involves reducing metal oxides with carbon and boron carbide, also conducted in electric furnaces. For example,

$$2TiO_2 + B_4C + 3C \rightarrow 2TiB_2 + 4CO$$

- Electrolysis of solutions of metal oxide and boron oxide mixtures:
  Alkaline and rare-earth borides are formed through an electrochemical boridization process in molten salt. This process involves reducing or extracting a particular cation from the molten salt while concurrently alloying it with boron to generate borides. The electrochemical alloying and selective reduction occur due to boron's semiconductive properties,

characterized by its high electronegativity and the formation of stable metal-boron (M-B) bonds. The approach is based on the one-pot ionothermal technique, where molten salts transition into a liquid state, acting as solvents and possibly as size templates for reactions. Both metal-rich and boron-rich metal borides can be produced by reacting metal halides with $NaBH_4$ at temperatures ranging from 400 to 1000°C within a LiCl/KCl mixture with a 45/55 wt % ratio.

- Reduction of metal hydrides by boron:
  Generally metal borohydrides like $Ti(BH_4)_3$, $Zr(BH_4)_3$, $RbBH_4$ etc. are modestly heated to decompose to form metal borides and to generate hydrogen.

$$Zr(BH_4)_3 \rightarrow ZrB_2 + B + 6H_2$$

- Reaction of boron halides and metal through the chemical vapor deposition (CVD) method:
  The preparation of pure metal boride materials through those abovementioned methods has proven challenging due to either the high temperatures involved or the inherent characteristics of the metal boride products themselves. An optimal approach for forming metal boride thin films *via* CVD would utilize economical and easily accessible starting materials, for instance metal halides and boranes (typically nido-pentaborane or decaborane), as CVD precursors, making it broadly applicable to diverse metal systems. During the CVD reactions, the formation of HCl as a by-product was anticipated. A very pure form LaB_6 film can be deposited at a relatively lower temperature of ~900°C.

9.3 Metal Borides Applications

Because of their unique molecular and crystal structures as mentioned above, metal borides offer superior properties in a very diverse field of applications because of their high hardness and mechanical strength, excellent electrical and thermal conductivity, extremely melting temperature, wear resilience, chemical non-reactivity, etc. Metal borides are utilized in aerospace for their lightweight yet durable properties, in automotive applications for their strength and wear resistance, in nuclear industries for their ability to withstand high temperatures and corrosive environments, and in energy production for their thermal conductivity and resistance to extreme conditions, in high temperature ceramic materials for high temperature resistance, in electronic devices for their unique electrical and thermal properties, , and in nuclear industries for high radiation resistance.

Metal borides indeed exhibit remarkable properties that make them suitable for demanding working conditions; for instance, those encountered in hypersonic aviation vehicles, lightweight armor, rocket engines refractories for high-speed cutting tools, materials in contact with molten metals, and fuel forms for advanced nuclear fission reactors. Additionally, they have potential applications in materials that come into contact with plasma for nuclear fusion reactors. Their high strength, heat resistance, corrosion resistance, and ability to withstand extreme environments make them invaluable in these advanced technological applications [29–32]. Titanium boride (TiB_2) is favored in aviation for its lightweight nature, high wear and corrosion resistance, as well as its ability to withstand high temperatures. It finds application in jet engines and space exploration due to these advantageous properties [33–35].

In the automotive industry, metal borides are commonly used in engine components that require high wear resistance. Specifically, chrome borides (CrB and Cr_2B) are preferred as coating materials to enhance the wear resistance of machine elements used in automotive applications [29, 32, 36]. Nuclear applications demand materials that can safely operate under challenging conditions such as high temperature, high corrosion, and high radiation. Metal borides like ReB_2, TaB_2 WB_2, DyB_6, and SmB_6 are among the preferred materials for nuclear applications due to their high temperature and radiation resilience, allowing them to operate reliably in nuclear environments [37]. Table 9.3 presents a comprehensive overview of the thermal, mechanical and nuclear properties of various metal borides. It reveals that metal borides generally exhibit high hardness and high melting points, along with relatively low densities ranging from 4 to 7 g/cm^3. This combination of properties positions metal borides as promising candidates for aerospace and aviation applications, where materials must withstand extreme conditions. Moreover, the nuclear properties listed in the table suggest potential uses of metal borides in nuclear applications, further expanding their utility [29, 38].

The hardness values of various metal borides at different temperatures are illustrated in Figure 9.1. HfB2 stands out with the highest hardness value of 38 GPa at room temperature. Additionally, all metal borides depicted in the figure have hardness values above 20 GPa at room temperature and surpass approximately 7 GPa at 900 °C. Notably, TiB2+MoSi2 exhibits the

Table 9.3 Thermal, mechanical and nuclear properties of boron and its derivatives [29].

	Mechanical features			Thermal	Nuclear
Material	**Den. g/cc**	**Hard. GPa**	**E, GPa**	**Melting point °C**	**Σ barns**
Boron	2.31–2.46	49	400	2181	759
TiB_2	4.52	25–35	560	3225	502
ZrB_2	6.10	22–26	300–350	3245	500
HfB_2	11.21	21–28	500	3380	538
NbB_2	6.97	18–19	505	2290	~500
MoB_2	7.87	24–25	569	2375	~500
TaB_2	12.50	25	550	3040	~510
W_2B_5	12.79	–	–	2365	~550
ReB_2	12.7	18	382	2400	~540
OsB_2	12.83	20	410	1870	~510
CrB_2	5.20	11–20	211	2200	502
LaB_6	4.72	19.7	479	2715	652
EuB_6	4.99	18–26	–	2580	1286
CeB_6	4.80	30	379	2550	651
GdB_4	6.47	18–21	380	2650	10.408

highest hardness value at 900°C. High-hardness materials are favored in applications prone to wear, including aviation, automotive, nuclear, and machining cutting tools [12].

Furthermore, machine components in the aerospace and automotive sectors operating at high temperatures demand materials with elevated hardness values for optimal performance. The notable hardness exhibited by these materials, particularly at high temperatures, underscores their suitability for high-temperature applications [12]. As previously discussed, metal borides offer a combination of traits encompassing high hardness, elastic modulus strength, and melting temperature, making them essential for industries like aerospace, automotive, and nuclear. Given their

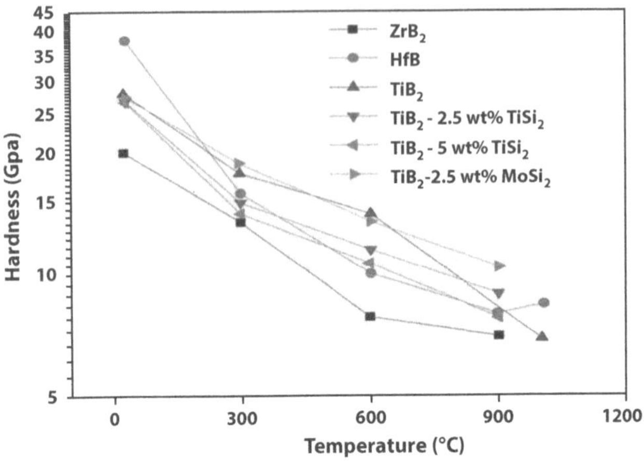

Figure 9.1 Temperature-dependent changes in hardness for different diboride-based ceramics, presented on a semi-logarithmic scale [12].

exceptional attributes, metal borides are poised for extensive use in applications with rigorous operating conditions. Industries such as nuclear, aerospace, and automotive, which face extreme environments, necessitate machine components capable of withstanding high temperatures and demonstrating superior oxidation resistance [12, 39].

Table 9.4 provides insights into the oxidation resistance properties of select metal borides and composite materials. Notably, compositions like ZrB2, HfB2, and TaB2, along with certain composites incorporating ceramics, exhibit outstanding oxidation resilience at elevated temperatures. Comparing the high-temperature oxidation resistance of these metal borides with that of composites reveals a significant enhancement in oxidation resistance for composites with ceramics added in specific proportions. This enhancement can be credited to the development of protective oxide layers on the surface, aided by the existence of ceramics [12, 39–41].

Utilization of metal borides in aerospace, automotive, and nuclear applications has been discussed, in detail in the subsequent sections.

9.3.1 Aerospace Applications of Metal Borides

The aerospace industry benefits from the use of metal borides due to their superior thermal stability, mechanical properties, lightweight nature, and resistance to extreme conditions, contributing to the advancement and reliability of aerospace technology. Metal borides find various applications

Table 9.4 Visualization of the oxidation resistance properties of specific metal borides and their composites.

Composition	Oxidation condition	Weight gain mg/ cm2	Film thickness (μm)	Oxide phase	Ref.
ZrB_2	1166 °C, 1 h	4.8	-	ZrO_2, $B2O_3$	[42]
ZrB_2	1500 °C, 3 h	14.9	150	m-ZrO_2	[43]
ZrB_2	1400 °C, 1 h	11.7	291	m-ZrO_2	[44]
ZrB_2	1500 °C, 3 h	29	1111	ZrO_2	[41]
ZrB_2	1400 °C, 2 h	18.7	300	ZrO_2	[45]
ZrB_2-4 mol.% W	1600 °C, 5 min	2.1	30	ZrO_2	[46]
ZrB_2-20 vol% $MoSi_2$	1400 °C, 30 h	6.5	-	SiO_2, ZrO_2, MoB	[47]
HfB_2	1400 °C, 1 h	8.80	248	-	[40]
HfB_2–20 vol% SiC	1400 °C, 1 h	0.60	12	HfO_2, HfOxCy, SiOxCy	[40]
HfB_2-20 vol% $MoSi_2$	1400°C,26 min	1.2	-	HfO_2, $HfSiO_4$, MoO, MoB	[48]
HfB_2–20 vol% SiC	1600 °C, 4 h	5.6	72	ZrO_2, SiO_2	[40]

(Continued)

Table 9.4 Visualization of the oxidation resistance properties of specific metal borides and their composites. (*Continued*)

Composition	Oxidation condition	Weight gain mg/cm2	Film thickness (μm)	Oxide phase	Ref.
TaB_2	1500 °C, 5 h	20	-	Ta_2O_5	[49]
TaB_2	1450 °C, 12 h	24	-	TaO, Ta_2O_5	[50]
TaB_2-10 vol% $MoSi_2$	1600°C,30 min	2.5	9	SiO_2, Ta_2O_5	[51]
ZrB_2-20 vol% SiC	1800°C,30 min	3.06	96	m-ZrO_2, t-ZrO_2	[39]
ZrB_2–40$MoSi_2$–5SiC (vol.%)	1500 °C, 10 h	3	148	ZrO_2, MoB, $ZrSiO_4$	[52]

in the aerospace industry because of their distinctive combination of characteristics. Some of the key applications include:

1. Turbine Blades: Metal borides are utilized in turbine blades for aircraft engines. Their high strength, hardness, and thermal durability make them suitable for withstanding the high temperatures and stresses experienced in the turbine section of jet engines.
2. Rocket Nozzles: Boride-based materials are employed in the construction of rocket nozzles. These materials can withstand the extreme temperatures and pressures generated during rocket propulsion, providing durability and reliability in space exploration missions.
3. Structural Components: Metal borides are used in the fabrication of lightweight structural components for spacecraft and satellites. Their low density contributes to reducing the overall weight of aerospace vehicles while maintaining structural integrity.
4. Heat Shields: Boride-based materials serve as effective heat shields in aerospace applications. They provide thermal insulation and protection against high temperatures during atmospheric re-entry and other heat-intensive phases of space missions.
5. High-Temperature Materials: Metal borides are utilized in the production of high-temperature materials for aerospace applications, including thermal protection systems, engine components, and hypersonic vehicle structures.
6. Corrosion Resistance: Boride coatings are applied to aerospace components to enhance their corrosion resistance in harsh environments, such as those encountered during long-duration flights or exposure to atmospheric pollutants.

Refractory metal borides, distinguished by their elevated melting points (>3000°C), exceptional thermal stability, conductivity, minimal thermal expansion coefficient, and resilience to oxidation, in addition exhibiting good thermal shock resilience and erosion resistance, stand out as special materials with potential applications in space. [6, 29, 34, 53, 54]. Metal borides are considered special materials with potential applications in aerospace especially for their high strength values and excellent heat resistance [32, 55, 56]. These special materials are preferred over Si-based ceramics and Ni-based superalloys, which cannot be used in components of hypersonic aerospace vehicles exceeding temperatures of 2000°C. The stringent stability requirements for high-temperature materials exceed the safe operating limits of Si-based ceramics and Ni-based alloys, rendering them unsuitable for aerospace applications involving extraordinary operating conditions. On the other hand, metal borides, with their high melting temperatures, high temperature resistance, and high hardness, emerge as materials that can safely operate in these applications. Titanium Diboride (TiB_2), distinguished among metal borides because of its high melting temperature, hardness, and excellent electrical conductivity, finds applications in aerospace including spacecraft and aircraft component coatings [29, 53].

Zirconium Diboride (ZrB_2) is extensively utilized in the aerospace industry because of its remarkable high temperature stability, thermal conductivity, and resistance to oxidation. It is preferred for components subjected to extreme temperatures in spacecraft. Similarly, Hafnium Diboride (HfB_2), known for its high temperature resistance and high radiation resistance, holds significance as a crucial material for aerospace components exposed to extreme conditions, such as those encountered in hypersonic vehicles. Tantalum Boride (TaB_2) is used in the construction of machine elements intended for space exploration studies, owing to high temperature resilience and hardness it has. Meanwhile, Niobium Boride (NbB_2), characterized by its high temperature stability and high hardness, is among the materials favored for aerospace components exposed to extreme conditions such as high temperature, corrosion, and wear [29, 38, 57]. Metal borides find extensive applications in aerospace and space industries due to their unique properties (Figure 9.2). Compared to Si, Ni, and Co-based alloys, which lack the necessary melting temperatures and exhibit low strength under extreme operating conditions, metal borides offer superior characteristics. With high melting temperatures (>2000°C), exceptional corrosion and wear resistance, high hardness, radiation resistance, low density, and oxidative resistance, metal borides are preferred for aerospace and space applications [29, 31, 32, 38, 57–59].

Aluminum, with its low density and high strength, is extensively employed in the aerospace industry through the widespread use of alloys. To enhance their mechanical properties and strength, various elements for instance ceramics, refractory metals, oxides, semimetals, and borides are added to aluminum. Metal borides, known for their low density, high hardness, yield strength, corrosion, and radiation resilience, are commonly added to aluminum. Loganathan and colleagues produced two composites by adding ZrB_2 + hBN and ZrB_2 + WS2 materials to AA7075 alloy [60]. The microhardness and yield strengths of these composites were investigated, revealing improvements over the base alloy. Kumar Naik *et al.* synthesized

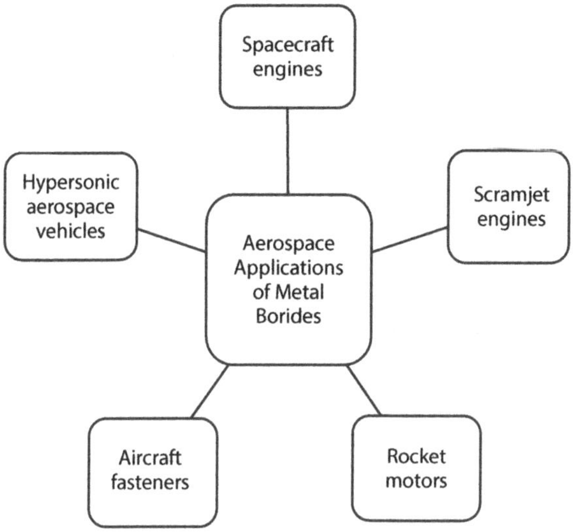

Figure 9.2 Some potential applications of metal borides in aerospace and space applications.

a ZrB2–B_4C composite with an outer layer of 70% ZrB_2 and an inner layer of 90% B_4C, providing high-temperature resistance and low density. The composite exhibited a hardness value of 30 GPa in the inner layer and 15 GPa at the outer surface, with a density of 3.84 g/cm^3, demonstrating its high potential for aerospace applications [61]. Furthermore, aside from the aforementioned properties and applications of metal borides, it has been noted that these materials harbor energy levels adequate for utilization as rocket propellants. In previous studies, it has been emphasized that AlB2 and MgB2 metal borides have the capacity to produce high energy. This indicates the potential for using metal borides as solid fuel in rockets [62].

9.3.2 Metal Borides Applications of Automotive Industries

Like aerospace industries, the automotive industries benefit from the use of metal borides as well owing to their advanced strength, wear resistance, lightweight nature, and corrosion resilience, leading to improved performance, efficiency, and durability of vehicles. Metal borides find several applications in the automotive industry due to these unique properties. Some of the key applications include:

1. Engine Components: Metal borides are utilized in the manufacturing of engine components such as pistons, valves,

and cylinder liners. Their high hardness and wear resistance contribute to improved engine performance and durability.

2. Lightweight Armor: Boride-based materials are used in the production of lightweight armor for vehicles. These materials offer excellent protection against ballistic threats while reducing the overall weight of the vehicle.
3. High-Speed Cutting Tools: Metal borides are employed in the fabrication of high-speed cutting tools used in automotive manufacturing processes. Their exceptional hardness and thermal stability enhance cutting performance and tool life.
4. Exhaust Systems: Boride coatings are applied to exhaust system components to advanced their corrosion resilience and thermal insulation properties. This helps in reducing emissions and extending the service life of exhaust systems.
5. Fuel Efficiency Enhancement: By reducing the weight of automotive components, metal borides contribute to enhancing fuel efficiency and reducing carbon emissions in vehicles.

High hardness and wear resistance properties of metal borides, especially titanium diboride (TiB_2), zirconium diboride (ZrB_2), and hafnium boride (HfB_2), make them ideal candidates for critical components in internal combustion engines, such as pistons, cylinder coatings, and valves, where wear-resistant coatings are essential. These applications are utilized to enhance durability and performance. Titanium boride (TiB) is used in cutting tools in machining processes to ensure effective and durable production of automotive components. Additionally, some metal borides exhibit good thermal conductivity, making them preferable in the areas where efficient heat dissipation is crucial, such as thermal management systems in electric vehicle applications [38, 58, 63, 64]. Metal borides of zirconium diboride (ZrB_2) and hafnium diboride (HfB_2) which exhibit high-temperature stability, are materials with the potential for use in exhaust systems. In the automotive sector, particularly in the pursuit of vehicle weight reduction, metal borides such as TiB2, ZrB2, and NbB2 emerge as notable due to their unique blend of lightweight properties and exceptional strength. These materials hold promise for various automotive applications, offering the potential to enhance fuel efficiency, improve performance, and meet stringent regulatory [6, 65, 66]. Figure 9.3 illustrates some potential applications of metal borides in the automotive industry. Upon examination of the figure, it becomes evident that metal borides offer potential applications

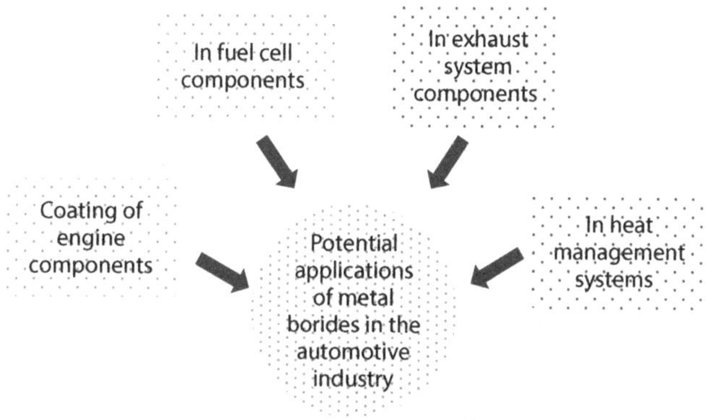

Figure 9.3 Some potential applications of metal borides in the automotive industry.

across a spectrum of areas, encompassing engine components, heat management and fuel cells. While cost remains a limiting factor for widespread adoption in industries like automotive, the enhanced efficiency, lifespan, and effectiveness of parts and systems underscore the viability of utilizing these materials in demanding applications such as aerospace, space exploration, automotive, and nuclear industries.

Metal borides exhibit a range of mechanical properties including high strength, hardness, corrosion and oxidation resistance, and low density, along with superconductivity, magnetic, optic, and high radiation resistance. In the automotive industry, they are utilized for coating substance to enhance strength, hardness, and wear resilience, with potential applications in thermal conduction components. Metal borides such as MgB_2, TaB_2, and NbB_2, known for their good thermal conductivity, hold promise for use in heat exchangers [31]. Metal borides stand out as powerful materials with the potential for various automotive components, ranging from increasing the wear resistance of engine components to enhancing the efficiency of cutting tools used in manufacturing [64, 67]. Xie and colleagues discovered that Mg4B6 metal boride demonstrates potential as an electrode material with high capacity in lithium-ion batteries, achieving a theoretical capacity of 4299 mAh [68]. Furthermore, the boriding process, employed to enhance the mechanical properties of steels and cast irons in various industrial applications including automotive, forms metal borides such as FeB, Fe_2B, and Fe_3B on the surface. These borides significantly enhance the strength, corrosion resistance, wear resistance, oxidation resistance, hardness, and fracture toughness of steel materials, making them suitable for use in automotive components such as heat exchangers, extrusion screws, gears, and casting molds [69, 70].

9.3.3 Metal Borides Nuclear Applications

Boron-containing materials are commonly utilized in areas where their high neutron absorption cross-section is required. Two stable isotopes of Boron are 10B and 11B, with natural abundances ranging from approximately 19.1% to 20.3% for 10B and 79.7% to 80.9% for 11B [71, 72]. Notably, the 10B isotope exhibits a significantly higher thermal neutron capture capability, with a cross-section of 3838 barns, or 3.838 x $10^{-21}cm^2$, compared to other boron isotopes. Consequently, boron carbide (B4C) pellets enriched in boron-10 are preferred materials for control mechanisms in nuclear reactors [73]. While boron carbide is commonly used in control rods because of its high neutron absorption cross-section, certain metal borides, such as hexaborides of Samarium and Dysprosium (SmB_6 and DyB_6), offer comparable or even higher neutron absorption capabilities [46]. Metal borides enriched with specific isotopes can be tailored for optimal neutron absorption performance in control rod applications. The uses of metal borides in nuclear industries are generally dependent on the specific application requirements, like thermal conductivity, mechanical properties, radiation resistance, and neutron absorption characteristics. Metal borides offer advantages in certain scenarios where these properties are critical for safe and efficient operation within nuclear reactors and related facilities. Metal borides are employed as radiation shielding materials in nuclear facilities. Their high neutron absorption cross-section makes them effective at attenuating radiation and protecting personnel and equipment from harmful nuclear radiation. The usage of effective metal borides as coatings on the metal plates like Al, Cu, Fe etc. has been successful to replace the use of toxic Lead as a protective material [74, 75] against both particles. For enhanced protection, the recent trend is to increase coating thickness of metal borides increasing the unit cost of coated metal plates.

These next-generation nuclear reactors, designed for high-energy production, require materials with high melting temperatures, high radiation resistance, and high corrosion, wear, and oxidation resistance. For that matter, metal borides are utilized in the fabrication of structural components for nuclear reactors, such as cladding materials and containment vessels. Their high temperature and corrosion resistance, high hardness, high radiation resistance, and superconductivity, make them suitable for withstanding the harsh operating conditions of nuclear reactors [37, 76]. Figure 9.4 depicts the mass attenuation characteristics of several metal borides, including GdB_6, NdB_6, and SmB_6, along with lead and vermiculite, as analyzed using both XCOM and EGS methods. A noticeable decrease in neutron attenuation across all materials is observed with increasing energy

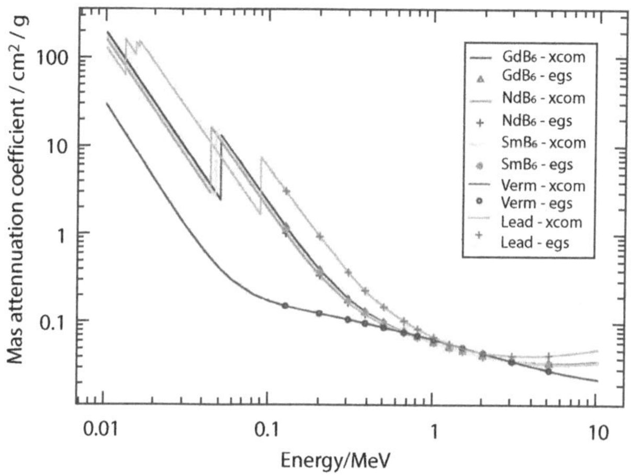

Figure 9.4 Differences in the mass attenuation coefficients of the analyzed samples as computed by XCOM and EGS [77].

levels from 0.01 to 0.3 MeV. Analysis of the figure reveals that metal borides exhibit significant neutron capture capabilities comparable to those of lead. Specifically, SmB_6, GdB_6, and NdB_6 metal borides demonstrate promising neutron attenuation performance while possessing a lightweight characteristic, with densities approximately half that of lead [77].

In another study, the mass and linear attenuation factors of some metal borides are discussed. Upon review of Table 9.5, it becomes apparent that the neutron capture properties of SmB_6 and DyB_6 are compared to those of B_4C, a material presently utilized in nuclear reactors. Notably, the table

Table 9.5 Comparison of neutron capture properties of SmB6 and DyB6 with the properties of B4C currently used in nuclear reactors [37].

Properties \ Compounds		B4C	SmB6	DyB6
The Mass Attenuation Factors	Thermal Neutrons (0.025 eV)	33.51172	29.58439	14.99297
	Thermal Neutrons (4 MeV)	0.07678	0.14001	0.17228
The Linear Attenuation Factors	Thermal Neutrons (0.025 eV)	84.44954	152.063	83.06
	Thermal Neutrons (4 MeV)	0.19349	0.71963	0.9544

illustrates that metal borides, particularly in terms of fast neutron attenuation, exhibit significantly superior neutron attenuation capabilities compared to B_4C. Dy_{B6} metal boride, in particular, demonstrates exceptional proficiency in capturing fast neutrons, surpassing the properties of other borides by a considerable margin. This positioning underscores DyB_6 as one of the most potent materials available for use in nuclear reactors [37].

9.3.4 Electromagnetic Properties of Metal Borides

As briefly mentioned in section 9.2, metal borides (MxBy) constitute a diverse class of compounds, boasting more than 150 distinct structure and multitude of compounds. The classifications broadly categorize into boron-rich (B:M ≥ 4:1) and metal-rich (B:M < 4:1) divisions. The unique arrangement and crystal form of boron atoms within these compounds play a crucial role in determining their physical properties. Metal borides exhibit diverse bonding arrangements, such as ionic M-B, covalent B-B, and metallic M-M/M-B bonds. In boron-rich borides, individual boron atoms are enveloped by metal atoms. As boron content increases, robust covalent B-B bonds become prominent, leading to diverse structural properties, such as one-dimensional boron chains, zigzags, two-dimensional sheets and ultimately three-dimensional boron networks. These bonding patterns and structural arrangements greatly influence the versatility of metal borides, impacting their physical and chemical properties. Understanding these intricate structures and bonding characteristics is essential for leveraging the potential applications of metal borides in various technological and material science fields, which requires diverse electronic and magnetic properties. Metal diborides can exhibit both metallic and ceramic characteristics as a result of their bonding nature, which arises from a combination of ionic and covalent bonds. Alkaline earth metal atoms with s^2 configuration possess some metallic properties, which can be tuned to become semiconductor with B addition to form borides. Examples include beryllium-rich compounds like Be_2B and Be_5B, exhibiting metallic conductivity, whereas, boron-rich compounds like BeB_6 and BeB_9 are semiconductors [78]. Borides belonging to the IV-VIII groups, which include transition metals, actinoids and rare earth metals, share a common classification. They are formed by elements with incomplete d or f subshells, specifically ds and fds elements [79]. Transition metal borides containing unoccupied d electron subshells hold a distinctive position within this category, primarily due to the partial confinement of valence electrons within individual metal atoms. These valence electrons can transition into p states. Borides associated with d and f transition metals, particularly spanning from Me2B to Me6B, demonstrate distinct properties influenced by the principal quantum number and the valence electron count of the metal atoms. Within this range, diborides are particularly noteworthy as they adopt sp2 configurations of boron atoms, resulting in a crystal structure reminiscent of graphite. Metals possessing stable d5 configuration atoms tend to form diborides, while those with fewer than five electrons within the isolated d shell, such as Fe, Co, Ni, Pd, and Pt, do not typically form diborides. Hexaborides are commonly formed by metals favoring the stabilization of f0, f7, and f14 configurations. The properties of hexaborides are largely influenced by the electron states of boron, with rare earth metals exhibiting donor activity through f-d transitions. As we move from diborides to tetraborides and then to dodecaborides, the influence of metallic states on boride properties decreases. The arrangement of electron states within the boron substructure significantly influences the creation of stable dodecaborides. Boron compounds involving p-block elements with s and p electrons fall into another category. The formation and stability of these compounds are affected by the interplay between s and p electron states, potentially leading to the establishment of stable sp3 and s2p6 configurations. Boron, being an electron acceptor, tends to adopt sp3 configurations. In substances such as BN or BP, the most basic alterations occur as atoms of boron acquire valence electrons from phosphorus or nitrogen atoms, resulting in both atoms within the compound assuming sp3 configurations. The robust B-B bonds present in BN, along with diagonally-oriented C-B-C type bonds, confer upon it semiconductor properties [80].

The majority of theoretical research and experimental efforts aimed at understanding magnetic properties focus on transition metal diborides and monoborides. Featuring a layered hexagonal structure reminiscent of AlB2, prominent compounds emerge as diborides formed by transition metals spanning groups 3 to 6, encompassing Sc, V, Hf, Ti, Zr, Nb, and others. The stability of these compounds is attributed to their diverse range of conduction band occupation numbers within their respective 3d, 4d, and 5d series [81]. Noteworthy is the observation that magnetic anisotropy displays a robust yet non-monotonic correlation with the occupation of the hybridized d-orbital band, peaking notably for diborides linked with group 4 metals. Through ab initio calculations of the susceptibility and electronic structure of these diborides, valuable insights are gleaned, suggesting that the magnetic anisotropy originates from the intricate interplay between competing phenomena of Van Vleck paramagnetism and the orbital diamagnetism of conduction electrons [82]. Transition metal borides are categorized as refractory hard metallic materials. The magnetic properties of these compounds exhibit significant variations, dependent on both the transition metal component and the composition of the boride. Among them, certain borides are Pauli-paramagnets, with weak paramagnetism. Some transition metal borides, in contrast, display robust ferromagnetic characteristics with ordering temperatures well surpassing room temperature. Intriguingly, a subset of these borides also demonstrates diamagnetic properties. Some bulk borides of Mn, Co and Fe are summarized in Table 9.6. As depicted in the table, MnB, FeB, Fe_2B, Co_2B and Co_3B are ferromagnetic with Curie point above the room temperature. Tetragonal Fe_2B has the highest saturation magnetization and Curie temperature which makes it a very promising magnetic material for use in harsh condition with its high Vickers hardness. Among them MnB is an interesting soft ferromagnetic material where the exchange interaction between manganese ions becomes positive by zigzag boron chains [83]. Recently, there has been a growing focus on the study of nanocrystalline transition metal boride ferromagnets, driven by their remarkable chemical stability. This research aims to tailor their physical properties, opening up possibilities for specialized applications, particularly in fields such as biomedical technology [84].

Boron readily reacts with a wide range of metals to produce metal borides, spanning from the most metal-rich Nd2Fe14B to the most boron-rich YB66. The most powerful permanent magnet, commonly referred to as neodymium magnets, is $Nd_2Fe_{14}B$. It exhibits exceptionally high remnant magnetization. Electric motors, speakers, hard drives, and along with all other device that needs a strong magnetic field all benefit from the employment of this rare earth magnets. Iron and neodymium, two ferromagnetic elements, are the source of the ferromagnetism where the crystalline arrangement of the boron contributes to the increased magnetism. However Nd and Dy (which is added to increase the Curie point of

Table 9.6 Ferromagnetic properties of some important borides.

	Crystal structure	Tc (K)	Ms (emu/g)	Hc (Oe)
MnB [83]	orthorhombic	566	130	near zero
FeB [85]	orthorhombic	590	80.5	-
Fe_2B [86]	tetragonal	1017	156.9	78.3
Co_2B [87, 88]	tetragonal	429	80	-
Co_3B [87, 89]	orthorhombic	747	91	1300
$Nd_2Fe_{14}B$ [90]	tetragonal	585	100	13000

the magnet) are categorized as critical elements by the U.S. Department of Energy, which prompts researchers to discovery of new rare-earth free alternatives [91, 92].

An interesting and relatively novel approach in the search of new magnets is manipulating the itinerant magnetism by using the 3d elements in metal-rich borides. Density functional theory calculations demonstrate that the introduction of a magnetically active element M (like Fe or Mn) to $A_3T_5B_2$ borides, where T atoms are filled by a valence electron-rich transition metal like Ir, Co or Rh, Co, results in a structural transformation to $A_2MT_5B_2$. This change leads to an M position with tetragonal coordination, as atoms typically occupying this position exhibit a smaller size compared to those in the A sites [93, 94]. Thus, the incorporation of a magnetic element into these compounds results in distinctive properties, showcasing the richness of magnetic phenomena. For instance, the compound $Sc_2MnIr_5B_2$ exhibits ferromagnetism whereas $Sc_2FeIr_5B_2$ displays antiferromagnetism and $Sc_2MnRh_5B_2$ exhibits metamagnetism [95]. These magnetic behaviors highlight the intricate interplay between the magnetic elements and the unique properties that emerge in these specific compounds. Recently, Shankhari *et al.*, successfully synthesized a novel quaternary boride known as $Hf_2FeIr_5B_2$ [96]. This compound was strategically designed to maximize the incorporation of 5d elements and achieve an optimal valence electron count. Notably, the synthesized phase exhibited hysteresis behavior at temperatures exceeding room temperature. The hard-ferromagnetic character of $Hf_2FeIr_5B_2$ makes it a very promising candidate for potential applications as a rare-earth free permanent magnetic material in future technologies. A large number of radiation shielding applications use boron-containing materials because of the element's high neutron capture cross-section [97, 98]. Integrating boron's neutron shielding capabilities with the high atomic number elements' ability to shield against high-energy radiation like X-rays and gamma rays represents an innovative approach for comprehensive radiation shielding applications [99, 100]. Avcıoğlu *et al.*, have recently reported that transition metal borides like WB_2, DyB_6, SmB_6 and TaB_2 are excellent materials for shielding against gamma and mixed neutron radiation [37].

9.4 Conclusion

Metal diborides have garnered interest for their applicability in extremely high-temperature environments. Boride-based UHTCs boast of high melting points and exceptional mechanical properties, though challenges such as poor fracture toughness have prompted the development of composites. Research focuses on refining processing techniques and microstructural development to enhance properties like fracture resistance. High-entropy ceramics (HECs), notably diborides, hold promise for hypersonic vehicle components and solar power concentrators. However, addressing engineering challenges requires comprehensive studies, blending experimental tests with computational simulations. Efforts have addressed processing challenges to enhance fracture properties, thermal shock resistance, and oxidation resistance. Customizing composition and microstructure is vital for compatibility with extreme conditions. Various production methods for metal borides offer diverse applications.

Metal borides play a crucial role in automotive and aerospace industries because of their exceptional features, including high strength, thermal stability, lightweight nature, and resistance to extreme conditions. In aerospace, they are utilized in turbine blades, rocket nozzles, structural components, heat shields, high-temperature materials, and corrosion-resistant coatings. These materials ensure durability and reliability in space exploration missions, advancing aerospace technology.

Metal borides are customizable for specific neutron absorption performance through isotopic enrichment. Their utilization in nuclear industries depends on diverse application requirements, including thermal conductivity, mechanical properties, radiation resistance, and neutron absorption characteristics. These borides serve as efficient radiation shielding materials in nuclear facilities, replacing hazardous substances such as lead.

Metal diborides possess both metallic and ceramic properties, rendering them versatile materials. Transition metal borides, particularly diborides, demonstrate predictable properties influenced by the principal quantum number and the number of valence electrons present in metal atoms. Research primarily focuses on understanding the magnetism exhibited by transition metal borides, which range from weak paramagnetism to robust ferromagnetism. Hexagonal diborides of transition metals, particularly those in groups 3–6, demonstrate stability and magnetic anisotropy.

References

1. Levine, S.R., Opila, E.J., Halbig, M.C., Kiser, J.D., Singh, M., Salem, J.A., Evaluation of ultra-high temperature ceramics foraeropropulsion use. *J. Eur. Ceram. Soc.*, 22, 2757–2767, 2002.

2. Bansal, N.P., *Handbook of ceramic composites*, Springer Science & Business Media, 2005, https://doi.org/10.1007/b104068, 978-0-387-23986-6.
3. Kaufman, L. and Clougherty, E.V., Investigation of Boride Compounds for Very High-Temperature Applications, ManLabs. Inc., Cambridge, Mass, 1963.
4. Kaufman, L., Thermodynamic properties of transition metal diborides, ManLabs, Inc., Cambridge, Mass, 1964.
5. Opeka, M.M., Talmy, I.G., Wuchina, E.J., Zaykoski, J.A., Causey, S.J., Mechanical, thermal, and oxidation properties of refractory hafnium and zirconium compounds. *J. Eur. Ceram. Soc.*, 19, 2405–2414, 1999.
6. Silvestroni, L. and Sciti, D., Densification of ZrB2–TaSi2 and HfB2–TaSi2 ultra-high-temperature ceramic composites. *J. Am. Ceram. Soc.*, 94, 1920–1930, 2011.
7. Monteverde, F. and Savino, R., ZrB2–SiC sharp leading edges in high enthalpy supersonic flows. *J. Am. Ceram. Soc.*, 95, 2282–2289, 2012.
8. Sonber, J. and Suri, A., Synthesis and consolidation of zirconium diboride. *Adv. Appl. Ceram.*, 110, 321–334, 2011.
9. Fahrenholtz, W. and Hilmas, G., Oxidation of ultra-high temperature transition metal diboride ceramics. *Int. Mater. Rev.*, 57, 61–72, 2012.
10. Yan, C., Liu, R., Zhang, C., Cao, Y., Long, X., A solid-state precursor route to MB2-Al2O3 composite powders. *Powder Technol.*, 301, 596–600, 2016.
11. Cotton, J., Ultra-high-temperature ceramics: ultra-high-temperature ceramics are composites designed to survive extremes of temperature and stress. *Adv. Mater. Processes*, 168, 26–29, 2010.
12. Golla, B.R., Mukhopadhyay, A., Basu, B., Thimmappa, S.K., Review on ultra-high temperature boride ceramics. *Prog. Mater Sci.*, 111, 100651, 2020.
13. Mukhopadhyay, A., Raju, G., Basu, B., Ultra high temperature ceramics: processing, properties, and applications, in: *MAX phases and ultra-high temperature ceramics for extreme environments*, pp. 49–99, 2013.
14. Fahrenholtz, W.G., Hilmas, G.E., Talmy, I.G., Zaykoski, J.A., Refractory diborides of zirconium and hafnium. *J. Am. Ceram. Soc.*, 90, 1347–1364, 2007.
15. Marschall, J., Chamberlain, A., Crunkleton, D., Rogers, B., Catalytic atom recombination on ZrB2/SiC and HfB2/SiC ultrahigh-temperature ceramic composites. *J. Spacecr. Rockets*, 41, 576–581, 2004.
16. Guo, S.-Q., Densification of ZrB2-based composites and their mechanical and physical properties: a review. *J. Eur. Ceram. Soc.*, 29, 995–1011, 2009.
17. Kong, D., Wang, Q., She, T., Wang, L., Effects of temperature on flexural behavior and flaw sensitivity of ZrB2-SiC-graphite composite. *J. Alloys Compd.*, 773, 905–912, 2019.
18. Suri, A., Subramanian, C., Sonber, J., Murthy, T.C., Synthesis and consolidation of boron carbide: a review. *Int. Mater. Rev.*, 55, 4–40, 2010.
19. Zhao, G., Huang, C., Liu, H., Zou, B., Zhu, H., Wang, J., Microstructure and mechanical properties of hot pressed TiB2–SiC composite ceramic tool

materials at room and elevated temperatures. *Mater. Sci. Eng., A*, 606, 108–116, 2014.

20. Wang, P., Kumar, R., Sankaran, E.M., Qi, X., Zhang, X., Popov, D., Cornelius, A.L., Li, B., Zhao, Y., Wang, L., Vanadium diboride (VB2) synthesized at high pressure: elastic, mechanical, electronic, and magnetic properties and thermal stability. *Inorg. Chem.*, 57, 1096–1105, 2018.
21. Clougherty, E.V., Kalish, D., Peters, E.T., Research and development of refractory oxidation resistant diborides, Part II, 5, 1968.
22. Zhu, S., Densification, microstructure, and mechanical properties of zirconium diboride based ultra-high temperature ceramics, 2008.
23. Tao, Q., Zhao, X., Chen, Y., Li, J., Li, Q., Ma, Y., Li, J., Cui, T., Zhu, P., Wang, X., Enhanced Vickers hardness by quasi-3D boron network in MoB 2. *RSC Adv.*, 3, 18317–18322, 2013.
24. Tao, Q., Ma, Y., Li, Y., Chen, Y., Ma, Y., Cui, T., Wang, X., Zhu, P., Exploring the high pressure behavior of 2D and quasi-3D boron layers in MoB 2. *RSC Adv.*, 4, 52878–52882, 2014.
25. Kiessling, R., Hagdahl, L., Sillén, L.G., Rottenberg, M., The Borides of Some Transition Elements. *Acta Chem. Scand.*, 4, 209–227, 1950.
26. Schwetz, K. and Lipp, A., Boron carbide, boron nitride, and metal borides, in: *Ullmann's Encyclopedia of Industrial Chemistry*, vol. 4, p. 295, 1985.
27. Baudis, U. and Fichte, R., Boron and boron alloys, in: *Ullmann's Encyclopedia of Industrial Chemistry*, 2000.
28. Pu, Z., Liu, T., Zhang, G., Liu, X., Gauthier, M.A., Chen, Z., Sun, S., Nanostructured Metal Borides for Energy-Related Electrocatalysis: Recent Progress, Challenges, and Perspectives. *Small Methods*, 5, 2100699, 2021.
29. Murthy, T.S., Sonber, J., Sairam, K., Majumdar, S., Kain, V., Boron-based ceramics and composites for nuclear and space applications: synthesis and consolidation, in: *Handbook of advanced ceramics and composites: Defense, security, aerospace and energy applications*, pp. 703–738, 2020.
30. Thompson, R., Production, fabrication and uses of borides, in: *The Physics and Chemistry of Carbides, Nitrides and Borides*, pp. 113–120, Dordrecht, Springer Netherlands, 1990.
31. Joshi, R. and Ningthoujam, R.S., Synthesis, characterization, physical properties and applications of metal borides, in: *Handbook on Synthesis Strategies for Advanced Materials: Volume-III: Materials Specific Synthesis Strategies*, pp. 251–305, 2021.
32. Ningthoujam, R., Synthesis and characterization of borides, carbides, and nitrides and their applications, in: *Materials Under Extreme Conditions*, S.B.A.K. Tyagi (Ed.), pp. 337–375, Elsevier, 2017.
33. Dandekar, D.P. and Benfanti, D.C., Strength of titanium diboride under shock wave loading. *J. Appl. Phys.*, 73, 673–679, 1993.
34. Millet, P. and Hwang, T., Preparation of TiB 2 and ZrB 2. Influence of a mechano-chemical treatment on the borothermic reduction of titania and zirconia. *J. Mater. Sci.*, 31, 351–355, 1996.

35. Murthy, T., Effect of sinter additives on the consolidation and properties of titanium diboride composites, 2012.
36. Telle, R., Sigl, L., Takagi, K., Boride-Based Hard Materials, in: *Handbook of Ceramic Hard Materials*, R. Riedel (Ed.), pp. 802–945, Wiley-VCH Verlag GmbH, 2008.
37. Avcıoğlu, C. and Avcıoğlu, S., Transition Metal Borides for All-in-One Radiation Shielding. *Materials*, 16, 6496, 2023.
38. Opeka, M.M., Talmy, I.G., Zaykoski, J., Oxidation-based materials selection for 2000 C+ hypersonic aerosurfaces: Theoretical considerations and historical experience. *J. Mater. Sci.*, 39, 5887–5904, 2004.
39. Xu, L., Yang, Y., Wang, S., Li, M., Xu, J., Qian, Y., Zuo, J., Zhang, D., Improved both mechanical and anti-oxidation performances of ZrB2-SiC ceramics with molybdenum disilicide addition. *Mater. Chem. Phys.*, 223, 53–59, 2019.
40. Zapata-Solvas, E., Jayaseelan, D.D., Brown, P., Lee, W., Effect of La2O3 addition on long-term oxidation kinetics of ZrB2–SiC and HfB2–SiC ultra-high temperature ceramics. *J. Eur. Ceram. Soc.*, 34, 3535–3548, 2014.
41. Zhao, J., Liu, H.-T., Liu, J.-X., Zhang, G.-J., ZrB2 ceramics doped with AlB2. *Ceram. Int.*, 40, 8915–8920, 2014.
42. Kuriakose, A.K. and Margrave, J., The oxidation kinetics of zirconium diboride and zirconium carbide at high temperatures. *J. Electrochem. Soc.*, 111, 827, 1964.
43. Zhang, S.C., Hilmas, G.E., Fahrenholtz, W.G., Improved oxidation resistance of zirconium diboride by tungsten carbide additions. *J. Am. Ceram. Soc.*, 91, 3530–3535, 2008.
44. Zapata-Solvas, E., Jayaseelan, D., Lin, H.-T., Brown, P., Lee, W., Mechanical properties of ZrB2-and HfB2-based ultra-high temperature ceramics fabricated by spark plasma sintering. *J. Eur. Ceram. Soc.*, 33, 1373–1386, 2013.
45. Talmy, I.G., Zaykoski, J.A., Opeka, M.M., High-temperature chemistry and oxidation of ZrB2 ceramics containing SiC, Si3N4, Ta5Si3, and TaSi2. *J. Am. Ceram. Soc.*, 91, 2250–2257, 2008.
46. Dehdashti, M.K., Fahrenholtz, W., Hilmas, G., Effects of temperature and the incorporation of W on the oxidation of ZrB2 ceramics. *Corros. Sci.*, 80, 221–228, 2014.
47. Sciti, D., Brach, M., Bellosi, A., Oxidation behavior of a pressureless sintered ZrB2–MoSi2 ceramic composite. *J. Mater. Res.*, 20, 922–930, 2005.
48. Sciti, D., Balbo, A., Bellosi, A., Oxidation behaviour of a pressureless sintered HfB2–MoSi2 composite. *J. Eur. Ceram. Soc.*, 29, 1809–1815, 2009.
49. Zhang, X., Hilmas, G.E., Fahrenholtz, W.G., Densification, mechanical properties, and oxidation resistance of TaC–TaB2 ceramics. *J. Am. Ceram. Soc.*, 91, 4129–4132, 2008.
50. Licheri, R., Musa, C., Orrù, R., Cao, G., Sciti, D., Silvestroni, L., Bulk monolithic zirconium and tantalum diborides by reactive and non-reactive spark plasma sintering. *J. Alloys Compd.*, 663, 351–359, 2016.

51. Silvestroni, L., Guicciardi, S., Melandri, C., Sciti, D., TaB2-based ceramics: microstructure, mechanical properties and oxidation resistance. *J. Eur. Ceram. Soc.*, 32, 97–105, 2012.
52. Guo, S., Mizuguchi, T., Ikegami, M., Kagawa, Y., Oxidation behavior of ZrB2–MoSi2–SiC composites in air at 1500° C. *Ceram. Int.*, 37, 585–591, 2011.
53. Murthy, T.C., Sonber, J., Sairam, K., Bedse, R., Chakarvartty, J., Development of refractory and rare earth metal borides & carbides for high temperature applications. *Mater. Today Proc.*, 3, 3104–3113, 2016.
54. Sonber, J., Murthy, T.C., Subramanian, C., Hubli, R., Fotedar, R., Suri, A., Effect of WSi2 addition on densification and properties of ZrB2. *Adv. Appl. Ceram.*, 113, 114–119, 2014.
55. Squire, T.H. and Marschall, J., Material property requirements for analysis and design of UHTC components in hypersonic applications. *J. Eur. Ceram. Soc.*, 30, 2239–2251, 2010.
56. Gasch, M.J., Ellerby, D.T., Johnson, S.M., Ultra high temperature ceramic composites, in: *Handbook of Ceramic Composites*, pp. 197–224, Springer US, Boston, MA, 2005.
57. Ohji, T., Singh, M., Singh, D., Salem, J., *Advanced processing and manufacturing technologies for structural and multifunctional materials III*, Ceramic Engineering and Science Proceedings, Daytona Beach, Florida, vol. 30, 2009.
58. Fahrenholtz, W.G., Binner, J., Zou, J., Synthesis of ultra-refractory transition metal diboride compounds. *J. Mater. Res.*, 31, 2757–2772, 2016.
59. Zhu, S., Densification microstructure and mechanical properties of zirco. pdf, Materials Science and Engineering, Doctor of Philosophy, Missouri University, Missouri University of Science and Technology, p. 194, 2008.
60. Loganathan, P., Gnanavelbabu, A., Rajkumar, K., Ayyanar, S., A comparative study of solid lubricant types on the microstructure and mechanical behaviour of AA7075-zirconium boride aerospace composites. *IOP Conf. Ser.: Mater. Sci. Eng.*, 764, IOP Publishing, 012001, 2020.
61. Naik, A.K., Nazeer, M., Prasad, D., Laha, T., Roy, S., Development of functionally graded ZrB2–B4C composites for lightweight ultrahigh-temperature aerospace applications. *Ceram. Int.*, 48, 33332–33339, 2022.
62. Gök, M.G. and Cihan, O., Energetic materials and metal borides for solid propellant rocket engines. *Int. J. Mater. Eng. Technol.*, 3, 109–119, 2020.
63. Sonber, J., Murthy, T.C., Subramanian, C., Hubli, R., Suri, A., Processing methods for ultra high temperature ceramics, in: *MAX Phases and Ultra-High Temperature Ceramics for Extreme Environments*, pp. 180–202, IGI Global, 2013.
64. Ortiz-Domínguez, M., Cruz-Avilés, A., Morales-Robles, A., Gómez-Vargas, O., Solís-Romero, J., Zuno-Silva, J., Cardoso-Legorreta, E., Hard layers based on metal borides: Microstructure and mechanical properties. *Microsc. Microanal.*, 27, 1556–1560, 2021.

65. Meléndez-Martınez, J., Domınguez-Rodrıguez, A., Monteverde, F., Melandri, C., De Portu, G., Characterisation and high temperature mechanical properties of zirconium boride-based materials. *J. Eur. Ceram. Soc.*, 22, 2543–2549, 2002.
66. Low, I.-M., *MAX phases and ultra-high temperature ceramics for extreme environments*, IGI Global, 2013.
67. Akopov, G., Pangilinan, L.E., Mohammadi, R., Kaner, R.B., Perspective: Superhard metal borides: A look forward. *APL Mater.*, 6, 2018.
68. Xie, M., Pang, J., Jin, W., Kuang, X., Theoretical Prediction of Two-Dimensional Metal Boride Mg4B6 as a High-Capacity Electrode Material for Lithium-Ion Batteries. *J. Phys. Chem. C*, 126, 17474–17481, 2022.
69. Campos-Silva, I.E. and Rodriguez-Castro, G.A., Boriding to improve the mechanical properties and corrosion resistance of steels, in: *Thermochemical Surface Engineering of Steels*, pp. 651–702, Elsevier, Woodhead Publishing, 2015.
70. Medvedovski, E., Jiang, J., Robertson, M., Boride-based coatings for protection of cast iron against wear. *Adv. Appl. Ceram.*, 115, 483–494, 2016.
71. Avcıoğlu, S., Buldu, M., Kaya, F., Kaya, C., Synthesis of nuclear-grade nano-sized boron carbide powders and its application in LDPE matrix composites for neutron shielding, in: *Composite Materials*, pp. 543–579, Elsevier, 2021.
72. Avcioğlu, S., LDPE matrix composites reinforced with dysprosium-boron containing compounds for radiation shielding applications. *J. Alloys Compd.*, 927, 166900, 2022.
73. Sinem, U., Investigation of boron element and its use in nuclear reactors. *J. Boron*, 8, 1, 25–30, 2023. https://doi.org/10.30728/boron.1131336
74. Kaur, T., Sharma, J., Singh, T., Review on scope of metallic alloys in gamma rays shield designing. *Prog. Nucl. Energy*, 113, 95–113, 2019.
75. Kharita, M., Yousef, S., AlNassar, M., Review on the addition of boron compounds to radiation shielding concrete. *Prog. Nucl. Energy*, 53, 207–211, 2011.
76. Pangilinan, L.E., Hu, S., Hamilton, S.G., Tolbert, S.H., Kaner, R.B., Hardening effects in superhard transition-metal borides. *Acc. Mater. Res.*, 3, 100–109, 2021.
77. Gülbiçim, H., Türkan, M.N., Aksu, M., Durmuş, S., A study on the investigation of gamma shielding properties of some metal borides. *Prog. Nucl. Energy*, 115, 107–114, 2019.
78. Serebryakova, T., Classification of borides. *J. Less-Common Met.*, 67, 499–503, 1979, 10.1016/0022-5088(79)90034-1.
79. Akopov, G., Yeung, M.T., Kaner, R.B., Rediscovering the Crystal Chemistry of Borides. *Adv. Mater.*, 29, 2017, 10.1002/adma.201604506.
80. Hayat, A., Sohail, M., Hamdy, M.S., Taha, T.A., AlSalem, H.S., Alenad, A.M., Amin, M.A., Shah, R., Palamanit, A., Khan, J., Nawawi, W.I., Mane, S.K.B., Fabrication, characteristics, and applications of boron nitride and their composite nanomaterials. *Surf. Interfaces*, 29, 2022, 10.1016/j.surfin.2022.101725.

81. Pluzhnikov, V.B., Svechkarev, I.V., Dukhnenko, A.V., Levchenko, A.V., Filippov, V.B., Chopnik, A., de Haas–van Alphen effect in the diborides ScB2, ZrB2, and HfB2. *Low Temp. Phys.*, 33, 350–354, 2007, 10.1063/1.2720083.
82. Grechnev, G.E., Fedorchenko, A.V., Logosha, A.V., Panfilov, A.S., Svechkarev, I.V., Filippov, V.B., Lyashchenko, A.B., Evdokimova, A.V., Electronic structure and magnetic properties of transition metal diborides. *J. Alloys Compd.*, 481, 75–80, 2009, 10.1016/j.jallcom.2009.03.123.
83. Simsek, T. and Ozcan, S., Structural and Magnetic Properties of MnxB100–x Alloys. *IEEE Trans. Magn.*, 51, 1–3, 2015, 10.1109/tmag.2015.2392715.
84. Şimşek, T., Avar, B., Şimşek, T., Yıldız, B., Chattopadhyay, A.K., Özcan, S., A single step synthesis by mechanical alloying and characterization of nanostructured Fe2B of high magnetic moment. *Ceram. Int.*, 47, 26119–26124, 2021, 10.1016/j.ceramint.2021.06.018.
85. Hamayun, M.A., Abramchuk, M., Alnasir, H., Khan, M., Pak, C., Lenhert, S., Ghazanfari, L., Shatruk, M., Manzoor, S., Magnetic and magnetothermal studies of iron boride (FeB) nanoparticles. *J. Magn. Magn. Mater.*, 451, 407–413, 2018, 10.1016/j.jmmm.2017.11.088.
86. Zhao, X., Li, L., Bao, K., Zhu, P., Tao, Q., Ma, S., Liu, B., Ge, Y., Li, D., Cui, T., Synthesis and characterization of a strong ferromagnetic and high hardness intermetallic compound Fe(2)B. *Phys. Chem. Chem. Phys.*, 22, 27425–27432, 2020, 10.1039/d0cp03380d.
87. Buschow, K.H.J., Magnetic Properties of Borides, in: *Boron and Refractory Borides*, Matkovich, V.I. (Ed.), 1977, 10.1007/978-3-642-66620-9_26.
88. Ener, S., Fries, M., Hammerath, F., Opahle, I., Simon, E., Fritsch, P., Wurmehl, S., Zhang, H., Gutfleisch, O., Magnetic and magnetocaloric properties of the Co2-xMn B system by experiment and density functional theory. *Acta Mater.*, 165, 270–277, 2019, 10.1016/j.actamat.2018.11.034.
89. Zieschang, A.M., Bocarsly, J.D., Schuch, J., Reichel, C.V., Kaiser, B., Jaegermann, W., Seshadri, R., Albert, B., Magnetic and Electrocatalytic Properties of Nanoscale Cobalt Boride, Co(3)B. *Inorg. Chem.*, 58, 16609–16617, 2019, 10.1021/acs.inorgchem.9b02617.
90. Givord, D., Li, H.S., Moreau, J.M., Magnetic properties and crystal structure of Nd2Fe14B. *Solid State Commun.*, 50, 497–499, 1984, 10.1016/0038-1098(84)90315-6.
91. Gutfleisch, O., Willard, M.A., Bruck, E., Chen, C.H., Sankar, S.G., Liu, J.P., Magnetic materials and devices for the 21st century: stronger, lighter, and more energy efficient. *Adv. Mater.*, 23, 821–842, 2011, 10.1002/adma.201002180.
92. Habib, K. and Wenzel, H., Exploring rare earths supply constraints for the emerging clean energy technologies and the role of recycling. *J. Clean. Prod.*, 84, 348–359, 2014, 10.1016/j.jclepro.2014.04.035.
93. Zhang, Y., Miller, G.J., Fokwa, B.P.T., Computational Design of Rare-Earth-Free Magnets with the Ti3Co5B2-Type Structure. *Chem. Mater.*, 29, 2535–2541, 2017, 10.1021/acs.chemmater.6b04114.

94. Shankhari, P., Bakshi, N.G., Zhang, Y., Stekovic, D., Itkis, M.E., Fokwa, B.P.T., A Delicate Balance between Antiferromagnetism and Ferromagnetism: Theoretical and Experimental Studies of A(2) MRu(5) B(2) (A=Zr, Hf; M=Fe, Mn) Metal Borides. *Chemistry*, 26, 1979–1988, 2020, 10.1002/chem.201904572.
95. Nagelschmitz, E.A. and Jung, W., Scandium Iridium Boride Sc3Ir5B2 and the Quaternary Derivatives Sc2MIr5B2 with M = Be, Al, Si, Ti, V, Cr, Mn, Fe, Co, Ni, Cu, Ga, or Ge: Preparation, Crystal Structure, and Physical Properties. *Chem. Mater.*, 10, 3189–3195, 1998, 10.1021/cm9802898.
96. Shankhari, P., Janka, O., Pottgen, R., Fokwa, B.P.T., Rare-Earth-Free Magnets: Enhancing Magnetic Anisotropy and Spin Exchange Toward High-T(C) Hf(2)MIr(5)B(2) (M = Mn, Fe). *J. Am. Chem. Soc.*, 143, 4205–4212, 2021, 10.1021/jacs.0c10778.
97. Uzun, S., Investigation of boron element and its use in nuclear reactors. *J. Boron*, 8, 1, 25–30, 2023, 10.30728/boron.1131336.
98. Pitto-Barry, A., Polymers and boron neutron capture therapy (BNCT): a potent combination. *Polym. Chem.*, 12, 2035–2044, 2021, 10.1039/d0py01392g.
99. Xiao, M., Qin, Q., He, X., Li, F., Wang, X., Shielding Capability Research on Composite Base Materials in Hybrid Neutron-Gamma Mixed Radiation Fields. *Materials (Basel)*, 16, 2023, 10.3390/ma16052084.
100. Daneshvar, H., Milan, K.G., Sadr, A., Sedighy, S.H., Malekie, S., Mosayebi, A., Multilayer radiation shield for satellite electronic components protection. *Sci. Rep.*, 11, 20657, 2021, 10.1038/s41598-021-99739-2.

10

Hard Machining and Characteristics for Industrial Cutting Tools and Coating Deposition Applications

Riyadh A. Al-Samarai[1]* and Yarub Al-Douri[2,3,4]

[1]*Electromechanical Engineering Department, Engineering College, University of Samarra, Samarra, Iraq*
[2]*College of Health and Medical Techniques, Al-Bayan University, Baghdad, Iraq*
[3]*Department of Mechanical Engineering, Faculty of Engineering, Piri Reis University, EflatunSk, Tuzla, Istanbul, Turkey*
[4]*Nanotechnology and Catalysis Research Centre, University of Malaya, Kuala Lumpur, Malaysia*

Abstract

Traditional grinding operations have totally taken the place of novel processes for modern processing requirements have exceeded the capabilities of older tools due to the rapid development of the manufacturing industry. This problem was largely solved with the advent of ceramic dishes. Tool coatings are constantly being improved to perform better in the harshest environments. Coated tools have superior corrosion resistance, superior thermal stability, superior hardness, and superior adhesive strength compared to conventional tools. This page primarily introduces the development of tool coating materials, coating preparation methods, and potential applications and developments of tool coatings.

***Keywords*:** Grinding operations, ceramic tools, tool coatings, corrosion resistance, thermal stability, coating materials, coating methods, cutting tool applications

**Corresponding author*: riyadh2824@yahoo.com

Inamuddin, Tariq Altalhi and Mohammad Abu Jafar Mazumder (eds.) Nanostructured Hard Materials: Industrial Applications, (317–342) © 2026 Scrivener Publishing LLC

10.1 Introduction

The term "hard machining" describes the process of machining parts with a hardness greater than 45 HRC; however, the most common application of this method is for parts with a hardness between 58 and 68 HRC. Materials used include hard chrome steel, tool steel, case-hardened steel, superalloys, nitride cast iron, heat-treated powder metallurgy components, and various hardened alloy steels [1]. Any machine shop can create dependable equipment, incredibly tough cutting materials, and hardened metal cutting required through sophisticated setups and specialized tool (holder) designs. Although hardened steel parts have historically been processed by grinding, hard parts can also be processed using a cutting instrument that provides a number of clear benefits.

They are compatible with a wide range of currently used applications, especially those related to milling, turning, and boring [1–3]. In modern manufacturing, this is no longer the case, although many research articles continue to support the outdated notion that machining can replace all grinding operations. Because the two processes work well together in a wide range of practical industrial applications, grinding spindles are commonly used in modern machine tools designed for harsh machining. Putting together a turning AISI 52100 steel with a 65.5 HRC, shown in Figure 10.1, is an example of industrial applications. The use of hard materials, ultra-precision cutting materials, and precision cutting materials has increased with the development of precision engineering and materials

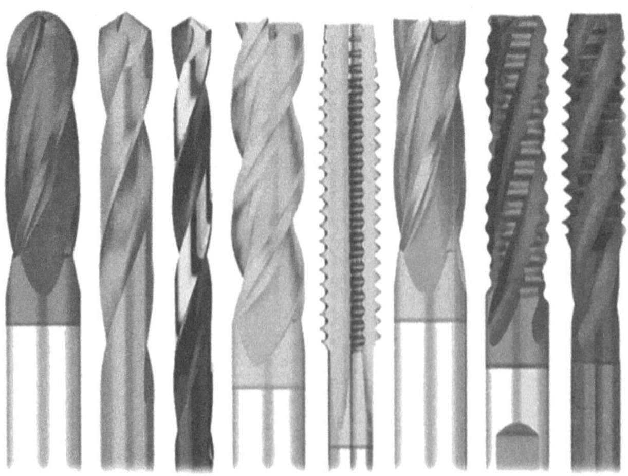

Figure 10.1 Putting together a turning AISI 52100 steel with a 65.5 HRC [9].

industries. In particular, the automotive, aviation, and engineering industries are growing at breakneck speed. As a result, more and more attention is being paid to materials and coatings of cutting tools [1, 2]. The basic design principles and characteristics of the cutting field have been used to describe the art of cutting tool design [3–6]. Particular attention is paid to tool matrix materials [10–12], tool coating materials [10–12] and recent advances [7–9]. Generally speaking, materials having a hardness greater than 45 HRC are considered difficult to machine.

A great place to start is a quick overview of the development of PVD coatings for cutting tools. Although the technology has been in use since the early 1970s, ultrathin titanium nitride (TiN) film was first commercially released as a tool coating in 1980. Perhaps business influences prevented its commercial use earlier. When this type of hard coating is used on the cutting edge of a machine tool, the tool life can be significantly increased. However, longer tool life leads to lower sales of new tools, which is bad news for tool manufacturers [4, 5]. Although end users should be aware of them, research papers and promotional materials generally do not address the limitations and disadvantages of hard processing:

- Unit costs for tools for machining are significantly higher than for grinding.
- In some cases, the geometry or size of the part makes hard turning impossible.

Generally speaking, unsupported parts should not have an L/D ratio greater than 4:1. Even though the tailstock is designed for long, thin workpieces, it vibrates when cutting under high pressure. Hard materials are often difficult to process without specialized rigid equipment. The rigidity of the machine determines the level of turning accuracy. When required part tolerances and surface finishes increase, machine tool rigidity becomes a more serious issue.

Coatings using nanostructures and nanolayers have also become a subject of recent research. They have been tested and the results are very promising: their tool life is on par with more advanced nano-coatings [8, 9]. When using PVD-coated TiN cemented carbide cutting inserts to machine steel, cutting speed and feed have a substantial impact on tool life [10]. Figure 10.2 Properties of advanced coating materials used in tools. Figure 10.3 shows the distinctive characteristics of the instrumentation material. The wear of PVD coatings is not impacted by the monolayer coating's thickness [11, 12]. PVD coatings have much greater density, greater hardness, and greater stickiness. Figure 10.4 illustrates the variables affecting

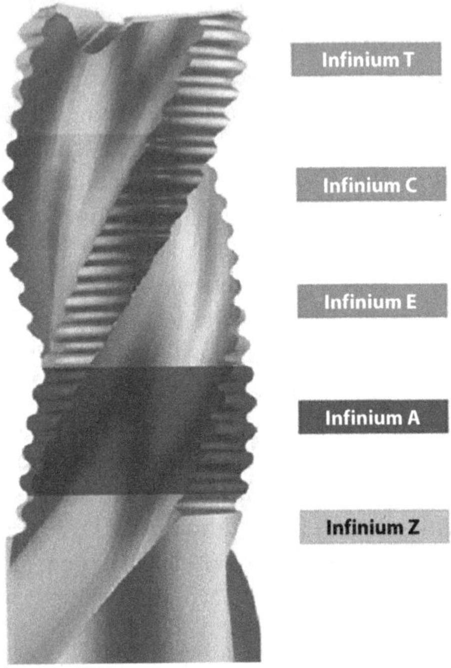

Figure 10.2 Properties of advanced coating materials used in tools [17].

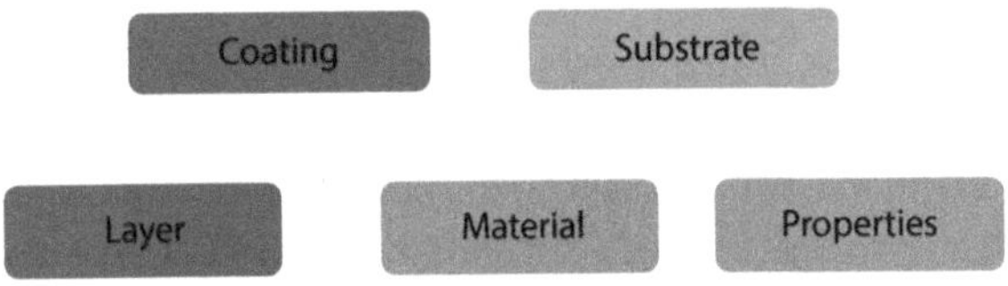

Figure 10.3 Factors influencing the cutting tool's efficacy.

the cutting tool's performance. A tool with a coating produces superior machining results. The way that CBN-TiN (coated) carbide tools respond to machining forces in numerous fields of machining research, notably in high-strength low-alloy steels. This work discusses various coating types, their general features, layering and applications. It also highlights technology shortcomings and difficulties with prospect research. The final section of this summary focuses on potential future research directions in the machining and tooling industries [7, 8].

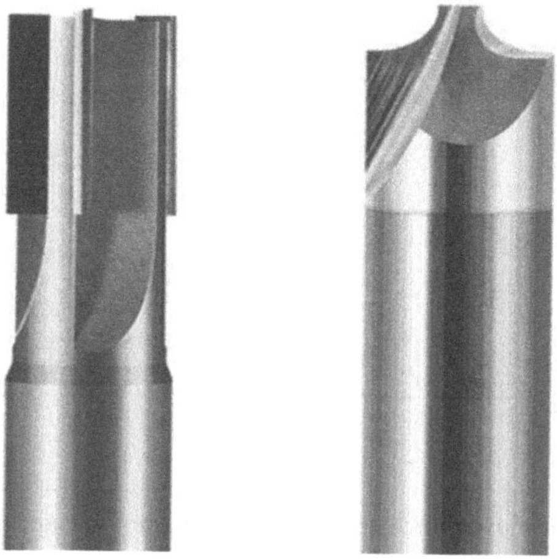

Figure 10.4 The surface quality deteriorates with tool wear. AISI 52100 steel with a 65.5HRC is turned.

The biggest mystery is what coolant is appropriate for heavy machining operations. Dry milling is the best method for milling objects with interrupted cuts, such as gears. The reason for this is the possibility of thermal stress failure during the process of entering and leaving the grooves. The high tool tip temperatures during dry turning cause the pre-cut area to be annealed (softened) during continuous cutting, which lessens the material's hardness and facilitates cutting. It is useful to expedite the process during dry cutting because of this feature. There are also definite economic benefits to dry cutting. On the other hand, it might be challenging to process, measure, and thermally deform parts during manufacture. Some applications are now required to use coolant due to recent issues. When employed, water-based coolant is required. It has been demonstrated that near-dry machining works well in demanding applications [9]. As illustrated in Figure 10.5, the workpiece's surface quality degrades even within the tool's lifetime due to wear. It is important to find solutions to the issues surrounding rigid equipment, industrial cutting tool properties, and their applications [13]. The coating deposition procedure, where the intense processing results in a white covering thickens with tool wear. It mostly affects parts like bearing strains, which are frequently formed on tools and are exposed to high-contact pressures. Eventually, the white coating can come off, which would cause the tool to fail. Most tool manufacturers have internal testing facilities to address this issue. The steel industry is able to carry out these tests. The attachment can precisely cut 400 parts; however, after 300 parts, it can get so dull that a white coating forms [28].

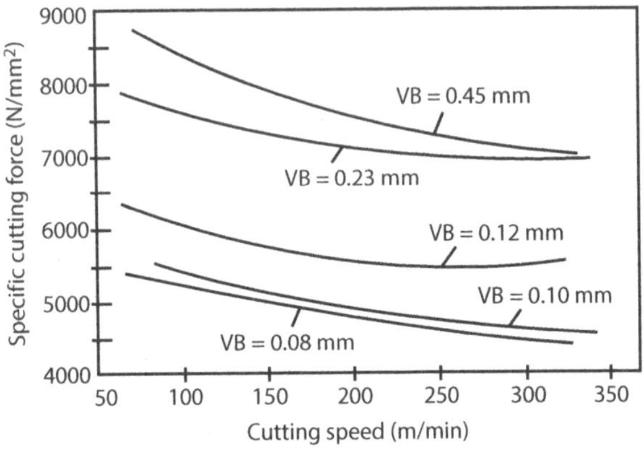

Figure 10.5 Changes in cutting force depending on how worn the side surfaces are. The alloy steel has a hardness of 52 HRC, a cutting depth of 0.25 mm, a cutting speed of 0.08 mm/rev, and CBN tool material.

10.2 Analysis of Hard Machining Research

Despite the significant amount of material on hard machining, there's only a small number of articles [9, 10]. Traditional and intensive processing neglect some basic truths that are well-known in practice. As a result, the nature of hard machining discussed in these articles is understood very differently by leading tooling companies. Unfortunately, it seems that not enough machining researchers are "hearing" this point of view. The first detailed work that attempted to summarize the hard machining experience was published by [12]. Studying the technological characteristics of these materials is necessary to create new technologies for processing hard metals. The results of [12], are summarized as the destruction of the material during hard machining, so-called sawtooth chips as shown in Figure 10.6. Thus, when the work material reaches maximum shear strain, a crack appears on the workpiece. Therefore, the process of creating a chip is determined by destruction. Although "segmented chips" produced by adiabatic shear in difficult-to-machine materials have the same cross-section as saw tooth chips, the two types of chips produced by hard machining are different because they arise from different processes. Compared to traditional machining, the shear angle is minimal. It increases sharply as the hardness of the material being processed increases.

- The disparity between the two components gets progressively larger. The working material's springing provides an explanation for this. In hard machining, a strong traction force is the primary cause of dimensional inaccuracies.

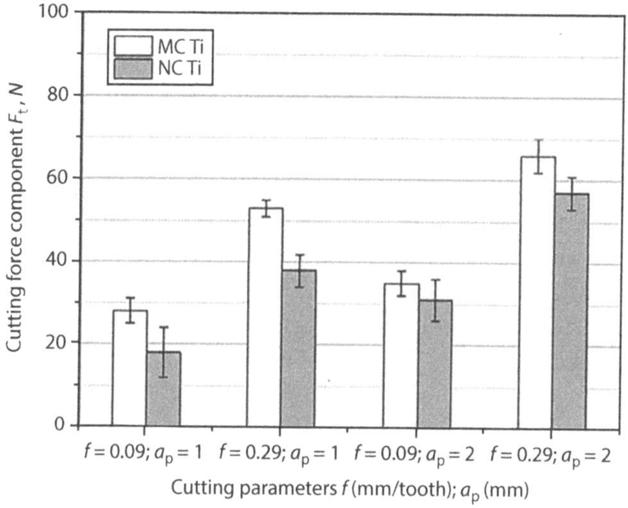

Figure 10.6 At a cutting speed vc = 40 m/min, the values of the tangential component of the cutting forces Ft were calculated for the selected cutting parameters (ap, f).

For instance, the bearing steel AISI 52100, one of the materials used most frequently for severe machining and hence testing, has a hardness range of 60 to 67 HRC. Grey cast iron has an elongation of greater than 7% and elongation at failure of 5% in their testing. Additionally, neither the model nor its analysis account for the energy involved [11, 14].

- Negative rake cutting appears to be an opposite of what is commonly practiced, with relatively low tangential (force) and radial (axial) cutting force components and a decrease with increasing workpiece hardness.
- Neither the cause of temperature during complex machining nor its effect or its results is investigated.

In a hard machining, which is an essence of CBN, ceramic and cermet tool materials [15–18].

10.3 Hard Machining Characteristics

Although this work analyzes the characteristics in great detail, there are several distinguishing characteristics that need to be considered when analyzing hard machining in order to set priorities for subsequent research and applications of this method. This can be achieved by comparing the energy balances of conventional and hard metal cutting methods [19–21]. AISI steel 52100 elongation-stress graphs during annealing and quenching are shown in Figure 10.7. Surprisingly, the energy required to plastically deform the removed layer is still significant during aggressive cutting, Figure 10.8. It shows cutting pens that have been coated with various materials, including titanium, tungsten, and other materials. No significant changes were observed in the range of Fr values for the radial components. This is because the so-called "cutting action" when machining MC titanium is different and in this case is more "passive" (as with a tool with a narrower rake angle).

10.3.1 Speed with Cutting Force

The Peclet criterion, a measure of similarity, is useful in understanding the role of thermal energy (heat) in hard machining [12, 25]. The Peclet

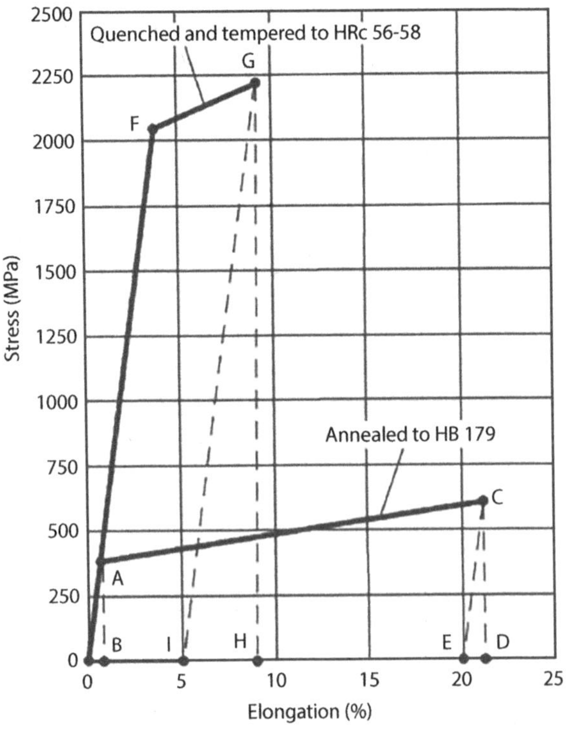

Figure 10.7 Elongation-stress graphs of AISI steel 52100 during annealing and quenching.

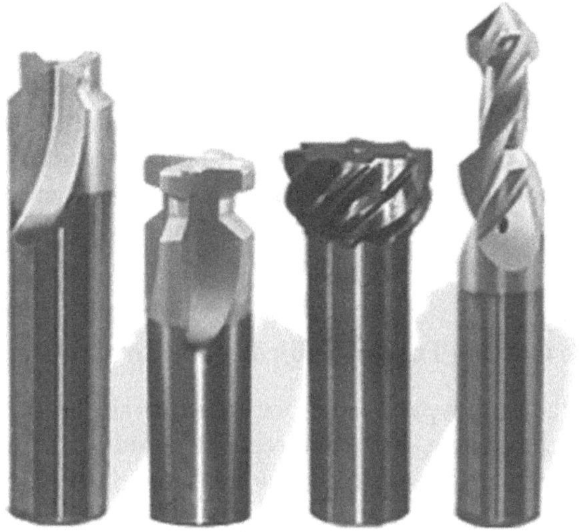

Figure 10.8 A hard turning: conventional, and high-speed.

criterion for metal cutting is illustrated using the following processing parameters [26, 27]:

$$Pe = \frac{vt_1}{w_w} \tag{10.1}$$

where w_w is thermal diffusivity of work material (m^2/s), and v is speed.

$$w_w = \frac{k_w}{(c_p \rho)_w} \tag{10.2}$$

where (c_p) w is specific of volume heat (J/(m^3 °C)) and (k_w) is the thermos-conductivity of work material (J/(ms°C)). The number Péclet that expresses how much the cutting regime (vt_l) has an impact on material's thermal characteristics (w_w). Abrasive sawtooth edges are formed due to high dynamic shear strength, creating notches on the tool [18].

The resistance of the processed material, which must be overcome during processing, is measured by the components of the cutting force. According to test data, titanium structure does not affect fundamental strength. Notable results were observed in the traction and pitching force scenarios. Regarding NK These forces for titanium are much less than for

MC Ti .The cutting force component was calculated both tangentially and axially using geometric relationships and dynamometer measurements (F_X, F_Y, and F_Z). During a fiber optic or other optical beam delivery, device is placed in front of tool to direct the laser beam onto the workpiece. The workpiece is softened by laser heat, which facilitates cutting. The heat is concentrated around the cut itself due to laser beam that is so precisely focused [29–33]. The chips dissipate heat, yet the physical characteristics of material being sliced remain unchanged as a result of heat [40].

10.3.2 The Cutting Forces Major Axial (Thrust) Components

Currently, the automotive and aerospace industries are the main consumers of titanium and its alloys. sports and recreation, industry, bioengineering, shipbuilding, industry, aerospace, and architecture. In addition, nanocrystalline materials (NCs) have special properties. physical and mechanical characteristics due to their special structure [2]. It is used in cutting tools and electronic devices to improve performance and protection, as well as in aviation for thermal barrier coatings and in the automotive industry for anti-corrosion coatings on engine components. Arc plasma deposition is the preferred coating method in many industries due to its versatility and efficiency [29, 30]. These characteristics provide them with great potential for energy applications and make them attractive for use in biological applications [1], such as orthopedic and dental implants. All of these applications must be processed using manufacturing procedures. make the corresponding parts. The most common industrial technology is mechanical processing, useful and practical metals. However, due to these, the same characteristics are considered difficult to work with. AISI 52100 steels are presented in Figure 10.9. If the applied stress is greater than the elastic limit also referred to as the yield limit, which for annealed 52100 steel is 370 MPa the material will display both elastic and plastic deformation.

It is shown in Figure 10.9 as a distance along the elongation axis 0 – D. However, after failure, the applied stress is released and the work material shows permanent extension, which is called elongation (shown as distance 0 – E in Figure 10.9). The total elongation after failure of the elastic component, shown in the E-D distance diagram, in the case under consideration, is less than 20%. This makes it easy to determine where point E draws a line parallel to line 0-A from point C. As a result, the E-D distance is sometimes called elastic recovery or elastic return in materials processing. The pressure at point of contact between the tool and workpiece increases sharply because this friction creates a relief, which leads to wear on the flank surface. This explains the significant radial (axial) component of cutting force,

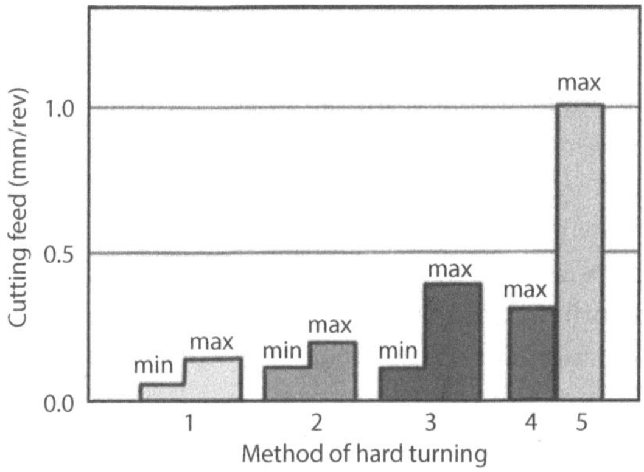

Figure 10.9 The minimum and maximum feed ranges for the following tools are as follows: tools with geometries f = 0.1–0.4 mm/rev can be found; two cylindrical; three - classical turning (f = 0.3 mm/rev); five - advanced turning (f up to 1.0 mm/rev).

and severe wear on the flank surface only happens when cutting something hard. However, prolonged heat reduces its flexibility.

10.3.3 Power Invested in Creation of Surfaces

The ability of solid materials and objects to restore their previous shape after removing loads or external forces that cause deformation is called elasticity. When an object regains its original size and shape after the load is removed, it is called elastic. Different materials behave differently for different physical reasons, which are influenced by their microscopic structure. For example, the stretching of polymer chains under the influence of applied force provides the elasticity of rubbers and polymers. Conversely, under the influence of external pressure, the crystal lattice cells that make up the structure of the metal material undergo changes in size and shape, which determine the elasticity of the metal.

The two factors that determine the elasticity of a material are the elastic limit and the elastic modulus. Materials that require large loads to produce significant deformation or those that are difficult to deform have a high modulus of elasticity. Steel strip is one example. Materials that easily deform under load, such as elastic bands, have a low modulus of elasticity. The material deforms permanently if the stress under load is too great, at which point it takes on a different shape and size rather than returning to its original state. The stress value above which a material ceases to behave elastically and instead permanently deforms is known as the elastic limit. The material with the lowest elastic modulus is the most elastic among materials with the same elastic limitations [12–18]. Surface engineering processes are constantly evolving, providing tremendous opportunities for a wide range of applications across multiple industries. From consumer products to electronics, medical devices, aerospace, and automotive components, product performance, and durability are greatly influenced by the adaptability of metals to their surface quality and process. Research into new materials, improved creative processes, and integration of surface technology into other manufacturing processes are enhanced by ongoing research and development in this field. This work's proportional influence on heavy machining (Figure 10.8) becomes much more substantial because it rises with the material's strength. In any model of chip creation, then, the labor necessary to generate a new surface must be taken into account in hard machining while balancing forces and/or energy.

10.3.4 Requirements for Rigid Processing

Models created using multibody tools in a real-time virtual environment require a trade-off between computational efficiency and accuracy. When complex multibody models come into contact with other parts of the system or with external surfaces, they are often deformed to some degree. Different modeling assumptions may be applied, resulting in different types of models depending on the intended use of the virtual environment. Although finite segments and methods offer computationally efficient ways to represent both large and small deformations, they are unable to convey the stresses and strains experienced by deformable components. Finite element analysis is a reliable method for describing the deformation of a system, including stress and strain fields, but requires more expensive computations. Appropriate contact models must be included in systems where touch or impact scenarios occur [31–34]. However, the driving force (friction changes) in the majority of sliding systems is frequently unknown or needs to be discovered by modeling. Frictional vibrations can, at most, be minimized; its frequency can be raised out of the undesirable range by altering the dynamics of the system.

10.4 Operations for Hard-Machining

The final manufacturing procedures of parts and components typically involve complex manufacturing such as boring, and drilling, and the required dimensions and surface finishes typically require stringent manufacturing standards and conditions.

10.4.1 Hard Turning

Hard turning (Figure 10.10) has the obvious benefit of eliminating grinding processes. The same machine, which "softly" spins the part before hardening it, enables this productivity and maximizes the equipment's effectiveness. However, some stores commit the error of initially selecting the incorrect tool insert for the job (one that is less expensive). Tool steel, bearing steel and case-hardened steel are examples of commonly used hard materials [34], this is untrue since these materials have hardness's that are substantially lower than 47 HRC, which means that the mechanism for how chips form during the operation, including the materials used for tools, differs greatly the best process predictability, from a metallurgical perspective. As previously stated, the L/D ratio for unsupported workpieces should generally not exceed 4:1. When cutting long and thin pieces, the tailstock is supported, but excessive cutting pressure can create vibration.

The level of turning accuracy is determined by the rigidity of the machine. Most modern machines have the rigidity needed to handle tasks that require heavy turning. In many cases, the overall condition of a car is more important. However, machine tool rigidity becomes an increasing concern as required part tolerances and surface finishes increase. The best option is to use specialized equipment and turning centers designed for serious machining. However, the secret to success in turning is not the rigidity of the machine, but the system. Unfortunately, the industry does not fully understand this simple concept. To maximize system rigidity, all overhang, tool, and part extensions must be minimized, such as by

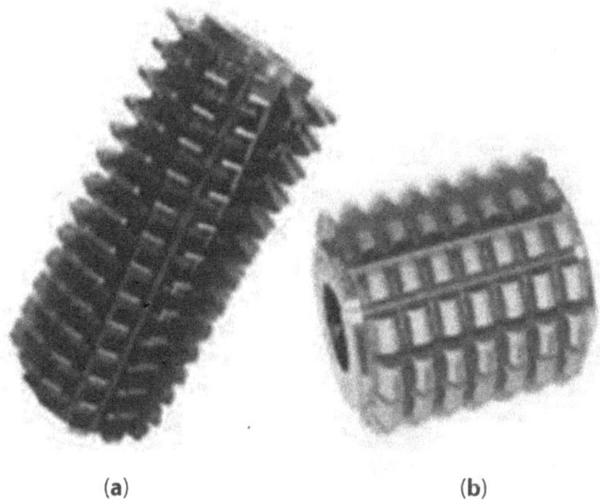

Figure 10.10 Carbide cutters (a) and two-tooth cutters (b).

eliminating spacers and shims. As already stated, the use of coolant is one of the major considerations when designing a turning process. Because dry cutting has obvious financial advantages, most turning operations are done dry. Parts can become hot, making it difficult to measure and handle components. Additionally, a surge in cherry red chips can lead to a number of other problems. Thus, high-pressure coolant through the tool is ideal for cooling the workpiece and hot chips while maintaining small chip size and shape control when coolant is required.

In light of this, if coolant is used, for obvious reasons it should be of low-concentration and water-based. When crushed by hand, the chilled chips practically fall apart, proving that the chips produce the right amount of heat. Depending on the required precision and surface polishing, the correct tool material (PCBN, cermet, or ceramic) must be selected to ensure process efficiency. Different tools, cutting inserts and tool manufacturers offer different brands. Optimal machining conditions, tool holders and insert shapes only complicate the multi-parameter optimization task.

10.4.2 Reaming and Hard Boring

Reaming is a widely used method for making holes in manufacturing. When it comes to machining smaller holes, reaming is more practical and cost-effective than internal grinding and fine boring. Boring is a machining technique in which a pre-made hole is enlarged using a cutting tool. In situations where deformation and dimensional changes caused by heat treatment create undue constraints in the "soft stage" of machining, hard boring will be a more economical alternative that reduces waste.

A brief description of hole machining includes boring, reaming, drilling, and honing. Adding additional cutting tools, more cutting tools select "More Cutting Tools" and we'll be here to support you every step of the way through the metal cutting process! In addition to being important surfaces on components such as discs, rings, couplings, brackets, and bushings, holes are also common surfaces encountered in the machining process. Cutting holes is a more complex task than cutting cylindrical surfaces while maintaining the same standards of accuracy and surface finish, resulting in lower productivity and higher costs. Figure 10.11 shows typical boring and reaming equipment for hard machining. Ceramic, metal-ceramic, or PSBN plates are attached to the tool body directly or in parts. Multi-bladed tools are used to achieve high performance. However, single-edge tools with backing plates are used when the hole being machined must have a high degree of accuracy. Clamping tiny inserts onto tiny boring bars becomes

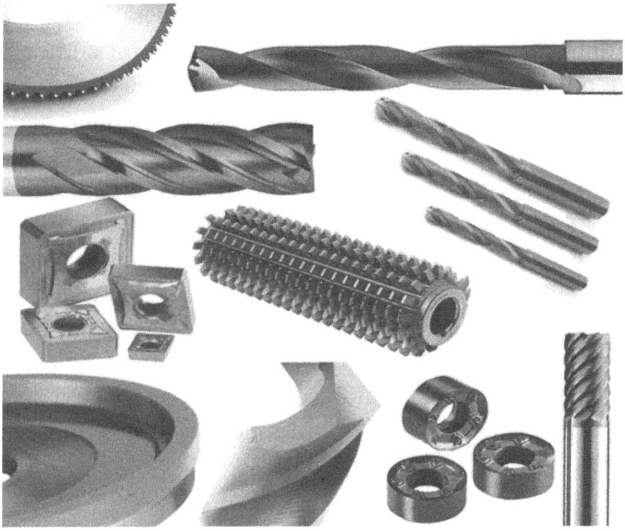

Figure 10.11 Tools with holes in them and a boring tool design for challenging machining.

more challenging for holes smaller than 12 mm in diameter. To solve this problem, a soldering tool is used.

10.4.3 Milling of Hard

Businesses frequently concentrate on creating dies and dies, which are used to cut materials like P20, H13, W5, and S7. Solid rods and cavities have traditionally been produced from these materials using electrical discharge devices. These materials can now be produced directly into hardened materials, thanks to the development of new technologies over time. This is often accomplished by utilizing new toolpath techniques and hard milling processes. Prominent mold manufacturers have realized that, in an increasingly competitive global environment, adjusting to new technologies may be essential to their survival. According to Zurek [39], the deployment of a system consisting of the system's most important element is the machine. The machine needs to be made for hard milling and also has some high-speed machining center features.

Hard milling demands a machine's fundamental design as well as specific machine parts like the motor, spindle and CNC. It is a crucial to have a study base with adequate vibration-dampening properties. Because polymer concrete bases often have vibration dampening that is 6 to 10 times better than cast iron, they are a viable option for high-speed, heavy-duty milling. Digital drives' quick acceleration and deceleration capabilities and superior contouring accuracy aid in cutting down on cutting tool wear. Depending on the application, each of these tools has different strengths and limitations. End mills made of solid carbide are often highly expensive, precisely ground and coated. Cutter sets with interchangeable carbide inserts are a second category of carbide cutters. They are not ideal for harsh milling tasks. Ceramic replacement inserts are a third type of strengthened ceramic inserts with whiskers. Faster cycle times and fewer operations per item are advantages of employing a milling system with changeable ceramic inserts. Using interchangeable ceramic inserts, ceramic cutters can safely and quickly grind everything from big face mills to small diameter end mills. Modern whisker-reinforced ceramics may run at speeds much above the point at which they melt, since they have a melting temperature of over 2000°C that fails with carbide tools.

10.4.4 Broaching of Hard

The broaching operation is the most cost-effective way for mass-producing parts with complex features in the field of metal cutting where high production rates and great component stability are required. One motion can be used to draw the part's profile. While exterior broaching is intended to machine the surface profile, internal broaching begins from a hole that has already been pre-machined. Both horizontal and vertical directions are open to firmware. Both linear and helical cutting movements are possible. It is feasible to use one of two rigid broaching techniques with this tool configuration:

- The cutting speed during typical broaching procedures is known to be between 0.5 and 6 m/min. Since that time, researchers and professionals have focused on rigid broaching's effectiveness and dependability, as well as productivity and surface quality. The outcomes thus far are outlined below:
- When it comes to the quality and dependability of broaching operations, coated carbide broaches, intended as side-cutting tools, perform well, obtaining IT profile quality of 7-8.
- When utilizing carbide broaches, an efficient cutting speed is identified as a crucial component in the broaching process. When compared to green (soft) feed, the cycle time at this speed is shorter since the feed stroke lasts between one and two seconds.

10.4.5 Operations for Manufacturing Hard-Gear

Cutting, forming, grinding, and heat treating are the four main steps in the gear manufacturing process required to produce gears that meet these requirements. Using techniques such as hobbing, forming, milling, and broaching, gear cutting produces precise tooth shapes on gears. One of the most important technologies in the production of gears is abrasive processing. Over the past 20 years, among the methods of abrasive processing of hardened gears, operations with cutting speeds of only 0.3–5 m/s have gained importance. Much work has been done to make these low-speed grinding operations more productive. The development of new abrasive tools is an integral part of this work, requiring a large amount of empirical research and, therefore, a significant investment of time and money [3, 4]. The method allows you to change the geometric surface of the side surfaces of the teeth and improve the surface quality. In addition, the noise level of the gearbox is reduced due to the characteristic surface roughness of the final gears. The most common process and wear pattern of milling tools is the development and propagation of cracks caused by thermal fatigue (high processing temperatures). The origin and wear patterns of coated tools are still the subject of recent research as they have a direct impact on tool life. Coating wear studies can help develop new coatings and improve processes.

Recent research has shown that using these methods can extend the life of coatings, and improve surface quality and overall process efficiency. However, machining with rigid gear machines places unique demands on the tool, machine, and overall process. Understanding these prerequisites is necessary for the successful implementation of this process. High-speed steel is the material most commonly used in the industry today to make cooktops. In addition, final machining cannot be done on a high-speed steel milling machine. Even these relatively soft materials, but this can only go so far before additional finishing with other dental instruments such as razors is required. The main reason is the incorrect angle of the lateral surface along the tooth profile. To improve the characteristics of finished surface, the cutting wedges of a form-cutting tool are often created with an inclination angle of 3–4° [19–22]. Since the tooth profile must be maintained within a limited tolerance when a cutter is sharpened repeatedly, substantially smaller flank angles are used in hob tools. Due to the high yield strength and high elasticity of processed materials, when working with hard materials, excessive friction occurs on the side surface. The workpiece and flange stick together due to high temperatures caused by this friction, reducing surface quality and shortening tool life. A carbide cutter with replaceable carbide inserts is shown in Figure 10.10 next to the cutter used for tiny modules. Tools with holes and a boring tool design for difficult machining are shown in Figure 10.11.

10.4.6 Skiving

10.4.6.1 Turning of Hard

Turning is a very effective and simple machining method that produces high-quality surfaces for rotating body parts made from a variety of materials. Processing efficiency for each unique situation, the best combination of workpiece and tool materials and cutting parameters is determined. circumstances, geometric characteristics of the cutting tool, condition of the apparatus, and other technical data. Machine parts consisting of hardened steel annealed and special cast iron, and other difficult-to-cut materials are successfully machined. The use of CBN tools for fine turning of workpieces composed of these materials often avoids the need for grinding, the traditional method of producing surfaces with a roughness less than Ra = 1.25 μm. Currently known as "hard turning," this processing method involves machining materials with a hardness greater than 47 HRC [11–14]. Figure 10.12 depicts the "sharpening" turning model proposed by Klimenko and Manokhin [35].

10.4.6.2 Skiving of Gear

Polishing hardened gears is known as gear hobbing. Using this method, errors in the gear ends and profiles associated with heat treatment are eliminated.

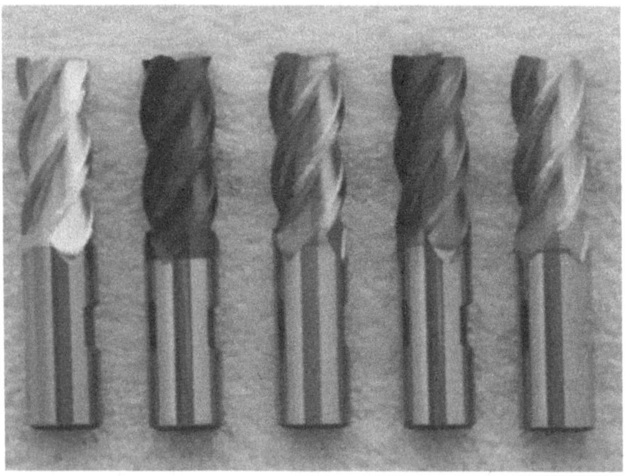

Figure 10.12 Characteristics of different coatings [8].

The end result is increased transmission accuracy. Hardened gears with a hardness of up to 63 HRC can be processed with carbide cutters [41–44]. The development of extremely sophisticated automated machine tools has made this possible, making the approach efficient and practical [45].

Figure 10.13 shows the hobbling of hardened spur gear with a diameter of 16 mm. The body of the worm gear tool is made of high-speed steel, with carbide cutting inserts soldered onto it. Each side surface of the gear is sharpened by removing a 0.5 mm thick allowance in one pass. Benefits of this technology include the ability to change gear flanks, such as cone and crown, over a range of 100 m or more in just one pass, as well as dramatic reductions in tooling costs and cycle times. The character has achieved DIN 8 quality, which is the most accurate indicator of process quality. Figure 10.13 illustrates the unique tooth geometry of a cutter. Many scientists believe that the main difference between sharpening teeth with worms and cutting teeth is the negative rake angle of the tool in the hand, caused by the displacement of the front surface of the tooth [43, 44]. The carbide cutter appears to be protected from impact and sudden overload by a negative rake angle. However, this is not entirely true as process modeling is sometimes too confusing, and important variables are ignored because the geometry of the tool teeth in the machine tool is not taken into account [17–19].

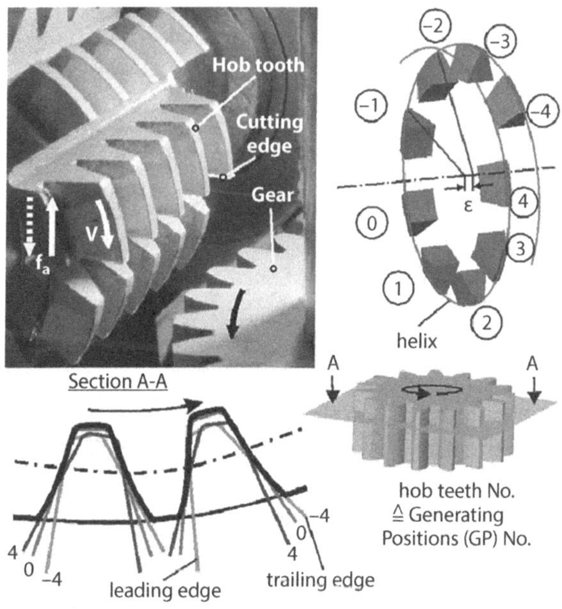

Figure 10.13 The tooth geometry of traditional and skiving hobs is compared.

10.5 Technology of Coating

High-performance carbide-coated tools are the result of significant advances in coating technology. This obviously exhibits improved qualities such as wear resistance, toughness, and hardness. Chemical coating techniques have advanced of new tactics and solutions for the selective coating of complicated surfaces. To enhance the performance of sol-gel coatings, new materials have been created and hybrid organic-inorganic coatings investigated [23, 24]. Developments in etchant formulation, process control, and masking techniques have proven advantageous in achieving high accuracy and reproducibility in chemical etching procedures.

10.5.1 Backdrop

This is cutting tools with hard protective coatings several micrometers thick. Up to 1980s, TiN, TiCN, and Al_2O_3 produced by CVD are used to make carbide tools. However, TiN is the first PVD coating to be used by cutting factories in 1980, followed by TiAlN [24]. Currently, much work is being done to improve cutting performance (CVD coatings) [25]. These layers can serve as a bridge between the desired characteristics of the top layer and the interfacial adhesion mechanism. Due to the increased friction created by this hardening property between the insert tip and the workpiece, higher temperatures are encouraged, resulting in diffusion and oxidation [44, 45].

10.5.2 Significance of Coating Layer

PVD and CVD coatings are applied using multiple cathodic arc deposition [46]. PVD instruments are more fragile than CVD instruments. PVD-coated tools offer more finishing options. On the other hand, CVD tools show better potential for roughing [47]. Substrates and layers of PVD coatings are being studied [48]. Tools coated with CVD (moderate temperature) materials are more durable and less likely to break [49]. Numerous studies have been conducted on different types of coatings and their application in specific situations [46, 48]. Figure 10.13 shows the characteristics of various coatings. Mechanical properties are determined by the properties of the substrates of various coatings [49, 36]. TiN is a durable substance that resists corrosion and wear well [50].

For better finishing, a number of coatings are applied to the tool surface, including TiN, TiCN, and TiAlN [20]. The impact of tool coatings on process efficiency is significant. Coated carbide tools can significantly reduce friction [21–23]. Tools used for dry cutting are usually made of carbide [24–25]. The production rate of CVD is higher than that of TiCN (single layer) [27, 37, 50]. Laser surface engineering methods take advantage of advances in laser technology, making surface modification more accurate and efficient [38]. Electroplating procedures have been improved using chemical methods, resulting in improved coating quality, higher deposition rates, and reduced environmental impact [39].

10.6 Future and Current Events

The release of innovative types of tools will occur much faster than in the past, as the time required to produce new tooling material will become shorter and shorter. However, the main obstacles in tool development continue to be hardness, wear resistance, and the balance of strength and toughness, which have been difficult to achieve. As a result, the future tool industry is expected to favor tool materials that combine the hardness and wear resistance of ultra-hard materials with the strength and toughness of high-speed steel and cemented carbide. This trend continues the tradition of creating and studying instruments. The most important point to note is that various coated instruments and composite structures can partially or completely overcome the above difficulties. Thus, tools coated with a composite structure have the best opportunity for development. Tooling materials will be in high demand and will face increasing challenges in the future regarding workpieces and production processes. In addition, the future tool industry will likely be characterized by the emergence of new types, shifts in the factors that determine the market share of each type, and situations where complementary and competing products exist.

10.7 Conclusion

The cutting tools market will focus on meeting the constant need for innovative tooling ideas. It is important to remember that cutting tool wear characteristics, tribological evaluation, tool life experiment, and mechanical behavior are the main factors when using a cutting tool to machine high-strength low-alloy steel. With the current state of coating technology, choosing the right coating structure and mixture can be important. Cutting factories have begun to use cement-coated carbide tools because of their excellent tribological properties, which have attracted the attention of scientists. Research is aimed at developing new coating methods and materials for cutting tools. In the manufacturing sector, economic and environmental factors are leading to ever-increasing cutting speeds, as well as decreasing use of coolants and lubricants. At the same time, the need to extend tool life requires constant further development of carbide technology. A crucial manufacturing procedure that boosts productivity is dry cutting. The tool will function more dynamically due to the continual trend towards better cutting processes and technological features. In general, mechanical engineering relies heavily on the utilization of nanostructured PVD coatings. When turning hardened steel, a modified tool design may be advantageous to provide better finishes on improved tool materials. Advanced tooling materials typically consist of a matrix-base material and a coating material applied on top of the matrix. As a result, extensive research is carried out to select materials suitable for tool bodies and coatings.

Since tool wear is a key indicator of tool life, longer tool life and improved machining productivity can be achieved. The materials from which metal-cutting tools are made must have the following qualities: heat resistance, wear resistance, strength, toughness, etc. Among them, the most important are hot hardness and strength. As a result, the best tool materials are expected to combine strength and hot hardness, another important goal of tool development in industry. Additionally, different tool materials can be used for different purposes due to their different qualities.

References

1. Gupta, K.M., Ramdev, K., Dharmateja, S., Sivarajan, S., Cutting Characteristics of PVD Coated Cutting Tools. *Mater. Today Proc.*, 5, 11260–11267, 2018.
2. Hosokawa, A., Hoshino, G., Koyano, T., Ueda, T., Cutting characteristics of PVD-coated tools deposited by filtered arc deposition (FAD) method. *CIRP Ann.*, 67, 83–86, 2018.
3. Grilli, M., Bellezze, T., Gamsjäger, E., Rinaldi, A., Novak, P., Balos, S., Piticescu, R., Ruello, M., Solutions for Critical Raw Materials under Extreme Conditions: A Review. *Materials*, *10*, 285, 2017.
4. Singla, A.K., Singh, J., Sharma, V.S., Processing of materials at cryogenic temperature and its implications in manufacturing: A review. *Mater. Manuf. Process.*, *33*, 1603–1640, 2018.
5. Bobzin, K., High-performance coatings for cutting tools. *CIRP J. Manuf. Sci. Technol.*, *18*, 1–9, 2017.
6. Ryoo, K.C., Song, Y.J., Kang, D.H., Jeong, C.W., Kong, J.H., Oh, J.H., Lim, D.W., Park, S.S., Kim, J.I., Kim, J.H., Park, J.H., Oh, Y.T., Kim, J.S., Shin, J.M., Park, J.H., Lee, K.W., Fai, Y., Koh, G.H., Jeong, G.T., Jeong, H.S., Kim, K.,

Optimization of Ring Type Electrode Process for High Density PRAM. *Ext. Abstr. 2006 Int. Conf. Solid State Devices Mater.*, 2015.

7. Byrne, G., Ahearne, E., Cotterell, M., Mullany, B., O'Donnell, G.E., Sammler, F., High Performance Cutting (HPC) in the New Era of Digital Manufactoring–A Roadmap. *Procedia CIRP, 46*, 1–6, 2016.
8. Vanegas, P., Durana, G., Zubia, J., De Ocariz, I.S., Advanced monitoring systems for smart tooling in aeronautical Industry 4.0, in: *Proceedings of the 14th Quantitative InfraRed Thermography Conference*, Berlin, Germany, 25–29 June 2018.
9. Das, S.R., Kumar, A., Dhupal, D., Surface roughness analysis of hardened steel using TiN coated ceramic inserts. *Int. J. Mach. Mach. Mater.*, 17, 22–38, 2015.
10. Cheng, K., *et al.*, Smart cutting tools and smart machining: development approaches, and their implementation and application perspectives. *Chin. J. Mech. Eng.*, 30, 1162–1176, 2017.
11. Bag, R., Panda, A., Sahoo, A.K., Kumar, R., A perspective review on surface integrity and its machining behavior of AISI 4340 hardened alloy steel. *Mater. Today Proc.*, 18, 3532–3538, 2019.
12. Brzezinka, T.L., Rao, J., Paiva, J.M., Azkona, I., Kohlscheen, J., Rabinovich, G.S.F., Veldhuis, S.C., Endrino, J.L., Facilitating TiB2 for filtered vacuum cathodic arc evaporation. *Coatings*, 10, 244, 2020.
13. Yousefi, S. and Zohoor, M., Effect of cutting parameters on the dimensional accuracy and surface finish in the hard turning of MDN250 steel with cubic boron nitride tool, for developing a knowledged base expert system. *Int. J. Mech. Mater. Eng.*, 14, 1–13, 2019.
14. Martinho, R.P., Silva, F.J.G., Martins, C., Lopes, H., Comparative study of PVD and CVD cutting tools performance in milling of duplex stainless steel. *Int. J. Adv. Manuf. Technol.*, 102, 2423–2439, 2019.
15. Airao, J., Chaudhary, B., Bajpai, V., Khanna, N., An Experimental Study of Surface Roughness Variation in End Milling of Super Duplex 2507 Stainless Steel. *Mater. Today Proc.*, 5, 3682–3689, 2018.
16. Roy, S., Kumar, R., Das, R.K., Sahoo, A.K., A Comprehensive Review on Machinability Aspects in Hard Turning of AISI 4340 Steel. *IOP Conf. Ser. Mater. Sci. Eng.*, 390, 12009, 2018.
17. Paulsen, T., Pecat, O., Brinksmeier, E., Influence of different machining conditions on the subsurface properties of drilled TiAl6V4. *Procedia CIRP, 46*, 472–475, 2016.
18. Li, J., Huang, Y., Meng, X., Xie, Y., A Review on High Entropy Alloys Coatings: Fabrication Processes and Property Assessment. *Adv. Eng. Mater., 21*, 1900343, 2019.
19. Zafar, M.Q., Uddin, G.M., Asim, M., Khan, A.A., Tahir, Z.U.R., Hayat, N., Ghufran, M., Jawad, M., Comparative analysis of low-temperature PVD-based TiN nano-thin-film-coated and -uncoated TNMG inserts in dry machining. *J. Chin. Inst. Eng. Trans. Chin. Inst. Eng. A*, 43, 143–152, 2020.

20. Li, S.K., Li, J.Q., Li, Y., Liu, F.S., Ao, W.Q., Dense pure binderless WC bulk material prepared by spark plasma sintering. *Mater. Sci. Technol.*, *31*, 1749–1756, 2015.
21. Badiger, P.V., Desai, V., Ramesh, M.R., Prajwala, B.K., Raveendra, K., Cutting Forces, Surface Roughness and Tool Wear Quality Assessment Using ANN and PSO Approach During Machining of MDN431 with TiN/AlN-Coated Cutting Tool. *Arabian J. Sci. Eng.*, 44, 7465–7477, 2019.
22. Sahoo, A.K. and Sahoo, B., Experimental investigations on machinability aspects in finish hard turning of AISI 4340 steel using uncoated and multilayer coated carbide inserts. *Meas. J. Int. Meas. Confed.*, 45, 2153–2165, 2012.
23. Sun, J., Zhao, J., Shen, X., Huang, Z., Yan, K., Xing, J., Gao, Y., Jian, Y., Yang, H., Liat, B., A Review on Binderless Tungsten Carbide: Development and Application. *Nano Micro Lett.*, *12*, 13, 2019.
24. Dressler, J., Open Source Platforms — Connected Car Becomes A Joint Project. *Auto Tech Rev.*, 5, 36–39, 2016.
25. Natasha, A.R., Ghani, J.A., Che Haron, C.H., Syarif, J., The influence of machining condition and cutting tool wear on surface roughness of AISI 4340 steel. *IOP Conf. Ser. Mater. Sci. Eng.*, 290, 12017, 2018.
26. Ginting, A., Skein, R., Cuaca, D., Herdianto, Pieter, Masyithah, Z., The characteristics of CVD- and PVD-coated carbide tools in hard turning of AISI 4340. *Meas. J. Int. Meas. Confed.*, 129, 548–557, 2018.
27. Bagcivan, N., Bobzin, K., Ludwig, A., Grochla, D., Brugnara, R.H., CrN/AlN nanolaminate coatings deposited *via* high power pulsed and middle frequency pulsed magnetron sputtering. *Thin Solid Films*, 572, 153–160, 2014.
28. Kumar, S., Singh, D., Kalsi, N.S., Surface quality evaluation of AISI 4340 steel having varying hardness during machining with TiN-coated CBN inserts. *Proc. Inst. Mech. Eng. Part J. J. Eng. Tribol.*, 231, 925–933, 2017.
29. Kumar, R., Sahoo, A.K., Mishra, P.C., Das, R.K., Comparative investigation towards machinability improvement in hard turning using coated and uncoated carbide inserts: part I experimental investigation. *Adv. Manuf.*, 6, 52–70, 2018.
30. Szutkowska, M., *et al.*, Fracture behaviour of WC-Co hardmetals with WC partially substituted by titanium carbide. *IOP Conference Series: Materials Science and Engineering*, vol. 329, no. 1, IOP Publishing, 2018.
31. Badiger, P.V., Desai, V., Ramesh, M.R., Prajwala, B.K., Raveendra, K., Effect of cutting parameters on tool wear, cutting force and surface roughness in machining of MDN431 alloy using Al and Fe coated tools. *Mater. Res. Express*, 6, 16401, 2019.
32. Mao, C., Ren, Y., Gan, H., Zhang, M., Zhang, J., Tang, K., Microstructure and mechanical properties of cBN-WC-Co composites used for cutting tools. *Int. J. Adv. Manuf. Technol.*, *76*, 2043–2049, 2015.
33. Fox, R.T. and Nilsson, R., Binderless tungsten carbide carbon control with pressureless sintering. *Int. J. Refract. Met. Hard Mater.*, *76*, 82–89, 2018.

34. Chen, K., Xiao, W., Li, Z., Wu, J., Hong, K., Ruan, X., Effect of Graphene and Carbon Nanotubes on the Thermal Conductivity of WC–Co Cemented Carbide. *Met. Open Access Metall. J.*, *9*, 377, 2019.
35. Fortunato, A., Valli, G., Liverani, E., Ascari, A., Additive Manufacturing of WC-Co Cutting Tools for Gear Production. *Lasers Manuf. Mater. Process.*, *6*, 247–262, 2019.
36. Goel, S., Luo, X., Agrawal, A., Reuben, R.L., Diamond machining of silicon: A review of advances in molecular dynamics simulation. *Int. J. Mach. Tools Manuf.*, *88*, 131–164, 2015.
37. Komanduri, R., Lee, M., Flom, D.G., Thompson, R.A., Jones, M.G., Douglas, R.J., General Electric Co, 1982. Pulse laser pretreated machining, U.S. Patent 4,356,376, 26 October 2020.
38. Kumar, J.P., Kishore, K.P., Kumar, M.R., Karthick, K.R.S., Gowtham, S.V., Experimental investigation on hard turning of AISI 4340 steel using cemented coated carbide insert. *IOP Conf. Ser. Mater. Sci. Eng.*, 314, 12004, 2018.
39. Bensouilah, H., Aouici, H., Meddour, I., Yallese, M.A., Mabrouki, T., Girardin, F., Performance of coated and uncoated mixed ceramic tools in hard turning process. *Meas. J. Int. Meas. Confed.*, 82, 1–18, 2016.
40. Suresh, R., Basavarajappa, S., Gaitonde, V.N., Samuel, G.L., Machinability investigations on hardened AISI 4340 steel using coated carbide insert. *Int. J. Refract. Met. Hard Mater.*, 33, 75–86, 2012.
41. Jena, J., Panda, A., Behera, A.K., Jena, P.C., Das, S.R., Dhupal, D., Modeling and Optimization of Surface Roughness in Hard Turning of AISI 4340 Steel with Coated Ceramic Tool. *Innov. Mater. Sci. Eng.*, pp. 151–160, 2019.
42. Ranjan Das, S., Panda, A., Dhupal, D., Hard turning of AISI 4340 steel using coated carbide insert: Surface roughness, tool wear, chip morphology and cost estimation. *Mater. Today Proc.*, 5, 6560–6569, 2018.
43. Patten, J., Micro Laser Assisted Machining. Google Patents US8933366B2, 13 January 2015.
44. Razavykia, A., Delprete, C., Baldissera, P., Correlation between Microstructural Alteration, Mechanical Properties and Manufacturability after Cryogenic Treatment: A Review. *Materials*, *12*, 3302, 2019.
45. Zhao, T., Zhou, J.M., Bushlya, V., Ståhl, J.E., Effect of cutting edge radius on surface roughness and tool wear in hard turning of AISI 52100 steel. *Int. J. Adv. Manuf. Technol.*, 91, 3611–3618, 2017.
46. Varghese, V., Akhil, K., Ramesh, M.R., Chakradhar, D., Investigation on the performance of AlCrN and AlTiN coated cemented carbide inserts during end milling of maraging steel under dry, wet and cryogenic environments. *J. Manuf. Process.*, *43*, 136–144, 2019.
47. Zhang, L., Dong, S., Du, J., Lu, Y. L., Zhao, H., Feng, L. First-principles forecast of gapless half-metallic and spin-gapless semiconducting materials: case study of inverse Ti2CoSi-based compounds. *App. Sci.*, 10, 3, 782, 2020.

48. Boing, D., de Oliveira, A.J., Schroeter, R.B., Limiting conditions for application of PVD (TiAlN) and CVD (TiCN/Al2O3/TiN) coated cemented carbide grades in the turning of hardened steels. *Wear*, 416–417, 54–61, 2018.
49. Dosbaeva, G.K., El Hakim, M.A., Shalaby, M.A., Krzanowski, J.E., Veldhuis, S.C., Cutting temperature effect on PCBN and CVD coated carbide tools in hard turning of D2 tool steel. *Int. J. Refract. Met. Hard Mater.*, 50, 1–8, 2015.
50. Borji, S., Ahangarkani, M., Zangeneh-Madar, K., Valefi, Z., Corrigendum to "The effect of sintering activator on the erosion behavior of infiltrated W-10 wt%Cu composite" [Int. J. Refract. Met. Hard Mater. 66 (2017) 150–157] (S0263436817300641)(10.1016/j.ijrmhm.2017.03.007). *Int. J. Refract. Met. Hard Mater.*, 74, 123, 2018.

11

Metal Carbides: Tools and Wear-Resaretant Coatings for Machining, Drilling and Mining Operations

Riyadh A. Al-Samarai[1]* and Yarub Al-Douri[2,3,4]

[1]*Electromechanical Engineering Department, Engineering College, University of Samarra, Samarra, Iraq*
[2]*College of Health and Medical Techniques, Al-Bayan University, Baghdad, Iraq*
[3]*Department of Mechanical Engineering, Faculty of Engineering, Piri Reis University, EflatunSk, Tuzla, Istanbul, Turkey*
[4]*Nanotechnology and Catalysare Research Centre, University of Malaya, Kuala Lumpur, Malaysia*

Abstract

This chapter covers cermet, sintered titanium carbonitride carbides, and titanium nitride. When exposed to aggressive, oxidizing, or abrasive environments, a special class of cermets known as TiN and TiCN compounds provides an exceptional combination of high-temperature resistance and hardness. Although TiN and TiCN coatings have several applications, the substrates on which they are applied influence their mechanical and tribological characteristics. The effectiveness of TiCN and TiN coatings on machined tool steel is the main focus of their work. The only innovation is the improvement of high stiffness at low hardness. Assessing improvements in hardness and toughness for use in cutting tools is critical to identifying trends and developments. In addition, it is expected that a very hard, binder-free TiCN phase will be required to overcome the disadvantages of using TiCN as a cutting tool at high temperatures.

***Keywords*:** Cermet, titanium carbonitride (TiCN), titanium nitride (TiN), high-temperature resistance, hardness, tool coatings, tribological properties, cutting tools

**Corresponding author*: riyadh2824@yahoo.com

Inamuddin, Tariq Altalhi and Mohammad Abu Jafar Mazumder (eds.) Nanostructured Hard Materials: Industrial Applications, (343–368) © 2026 Scrivener Publishing LLC

11.1 Introduction

Today's manufacturing industries are constantly improving their tooling and wear-resistant coatings for machining, drilling, and mining applications to enhance equipment performance and reliability and reduce maintenance costs. Manufacturers' technologies that can be used to create better products, remain competitive and ultimately increasing profits. To achieve the mentioned goals, materials should ideally have a number of critical characteristics including excellent wear and corrosion resistance, good hardness and toughness, environmental friendliness, high performance at high temperatures, and relatively low production costs [1]. The part must have high wear resistance and corrosion resistance to withstand long-term heavy wear. A part that performs reliably under such conditions is deemed suitable, contributing to enhanced safety and reduced maintenance costs [2].

While having a material that is highly resistant to corrosion is critical, it is equally important to have a material that is mechanically strong. Numerous types of metal-ceramic systems have become the subject of in-depth research. These include titanium carbonitride and titanium carbide (TiC). Ceramic stages (Ti(C,N)). The purpose of conventional "carbide" compounds, consisting of tungsten carbide (WC) embedded in a cobalt link, is the same as that of cermets, but their high density and relatively low resistance to oxidation and corrosion preclude their use in most applications. structural applications involving high temperatures. With a combination of high strength, toughness, and hardness that provides good wear resistance, they are mainly used for cutting and machining [3–8]. Metal carbides and nitrides have historically been prepared using a variety of methods, with solid-phase reaction methods being the most efficient and inexpensive methods [9]. The hardness of material and cermet can be adjusted depending on Ti(C, N). There is a large amount of TiC and TiN due to the hardness and toughness increase with increasing TiC content and TiN ratio, respectively. Nickel, cobalt, stainless steel and nickel aluminide (Ni_3Al) are typical metal binder materials used in these cermets. Due to its high heat resistance and good strength retention at high temperatures, Ni3Al is of particular interest [11–13]. However, TiB2 low self-diffusion coefficient and poor fracture toughness make it difficult to compact, limiting its application [4]. Among other benefits, the combination of TiB2 and AlN provides increased impact resistance due to unique strengthening mechanisms. When it comes to hardening ceramic materials, these mechanisms are critical [14–16]. A class of extremely hard, refractory, wear-resistant alloys produced by powder metallurgy are known as carbides and cermets. These materials are widely used as high-pressure components, cutting tools, wear parts in mining and industrial applications, and drill inserts. Properties including toughness, microstructure, hardness, strength, corrosion, and wear resistance have a significant impact on component life. While the binder (or matrix) is the dominant phase in MMC, the hard carbide phase predominates in carbide alloys and cermets. The long-term plan has two parts. First, the relationship between properties, microstructure, composition, production conditions, carbides, and cermets may perform better. Second, to gain new insights and expand potential markets for these materials, new carbides, ceramics, and hard coatings are being developed using primary resources from South Africa [1]. Co3O4 and WO3 [2–4], two wear products of WC-Co, are registered under REACH, the European Union's system of chemical registration, evaluation, authorization, and restriction [5]. A significant barrier to thermal and mechanical applications of ceramic materials is their low fracture toughness, as shown by KIC tests [2]. Among these applications, objects of high hardness, such as those made of hardened steel, require the use of ceramics exclusively as cutting tools.

11.2 Structural Formation

11.2.1 Nitride Titanium (TiN)

TiN can be produced by reducing the proportion of "rutile" variety of titanium dioxide in the ore. Large homogeneous mononitride and nitride forms of various stoichiometries, including sodium nitride (Na3N), have been discovered. For example, in the TiN0.1 and TiN0.5 phases, the nitrogen atom in titanium nitride can be easily replaced by an oxygen or carbon atom in the interstitial position. There results in the formation of individual atoms (C, O, and N) as shown in Figure 11.1 and in Figure 11.2, TiN is remarkably stable at temperatures below 1300°C. The mechanical properties of product materials, in particular their hardness, are improved. Drilling, turning, and milling are examples of older or traditional machining operations in which the hardness of the cutting tool exceeds the hardness of the material being cut. Due to progressively decreasing hardness, cutting tools used in various machining operations have a limited-service life, which increases the cost of machining [4–7].

For example, when the temperature increased to 500°C, the curve took a parabolic shape [20]. Refractory carbides and nitrides like titanium carbide and titanium nitride are typically not found in normal conditions. They frequently take a long time and cost a lot of money to install. These materials

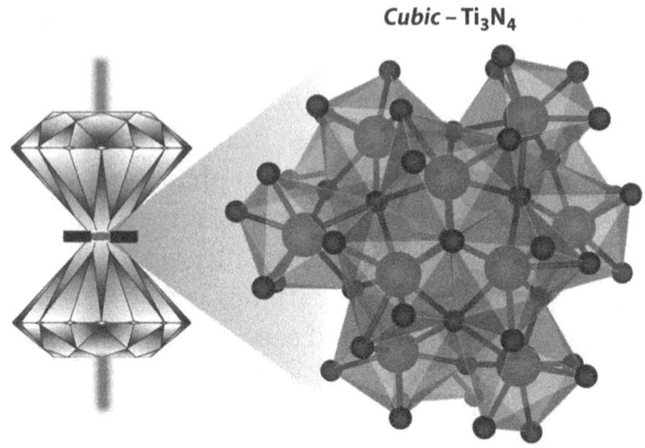

Figure 11.1 Titanium nitride's (TiN) crystal structure (VESTA software are used to calculate the crystal structure of titanium nitride) [17].

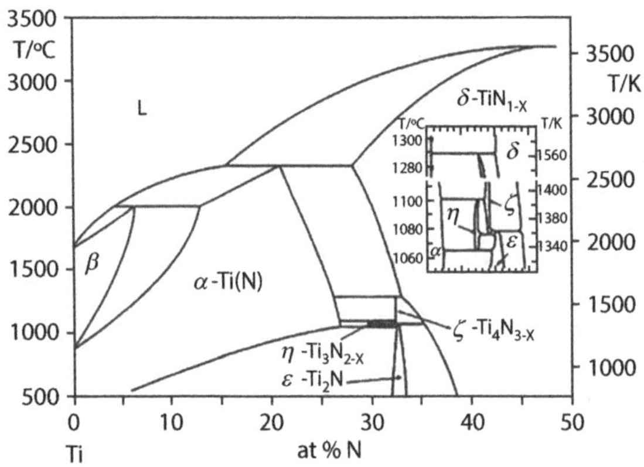

Figure 11.2 Titanium Nitride (TiN) phase diagram, reproduced with consent [17].

feature robust covalent bonding due to their excellent heat resistance and inertness. Therefore, it cannot be acquired through conventional means. This rescue prevented it from being widely used before now [21].

11.2.2 Carbonitride of Titanium (TiCN)

Refractory carbides and nitrides are commonly found in powders, monolithic structures, coatings, fibers or filaments [19]. However, later in studying powdered TiC, TiN, and Ti (C,N) materials that are currently produced using different processes will be considered. Due to its wear resistance and the ability to be used in the creation of complex ceramic composites or refractory materials, TiCN is widely recognized as an interesting material for cutting tools [22–24]. The wide dispersion of titanium carbonitride ceramic particles may be caused by the presence of carbon or nitrogen atoms in the interstices. The carbon and nitrogen atoms are randomly contributed in the octahedral lattice site. Cubic structural categorization TiCN are still face-centered cubic, in contrast to TiN and TiC [24–26]. The elements can be directly carbonized or nitrided to produce TiC, TiN, and Ti(C, N). Because it requires energy, this process consumes a lot of energy. The time ranges from 5 to 20 hours, with temperatures ranging between 1900°C and 2900°C [27]. Finally, Finarhyd is a very clumpy powder that needs to be ground more finely. To remove contaminants, additional chemical cleaning is required due to the small particle size. This is constructed from milling machines.

As a result, the procedure is very expensive due to the cost of pure titanium and because it uses a lot of energy [18], which leads to an ordered arrangement of titanium atoms and a bold formation in the presence of N and C atoms.

11.3 Coating of TiCN and TiN

The creation of titanium nitride, as well as their various applications in engineering services, are now being studied by researchers. Due to its adaptability in a wide range of study domains, solid coatings continue to be the center of attention [1–3]. The mechanical compatibility of coating materials like titanium nitride (TiN) and titanium carbonitride (TiCN) with a variety of metal substrates has garnered a lot of interest. However, due to variations in mechanical properties, particularly tribological and mechanical features [1–6], the manufacturing of these ceramics on metal substrates presents difficulties. As a result, extensive research has been done to enhance the qualities of these coatings and broaden their applicability [32, 33]. To improve mechanical qualities, particularly during long-term high-speed machining, cutting tool development is essential. However, there are a number of issues with how cutting tools are made. Prior to being used, the material must fulfill the standards for the required mechanical qualities. Second, the cost of the finished product is a reflection of the limits of traditional sintering techniques, which include the quantity of materials used and the time needed to finish the sintering cycle. This presents significant challenges for the manufacture of these instruments [8]. Furthermore, the materials used to make cutting tools need to be able to withstand abrasive cutting environments, like high temperatures and friction between the cutting tool's surface and the workpiece. Cutting tool materials need to have several characteristics to achieve these criteria [9].

TiCN coatings provide strong mechanical and wear-resistant qualities as well as great thermal stability. They also demonstrated minimal internal stresses and great resistance to corrosion [4, 6, 7]. Additionally, are still utilized today to safeguard mechanical components and the materials used in cutting tools [28–31] due to its resistance to damage during transportation [5, 8–10].

11.4 Challenges and Restrictions of TiN and TiCN

Despite this, the service life of H13 tools is impacted by a number of failure modes because of the harsh operating circumstances, including high load, temperature, and stability time. Thermal fatigue and abrasive and adhesive tool surface degradation are the principal effects of such working circumstances. As a result, working tools must be replaced after a given amount of time, which has an impact on the total cost of production operations [12, 13]. Only the stoichiometric arrangement of these components determines the melting point of these composites [38–41]. Refractory linings for furnaces, cutting tools and hypersonic vehicles, as well as molten metal crucibles are reported to be manufactured with these defining characteristics from TiN/TiCN. TiN/TiCN which has disadvantages in all applications, including low fracture toughness and poor oxidation resistance. In addition, important elements such as a long sintering period can limit grain formation and increase particle size. As a result, the performance and overall efficiency of the sintered body may also be reduced indirectly [42–47]. In any case, it is critical to improve the synthesis of TiN and TiCN composites, and properties such as tensile strength, elastic modulus, fracture toughness, and oxidation resistance that require urgent research [35, 36].

11.5 Treatment Techniques

These pertain to morphology of materials and components or microstructural integrity. Together, they influence the processing and performance of composites [34]. Since they are both extremely high-temperature ceramics, the processing methods and sintering [48, 49]. Spark plasma (SPS), pseudo-thermal pressure (PHIP), quasi-pneumatic pressure (QIP), and reaction spark pressure. For some modern applications such as semi-finished manufacturing and final manufacturing, TiCN-based cermet is a promising material [50, 51]. However, its low wear resistance does not allow it to be used in the future, for example, in high-speed cutting industries. Due to the increased hardness and the absence of a ductile metal binder phase contrasting with the entire cermet, the surface layer containing a large amount is called wear, as it can lead to improved cutting operations when the carbide phase is formed in the nitride environment layers of repetition [52–55].

Carbide-containing cermets are often manufactured mostly from titanium carbides, tungsten carbides, and possibly cermets, or solid materials based on titanium and tungsten carbides, in small quantities [55, 56]. Therefore, the harder the material, which reduces the wear rate and increases wear resistance, the greater the number of binder phases [56].

11.6 Factors That Improve Sintering Kinetics

- In comparison to other conventional sintering techniques, spark plasma sintering's diffusion kinetics are primarily improved. It is widely known that using pressure to produce plastic can increase the compaction of any composite specimen [57].The high pulsed energy source is what causes the spark plasma sintering to generate [58, 59]. According to studies, applying a continuous pulsed current is essential for achieving significant densification of any sintered product in IPS [60].
- According to a recent work by Chandran *et al.* [61], high heating rates are necessary to produce the best densification when producing nanocrystalline Al_2O_3 composites from the resultant powders [62–66].
- Additionally, studies have revealed that DC pulsing increases the titanium carbonitride's degree of condensation.

11.7 Binderless

Binder-free WC, in particular, exhibits exceptional wear resistance compared to titanium carbonitride and some other materials. Binderless carbide systems are beneficial and important in many manufacturing industries, such as high-speed cutting tools, etc., due to their exceptional wear resistance [72, 75]. Bondless carbides and cermets are gradually replacing traditional cemented carbides in industry due to the deterioration of the binder phase in some harsh environments. Bondless WC-MoxC carbides are the most productive and have good combination properties [79].

In addition, fracture toughness is reduced compared to previous sintering methods. Figure 11.3 Grain distribution of metal and ceramic materials is used at high temperatures, either with or without binders [78]. It has two intertwined fcc lattices, one of which has a comparable crystal structure and is composed of carbon and metal. Two-dimensional (2D) nanomaterials have garnered a lot of interest as a novel kind of nanomaterial. There are ultrathin 2D nanomaterials that have a special planar structure, in which the side measurements are larger than sub-microns including its lengthwise dimensions. Additionally, it is restricted to atomic or nanoscale levels. Frameworks are typically a bond created by combining d orbitals with 2s and 2p orbitals.

The structure of sp orbitals changes with increasing concentration of electron clouds. The bulk modulus and lattice constants of the carbides are theoretically analyzed and correlated using the Murnaghan equation. Building on individual two-dimensional transition metal atoms, this study explores 66 transition metal carbides. Owing to their high diversity, over 800 distinct computational configurations of MAX phases have been identified.

Zhou *et al.* [55], have conducted analysis of effect of toughness at different metal of Ti(C,N). It has been demonstrated that when using metal bonds ranging from 4–5 and 2–3 μm, respectively, the fracture toughness of TiCN approaches 20% and 25%. Consequently, as the proportion of binder metal increases, these values begin to decrease.

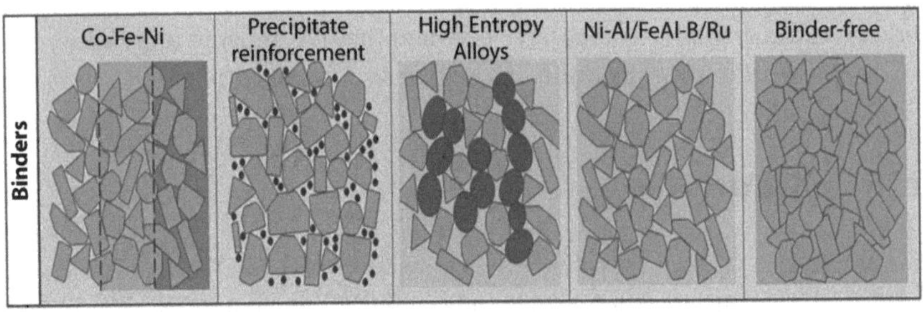

Figure 11.3 High-temperature granular dispersion of ceramic, with or without binders [70].

11.8 Prior Studies on Titanium Carbonate Nitride and Titanium Nitride without a Binder Used Sintering Techniques

Unbonded carbides are developed to replace the traditional carbides and supplement the old typical carbide cermet composition [79]. Interestingly, using outdated conventional sintering techniques, it is quite challenging to fuse any materials including tungsten carbides with titanium carbides [68, 69]. Due to its exceptional qualities like refractoriness, as well as thermal, Rosias, Cardinal and Young.

Superior tolerance control is possible with titanium carbonitride cermet enhanced surface quality, feed rate, and any complex machine part [72, 73]. Hard coatings made of Ti(C,N), TiN and TiC are applied to the surface of hard alloys. These coating materials have recently undergone modifications to better suit particular applications in sectors like [64]. The various structural distributions of hard alloys are shown in Figure 11.4 The majority of materials used in friction and wire drawing applications develop these microstructures. Different grain is produced by various reinforcing chemicals buildings, as depicted in Figure 11.5. However, it has been demonstrated

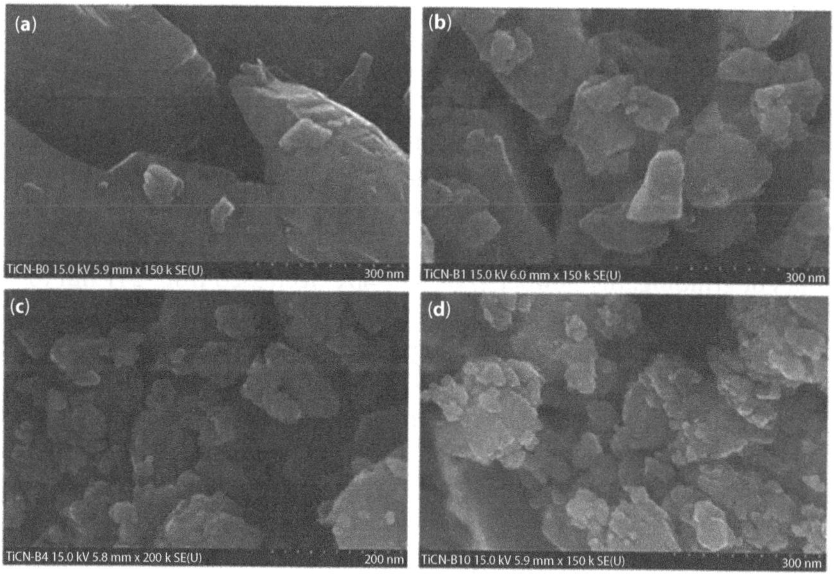

Figure 11.4 FESEM microstructures with different grinding times on TiCN without binder, including (a) as received, (b) 1 hour, (c) 4 hours and (d) 10 hours [70].

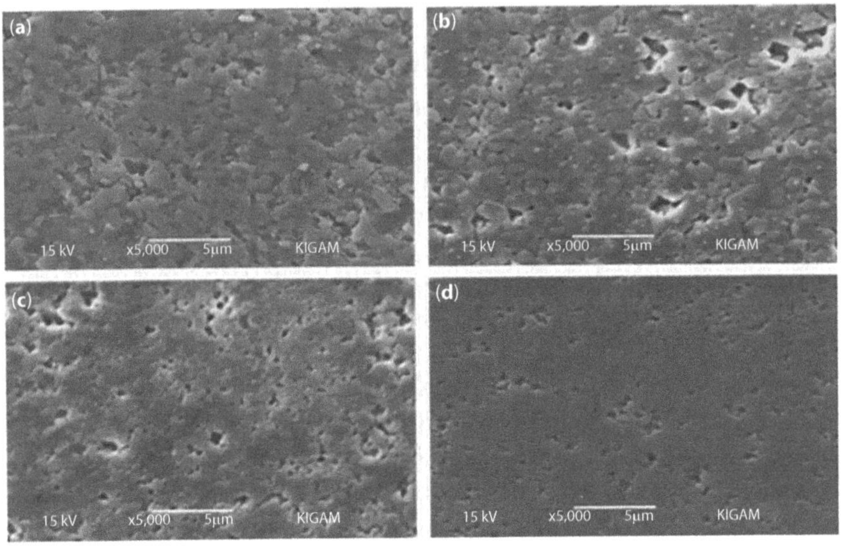

Figure 11.5 FESEM microstructures on sintered TiCN without binder after milling for (a) 1 hour, (b) 1 hour, (c) 4 hours, and (d) 10 hours [70].

that TiC inclusions can enhance the processing capabilities of high-speed cutting tools [75].

Zhang *et al.* [72], investigated the mechanism of tungsten carbide created during the initial alloying with 0.5% Cr3C2-0.3%VC. According to Kim *et al.* [73], binder-free ($TiC_{0.5}N_{0.5}$) titanium carbonate nitride, which are ball at different times and then sintered at 1170°C using high-frequency induction heating methods, showed good results. The compaction process has been significantly improved by developing the crystalline structure of the ball milled powders. Shredding is reported to be decreasing. After 10 hours of grinding, the powder crystallite size decreased from 241 nm to 9 nm as illustrated in Figure 11.6.

After sintering, the rapid sintering procedure preserved the fine grain microstructure. The relationship between variable crystallites and relative density and fracture toughness are insignificant. It is estimated that the crystallite size decreased by 231-9 nm after treatment for 10 hours. HFIHS, sometimes called fast combined high frequency induction thermal sintering [74, 75]. After sintering, the process retained a fine microstructure with acceptable mechanical properties, as shown in Figure 11.7 Graphite. According to the results, 99% compaction are achieved after 4 minutes of exposure.

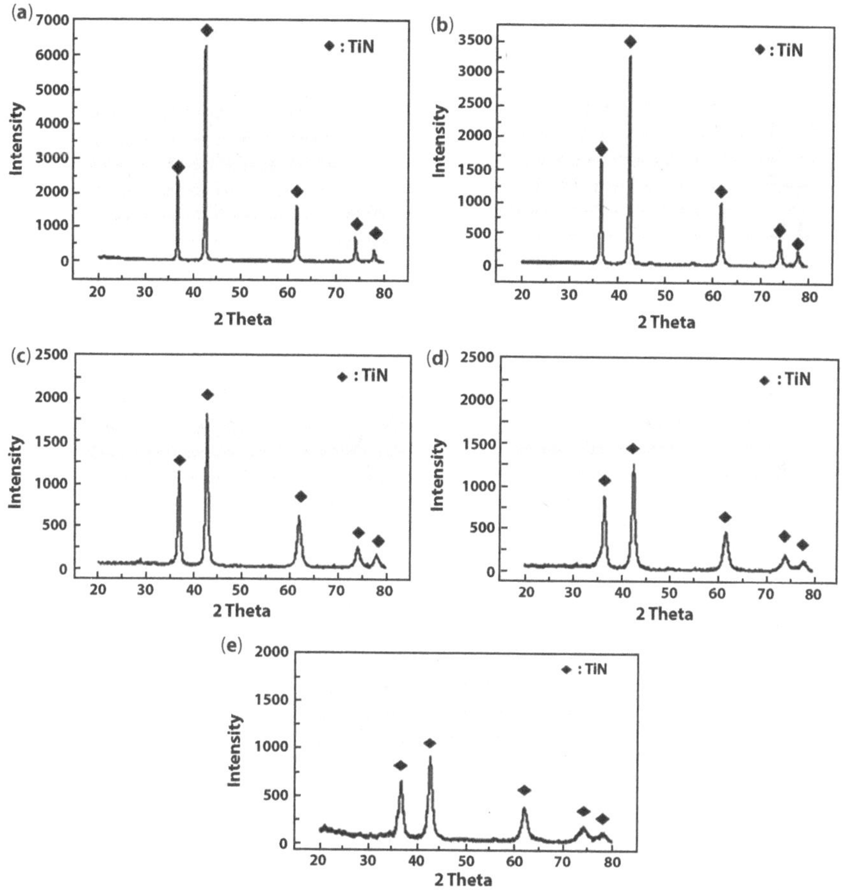

Figure 11.6 The XRD patterns below illustrate titanium nitride to milling durations without the use of binders: Cp-TiN, TiN-1 h, TiN-4 h, TiN-10 h, TiN-40 h, and TiN-1 h [70]).

Relative density and microhardness are in direct inverse relationship, since the grinding time reached 40 hours. Using HEBM high-energy ball milling technology, the microstructure of binder-free titanium nitride particles has been thoroughly improved, as shown in Figure 11.8. The X-ray patterns show after 40 hours of grinding, broadening due to deformation of the crystallites and a decrease in crystallite size, noticeable after 1 hour of grinding. The ball mills used have, due to the constant contact of the granules with the powder in the grinding bottle, a tendency for unwanted particles or contaminants to enter. In Figure 11.9, it clearly shows the X-ray of grinding. The peak broadening of titanium nitride decreases as a result of grain development during the sintering process.

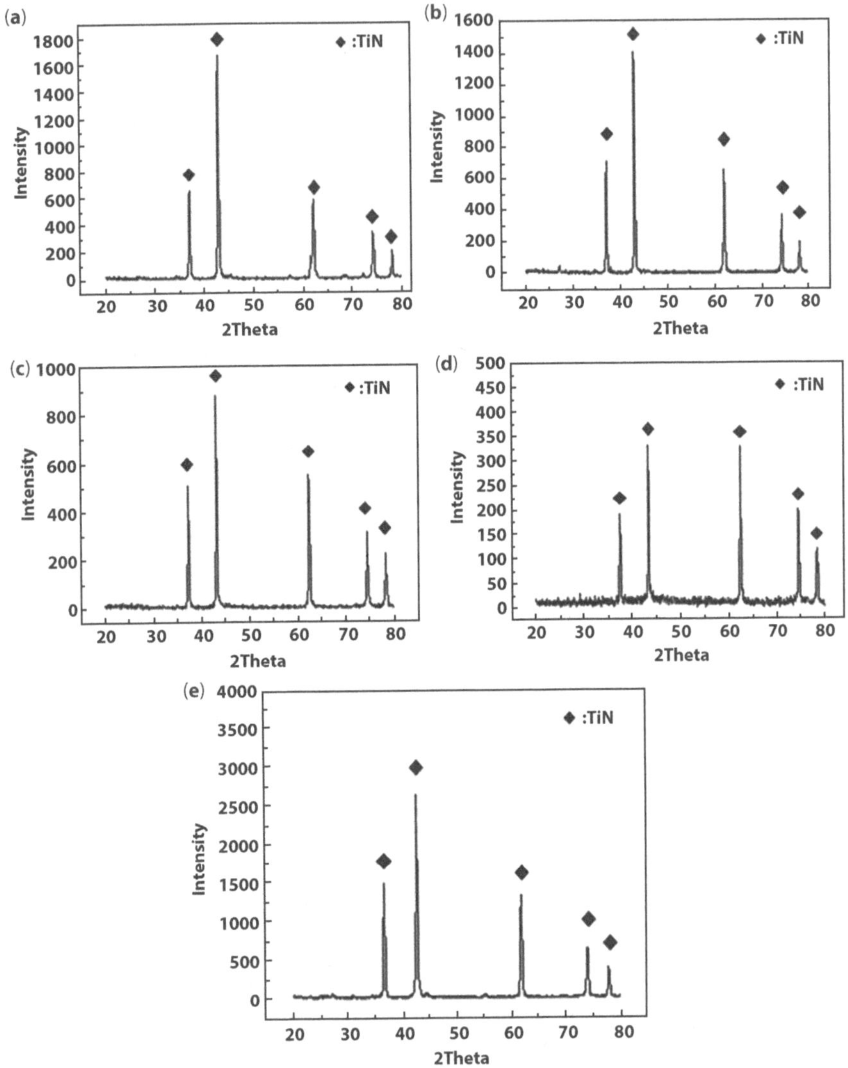

Figure 11.7 Demonstrates the results through the X-ray diffraction patterns of titanium nitride to milling periods (without binders): 40 h, 40 h, 10 h, 1 h, and Cp- TiN [70].

As grinding time progresses, as shown in Figure 11.8. The formula developed determines the final size of each particle. After long milling segments of 1 hour, 4 hours, 10 hours, and 40 hours, Norton determined them to be 56 nm, 21 nm, 15 nm, and 10 nm. When TiN is added, the concentration of Mo in the bond phase increases, which greatly increases the corrosion resistance of cermet. The grain size and edge thickness of solid particles generally

Figure 11.8 Titanium nitride powder SEM pictures for (a) Cp-TiN, (b) 1 h, (c) 4 h, (d) 10 h, and (e) 40 h, respectively, were obtained [70].

decrease with increasing nitrogen content in cermet. In this regard, TiCN-based cermet shows interesting properties, including low metal sliding friction, excellent chemical stability, high hardness at high temperatures, and good wear resistance. In addition to better performance in terms of wear resistance and good surface quality of mechanical workpieces, it can replace traditional carbide alloys. The effect of adding nanometric or micrometric TiN on the mechanical properties and microstructure of TiC-based cermet has been the subject of recent research [6]. These results demonstrated that the addition of Nano TiN could stop the melting of TiC grains and that the mechanical performance are improved when TiN are added at a weight percentage of less than 6 wt%. It would be interesting to investigate how the TiN

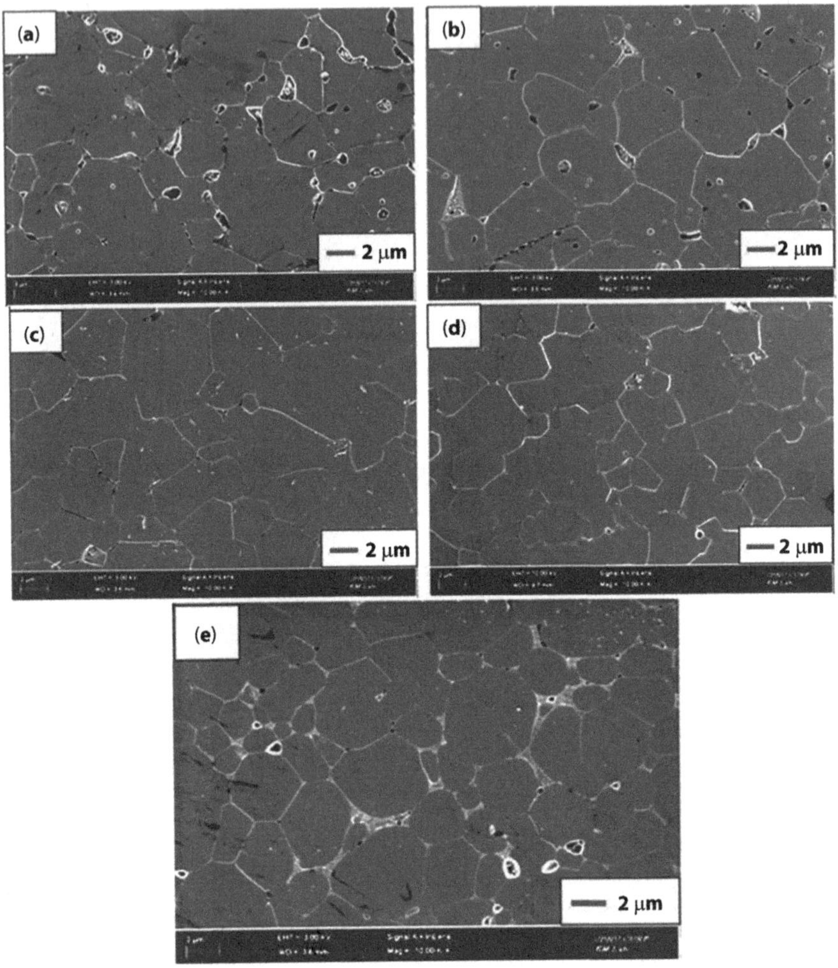

Figure 11.9 A scanning microscopy (SEM) of titanium nitride taken at milling times for the following Cp-TiN, (a) 1 h, (b) 4 h, (d) 10 h, and (e) 40 h [70] are the times.

particles generated by SHS affect the final mechanical properties of the bulk cermet in this case and compare its frictional behavior with that of other solid materials. Nitrogen addition to TiC-Ni-Mo ceramic alloys has been the subject of numerous studies aimed at optimizing the alloys' microstructure, particularly the solids' grain size. These days, titanium carbide (TiC)-based modified cermets are frequently utilized as cutting tools due to their superior hardness and wear resistance compared to TiC–Mo_2C–Ni cermets [3–5].

As the TiC content areas, so does the hardness. To ensure that the metal binder adequately wets the ceramic phase and forms a mixed carbide (Ti, Mo, C) ring around the TiC grains, molybdenum is often supplied as Mo_2C.

Harder surfaces are attributed to nickel. The parent combination forms a complex microstructure after sintering, with the development of the distinctive fundamental edge structure of the grains (Ti, Mo) (C, N). The edge is enriched with the heaviest element Mo and shares the same crystal structure as the nucleus; whereas, the center is virtually unresolved TiC or TiCN.

11.9 Niobium Carbide

The search for new materials possessing exceptional qualities such as chemical inertness, hardness, high incompressibility, and thermal stability has regained momentum recently. In this context, a variety of materials have been investigated, including trimetallic carbides, diamonds, nitrides, and boron derivatives. Studying, creating, and improving materials for strength-related applications are the main goals of these endeavors. In the past, indentation tests such as the Brinell and Rockwell tests were the main methods used to characterize these materials [18–22]. By optimizing various properties such as hardness, heat resistance, fracture toughness, strength, and wear resistance, it is critical to ensure good performance and long component life for all these applications [23]. While other parameters, like use and tool constants, are initially ignored in theory, they are crucial; most theoretical theories rely solely on the overall material unit that is extracted from the radioisotope material. These materials are the only ones that use conventional glass materials like Almaz and C-BN; however, even these materials have certain uses and needs [23].

11.10 NbC Cermets

NbC are fragile without a binder; therefore, cermets and metal matrix composites combine the benefits of, say, the ceramic's hardness and the binder's fracture toughness. Numerous parameter studies have been carried out to look at different metal-bonded NbC ceramics, including binders based on cobalt, nickel, and iron. Either a percentage by weight (wt%) or a percentage by volume (vol%) are used to represent the binder content.

11.11 NbC-Co Cermets

Cho *et al.* [30], completed the first fundamental studies on the development of NbC grains in a Co matrix during sintering in 1968, 2001 and

2004, respectively. Due to the fact that a cobalt binder, namely WC-6Co, are used for the production and use of cemented WC carbides, Co-bonded NbC cermets with a binder content of 8 or 12 wt.% Co is created [8].

11.12 NbC-Ni Cermets

Co are being replaced by a Ni binder because the use of Co is associated with health problems and the transition from WC to NbC results in rather poor binder wettability [17, 37]. Compared to NbC-Co cermets, NbC-Ni cermets are stronger but less hard [4]. Metal matrix composites (MMCs), called cemented carbides, consist of a ductile metal binder enclosing a hard carbide phase, typically a tungsten carbide (WC) monolith.

The hard carbide phase predominates in hard alloys and ceramics, while the binder (or matrix) is the main phase in MMC. The word "cermet", a combination of the "metal" and "ceramic", refers to a subgroup of cemented carbides where TiC or NbC, rather than WC, are the predominant hard carbide phase and are embedded in a carbide matrix. Co, Ni or Fe bonds. Until now, 70% of cutting tools produced are made from tungsten carbide (WC), the material of choice for cemented carbide [76, 77].

11.13 Secondary Carbides' Impact

In the hydropower, chemical, oil, and petroleum refining industries, corrosion are a serious problem that occurs frequently [4–6]. There is a leading cause of failure of flow components, including pumps and valves used in the transport of slurries or marine oil. Ceramic-metal composites provide an exceptional combination of high strength and high hardness to effectively combat aggressive media in such harsh environments. Numerous fundamental studies have been carried out on the corrosion behavior of metal and ceramic composites, and some important results have already been obtained [38, 39].

11.14 Mechanical Properties

The materials are easily destroyed regardless of the joining method due to porosity, showing poor mechanical strength and toughness. According to the study [42], carbides become ductile only when there is a large stress near the crack source. They say that the only way to increase the strength of carbides is to stop fractures from occurring. Surface porosity, microstructure grain size and grain boundary phases influence the hardness [43]. Much attention is paid to the development and study of NbC ceramics. NbC-Co metal ceramics have been widely researched between 2000 and 2014. Similar to WC cemented carbide alloys in terms of hardness, NbC-Co is less durable. The hardness of NbC0.88-12Co has been increased by replacing NbC1.0. More recently, NbC cermets with iron bonds have been studied. In contrast to NbC-Ni cermets, the hardness-to-fracture toughness ratio is often worse. The durability and endurance of cutting tools, particularly those used for heavy lifting and heavy-duty applications, are now subject to stronger limitations due to recent technological advancements [1]. Enhancing the mechanical and physical characteristics of diamond composites, which are utilized to create cutting tools for a variety of technical applications, is crucial in this regard. High physical, mechanical, and performance characteristics, including wear resistance, heat and friction resistance, hardness, elastic modulus, fracture toughness, thermal conductivity, ductility, and high diamond holding capacity, must all be present in these materials [2–4]. It is commonly known that the wear resistance of diamond composites is influenced by the elemental composition and matrix properties, whether they are metallic or carbide [5–7].

In this regard, a large number of studies have focused on nickel-bonded NbC ceramics with varying concentrations of secondary carbides. Metal ceramics are produced using numerous technological methods. The main focus are metal powder production including mechanical grinding, colloidal powder, hot pressing and sintering, SPS, LPS and further compression using GPS or HIP. Casting Fe3Al-NbC metal matrix composites has also been successful in some cases. Table 11.1 Describes a few ceramic nanocomposites' properties [79]. The study of wear, lubrication, and friction is known as tribology. Materials can become worn out either by contact between two surfaces or through contact between an exposed surface and abrasive particles, liquids, chemicals, or heat [67].

All in all, there are two basic categories of wear mechanisms: mechanical and all other cases that include an additional degree of active chemistry, like oxidation or corrosion. Four categories can be used to separate the different types of mechanical wear. Categories include surface fatigue, adhesion, abrasion, and erosion. Additionally, there is also a fretting mechanism operating under similar wear circumstances, which combines adhesive, abrasive, and occasionally corrosive wear. From an industrial perspective, all of the aforementioned modes result in loss of dimensional tolerances, higher operating costs, and supplementary costs like part replacement [67, 68]. In addition, according to some studies, the main wear mechanisms include brittle pull-out or separation of carbon nitride grains, as well as extrusion and spreading

Table 11.1 Some ceramic nanocomposites' characteristics [79].

Ceramic composite	Properties
SiN/SiC	increased toughness and fracture strength
Al_2O_3/SiC	durability, strength at high temperatures,
Al_2O_3/Co	creepmagnetic, electrical, and optical qualities
Al_2O_3/Ni	thermal conductivity, electrical, and electronic qualities.
Fe_2O_3/CNT	enhanced electrical conductivity, microhardness, friction/wear coefficients, and fracture toughness
Al_2O_3/CNT	superior strength and toughness to fracture over monolithic SiO2
SiO_2/CNT	enhanced fracture toughness, optical qualities, and tensile strength
SiC/CNT	enhanced storage modulus, reduced flammability, and increased
PE/TiO_2	tensile Young's modulus and robust matrix-silica layer interaction
Nylon-6/layered-silicates	enhanced stiffness, Young's modulus, strain at break, tensile stress, and tensile strength
Nylon-6/montmorillonite	improved thermal stability and glass transition
PP/montmorillonite	Improved heat stability
PVA/Ag	bending strength, break, bending modulus, and storage modulus
PPX/Pd or Sn NPs	
PMMA/montmorillonite	Increased modulus and tensile strength
Polylactide/layered-silicates	Strength and Young's modulus improvements
Epoxy/layered-silicates	enhanced strain to failure, and strength a decrease in ultimate
Epoxy/CNTs	strain and tensile strength and an increase in Young's modulus

(Continued)

Table 11.1 Some ceramic nanocomposites' characteristics [79]. (*Continued*)

Ceramic composite	Properties
Epoxy/CNTs	increased ductility and strain energy density
Epoxy/DWCNT	heightened yielding and breaking energy
Polyimide/MWCNT	increases in strength and stiffness
PE/PAN/MWCNT	Growth in electrical conductivity, tensile strength, and Young's modulusSignificant increase in hardness
Silicone/elastomer SWCNT	strengthened microhardness Higher hardness and Young's modulus
PVC/MWCNT	
a-Fe/$Fe_{23}C_6$/Fe_3B	
Al/SiC	
Cu/Nb	

of the binder phase. From a mechanical point of view, "water" and "dry" corrosion have certain similarities [71]. However, in solid and liquid corrosive environments, the fluid effect and more complex particle trajectories lead to more complex material failure mechanisms [12].

11.15 Conclusion

UHTC materials fill a gap in many advanced industrial manufacturing applications. It can be achieved with super-hard ceramic materials that withstand the high temperatures of high-speed cutting tools, molten metal crucibles, etc. due to the oxidation resistance of TiCN/TiN-based cermet. However, the main disadvantages are poor sintering and insufficient mechanical stability of properties. Despite its high-temperature properties, TiCN/TiN are brittle, which may limit its use in some applications, especially in the aerospace industry and in high-speed cutting tools. Manufacturing processes are the most important factor in the use and standardization of TiCN/TiN composites. Titanium nitride, one of several compounds used as an additional reinforcement to improve the properties of TiCN, can impart special properties to TiCN and TiN that make them very attractive as advanced technologies.

When many carbon allotropes are combined, TiCN performs better. Enhancing TiCN's qualities, such as its oxidation resistance and fracture toughness, is also crucial. The production of binder-free titanium carbonitride is examined in further detail in this review to clarify the impacts of several additions, such as WC, WC-Cr2C3, and TiC. It demands a great deal of attention in research labs and academic institutions.

Additionally, the computed wear factors were found to be incredibly low—on the scale of 10-8. For both coatings, abrasive wear is the primary source of wear, though. Furthermore, some plastic deformation was found by the wear trace profiles. Energy-dispersive X-ray spectroscopy (EDS) and scanning electron microscopy (SEM) analyses reveal that the coating maintains its integrity under lubricated conditions, while secondary carbides demonstrate notable microstructural features.

Methods for producing binder-free TiCN by powder metallurgy are not addressed in the literature, nor are starting powders, powder impurity preparation, dispersion techniques, equipment requirements (mixing, grinding), sintering method, sintering parameters, or best methods for characterizing composite materials. Last, but not least, it is very important to characterize the mechanical behavior (varicosity and hardness) of the obtained titanium carbonitride cermet without binder.

References

1. Belgacem, L., Ouinas, D., Viña Olay, J.A., Amado, A.A., Experimental Investigation of Notch Effect and Ply Number on Mechanical Behavior of Interply Hybrid Laminates (Glass/Carbon/Epoxy). *Compos. Part B Eng.*, *145*, 189–196, 2018.
2. Yu, J., Pan, Z., Ye, W., Li, Q., Wu, Z., Dynamic Temperature Field and Drilling Damage Mechanism of Plain Woven Carbon/Glass Hybrid Composites. *Compos. Struct.*, *304*, 116375, 2023.
3. Gao, M., Li, S., Guan, W., Xie, H., Wang, X., Liu, J., Wang, H., Excellent thermal shock resistance of NiCrAlY coatings on copper substrate *via* laser cladding. *J. Mater. Sci. Technol.*, 130, 93–102, 2022.
4. Geier, N., Davim, J.P., Szalay, T., Advanced Cutting Tools and Technologies for Drilling Carbon Fibre Reinforced Polymer (CFRP) Composites: A Review. *Compos. Part A Appl. Sci. Manuf.*, *125*, 105552, 2019.
5. Nagy, S.R., Pop, A.B., Titu, A.M., Determination of processing precision of hole in industrial 2 plastic materials. *Polymers*, *15*, 347, 2023.
6. Sorrentino, L., Turchetta, S., Bellini, C., A new method to reduce delaminations during drilling of FRP laminates by feed rate control. *Compos. Struct.*, *186*, 154–164, 2018.

7. Zeng, W., Gan, X., Li, Z., Zhou, K., Effect of TiC addition on the microstructure and mechanical properties of TiN-based cermets. *Ceram. Int.*, 43, 1, 1092–1097, 2017.
8. Imasato, S., Tokumoto, K., Kitada, T., Sakaguchi, S., Properties of ultra-fine grain binderless cemented carbide 'RCCFN'. *Int. J. Refract. Met. Hard Mater.*, 13, 5, 305–312, 1995.
9. Kim, H., Kim, J., Kwon, Y., Mechanical properties of binderless tungsten carbide by spark plasma sintering. *Proceedings. The 9th Russian-Korean International Symposium on Science and Technology, KORUS 2005*, IEEE, pp. 458–461, 2005.
10. Al-wandi, S., Ding, S., Mo, J., An approach to evaluate delamination factor when drilling carbon fiber-reinforced plastics using different drill geometries: Experiment and finite element study. *Int. J. Adv. Manuf. Technol.*, 93, 4043–4061, 2017.
11. Demirskyi, D., Agrawal, D., Ragulya, A., Comparareons of grain size-density trajectory during microwave and conventional sintering of titanium nitride. *J. Alloys Compd.*, 581, 498–501, 2013.
12. Karimi, N.Z., Heidary, H., Minak, G., Critical thrust and feed prediction models in drilling of composite laminates. *Compos. Struct.*, *148*, 19–26, 2016.
13. Fernández-Pérez, J., Cantero, J.L., Díaz-Álvarez, J., Miguélez, M.H., Hybrid Composite-Metal Stack Drilling with Different Minimum Quantity Lubrication Levels. *Materials*, 12, 448, 2019.
14. Gaugel, S., Sripathy, P., Haeger, A., Meinhard, D., Bernthaler, T., Lissek, F., Kaufeld, M., Knoblauch, V., Schneider, G., A comparative study on tool wear and laminate damage in drilling of carbon-fiber reinforced polymers (CFRP). *Compos. Struct.*, 155, 173–183, 2016.
15. Qiu, X., Li, P., Niu, Q., Chen, A., Ouyang, P., Li, C., Ko, T.J., Influence of machining parameters and tool structure on cutting force and hole wall damage in drilling CFRP with stepped drills. *Int. J. Adv. Manuf. Technol.*, 97, 857–865, 2018.
16. Xu, J., Li, C., Mi, S., An, Q., Chen, M., Study of drilling-induced defects for CFRP composites using new criteria. *Compos. Struct.*, 201, 1076–1087, 2018.
17. Che, D., Saxena, I., Han, P., Guo, P., Ehmann, K.F., Machining of carbon fiber reinforced plastics/polymers: A literature review. *J. Manuf. Sci. Eng.*, 136, 034001, 2014.
18. Su, F., Zheng, L., Sun, F., Wang, Z., Deng, Z., Qiu, X., Novel Drill Bit Based on the Step-Control Scheme for Reducing the CFRP Delamination. *J. Mater. Process. Technol.*, 262, 157–167, 2018.
19. Hocheng, H. and Tsao, C.C., Effects of Special Drill Bits on Drilling-Induced Delamination of Composite Materials. *Int. J. Mach. Tools Manuf.*, 46, 1403–1416, 2006.
20. Feito, N., Muñoz-Sánchez, A., Díaz-Álvarez, A., Miguelez, M.H., Multi-Objective Optimization Analysis of Cutting Parameters When Drilling

Composite Materials with Special Geometry Drills. *Compos. Struct.*, 225, 111187, 2019.

21. Bilgin, S., Güler, O., Alver, U., Erdemir, F., Aslan, M., Çanakçı, A., Effect of TiN, TiAlCN, AlCrN, and AlTiN ceramic coatings on corrosion behavior of tungsten carbide tool. *J. Australas. Ceram. Soc.*, 57, 263–273, 2021.
22. Dabees, S., Mirzaei, S., Kaspar, P., Holcman, V., Sobola, D., Characterization and Evaluation of Engineered Coating Techniques for Different Cutting Tools—Review. *Materials*, 15, 5633, 2022.
23. Taheri, R., Jalali, M., Al-Yaseri, A., Yabesh, G., Improving TC drill bit's efficiency and resistance to wear by graphene coating. *Pet. Res.*, 7, 430–466, 2022.
24. Aamir, M., Tolouei-Rad, M., Giasin, K., Nosrati, A., Recent advances in drilling of carbon fiber–reinforced polymers for aerospace applications: A review. *Int. J. Adv. Manuf. Technol.*, 105, 2289–2308, 2019.
25. Wang, X., Shen, X., Zeng, C., Sun, F., Combined influences of tool shape and as-deposited diamond film on cutting performance of drills for CFRP machining. *Surf. Coat. Technol.*, 347, 390–397, 2018.
26. Iliescu, D., Gehin, D., Gutierrez, M.E., Girot, F., Modeling and tool wear in drilling of CFRP. *Int. J. Mach. Tools Manuf.*, 50, 204–213, 2010.
27. Hrechuk, A., Bushlya, V., M'Saoubi, R., Ståhl, J.E., Experimental investigations into tool wear of drilling CFRP. *Procedia Manuf.*, 25, 294–301, 2018.
28. Karpat, Y., Deger, B., Bahtiyar, O., Drilling thick fabric woven CFRP laminates with double point angle drills. *J. Mater. Process. Technol.*, 212, 2117–2127, 2012.
29. Cavatorta, M.P., A comparative study of the fatigue and post-fatigue behavior of carbon-glass/epoxy hybrid RTM and hand lay-up composites. *J. Mater. Sci.*, 42, 8636–8644, 2007.
30. Nagaraja, K.C., Rajanna, S., Prakash, G.S., Koppad, P.G., Alipour, M., Studying the effect of different carbon and glass fabric stacking sequence on mechanical properties of epoxy hybrid composite laminates. *Compos. Commun.*, 21, 100425, 2020.
31. Dong, C. and Davies, I.J., Flexural and tensile strengths of unidirectional hybrid epoxy composites reinforced by S-2 glass and T700S carbon fibres. *Mater. Des.*, 54, 955–966, 2014.
32. Tan, C.L., Azmi, A.I., Mohamad, N., Performance Evaluations of carbon/glass hybrid Polymer composites. *Adv. Mater. Res.*, 980, 8–12, 2014.
33. Selmy, A., Abd El-baky, M.A., Hegazy, D.A., Mechanical properties of inter-ply hybrid composites reinforced with glass and polyamide fibers. *J. Thermoplast. Compos. Mater.*, 32, 267–293, 2019.
34. Attia, M.A., Abd El-baky, M.A., Alshorbagy, A.E., Mechanical performance of intraply and inter-intraply hybrid composites based on e-glass and polypropylene unidirectional fibers. *J. Compos. Mater.*, 51, 381–394, 2017.
35. Abd El-baky, M.A., Evaluation of mechanical properties of jute/glass/carbon fibers reinforced hybrid composites. *Fibers Polym.*, 18, 2417–2432, 2017.

36. Franz, G., Vantomme, P., Hassan, M.H., A Review on Drilling of Multilayer Fiber-Reinforced Polymer Composites and Aluminum Stacks: Optimization of Strategies for Improving the Drilling Performance of Aerospace Assemblies. *Fibers*, 10, 78, 2022.
37. Gemi, L., Köklü, U., Yazman, S., Morkavuk, S., The Effects of Stacking Sequence on Drilling Machinability of Filament Wound Hybrid Composite Pipes: Part-1 Mechanical Characterization and Drilling Tests. *Compos. Part B Eng.*, 186, 107787, 2020.
38. Sonkar, V., Abhishek, K., Datta, S., Mahapatra, S.S., Multi-Objective Optimization in Drilling of GFRP Composites: A Degree of Similarity Approach. *Procedia Mater. Sci.*, 6, 538–543, 2014.
39. Tan, C.L., Azmi, A.I., Muhammad, N., Delamination and Surface Roughness Analyses in Drilling Hybrid Carbon/Glass Composite. *Mater. Manuf. Process.*, 31, 1366–1376, 2016.
40. Krishnaraj, V., Prabukarthi, A., Ramanathan, A., Elanghovan, N., Senthil Kumar, M., Zitoune, R., Davim, J.P., Optimization of Machining Parameters at High Speed Drilling of Carbon Fiber Reinforced Plastic (CFRP) Laminates. *Compos. Part B Eng.*, 43, 1791–1799, 2012.
41. Gaitonde, V.N., Karnik, S.R., Rubio, J.C., Correia, A.E., Abrão, A.M., Davim, J.P., Analysis of Parametric Influence on Delamination in High-Speed Drilling of Carbon Fiber Reinforced Plastic Composites. *J. Mater. Process. Technol.*, 203, 431–438, 2008.
42. Campos Rubio, J., Abrao, A.M., Faria, P.E., Correia, A.E., Davim, J.P., Effects of High Speed in the Drilling of Glass Fibre Reinforced Plastic: Evaluation of the Delamination Factor. *Int. J. Mach. Tools Manuf.*, 48, 715–720, 2008.
43. Rawat, S. and Attia, H., Wear Mechanisms and Tool Life Management of WC–Co Drills during Dry High Speed Drilling of Woven Carbon Fibre Composites. *Wear*, 267, 1022–1030, 2009.
44. Navarro-Mas, M.D., Meseguer, M.D., Lluch-Cerezo, J., García-Manrique, J.A., Comparison of Different Parameters to Evaluate Delamination in Edge Trimming of Basalt Fiber Reinforced Plastics (Bfrp). *Materials*, 13, 5326, 2020.
45. Mkaddem, A., Soussia, A.B., El Mansori, M., Wear resistance of CVD and PVD multilayer coatings when dry cutting fiber reinforced polymers (FRP). *Wear*, 302, 946–954, 2013.
46. Meinhard, D., Haeger, A., Knoblauch, V., Drilling induced defects on carbon fiber-reinforced thermoplastic polyamide and their effect on mechanical properties. *Compos. Struct.*, 256, 173–183, 2021.
47. Heisel, U. and Pfeifroth, T., Influence of point angle on drill hole quality and machining forces when drilling CFRP. *Proc. CIRP*, 1, 471–476, Jan. 2012.
48. Poulachon, G., Outeiro, J., Ramirez, C., André, V., Abrivard, G., Hole surface topography and tool wear in CFRP drilling. *Proc. CIRP*, 45, 35–38, 2016.
49. Findenig, G., Buchegger, C., Lengauer, W., Veitsch, C., Demoly, A., Investigation of the main influencing parameters on the degassing behavior

of titanium carboni-trides using mass spectrometry. *Int. J. Refract. Met. Hard Mater.*, 63, 38–46, 2017.

50. Suib, S.L., A review of nanoceramic materials for use in ceramic matrix composites, in: *Sol-gel Based Nanoceramic Materials: Preparation, Properties and Applications*, pp. 185–230, 2017.
51. Nguyen, T.K., Fundamental Tool Wear Study in Turning of Ti-6Al-4V Alloy (Ti64) and Nano-Enhanced Minimum Quantity Lubrication (MQL) Milling, Michigan State University, East Lansing, MI, USA, 2015.
52. Lagarde, Q., Wagner, V., Dessein, G., Harzallah, M., Effect of Temperature on Tool Wear During Milling of Ti64. *J. Manuf. Sci. Eng.*, *143*, 071007, 2021.
53. Veiga, C., Davim, J.P., Loureiro, A., Properties and applications of titanium alloys: A brief review. *Rev. Adv. Mater. Sci.*, *32*, 14–34, 2012.
54. Pramanik, A., Islam, M.N., Basak, A., Littlefair, G., Machining and Tool Wear Mechanisms during Machining Titanium Alloys, in: *Engineering Materials and Application*, pp. 338–343, Trans Tech Publications Ltd., Bäch, Switzerland, 2013.
55. Ulutan, D. and Ozel, T., Machining induced surface integrity in titanium and nickel alloys: A review. *Int. J. Mach. Tools Manuf.*, *51*, 250–280, 2011.
56. Amin, A.K.M.N., Ismail, A.F., Khairusshima, M.K.N., Effectiveness of uncoated WC–Co and PCD inserts in end milling of titanium alloy—Ti–6Al–4V. *J. Mater. Process. Technol.*, *192–193*, 147–158, 2007.
57. Sahoo, P., Singh, V.K., Patra, K., Influences of tribological and mechanical properties of (Ti/Al)N and diamond like carbon coated tungsten carbide tool on machining and dynamic stability in micro milling of Ti6Al4V. *J. Manuf. Process.*, *85*, 915–934, 2023.
58. Rahman, M., Wang, Z., Wong, Y., A Review on High-Speed Machining of Titanium Alloys. *JSME Int. J. Ser. C Mech. Syst. Mach. Elem. Manuf.*, *49*, 11–20, 2006.
59. Jianxin, D., Yousheng, L., Wenlong, S., Diffusion wear in dry cutting of Ti–6Al–4V with WC/Co carbide tools. *Wear*, *265*, 1776–1783, 2008.
60. Hosseini, A., Kishawy, H.A., Hussein, H.M., Machinability of Titanium and Its Alloys, in: *Machinability of Advanced Materials*, pp. 95–118, John Wiley & Sons, Inc., Hoboken, NJ, USA, 2014.
61. Umbrello, D., Finite element simulation of conventional and high speed machining of Ti6Al4V alloy. *J. Mater. Process. Technol.*, *196*, 79–87, 2008.
62. Kumar, K.M., Mathew, N.T., Baburaj, M., Sustainable milling of Ti-6Al-4V super alloy using AlCrN and TiAlN coated tools. *Mater. Today Proc.*, *50*, 1732–1738, 2022.
63. Hosseini, A. and Kishawy, H.A., Cutting Tool Materials and Tool Wear, in: *Machining of Titanium Alloys*, J.P. Davim, (Ed.), pp. 31–56, Springer, Berlin/Heidelberg, Germany, 2014.
64. Nabhani, F., Machining of aerospace titanium alloys. *Robot Comput. Integr. Manuf.*, *17*, 99–106, 2001.

65. Soroka, O.B., Evaluation of residual stresses in PVD-coatings. Part 1. Review. *Strength Mater.*, *42*, 287–296, 2010.
66. Abadias, G., Chason, E., Keckes, J., Sebastiani, M., Thompson, G.B., Barthel, E., Doll, G.L., Murray, C.E., Stoessel, C.H., Martinu, L., Review Article: Stress in thin films and coatings: Current status, challenges, and prospects. *J. Vac. Sci. Technol., A*, *36*, 20801, 2018.
67. Patro, S.S., Roy, S., Sahoo, A.K., Investigation of influence of design on residual stress at free edge of multilayer Ti/AlN coatings by PVD using FEM. *Int. J. Interact. Des. Manuf. JIDeM*, *17*, 2987–2998, 2023.
68. Liu, J., Wang, Y., Liu, G., Hua, J., Deng, X., Properties and Performance of TiAlSiN and AlCrN Monolayer and Multilayer Coatings for Turning Ti-6Al-4V. *Coatings*, *13*, 1229, 2023.
69. Mishra, S.K., Ghosh, S., Aravindan, S., Characterization and machining performance of laser-textured chevron shaped tools coated with AlTiN and AlCrN coatings. *Surf. Coat. Technol.*, *334*, 344–356, 2018.
70. Reiter, A.E., Derflinger, V.H., Hanselmann, B., Bachmann, T., Sartory, B., Investigation of the properties of Al1−xCrxN coatings prepared by cathodic arc evaporation. *Surf. Coat. Technol.*, *200*, 2114–2122, 2005.
71. Mo, J.L., Zhu, M.H., Leyland, A., Matthews, Impact wear and abrasion resistance of CrN, AlCrN and AlTiN PVD coatings. *Surf. Coat. Technol.*, *215*, 170–177, 2013, [Google Scholar] [CrossRef].
72. Feng, Y., Zhang, L., Ke, R., Wan, Q., Wang, Z., Lu, Z., Thermal stability and oxidation behavior of AlTiN, AlCrN and AlCrSiWN coatings. *Int. J. Refract. Met. Hard Mater.*, *43*, 241–249, 2014.
73. Zhang, J., Ling, L., Luo, D., Deng, C., Huang, X., Tao, G., Cao, H., Cutting performance and surface quality of Ti-6Al-4V by longitudinal ultrasonic vibration-assisted high-speed dry milling with coated carbide tools. *Int. J. Adv. Manuf. Technol.*, *126*, 5583–5596, 2023.
74. Kim, D., Swan, S.R., He, B., Khominich, V., Bell, E., Lee, S.-W., Kim, T.-G., A study on the machinability of advanced arc PVD AlCrN-coated tungsten carbide tools in drilling of CFRP/titanium alloy stacks. *Carbon Lett.*, *31*, 497–507, 2021.
75. ISO 19252:2008, Plastics—Determination of Scratch Properties, Technical Committee ISO/TC 61, Plastics, Subcommittee SC 2, Mechanical Properties. ISO, Geneva, Switzerland, 2008.
76. ISO 8688-2:1989, Tool Life Testing in Milling—Part 2: End Milling, ISO: Geneva, Switzerland, 1989.
77. Panjan, P. and Cekada, M., Growth defects in PVD hard coatings Growth defects in PVD hard coatings. *Vaccum*, *84*, 209–214, 2009.
78. Polcar, T. and Cavaleiro, A., High temperature properties of CrAlN, CrAlSiN and AlCrSiN coatings—Structure and oxidation. *Mater. Chem. Phys.*, *129*, 195–201, 2011.
79. Kumar, T.S., Prabu, S.B., Manivasagam, G., Padmanabhan, K.A., Comparison of TiAlN, AlCrN, and AlCrN/TiAlN coatings for cutting-tool applications. *Int. J. Miner. Metall. Mater.*, *21*, 796–805, 2014.

12

Industrial Applications of Hard Materials: Wear-Resistant Coatings, Cutting Tools, and Advanced Manufacturing Techniques

Riyadh A. Al-Samarai[1*] and Yarub Al-Douri[2,3,4]

[1]*Electromechanical Engineering Department, Engineering College, University of Samarra, Samarra, Iraq*
[2]*College of Health and Medical Techniques, Al-Bayan University, Baghdad, Iraq*
[3]*Department of Mechanical Engineering, Faculty of Engineering, Piri Reis University, EflatunSk, Tuzla, Istanbul, Turkey*
[4]*Nanotechnology and Catalysis Research Centre, University of Malaya, Kuala Lumpur, Malaysia*

Abstract
Traditional grinding operations have replaced novel processes for turning hard components since introducing new ultra-cutting hard tool materials. Hard machining has significantly improved various machining operations, including milling, boring, broaching, hobbing, and others. Cutting performance is significantly impacted by coatings. The performance of cutting tools has gradually increased along with cutting technology through the development of thin layers that provide the properties required for each cutting condition. To increase wear resistance, the layer on the tool is crucial. A coating layer can be used to enhance the cutting tool's performance by either chemical or physical vapor deposition (PVD) comprised of titanium nitride (TiN), titanium carbonitride (TiCN), and aluminum chromium nitride (AlCrN). The characteristics of PVD and CVD multilayer coatings are outlined in this work, together with additional research issues relating to tribological phenomena from the viewpoint of cutting tool technology.

Keywords: Hard machining, ultra-hard tool materials, cutting performance, tool coatings, wear resistance, PVD and CVD, titanium nitride (TiN), tribological behavior

**Corresponding author*: riyadh2824@yahoo.com

Inamuddin, Tariq Altalhi and Mohammad Abu Jafar Mazumder (eds.) Nanostructured Hard Materials: Industrial Applications, (369–394) © 2026 Scrivener Publishing LLC

12.1 Introduction

Hard machining refers to the machining of parts with a hardness of more than 45 HRC, while the method is most frequently used for parts with a hardness between 58 and 68 HRC. Some of the materials that are used are hardened alloy steels, heat-treated powder metallurgy components, and nitride casting irons [1]. The creation of appropriate rigid types of machinery, extremely tough cutting materials, and cutting metals with hardened pieces is a task for any machine shop, thanks to specialized tool (holder) designs and intricate setups. Although grinding has traditionally been used to finish hardened steel parts, machining hard parts using a cutting tool has several obvious advantages.

They support a large number of currently active applications, especially those related to turning, boring, and milling [1–3]. Even though many research articles still uphold the antiquated belief that all grinding activities can be replaced by machining, this is no longer the case in modern manufacturing. Advanced manufacturing has increased the demands and challenges placed on cutting tools due to the development of modern manufacturing technologies. The hard coating of the cutting tool system provides superior mechanical properties and friction characteristics, which improves overall tool performance. These two processes complement one another in a variety of real-world industrial applications, which is why grinding spindles are frequently included in modern machine tools for hard machining. Support shaft assembly is a typical automotive component shown in Figure 12.1. After the hard (42 HRC) cast iron flange and hardened (48–50 HRC) support shaft have been pressed together, machining operations are carried out to guarantee the flange end perpendicular to the shaft shoulders [4, 5]. Research papers and promotional materials generally do not address the constraints and downsides of hard processing, even though end users should be aware of them:

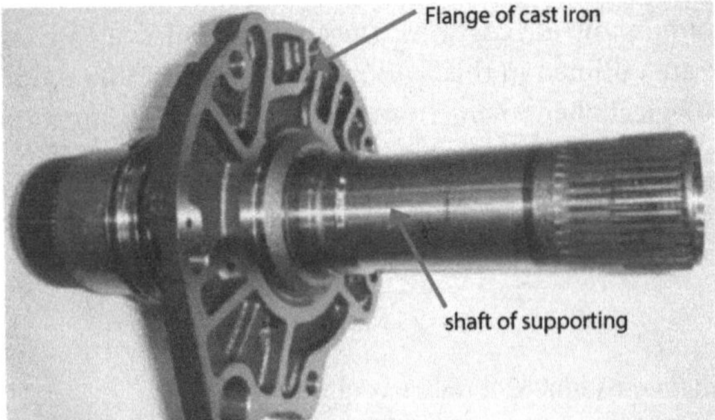

Figure 12.1 Assembling a shaft for support [9].

- Compared to grinding, machining requires much higher unit costs per tool.
- Sometimes the size or geometry of a part simply does not allow for hard turning.

The L/D ratio for unsupported workpieces should generally not exceed 4:1. Although the tailstock is designed for long and thin parts, the vibration occurs at high cutting pressures. Successful machining of hard materials often requires specialized and rigid machines. The level of turning accuracy depends on the rigidity of the machine. Machine tool rigidity becomes a more serious issue when required part tolerances and surface finishes increase. For applications requiring harsh machining, machining systems must include several elements to enhance stiffness and damping properties. All overhangs, tools, and shims must be kept to a minimum to increase system rigidity. The goal is to keep everything as close to the tower as possible. Additionally, coatings based on nitride include TiN, TiCN, TiAlN, and AlCrN [1–3], and multilayer carbide tools with an exterior TiN coating work well at faster cutting rates. In many different industries, surface finishing technologies are essential to improve the functionality and performance of metallic materials. The ability to change the surface characteristics of metals has created new opportunities to improve wear resistance, corrosion protection, and mechanical strength [1, 2].

Figure 12.2 illustrates the elements affecting the selection of materials for cutting tools. When cutting cast iron and steel, the technological effectiveness of ceramic (silicon nitride) coated inserts outperforms uncoated inserts [4, 5]. Tools built of TiAlN carbide alloy with a multilayer (coating) have longer service lives compared to instruments with a single layer covering by a factor of 75% [6].

Coatings using nanostructures and nanolayers have also become a subject of recent research. They have been tested and the results are very promising: their tool life is on par with more advanced nano-coatings [8, 9]. When using PVD-coated TiN cemented carbide cutting inserts to machine steel, cutting speed and feed have a substantial impact on tool life [10]. Figure 12.3 shows the distinctive characteristics of the instrumentation material. The wear of PVD coatings is not impacted by the monolayer coating's thickness [11, 12]. PVD coatings have much greater density, greater hardness, and greater stickiness. Figure 12.4 illustrates the variables affecting the cutting tool's performance. A tool with a coating produces superior machining results. The way that CBN-TiN (coated) carbide tools respond to machining forces in numerous fields of machining research, notably in high-strength low-alloy steels. This work discusses various coating types, their general features, layering, and applications. It also highlights technology shortcomings and difficulties

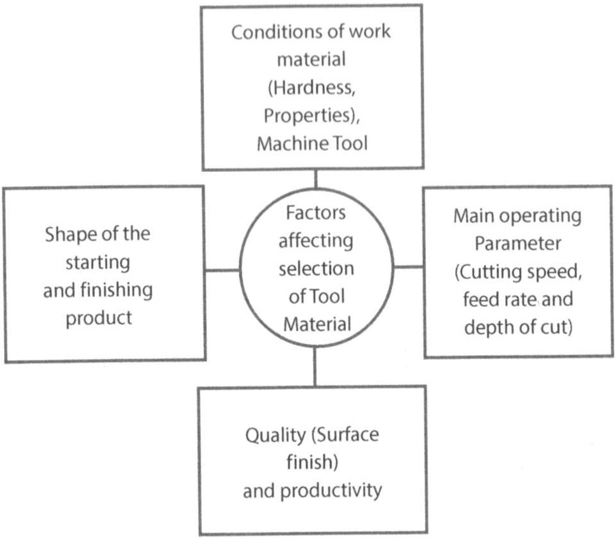

Figure 12.2 Elements influencing the choice of tool material [17].

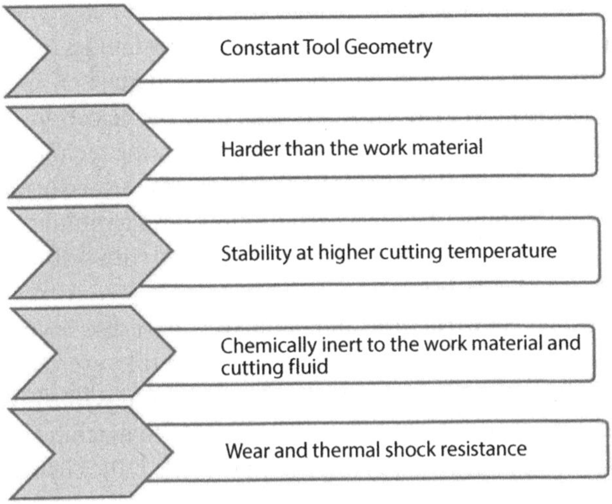

Figure 12.3 Characteristics of tool materials [17].

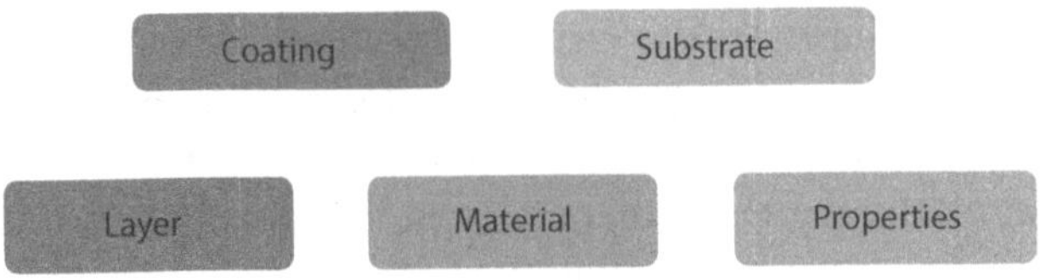

Figure 12.4 Factors affecting the effectiveness of the cutting tool.

with prospect research. The final section of this summary focuses on potential future research directions in the machining and tooling industries [7, 8].

The biggest mystery is what coolant is appropriate for heavy machining operations. Dry milling is the best method for milling objects with interrupted cuts, such as gears. The reason for this is the possibility of thermal stress failure during the process of entering and leaving the grooves. The high tool tip temperatures during dry turning cause the pre-cut area to be annealed (softened) during continuous cutting, which lessens the material's hardness and facilitates cutting. It is useful to expedite the process during dry cutting because of this feature. There are also definite economic benefits to dry cutting. On the other hand, it might be challenging to process, measure, and thermally deform parts during manufacture. Some applications are now required to use coolant due to recent issues. When employed, water-based coolant is required. It has been demonstrated that near-dry machining works well in demanding applications [9]. As illustrated in Figure 12.5, the workpiece's surface quality degrades even within the tool's lifetime due to wear. It is important to find solutions to the issues surrounding rigid equipment, industrial cutting tool properties, and their applications. The coating deposition procedure, with intense processing results in a white covering that thickens with tool wear. It mostly affects parts like bearing strains, which are frequently formed on tools and are exposed to high contact pressures. Eventually, the white coating can come off, which would cause the tool to fail. Most tool manufacturers have internal testing facilities to address this issue. The steel industry can carry out these tests. The attachment can precisely cut 400 parts; however, after 300 parts, it can get so dull that a white coating forms [13].

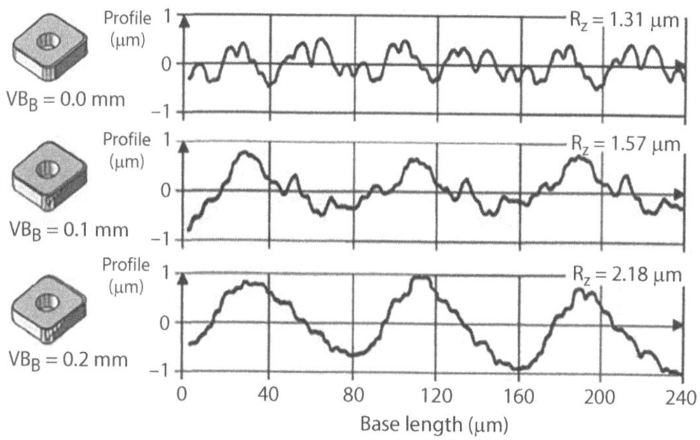

Figure 12.5 As the tool wears, the surface quality deteriorates. Turning AISI 52100 steel (65.5HRC).

12.2 A Brief Analysis of Hard Machining Research

Despite the significant amount of material on hard machining, only a small number of articles [9, 10] focus on advanced coating technologies for cutting tools. Traditional and intensive processing, neglect some basic truths that are well-known in practice. As a result, the nature of hard machining discussed in these articles is understood very differently by leading tooling companies. Unfortunately, it seems that not enough machining researchers are "hearing" this point of view. The first detailed work that attempted to summarize the hard machining experience was published by [12]. Studying the technological characteristics of these materials is necessary to create new technologies for processing hard metals. The results of [12], are summarized as the destruction of the material during hard machining, so-called sawtooth chips are formed as shown in Figure 12.6. Thus, when the work material reaches maximum shear strain, a crack appears on the workpiece. Therefore, the process of creating a chip is determined by destruction. Although "segmented chips" produced by adiabatic shear in difficult-to-machine materials have the same cross-section as saw tooth chips, the two types of chips produced by hard machining are different because they arise from different processes. Compared to traditional machining, the shear angle is minimal. It increases sharply as the hardness of the material being processed increases. When milling an alloy steel with a hardness of 52 HRC and a tensile strength of 1900 MPa, with a depth of cut of 0.25 mm and a rotary speed of 0.08 mm/rev, Figure 12.7 shows that the cutting force increases with the wear of the CBN tool face. The tool becomes duller as the wear of the tool face increases, affecting surface quality, cutting resistance, energy consumption, and tool wear.

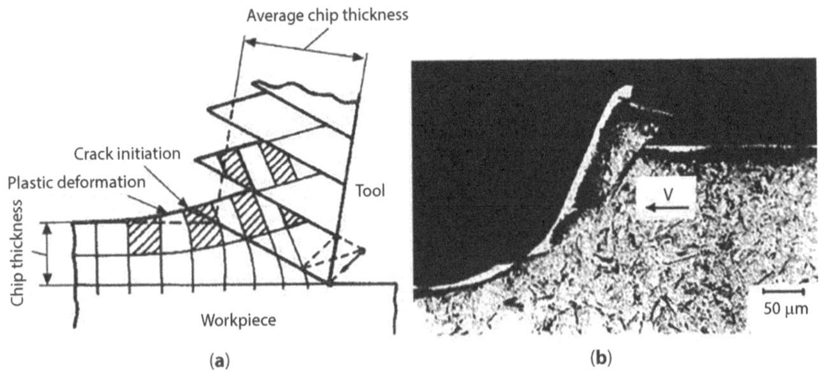

Figure 12.6 Nakayama *et al.*'s (a) model for chip creation and König *et al.*'s (b) experimental confirmation.

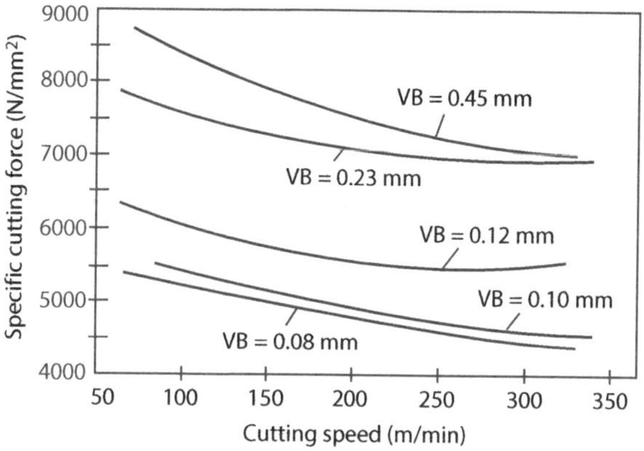

Figure 12.7 Changes in cutting force with different degrees of flank wear. Alloy steel with hardness 52 HRC, 1900 MPa, CBN tool material, cutting depth 0.25 mm, cutting speed 0.08 mm/rev.

- The disparity between the two components gets progressively larger. The working material's springing explains this. In hard machining, a strong traction force is the primary cause of dimensional inaccuracies.
- For instance, the bearing steel AISI 52100, one of the materials used most frequently for severe machining and hence testing, has a hardness range of 60 to 67 HRC. Grey cast iron has an elongation of greater than 7% and elongation at failure of 5% in their testing. Additionally, neither the model nor its analysis accounts for the energy involved in the creation of new surfaces as a result of crack propagation [11, 14].
- Negative rake cutting appears to be the opposite of what is commonly practiced, with relatively low tangential (force) and radial (axial) cutting force components and a decrease with increasing workpiece hardness.
- Neither the cause of temperature during complex machining nor its effect or its results is investigated.
- In hard machining, which primarily involves materials such as CBN and ceramics, advanced coatings play a crucial role in enhancing tool performance and durability [15–18].

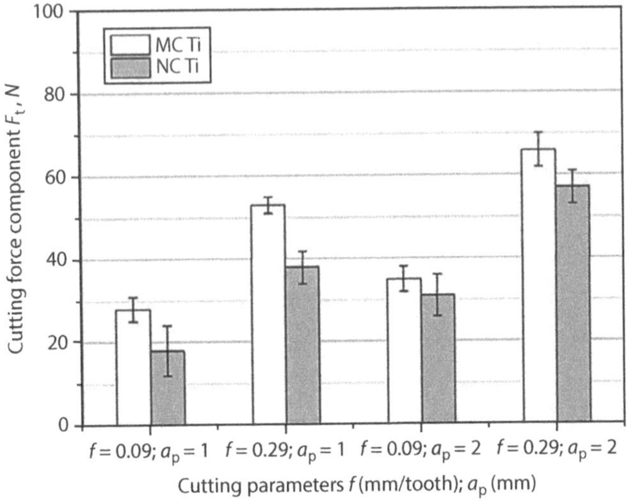

Figure 12.8 At a cutting speed of vc = 40 m/min, the values of the tangential component of the cutting forces Ft were determined for the chosen cutting parameters (ap, f).

12.3 Characteristics of Hard Machining

Although this work analyzes the characteristics in great detail, several distinguishing characteristics need to be considered when analyzing hard machining to set priorities for subsequent research and applications of this method. This can be achieved by comparing the energy balances of conventional and hard metal cutting methods [19–21]. Surprisingly, the power required to plastically deform the removed layer is still significant during aggressive cutting (Figure 12.8). The obtained results show that compared with MC titanium, no noticeable changes were observed in the range of Fr values for the radial components. This is because the so-called "cutting action" when machining MC titanium is different and in this situation is more "negative" (as is the case with a tool with a narrower rake angle) [28].

12.3.1 Reduction in Cutting Force with Speed

The Peclet criterion, a measure of similarity, is useful in understanding the role of thermal energy (heat) in hard machining [12, 25]. The Peclet criterion for metal cutting is illustrated using the following processing parameters [26, 27]:

$$Pe = \frac{vt_1}{w_w} \tag{12.1}$$

where w_w is thermal diffusivity of work material(m^2/s), and v is speed of a moving heat source (the cutting speed),

$$w_w = \frac{k_w}{(c_p \rho)_w} \tag{12.2}$$

where ($c\rho$)w is specific of volume heat (J/(m^3 °C)) and (k_w) is the thermos-conductivity of work material (J/(ms°C)). The number Péclet that expresses how much is the cutting regime (vt_1) has an impact on material's thermal characteristics (w_w). Abrasive sawtooth edges are formed due to high dynamic shear strength, creating notches on the tool [18].

The resistance of the processed material, which must be overcome during processing, is measured by the components of the cutting force. According to test data, titanium structure does not affect fundamental strength. Notable results were observed in the traction and pitching force scenarios. Regarding NK these forces for titanium are much less than for MC Ti. The cutting force component was calculated both tangentially and axially using geometric relationships and dynamometer measurements (FX, FY, and FZ). During use, a fiber optic or other optical beam delivery device is placed in front of the tool to direct the laser beam onto the workpiece. The workpiece is softened by laser heat, which facilitates cutting. The heat is concentrated around the cut itself due to the laser beam being so precisely focused [29–33]. The chips dissipate heat, yet the physical characteristics of the material being sliced remain unchanged as a result of heat.

12.3.2 The Cutting Forces Major Axial (Thrust) Components

Currently, the automotive and aerospace industries are the main consumers of titanium and its alloys—sports and recreation, industry, bioengineering, shipbuilding, industry, aerospace, and architecture. In addition, nanocrystalline materials (NCs) have special properties. physical and mechanical characteristics due to their special structure [2]. It is used in cutting tools and electronic devices to improve performance and protection, as well as in aviation for thermal barrier coatings and in the automotive industry for anti-corrosion coatings on engine components. Arc plasma deposition is the preferred coating method in many industries due to its versatility and efficiency [29, 30]. These characteristics provide them with great potential for energy applications and make them attractive for use in biological applications [1], such as orthopedic and dental implants. All of these applications must be processed using manufacturing procedures in making the corresponding parts. The most common industrial technology is mechanical processing through useful and practical metals. However, due to these same characteristics, materials such as titanium and its alloys are considered difficult to work with. AISI 52100 steels are presented in Figure 12.9. If the applied stress is greater than the elastic limit also referred to as the yield limit, which for annealed 52100 steel is 370 MPa, the material will display both elastic and plastic deformation.

It is depicted in Figure 12.9 as a distance along the elongation axis 0–D. However, after failure, the applied stress is released and the working material exhibits a permanent extension, which is called elongation (shown by a distance 0-E in Figure 12.9). The total elongation after failure of the elastic component, illustrated in the diagram by a distance E-D, in the situation under consideration, is less than 20%. This makes it easy to determine where to point E is drawing a line parallel to line 0-A from point C. As a result, the E-D distance is sometimes called elastic recovery or elastic return in materials processing. The pressure at the point of contact between the

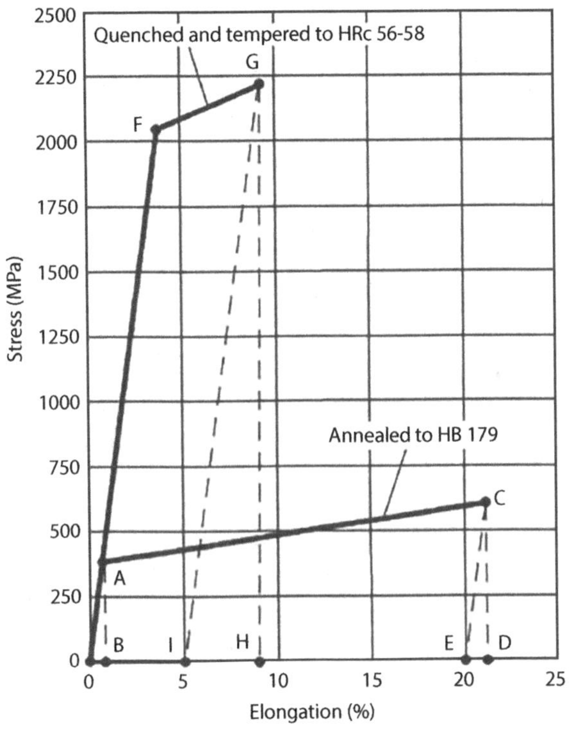

Figure 12.9 AISI steel 52100 elongation-stress diagrams after being annealed and quenched.

tool and workpiece increases sharply because this friction creates a relief, which leads to wear on the flank surface. This explains the significant radial (axial) component of cutting force and severe wear on the flank surface which only happens when cutting something hard. However, prolonged heat reduces its flexibility, which leads to the formation of a white coating.

12.3.3 Large Power Invested in Creation of Surfaces

The ability of solid materials and objects to restore their previous shape after removing loads or external forces that cause deformation is called elasticity. When an object regains its original size and shape after the load is removed, it is called elastic. Different materials behave differently for different physical reasons, which are influenced by their microscopic structure. For example, the stretching of polymer chains under the influence of applied force provides the elasticity of rubbers and polymers. Conversely, under the influence of external pressure, the crystal lattice cells that make up the structure of the metal material undergo changes in size and shape, which determine the elasticity of the metal [40].

The two factors that determine the elasticity of a material are the elastic limit and the elastic modulus. Materials that require large loads to produce significant deformation or those that are difficult to deform have a high modulus of elasticity. Steel strip is one example. Materials that easily deform under load, such as elastic bands, have a low modulus of elasticity. The material deforms permanently if the stress under load is too great, at which point it takes on a different shape and size rather than returning to its original state. The stress value above which a material ceases to behave elastically and instead permanently deforms is known as the elastic limit. The material with the lowest elastic modulus is the most elastic among materials with the same elastic limitations [12–18]. Surface engineering processes are constantly evolving, providing tremendous opportunities for a wide range of applications across multiple industries. From consumer products to electronics, medical devices, aerospace, and automotive components, product performance, and durability are greatly influenced by the adaptability of metals to their surface quality and process. Research into new materials, improved creative processes, and integration of surface technology into other manufacturing processes are enhanced by ongoing research and development in this field. This work's proportional influence on heavy machining (Figure 12.8) becomes much more substantial because it rises with the material's strength. In any model of chip creation, then, the labor necessary to generate a new surface must be taken into account in hard machining while balancing forces and/or energy.

12.3.4 Requirements for Rigid Processing

Models created using multibody tools in a real-time virtual environment require a trade-off between computational efficiency and accuracy. When complex multibody models come into contact with other parts of the system or with external surfaces, they are often deformed to some degree. Different modeling assumptions may be applied, resulting in different types of models depending on the intended use of the virtual environment. Although finite segments and methods offer computationally efficient ways to represent both large and small deformations, they are unable to convey the stresses and strains experienced by deformable components. Finite element analysis is a reliable method for describing the deformation of a system, including stress and strain fields, but requires more expensive computations. Appropriate contact models must be included in systems where touch or impact scenarios occur [31–34]. However, the driving force (friction changes) in the majority of sliding systems is frequently unknown or needs to be discovered by modeling. Frictional vibrations can, at most, be minimized its frequency can be raised out of the undesirable range by altering the dynamics of the system.

12.4 Operations for Hard-Machining

The final manufacturing procedures of parts and components typically involve complex manufacturing processes such as turning, milling, boring, and drilling, and the required dimensions and surface finishes typically require stringent manufacturing standards and conditions.

12.4.1 Hard Turning

Hard turning (Figure 12.10) has the obvious benefit of eliminating grinding processes. The same machine, which "softly" spins the part before hardening it, enables this productivity and maximizes the equipment's effectiveness. However, some stores commit the error of initially selecting the incorrect tool insert for the job (one that is less expensive). Tool steel, bearing steel, and case-hardened steel are examples of commonly used hard materials [34], this is untrue since these materials have a hardness that is substantially lower than 47 HRC, which means that the mechanism for how chips form during the operation, including the materials used for tools, differs greatly the best process predictability, from a metallurgical perspective. As previously stated, the L/D ratio for unsupported

(a) (b)

Figure 12.10 A hard turning: (a) conventional, and (b) high-speed.

workpieces should generally not exceed 4:1. When cutting long and thin pieces, the tailstock is supported, but excessive cutting pressure can create vibration.

The level of turning accuracy is determined by the rigidity of the machine. Most modern machines have the rigidity needed to handle tasks that require heavy turning. In many cases, the overall condition of a car is more important. However, machine tool rigidity becomes an increasing concern as required part tolerances and surface finishes increase. The best option is to use specialized equipment and turning centers designed for serious machining. However, the secret to success in turning is not the rigidity of the machine, but the system. Unfortunately, the industry does not fully understand this simple concept. To maximize system rigidity, all overhang, tool, and part extensions must be minimized, such as by eliminating spacers and shims. As already stated, the use of coolant is one of the major considerations when designing a turning process. Because dry cutting has obvious financial advantages, most turning operations are done dry. Parts can become hot, making it difficult to measure and handle components. Additionally, a surge in cherry red chips can lead to some other problems. Thus, high-pressure coolant through the tool is ideal for cooling the workpiece and hot chips while maintaining small chip size and shape control when coolant is required.

In light of this, if coolant is used, for obvious reasons it should be of low-concentration and water-based. When crushed by hand, the chilled chips practically fall apart, proving that the chips produce the right amount of heat. Depending on the required precision and surface polishing, the correct tool material (PCBN, cermet, or ceramic) must be selected to ensure process efficiency. Different tools, cutting inserts, and tool manufacturers offer different brands. Optimal machining conditions, tool holders and insert shapes only complicate the multi-parameter optimization task. Success requires knowledge, understanding of the basics of heavy machining, and experience.

12.4.2 Reaming and Hard Boring

To enlarge the inside diameter of an existing hole, two techniques are used; boring is known as internal turning and reaming. The first hole is drilled into the casting using a drill if necessary. Reaming and boring accomplish three main objectives: (a) calibration - boring and reaming ensure the required surface quality while also bringing the hole to required diametrical accuracy with tight tolerances; (b) straightness – boring. Hard boring will be a cost-effective replacement that minimizes waste when distortion and dimensional changes caused by heat treatment place undue restrictions on the "soft stage" of machining. The benefits of rigid boring increase with a length-to-diameter ratio and degree of deformation (for example, due to part asymmetry).

By replacing internal grinding with block boring when cutting 63 HRC hardened steel, cycle time is reduced from 26 minutes to 2 minutes, demonstrated by a leading tool manufacturer in 20 seconds. Figure 12.11 shows typical boring and reaming equipment for hard machining. Ceramic, metal-ceramic, or PSBN plates are attached to the tool body directly or in parts. Multi-bladed tools are used to achieve high performance. However, single-edge tools with backing plates are used when the hole being machined must have a high degree of accuracy. Clamping tiny

(a) (b)

Figure 12.11 (a) Hole- tools, and (b) a boring tool design for difficult machining.

inserts onto tiny boring bars becomes more challenging for holes smaller than 12 mm in diameter. To solve this problem, a soldering tool is used.

12.4.3 Milling of Hard

Companies often focus on the production of dies and dies, which are used to cut materials such as P20, H13, W5, and S7. Traditionally, electrical discharge machines have been used to create solid rods and cavities from these materials. With the advancement of new technologies over time, it is now possible to manufacture these materials straight into hardened materials, frequently by using hard milling and new toolpath techniques. Leading mold producers have understood that adapting to new technology might be one of the keys to surviving in a highly competitive climate on a worldwide scale. According to Zurek [39], the deployment of a system consisting of the machine, cutting tools and tool holders, and a computer-aided design/manufacturing system leads to successful hard milling. The system's most important element is the machine. The machine needs to be made for hard milling and also has some high-speed machining center features.

Hard milling demands a machine's fundamental design as well as specific machine parts like the motor, spindle, and CNC. It is crucial to have a sturdy base with adequate vibration-dampening properties. Because polymer concrete bases often have vibration dampening that is 6 to 10 times better than cast iron, they are a viable option for high-speed, heavy-duty milling. Digital drives' quick acceleration and deceleration capabilities and superior contouring accuracy aid in cutting down on cutting tool wear. Depending on the application, each of these tools has different strengths and limitations. End mills made of solid carbide are often highly expensive, precisely ground and coated. Cutter sets with interchangeable carbide inserts are a second category of carbide cutters. They are not ideal for harsh milling tasks. Ceramic replacement inserts are a third type of strengthened ceramic inserts with whiskers. Faster cycle times and fewer operations per item are advantages of employing a milling system with changeable ceramic inserts. Using interchangeable ceramic inserts, ceramic cutters can safely and quickly grind everything from big face mills to small diameter end mills. Modern whisker-reinforced ceramics may run at speeds much above the point at which they melt since they have a melting temperature of over 2000°C that fails with carbide tools. In actuality, ceramics reinforced with whiskers operate more effectively at temperatures higher than the melting point of carbide inserts.

12.4.4 Broaching of Hard

The broaching operation is the most cost-effective way for mass-producing parts with complex features in the field of metal cutting where high production rates and great component stability are required. One motion can be used to draw the part's profile. While exterior broaching is intended to machine the surface profile, internal broaching begins from a hole that has already been pre-machined. Both horizontal and vertical directions are open to the firmware. Both linear and helical cutting movements are possible. It is feasible to use one of two rigid broaching techniques with this tool configuration:

- The cutting speed during typical broaching procedures is known to let between 0.5 and 6 m/min. Since that time, researchers and professionals have focused on rigid broaching's effectiveness and dependability, as well as productivity and surface quality. The outcomes thus far are outlined below:
- When it comes to the quality and dependability of broaching operations, coated carbide broaches, intended as side-cutting tools, perform well, obtaining IT profile quality of 7-8.
- When utilizing carbide broaches, an efficient cutting speed of 60–63 m/min is identified as a crucial component in the broaching process. When compared to green (soft) feed, the cycle time at this speed is shorter since the feed stroke lasts between one and two seconds.

12.4.5 Operations for Manufacturing Hard-Gear

The same kinematic principles of abrasive technology are described by such terms as "gear honing", "mechanical honing", "spherical honing", "neck grinding" and others. These principles are based on the meshing contact of geared tools with a working mechanism in the form of a helical gear pair [1]. The highest temperatures in the tribological contact zone are believed to be between 100 and 300°C due to the low cutting speeds [2]. The fact that thermal damage to workpieces has never been observed with these methods can be explained by such low temperatures combined with contact periods of a few microseconds. Throughout the entire service life of the transmission, high compressive residual stresses are detected in the near-surface layers of the side surfaces of low-speed transmissions, which can be detected at significantly reduced amplitudes [3, 4]. The method allows you to change the geometric surface of the side surfaces of the teeth and improve the surface quality. In addition, the noise level of the gearbox is reduced due to the characteristic surface roughness of the final gears. The most common process and wear pattern of milling tools is the development and propagation of cracks caused by thermal fatigue (high processing temperatures). The origin and wear patterns of coated tools are still the subject of recent research as they have a direct impact on tool life. Coating wear studies can help develop new coatings and improve processes.

Recent research has shown that using these methods can extend the life of coatings, and improve surface quality and overall process efficiency. However, machining with rigid gear machines places unique demands on the tool, machine, and overall process. Understanding these prerequisites is necessary for the successful implementation of this process. High-speed steel is the material most commonly used in the industry today to make cooktops. In addition, final machining cannot be done on a high-speed steel milling machine. Even these relatively soft materials, can only go so far before additional finishing with other dental instruments such as razors is required. The main reason is the incorrect angle of the lateral surface along the tooth profile. To improve the characteristics of the finished surface, the cutting wedges of a form-cutting tool are often created with an inclination angle of 3–4° [19–22]. Since the tooth profile must be maintained within a limited tolerance when a cutter is sharpened repeatedly, substantially smaller flank angles are used in hob tools. Due to the high yield strength and high elasticity of processed materials, when working with hard materials, excessive friction occurs on the side surface. The workpiece and flange stick together due to high temperatures caused by this friction, reducing surface quality and shortening tool life. A carbide cutter with replaceable carbide inserts is shown in Figure 12.12 next to the cutter used for tiny modules.

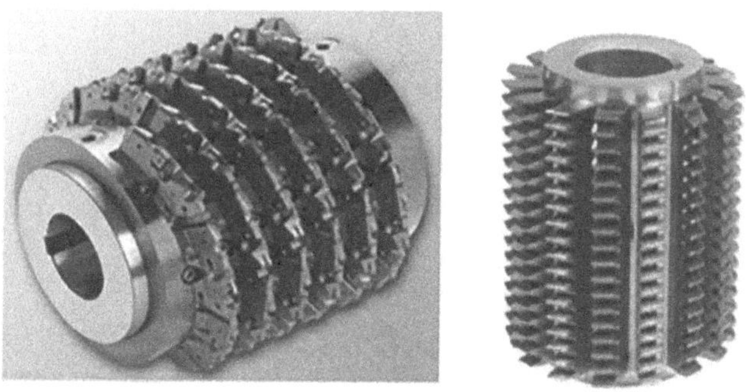

(a) (b)

Figure 12.12 (a) Carbide hobs, and (b) double- gear hobs.

12.4.6 Skiving

12.4.6.1 Turning of Hard

Turning is a very effective and simple machining method that produces high-quality surfaces for rotating body parts made from a variety of materials. Processing efficiency for each unique situation, the best combination of workpiece and tool materials and cutting parameters is determined. circumstances, geometric characteristics of the cutting tool, condition of the apparatus, and other technical data. Machine parts consisting of hardened steel annealed and special cast iron, and other difficult-to-cut materials are successfully machined. The use of CBN tools for fine turning of workpieces composed of these materials often avoids the need for grinding, the traditional method of producing surfaces with a roughness less than Ra = 1.25 μm. Currently known as "hard turning," this processing method involves machining materials with a hardness greater than 47 HRC [11–14]. Figure 12.13 depicts the "sharpening" turning model proposed by Klimenko and Manokhin [35].

12.4.6.2 Skiving of Gear

Polishing hardened gears is known as gear hobbing. Using this method, errors in the gear ends and profiles associated with heat treatment are eliminated. The end result is increased transmission accuracy. Hardened gears with a hardness of up to 63 HRC can be processed with carbide cutters [41–44]. The development of extremely sophisticated automated machine tools has made this possible, making the approach efficient and practical [45].

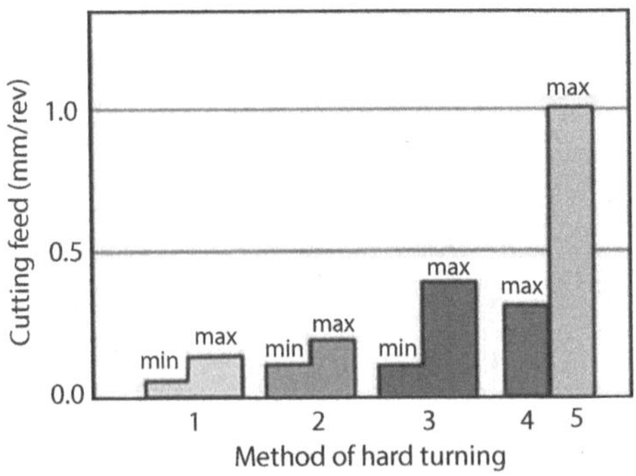

Figure 12.13 The minimum and maximum feeds: tools (f=0.05–0.12mm/rev) (1), a cylindrical (f=0.1–0.2mm/rev) (2), tools with a geometry (f=0.1–0.4mm/rev) (3), traditional turning (f=0.3mm/rev) (4), and advanced turning (f up to 1.0mm/rev) (5).

Figure 12.14 Skipping and buckling.

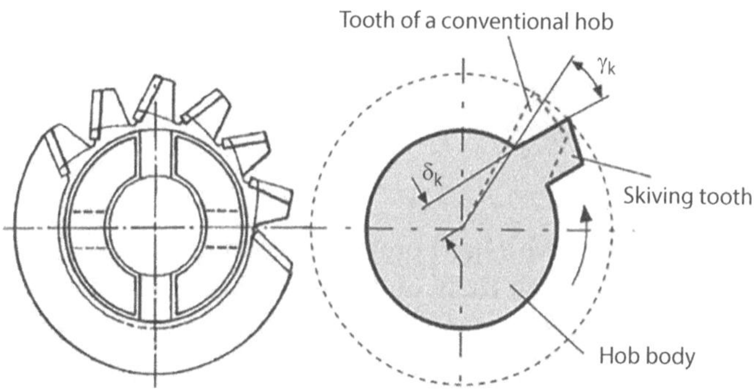

Figure 12.15 The tooth geometry of traditional and skiving hobs is compared.

Figure 12.14 shows the hobbling of hardened spur gear with a diameter of 16 mm. The body of the worm gear tool is made of high-speed steel, with carbide cutting inserts soldered onto it. Each side surface of the gear is sharpened by removing a 0.5 mm thick allowance in one pass. Benefits of this technology include the ability to change gear flanks, such as cone and crown, over a range of 100 m or more in just one pass, as well as dramatic reductions in tooling costs and cycle times. The character has achieved DIN 8 quality, which is the most accurate indicator of process quality. Figure 12.15 illustrates the unique tooth geometry of a cutter. Numerous researchers believe that the main difference between worms and cutting teeth for sharpening is the negative rake angle of the tool in the hand, which is caused by the displacement of the front surface of the tooth [43, 44]. The carbide cutter appears to be protected from impact and sudden overload by a negative rake angle. However, this is not entirely true as process modeling is sometimes too confusing, and important variables are ignored because the geometry of the tool teeth in the machine tool is not taken into account [17–19].

12.5 Coating Technology

High-performance carbide-coated tools are the result of significant advances in coating technology. These exhibit improved qualities such as wear resistance, toughness, and hardness. Chemical coating techniques have advanced new tactics and solutions for the selective coating of complicated surfaces. To enhance the performance of sol-gel coatings, new materials have been created and hybrid organic-inorganic coatings investigated [23, 24]. Developments in etchant formulation, process control, and masking techniques have proven advantageous in achieving high accuracy and reproducibility in chemical etching procedures.

12.5.1 Backdrop

These are cutting tools with hard protective coatings several micrometers thick. Up to the 1980s, TiN, TiCN, and Al2O3 produced by CVD were used to make carbide tools. However, TiN was the first PVD coating to be used by cutting factories in 1980, followed by TiAlN [24]. Currently, much work is being done to improve cutting performance (CVD coatings) [25]. These layers can serve as a bridge between the desired characteristics of the top layer and the interfacial adhesion mechanism. Due to the increased friction created by this hardening property between the insert tip and the workpiece, higher temperatures are encouraged, resulting in diffusion and oxidation.

12.5.2 Significance of Coating Layer

Multiple cathodic arc deposition is used to apply PVD and CVD coatings [46]. PVD instruments have been demonstrated to have higher brittleness than CVD instruments. PVD-coated tools provide greater finishing potential. However, CVD tools show improved roughing potential [47]. Layers and substrates of PVD coatings are studied [48]. Tools with CVD (moderate temperature) coatings have increased durability, which reduces the

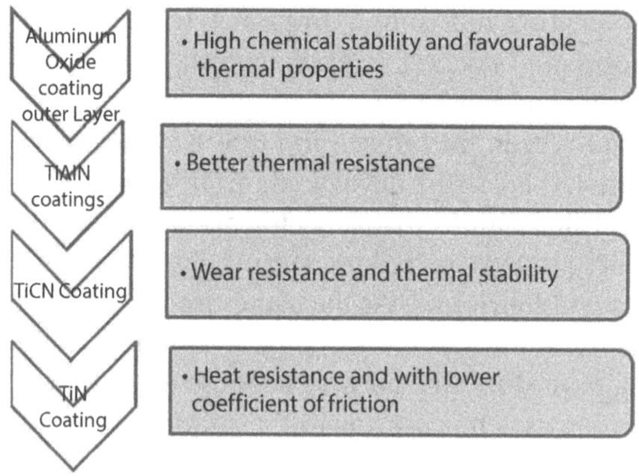

Figure 12.16 Characteristics of different coatings [8].

likelihood of cracking [49]. There is extensive research into the different types of coatings and how they are used in specific situations [46, 48]. A reduced nose radius results in a reduced contact area between the tool and the workpiece, which increases the force and wear of the cutting tool flank edge [48]. Figure 12.16 shows the features of various coatings. The substrate determines the mechanical characteristics of various coatings [36, 49]. TiN is a durable material with excellent wear and corrosion resistance [50].

For better finishing, several coatings are applied to the tool surface, including TiN, TiCN, and TiAlN [20]. The impact of tool coatings on process efficiency is significant. Coated carbide tools can significantly reduce friction [21–23]. Tools used for dry cutting are usually made of carbide [24, 25]. The production rate of CVD is higher than that of TiCN (single layer) [27, 37, 50]. Laser surface engineering methods take advantage of advances in laser technology, making surface modification more accurate and efficient [38]. Electroplating procedures have been improved using chemical methods, resulting in improved coating quality, higher deposition rates, and reduced environmental impact [39].

12.6 Concluding Remarks

The cutting tool industry will focus on the ongoing demand for creative tooling concepts. It is significant to note that the primary aspects to be taken into account when utilizing cutting tools for machining high-strength low-alloy steel are cutting tool wear characteristics, tribological evaluation, tool life experiment, and mechanical behavior. Selection of an appropriate coating mix and appropriate structure may be necessary from the standpoint of coating technology today. Due to its excellent tribological phenomena, cement-coated carbide tools have received scientific attention and have been introduced in cutting plants. New coating techniques and materials are being developed through research for cutting tool applications. In the manufacturing sector, economic and environmental factors are leading to ever-increasing cutting speeds, as well as decreasing use of coolants and lubricants. At the same time, the need to extend tool life requires constant further development of carbide technology. A crucial manufacturing procedure that boosts productivity is dry cutting. The tool will function more dynamically due to the continual trend towards better cutting processes and technological features. In general, mechanical engineering relies heavily on the utilization of nanostructured PVD coatings. When turning hardened steel, a modified tool design may be advantageous to provide better finishes on improved tool materials.

References

1. Liu, J., Khan, A.S., Takacs, L., Meredith, C.S., Mechanical behavior of ultra-fine-grained/ nanocrystalline titanium synthesized by mechanical milling plus consolidation: experiments, modeling and simulation. *Int. J. Plast.*, 64, 151–163, 2015, https://doi.org/10.1016/j.ijplas.2014.08.007.
2. Ding, H. and Shin, Y.C., Dislocation density-based grain refinement modeling of orthogonal cutting of titanium. *J. Manuf. Sci. Eng.*, 136, 41003–41011, 2014.
3. Melkote, S.N., Liu, R., Fernandez-Zelaia, P., Marusich, T., A physically based constitutive model for simulationof segmented chipformation in orthogonal cutting of commercially pure titanium. *CIRP Ann. Manuf. Technol.*, 64, 65–68, 2015, https://doi.org/10.1016/j.cirp.2015.04.060.
4. Subbaiah, K.V., Raju, C., Pawade, R.S., Suresh, C., Machinability investigation with wiper ceramic insert and optimization during the hard turning of AISI 4340 steel. *Mater.Today Proc.*, 18, 445–454, 2019.
5. Das, R.K., Sahoo, A.K., Mishra, P.C., Kumar, R., Panda, A., Comparative machinability performance of heat treated 4340 Steel under dry and minimum quantity lubrication surroundings. *Procedia Manuf.*, 20, 377–385, 2018.
6. Markopoulos, A.P., Habrat, W., Galanis, N.I., Karkalos, N.E., Modelling and optimization of machining with the use of statistical methods and soft computing, in: *Design of Experiments in Production Engineering*, J.P. Davim (Ed.), pp. 39–88, Springer International Publishing, Champp, 2016, https://doi.org/10.1007/978-3-319-23838-8_2.
7. Goel, S., Martinez, F.D., Chavoshi, S.Z., Khatri, N., Giusca, C., Molecular dynamics simulation of the elliptical vibration-assisted machining of pure iron. *J. Micromanuf.*, *1*, 6–19, 2018.

8. Sert, A. and Celik, O.N., Characterization of the mechanism of cryogenic treatment on the microstructural changes in tungsten carbide cutting tools. *Mater. Charact.*, *150*, 1–7, 2019.
9. Swamini, A., Chopra, V., Sargade, G., Metallurgy behind the Cryogenic Treatment of Cutting Tools: An Overview. *Mater. Today Proc.*, *2*, 1814–1824, 2015.
10. Dhananchezian, M., Study the machinability characteristics of Nicked based Hastelloy C-276 under cryogenic cooling. *Measurement*, *136*, 694–702, 2019.
11. Davis, B., Dabrow, D., Newell, R., Miller, A., Schueller, J., Xiao, G., *et al.*, Chip morphology and chip formation mechanisms during machining of ECAE- processed titanium. *J. Manuf. Sci. Eng.*, 140, 3, 031008, 2017, https://doi.org/10.1115/1.4038442.
12. Habrat, W., Motyka, M., Topolski, K., Sieniawski, J., Nowotnik, A., Machinability of micro-and nanocrystalline titanium in milling process, in: *Proceedings of the 13th World Conference on Titanium*, pp. 999–1005, 2015, https://doi.org/10.1002/9781119296126.ch170.
13. Khan, A. and Maity, K., Comparative study of some machinability aspects in turning of pure titanium with untreated and cryogenically treated carbide inserts. *J. Manuf. Process.*, 28, 272–284, 2017. https://doi.org/10.1016/j.jmapro.2017.05.018.
14. Sivalingam, V., Sun, J., Yang, B., Liu, K., Raju, R., Machining performance and tool wear analysis on cryogenic treated insert during end milling of Ti-6Al-4V alloy. *J. Manuf. Process.*, 36, 188–196, 2018.
15. Montenegro, P., Gomes, J., Rego, R., Borille, A., Potential of niobium carbide application as the hard phase in cutting tool substrate. *Int. J. Refract. Met. Hard Mater.*, 70, 116–123, 2018.
16. Saini, A., Pabla, B.S., Dhami, S.S., Developments in cutting tool technology in improving machinability of Ti6Al4V alloy: a review. *Proc. Inst. Mech. Eng. B J. Eng. Manuf.*, 230, 1977–1989, 2016, https://doi.org/10.1177/0954405416640176.
17. Bag, R., Panda, A., Sahoo, A.K., Kumar, R., Cutting tools characteristics and coating depositions for hard part turning of AISI 4340 martensitic steel: A review study. *Mater. Today Proc.*, 26, 2073–2078, 2020.
18. Sumiya, H. and Harano, K., Innovative ultra-hard materials: Binderless nanopolycrystalline diamond and nano-polycrystalline cubic boron nitride. *SEI Tech. Rev.*, 82, 21–26, 2016.
19. Ji, W., Zou, B., Zhang, S., Xing, H., Yun, H., Wang, Y., Design and fabrication of gradient cermet composite cutting tool, and its cutting performance. *J. Alloys Compd.*, 732, 25–31, 2018.
20. VakkasYıldırım, C., Experimental comparison of the performance of nanofluids, cryogenic and hybrid cooling in turning of Inconel 625. *Tribol. Int.*, *137*, 366–378, 2019.
21. Cai, X.C., Song, J., Yang, T.T., Peng, Q.M., Huang, J.Y., Shen, T.D., A bulk nanocrystalline Mg–Ti alloy with high thermal stability and strength. *Mater. Lett.*, 210, 121–123, 2018, https://doi.org/10.1016/j.matlet.2017.09.021.

22. Biswal, B., Sarkar, B., Mahanta, P. (Eds.), Characteristics During Hard Turning of Alloy Steel with Untreated and Cryotreated Cermet Inserts, in: *Advances in Mechanical Engineering*, Lecture Notes in Mechanical Engineering, Springer, Singapore, 2018.
23. Naseri, R., Kadkhodayan, M., Shariati, M., Static mechanical properties and ductility of biomedical ultrafine-grained commercially pure titanium produced by ECAP process. *Trans. Nonferrous Met. Soc. China (English Ed.)*, 27, 1964–1975, 2017, https://doi.org/10.1016/S1003-6326(17)60221-8.
24. Navarro-Devia, J.H., Amaya, C., Caicedo, J.C., Aperador, W., Performance evaluation of HSS cutting tool coated with hafnium and vanadium nitride multilayers, by temperature measurement and surface inspection, on machining AISI 1020 steel. *Surf. Coat. Technol.*, 332, 484–493, 2017.
25. Santhanakrishnan, M., Venkateshwaran, N., Rajkumar, M., Vignesh, T., Performance evaluation of Ni/Nano SiC coated tool insert for machining SS316l using Response Surface Methodology (RSM). *Mater. Today Proc.*, 47, 4671–4675, 2021.
26. Swain, P.K., Mohapatra, K.D., Swain, P.K., Advances in machine technology and improvements in Al-SiCp nano composite materials by using multilayer coated carbide tool. *Mater. Today Proc.*, 44, 1071–1078, 2021.
27. Habrat, W., Motyka, M., Topolski, K., Sieniawski, J., Evaluation of the cutting force components and the surface roughness in the milling process of micro- and nanocrystal-line titanium. *Arch. Metall. Mater.*, 61, 1033–1038, 2016, https://doi.org/10.1515/ amm-2016-0226.
28. Sui, H., Zhang, X., Zhang, D., Jiang, X., Wu, R., Feasibility study of high-speed ultrasonic vibration cutting titanium alloy. *J. Mater. Process. Technol.*, *247*, 111–120, 2017.
29. Ultrasonically Assisted Machining of Titanium Alloys. Available online: https://hdl.handle.net/2134/15909 (accessed on 17 March 2020).
30. Joshi, S., Tewari, A., Joshi, S.S., Microstructural characterization of chip segmentation under different machining environments in orthogonal machining of Ti6Al4V. *J. Eng. Mater. Technol.*, *137*, 011005, 2015.
31. Sousa, V.F.C. and Silva, F.J.G., Recent Advances on Coated Milling Tool Technology—A Comprehensive Review. *Coatings*, 10, 235, 2020.
32. Dumitrescu, P.K., Koshy, P., Stenekes, J., Elbestawi, M.A., High-power diode laser assisted hard turning of AISI D2 tool steel. *Int. J. Mach. Tools Manuf.*, 46, 15, 2009–2016, 2006.
33. Sousa, V.F.C. and Silva, F.J.G., Recent Advances in Turning Processes Using Coated Tools—A Comprehensive Review. *Metals*, 10, 235, 2020.
34. Muhammad, R., Mistry, A., Khan, W.S., Ahmed, N., Roy, A., Silberschmidt, V.V., Analysis of tool wear in ultrasonically assisted turning of Iranica. *B Trans. Eng.*, *23*, 1800–1810, 2016.
35. Tamerabeta, Y., Briouaa, M., Tamerabeta, M., Khoualdia, S., Experimental Investigation on Tool Wear Behavior and Cutting Temperature during Dry

Machining of Carbon Steel SAE 1030 Using KC810 and KC910 Coated Inserts. *Tribol. Ind.*, *40*, 52–65, 2018.

36. Vereschaka, A., Kataeva, E., Sitnikov, N., Aksenenko, A., Oganyan, G., Sotova, C., Influence of Thickness of Multilayered Nano-Structured Coatings Ti-TiN-(TiCrAl)N and Zr-ZrN-(ZrCrNbAl)N on Tool Life of Metal Cutting Tools at Various Cutting Speeds. *Coatings*, *8*, 44, 2018.
37. Gao, M., Li, S., Guan, W., Xie, H., Wang, X., Liu, J., Wang, H., Excellent thermal shock resistance of NiCrAlY coatings on copper substrate via laser cladding. *J. Mater. Sci. Technol.*, 130, 93–102, 2022.
38. Klocke, F., Brinksmeier, E., Weinert, K., Capability profile of hard cutting and grinding processes. *CIRP Ann.-Manuf. Technol.*, 54, 2, 552–580, 2005.
39. Vereschaka, A.A., Grigoriev, S.N., Sitnikov, N.N., Oganyan, G.V., Batako, A., Working efficiency of cutting tools with multilayer nano-structured Ti-TiCN-(Ti,Al)CN and Ti-TiCN-(Ti,Al,Cr)CN coatings: Analysis of cutting properties, wear mechanism and diffusion processes. *Surf. Coat. Technol.*, *332*, 198–213, 2017.
40. Kuo, C., Wang, C., Ko, S., Wear behaviour of CVD diamond-coated tools in the drilling of woven CFRP composites. *Wear*, *398–399*, 1–12, 2018.
41. Chang, Y.-Y., Yang, Y.-J., Weng, S.-Y., Effect of interlayer design on the mechanical properties of AlTiCrN and multilayered AlTiCrN/TiSiN hard coatings. *Surf. Coat. Technol.*, 389, 125637, 2020.
42. Astakhov, V.P., *Geometry of Single-Point Turning Tools and Drills, Fundamentals and Practical Applications*, Springer, London, 2010.
43. Astakhov, V.P. and Davim, P.J., Tools (geometry and material) and tool wear, in: *Machining: fundamentals and recent advances*, P.J. Davim (Ed.), pp. 29–58, Springer, London, 2008.
44. Guimarães, B., Guedes, A., Fernandes, C.M., Figueiredo, D., Bartolomeu, F., Miranda, G., Silva, F.S., WC-Co/316L stainless steel joining by laser powder bed fusion for multi-material cutting tools manufacturing. *Int. J. Refract. Met. Hard Mater.*, 112, 106140, 2023.
45. Chandrasekhar, U., Yang, L.-J., Gowthaman, S., *Innovative Design, Analysis and Development Practices in Aerospace and Automotive Engineering (I-DAD 2018)*, vol. 2, Springer Singapore, 2019.
46. Ye, F., Li, Y., Sun, X., Yang, Q., Kim, C.-Y., Odeshi, A.G., CVD diamond coating on WC-Co substrate with Al-based interlayer. *Surf. Coat. Technol.*, *308*, 121–127, 2016.
47. Chandran, M., Sammler, F., Uhlmann, E., Akhvlediani, R., Hoffman, A., Wear performance of diamond coated WC-Co tools with a CrN interlayer. *Diamond Relat. Mater.*, *73*, 47–55, 2017.
48. An, S., Foest, R., Fricke, K., Riemer, H., Fröhlich, M., Quade, A., Schäfer, J., Weltmann, K.-D., Kersten, H., Pretreatment of cutting tools by plasma electrolytic polishing (PEP) for enhanced adhesion of hard coatings. *Surf. Coat. Technol.*, 405, 126504, 2021.

49. Ramasubramanian, K., Arunachalam, N., Rao, M.S.R., Wear performance of nano-engineered boron doped graded layer CVD diamond coated cutting tool for machining of Al-SiC MMC. *Wear*, *426–427*, 1536–1547, 2019.
50. Nayak, R.K., Pradhan, M.K., Sahoo, A.K., Machining of nanocomposites. *Int. J. Automot. Mech. Eng.*, 16, 4, 7315–7340, CRC Press, 2022.

13

Biomedical Applications of Hard Materials: Implants, Coatings and Drug Delivery Systems for Medical Devices

Fatema Tuz Zohera[1†], Abul Kalam Azad[2†] and Madhusmruti Khandai[3*†]

[1]International Islamic University Malaysia, Kuantan, Pahang, Malaysia
[2]University College of MAIWP International (UCMI), Kuala Lumpur, Malaysia
[3]Royal College of Pharmacy and Health Sciences, Berhampur, Odisha, India

Abstract

Hard materials are of utmost importance in the biomedical area, serving either as toxins or treatments. Additionally, they are employed in the field of medicine for the purposes of diagnosis and therapy, with a notable multitude of applicability. In addition to established therapies, a range of novel application possibilities are appearing and garnering the attention of academics worldwide. Materials represent a revolutionary change in the realm of delivery of drugs including their use for healthcare purposes. Being able to be used in many ways and adjust to different situations has not only improved the results of therapy but also greatly lessened the negative consequences. A wide range of hard materials are utilized in the domains of dental and medical care. It provides an explanation of the present metallic, ceramic, and polymer materials used in healthcare, along with the associated challenges. Additionally, it explores conceivable future materials that could address these issues. The term "hard biomaterial" is initially contrasted with "soft Biomaterial", and the importance of smart link among hard materials with biological tissue, known as biosis-abiosis, is illustrated. The discussion focuses on the tissue compatibility of titanium, as well as the surface treatment methods used to enhance its ability to interact with hard tissues. The present challenges in this area are also addressed, along with the need for more study in materials science.

**Corresponding author*: drmskhandai@gmail.com
†Authors have equally contributed
Madhusmruti Khandai: Orcid: https://orcid.org/0000-0003-4024-9735

Inamuddin, Tariq Altalhi and Mohammad Abu Jafar Mazumder (eds.) Nanostructured Hard Materials: Industrial Applications, (395–414) © 2026 Scrivener Publishing LLC

Additionally, the use of hard materials in the field of regenerative healthcare is explored. Furthermore, the purpose of the article represents to examine the promising developments and potential in biomaterials, which anticipating medical settings in which they enable accurate, specific and individualized drug administration. Biomaterials possess the capacity to fundamentally transform the medical field through the provision of personalized therapies that cater to the unique requirements of each patient. It presents significant potential for enhanced therapeutic effectiveness accompanied by reduced adverse effects, thereby securing greater prospects for pharmaceuticals.

Keywords: Hard materials, metals, ceramics, polymers, biomaterials, surface treatment

13.1 Introduction

Novel elements play a crucial role in driving progress in health care technology, providing fresh and inventive alternatives for both design and implementation [1]. Advancements in software, electronic devices and robots have also played a pivotal role in the progress of healthcare. The creation and utilization of novel hard materials are crucial in determining the trajectory of medical advancements. Investigative instruments, implantation equipment, therapeutic platforms and regenerative therapy are all touched by this progress, which surpasses earlier boundaries. Figure 13.1 focuses on the basic demarcation between hard and soft materials.

Metal plates were initially developed by Lane in 1985 for the purpose of fixing bone fractures, marking the beginning of contemporary histories with the use of metals used in making implants [2]. Metallic implantation

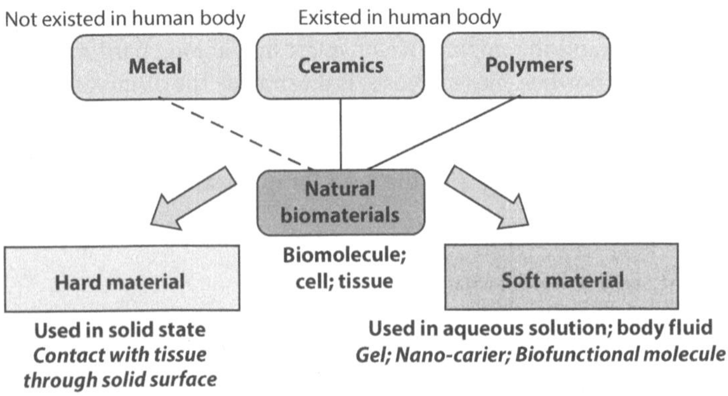

Figure 13.1 Concept of hard biomaterials and soft biomaterials.

devices have issues with rusting and lack of sturdiness throughout their early stages of research [3]. Following the development of 18-8 stainless steel in the 1920's, which exhibited significantly better endurance against rusting compared to other materials at that moment, it quickly captured the attention of doctors. Subsequently, metal implants saw significant advancement and were utilized in medical practices.

Research has revealed, metallic implants releases ions of metal, having damaging effects on the organs; and therefore, do not effectively sustain the individual's health over an extended period [4]. Various factors can hinder the functionality as well as activation of the implants, including inadequate osteointegration, inflammatory conditions, infections due to bacteria, person's age systemic illness and implant dimensions along with design. Various features and unexpected relationship between cells are seen on the outermost layers of implant components. It is common for microbes to invade implants while undergoing surgical procedures, which hinders the surface's ability to integrate with the body [5]. The most appropriate choices for implants for various regions of human body tend to be stainless steel, commercial pure titanium (Cp-Ti) Ti alloys, aluminum compounds, cobalt metals, gold compounds, zirconium and others. This is due to their outstanding biological compatibility and low foreign body engagement in contrast with alternative products [6]. Implants made of titanium (Ti) exhibit favorable biocompatibility characteristics. However, their inadequate resilience to wear might result in allergy symptoms leading to discomfort in specific site of action of the implant. The discomfort further weakens the joint and potentially trigger bacteria borne infection and thus dislocation of the implant within few years of being implanted [7].

Multiple events can occur involving the device and connective tissue following the implantation process, including ion transfer, protein absorption, coagulation of blood as well as cell adhesion [8]. Osteointegration of native muscle and tissues by implants is challenging. Bioactive coatings like osteoconductive, environment friendly, tough antimicrobial coatings that promote long-term antibiotic discharge, antimicrobial paints within clinical settings and ant-corrosion coatings can help implant components attain desired chemical, physiological, and multifaceted characteristics.

13.2 Hard Materials for Biomedical Application

Hard materials refer to metallic substances, crystallized ceramics, glasses and rigid polymeric substances. A crucial quality that hard materials must possess is the ability to maintain their dimensions without any changes in the solid state along with the following attributes:

1. Undistorted,
2. Intact, and
3. ensuring the durability of solid-state over an extended period of time
4. Biochemical processes and tissue connectivity like growth of bones, safeguarding bone integration, soft tissue attachment, thrombus avoidance along with inhibition of biofilm
5. substantial rigidity to avoid stress while restoration and biological tissue-like distortion afterwards therapy and
6. frictional erosion does not do any harm to healthy tissue
7. Put simply, the necessary qualities are outstanding durability, fracture resilience, fatigue endurance and anti-corrosion properties.

13.2.1 Drug-Delivery Systems

Biomaterials are extensively utilized to develop methods for delivering drugs, which is one of their most notable uses [9]. Such objects are capable of being engineered to enclose and dispense pharmacological substances in a regulate ad precise approach. The technique provides numerous benefits, such as enhanced medicine effectiveness, less adverse reactions, and extended discharge for long-term illness.

13.2.2 Tissue Engineering

Scientists created substrates composed of biologically attuned substances such as hydrogels, ceramics and polymers to facilitate cell development and regrowth. These scaffolding structures serves as a basis of support for cells to adhere, multiply and specialize, resulting in the development of fully functioning substitute organs and tissues, including bones [10, 11]. Tissue engineers utilizes biomaterials to fabricate functioning tissues primarily the purposes of regeneration therapy and transplanting.

13.2.3 Implantable Devices

Biomaterials play a crucial role for the advancement of medical implants, including prosthetic connections, pacemakers, and stents. Such substances demand biocompatibility so that they may effortlessly merge within the cells of the body so as to prevent eliciting an allergic response. Metallic titanium and polymers with biodegrability, which are biomaterials have significantly improved the standards lifestyle of individuals requiring such medical equipment [12–15].

13.2.4 Diagnostic Tools

Hard materials are additionally crucial towards the advancement of diagnosis equipment. The initial stages of illness detection might be achieved by utilizing nanoparticles that are encapsulated using particular biomolecules. For instance, when antibody is attached to magnetized nanoparticles, they're able to specifically attach to cancer cells, enabling the cells to separate using magnets and further detection [16, 17].

13.3 Types of Implant Coatings

13.3.1 Biocompatible Coatings

There is a continual interaction between the surrounding living tissues and the orthopedic implantation device. The external appearance of the material has a major impact on the biological body's reaction to the implanted device. Any adverse response, including discomfort and swelling, may occur if the outer layer of the implant becomes subjected to corrosion and release charges into the bloodstream [18–20].

The implants necessitate exterior interface alterations to enhance their characteristics, hence preventing premature removal as well as extending duration of their active life-span. Calcium orthophosphates function as osteoconductive coatings which are employed as scaffolding or adhesive agents [21–25]. The combination of Chitosan and gelatin matrices is highly suitable for usage as scaffolding elements to coat the implantable devices due to its robust physico-chemical features, lack of toxicities, compliance with living tissues, capacity to degrade naturally, and very little or minimum immune response [26].

13.3.1.1 Calcium Based Apatite Coatings

Noncalcified edges characterize implanted elements. Those substances include elements such as ceramics, metallic materials and polymeric components. Calcium phosphate or apatite is extensively biocompatible and osteoconductive according to research [27]. Implants with a blast apatite coating are highly biocompatible with the bones. Hydrothermal hot pressing of titanium along with hydroxyapatite ceramic coatings enhances the core density of apatite and was used to modify its surface at the same time [28].

13.3.1.2 Bone Morphogenetic Protein Coatings (BMP)

The body's own BMP protein facilitates the spontaneous repair of cartillage as well as bone tissue structures. Implant content fail to meet expectation when it comes to their ability to promote the growth of bone structures. To restore the affected areas, bony augmenting is a must. A study conducted by Yeo *et al.*, found that, acid etching and sandblasting on the external contour of Ti implants greatly improved bone proximity, leading to a spike in BMP functionality which was very impressive [29]. As a result of its bone resorption characteristics, hydroxyapatite (HAp) coatings helps to increase the biological functionality of the implants over the time. One possible way to improve mineralization and at the same time stimulate BMP output is to cover the edges of the implant with granular hydroxyapatite.

13.3.1.3 RGD Peptide Based Coatings

The 3-amino acid pattern known as arginyl-glycyl-aspartic acid (RGD) peptide facilitates cell adherence to implanted devices using integrin receptor components and thus transmits impulses to the center of the cell vi cytoplasmic signaling routes [30]. One in ten orthodontic implant device failures is attributable to premature osseointegration and failures caused by insufficient host responses (e.g. the incorporation of extracellular matrix (ECM) proteins, interfacial contacts, tissue development etc.). the outermost layer of the implant that holds the tooth can become reactive during the process that accelerates bone matrix creation by combining it with peptides that are biologically functional that are present in ECM proteins [31].

13.3.1.4 Mg Based Coating

One of the most important traces of elements highly indispensable to bones of an individual's metabolic function is called magnesium (Mg). Because of these protective coatings improve osseointegration along with high adherence, they find use in orthopedics and orthodontics [32]. Pardun *et al.*, investigated briefly the impact of zirconia-calcium phosphate coated coverings on magnesium accumulation [33]. Yttria stabilized zirconia (YSZ) and hydroxyapatite (HPa) having been mixed with various concentrations of magnesium oxide or fluoride to create this specific type of specialized coating. According to Tong *et al.*, the amount of magnesium incorporated as the protective layer affected its exterior shape, stiffness, as well as the rate at which calcium is dissolved [34].

13.3.1.5 ZnO Based Coatings

Zinc oxide (ZnO) has become a wonderful substance because it's usually non-toxic, has outstanding photocatalysis capacity to kill a wide range of infections, also inexpensive as well as biocompatible against body parts [35]. The biocompatibility along with enhanced antimicrobial properties as well as resistant to corrosive activity nature of ZnO make them useful against a wide range of natural microbes. In this way ZnO coating caused biofilms of bacterial organisms to accumulate once the microorganisms are attached onto the implanted device's exterior; however, it is usually deteriorating after 24 hours of exposition.

13.3.1.6 TiO_2 Based Coatings

Because of its consistency, responsiveness, recycling, resilience and relatively low inexpensive price, titanium dioxide also known as TiO_2 is a wonderful coating material with excellent antimicrobial properties that are currently accessible. Several biomedical appliances are made possible by TiO_2's biological compliance, biomechanical robustness, and exceptional resistance power to corrosive activity. Hou *et al.*, effectively applied a 400 nm-thick TiO_2 thin layer of coating to the surface of a MgZn composite [36, 37]. The degree of corrosiveness of MgZn material turned out to have substantially diminished upon the implementation of TiO_2 coating. Additionally, the material was observed to be considered to a more compact along with granular grade, according to their investigation's outcome. In an earlier investigation, Yu *et al.*, employed microarc oxidation to encapsulate TiO_2 nanotubes alongside biologically active calcium and phosphorus in addition to antimicrobial activity of Ag [38]. In addition, cellular investigations provided extra proof that the manufactured layers enhanced the adherence along with survivability of endothelial cells and exhibited a widely significant level of biocompliance.

13.3.1.7 Carbon Based Coatings

The biomedical community has paid considerable focus on durable lacquer coating (DLC) since they possess excellent biocompliance, that is crucial for DLC implementations [39]. The utilization of DLCs in biomedical pressure monitoring device has been contemplated due to their respective positions elevated piezo resistivity [40]. Prosthetic cardiovascular valves, surgical cords, prosthetic joint devices, as well as circulatory catheters have all been widely coated with DLCs. Coating with aluminum along with its corresponding alloys *via* DLCs makes them biomedically useful. Because of the necessity of preserving individual's extensive degrees of movement damage and friction are unavoidable in orthopedic replacement of prosthetic joints. Metallic implants with a DLC are ideal for insertion into the hip and knee because they are resistant to oxidation and have a relatively small coefficient of friction, two factors that are problematic inside the body of any person. Since there is a negative correlation among energy and bacterial adherence, DLC coatings may find utility in surgical equipment, thanks to their biocompatible nature and specifically their antibacterial properties [41].

13.3.1.8 TiN and CrN Based Coatings

Nitride-based coating are being utilized to improve the physiological effectiveness of mechanical implant components by enhancing their exterior features [42]. Titanium nitride (TiN) is commonly used as a covering for orthopedic devices due to its biocompatibility, minimal coefficient of wear and abrasion, excellent durability and the ability to resist oxidation [43]. By applying a TiN-ceramic covering onto the exterior part of the implant, diffusion of the ionized metals into the living tissue is reduced, hence impeding proliferation of microbes. Ceramic coatings not only provide an anatomical barrier separating the surface being coated along with the surroundings, but they also enhance protection against corrosion [44, 45]. Yet another improved feature pertains to the materials wearing toughness. Considering the nearly nonexistent hemolysis %, TiN demonstrates promising compatibility with the blood vessel characteristics [45]. Consequently, cardiologists utilize TiN-coatings to create pacemaker leads along with ventricular assistance equipment for individuals with coronary diseases. According to Cogan *et al.*, neurologists are investigating the use of TiN-coated electrodes while designing the surgical implants that can be used persistently for the management of spinal cord injuries [46].

13.3.2 Polymer Based Antimicrobial Coatings

Microbial linked implantation can lead to contamination which may pose a formidable challenge to the surgical implant sector, resulting in permanent destabilization, persistent discomfort, bone degradation at the afflicted region, along with various other problems. After being implanted, the microorganisms that have attached towards the external layer of the implant move quickly around the immediate vicinity and establish a colony. Nevertheless, numerous uses are constrained by the poor immediate effectiveness of antimicrobial agents, which exhibit limited susceptibility as well as toxicological properties [47]. These agents garnered significant interest owing to the following characteristics

(a) extensive resistance to disease that affect humans,
(b) sustained effectiveness,
(c) resilience in challenging settings,
(d) harmony within the natural world,
(e) simplicity in production, and
(f) financial efficiency [48].

13.3.2.1 Polymer Coatings

Polymers adaptability, affordability, and simplicity of surface alteration garnered significant interest in a range of domains [49]. Medication transportation tissue science, bioinert coatings, wound reconstruction, as well as bone grafting technologies are just a few of the many uses for polymers. Biomedical uses of polymers with inherent antibacterial properties. 3D granular frameworks enabling neural development in periphery as well as the creation of engineered circulatory systems can be progressively fabricated using biodegradable polymers such as PLGA, PEE, PGA, PBT and PDLLA, [50]. Perforated composite frameworks made of gelatin, PLGA, PLLA-CL, and silk fibroin, PEVA, all have been utilized to connective tissue cartilage, bone, blood vessel, as well as neuronal cell related tissue engineering applications [51]. These polymers also help naturally to regenerate cartilage tissue [52].

13.3.2.2 Coatings for Sustainable Antibiotic Release

Antibiotic-resistant biofilm is produced by prosthetic joint infections in the medical field, and this biofilm frequently produces antibiofouling or antimicrobial surfaces for the prosthetic therapy. Conventionally, a number of antibiotics are prescribed to treat diseases such biliary tract infections, endocarditis, cellulitis, and pneumonia. It must be sought through the application of conventional procedures, such as systemic antibiotic administration for the treatment and prevention of liver and renal problems. Nevertheless, insufficient antibiotic penetration into impacted areas necessitates inpatient observation [53]. It is especially important to release antibiotics in a more efficient manner; hence, several *in situ* drug-dispensing beds are being considered to build drug-eliminating surfaces. The capacity to provide the therapeutic dosage at or near the target locations, enhancing pharmacological effectiveness and limiting adverse effects, is the main advantage of *in situ* drug release. A number of methods are used to achieve sustainable medication release: (a) biocidal agents are trapped inside the device; (b) antibiotics are adsorbed in porous structures; (c) drug/antimicrobial substances are put onto the surface; (d) antibiotics are loaded into interior cavities [54].

13.3.2.3 The Properties of Coatings Systems and Their Resistivity to Corrosion

After implantation, metal ions will infiltrate into the surrounding tissues from the surface of the implant materials—a situation that can be scientifically defined as surface deterioration can be avoided.

The blood plasma contains a variety of ions that regularly interact with the implant surface, including Na^+, K^+, Cl^-, and SO_4^-. It is common knowledge that each material has a specific potential (voltage) that determines how corrosion protection develops. Pitting corrosion may occur when the body current is above the limit, perhaps leading to the breakdown of the passive layer on the implants. A microhole that forms on the implant's surface and causes metal ions to be dramatically liberated and deposited in the surrounding tissues is known as pitting. T An immunogenic response is often elicited by the deposition of metal ions, and subsequently, immune cells are drawn to the site of the inflammation. Serious health problems like pain, bacterial infections, morbidity, and mortality are brought on by the aforementioned circumstance. Therefore, it is necessary to alter the implant's surface to enhance corrosion resistance. However, surface alteration shouldn't have an impact on the mechanical behavior of the material. Various materials have been employed to improve the performance of implant materials, such as metal oxide coatings, biopolymers, metal nitrides, and ceramics. Biocompatible ceramic coatings increase corrosion resistance, mineralization, and surface hardness. Zirconia (ZrO_2) in particular showed resilience and corrosion resistance to bodily fluids such fake saliva and blood plasma [55]. However, taking antibiotics for an extended period of time may harm your organs, which increases morbidity and death. Producing antimicrobial treatments using ceramic nanocomposite coatings that show favorable response even to multidrug resistant bacterial strains can minimize the need of antibiotics and their associated negative effects [56].

13.4 Ceramics

Ceramics are produced from inorganic material at elevated temperatures. Typical examples of ceramics based on calcium and silica include hydroxyapatite (HAp), zirconium oxide (ZrO_2), aluminum oxide (Al_2O_3) and bioactive glass (BG). Because of their exceptional biocompatibility, these glazed systems are a fantastic option for applications involving biomedical implants. Based on how potentially these systems react with the human body, ceramic implants can be classified majorly in to three different classes namely as bioactive, bioinert, or bioresorbable [57]. Utilizing bioactive ceramics allows implant sites to be more reactive and to engage with surrounding cells. Fluorapatites and HA are typical examples of bioactive ceramics [58]. On the other hand, when bioinert ceramics are implanted, they physically link with the host tissues without displaying any response [59].

The primary benefits of ceramics are their immunogenicity post-implantation and their lack of disease transmission risk [60]. Other notable advantages are low toxicity, enhanced resistivity to compressive stress, and encouragement for the growth of new hard tissue. Ceramics is known for its exceptional toughness and stress-resistant qualities because of its high elastic moduli. They also have a sluggish rate of crack initiation and propagation over time, which considerably reduces implant dependability [61]. Their application as bioimplants is further restricted by their brittleness, fracture toughness, and composition. The brittleness factor common to all ceramics restricts its performance for sustaining load (hip implants). A significant discrepancy associated with the characteristics between glazed system and bone can prevent the burden from passing through the latter, ultimately resulting in bone failure [62].

13.5 Commercially Available Multifunctional Coatings

In the last fifty years, the biomedical sector has employed a diverse array of implants for a varied range of applications. Nevertheless, the market has seen very little commercially available materials and compounds. Among these was the TiN-coated Co-based alloy created by Sree Chitra Tirunal Institute for Medical Sciences and Technology (SIRTRA), in India, which also generated a low-cost heart valve [63]. For the purpose of reconstructing cranial bones, Marcin Kozakiewicz and colleagues created specialized zirconium oxide implants [64]. A multipurpose TiN coating that is resistant to corrosion and wear has been developed by the company "Copper Development Association Inc." Additionally, to prevent restenosis and excessive neointimal development, clinical professors Cassignol and Saad created metallic stents coated with sirolimus, a drug-eluting polymer. According to Cassignol *et al.*, the United States authorized the drug-eluted stents that were created for use in patients undergoing percutaneous coronary intervention (PCI) [65]. According to Houssam Sahwil, DDS Laboratories created metal-free, FDA-approved zirconia dental implants with low elastic modulus, white color, and strong flexural strength to increase longevity. A hydroxyapatite-based ocular implant was created by Gradinaru *et al.* specifically for patients who have been disemboweled [66]. Furthermore, a number of metal oxide polymers and nanocomposites are waiting to be commercialized.

13.6 Surface Modification

Any substance inserted into the body during an implant process is regarded as foreign material for human body system. In case the foreign object that has been implanted is not compatible, then layers of fibrous tissues, often known as scar tissues, begin to grow between the implanted device and bio tissues. Implant failure eventually results from the implant's inability to get integrated with the osteo system of the host bone owing to the formation of scar tissue. Thus, full integration of bioimplants with host body system is main obligation for a successful implementation of the implants [67]. Biomedical gadgets' surface morphology and chemistry controls the bio-responses and reaction of these devices along with their working. To improve osteoconductivity and biocompatibility, it has been suggested that surface modification of biometallic materials be used to obtain the desired characteristics and raise the implant success rate. Effective surface modification will have a long-term negative impact on the biomedical implant device's bulk functionality and attributes [68, 69]. The benefit of integrating these devices with biological system and biomaterials' capacity to support loads can significantly raise the success rate of bioimplants.

Through surface modification of these devices that get adjoined to human environment, researchers have attempted to improve the bio-consolidation of implants. There are 2 methods taken into consideration for altering the implants' surface. The first method involves enclosing the metallic strata with coatings made of organic or inorganic substrates without changing the implant stratum [70]. The second method demands for engaging the surface modified layers, also known as conversion coatings, in which a substrate's strata is altered to produce a modest enhancement in thickness by chemical method. In this instance, creating conversion coatings involves the substrate constituents. To improve the mechanical interlocking of coatings, the bed should be grinded and polished since it's necessary before going for conversion of coating. It is advised to carry modification of the surface by employing an overlay coating because this procedure is crucial [71]. Recently, surface modification and thin film deposition have been coupled to achieve the synergy of both properties.

13.7 Conclusion

Fast osseointegration makes the qualities of the interface between the implant and bone important. Multifunctional coatings are applied based on implant application to enhance both the strength and quality of bone-implant interaction. Superior mechanical and corrosion resistance qualities characterize ideal implants; nevertheless, to attain the superior surface qualities, coatings with a polymer-based antibacterial agent, osteoconductive coating, and biocompatible coating work well to prolong the life of the implant. To enhance the execution of primary materials, such as alloys of metallic origin or metals, polymers, ceramics, etc., surface modification is a necessary step. Surface modification is mostly used in the biomedical industry to improve a material's mechanical stability, bioactivity, and biocompatibility. The three main uses of materials in the human body mentioned above involve various kinds of materials. Zirconia and its allotropes as well as alumina coatings were applied to the surface of Ti implants for usage in the skull. The three main uses of materials in the human body mentioned above involve various kinds of materials. For example, TiN/TiO_2, ZrN, and TiON on orthodontic wire are examples of coatings that are utilized in oral applications to improve resistivity to corrosive reactions and prevent microbial especially bacteria invasive reactions in bionic saliva solutions. Dental abutments with hydroxyapatite derived coatings have better osseointegration characteristics and a lower chance of bacterial infection. To improve the longevity of dental implants, cardiac stents, and orthopedic implants, drug-eluting polymer coatings are also successfully used in these devices to release antibiotics and deliver morphogenetic proteins [72]. To protect the implant material from wear and degradation, wear resistance coatings have also received a lot of attention. Ankle implants, total hip and knee replacements, and their multilayer coatings all had their surfaces altered using TiC, ZrC, AlCrN, and TiZrAlN. Furthermore, coatings made of metal oxide nanocomposites and HAp are used to enhance the performance of HAp and lessen its dissolution behavior. Simultaneously, ceramic coatings including silver, zinc, and copper significantly decreased both bacterial invasion and biocompatibility. Its surface modification technique may help prevent prolonged use and the negative effects that follow. Numerous surface modification techniques are employed, including spin, dip, sol-gel coating etc., vapor deposition utilizing chemicals or physical methods, electrochemical deposition, electrodeposition, and electroless deposition. Based on our review of the literature, we conclude that the vapor deposition by physical method is a useful technique for strata modification because of its inherent qualities, which include higher crystalline fill without defects on the coating substrate along with enhanced adhesion strength. These attributes improve the execution of the surface-coated implant even in a hostile setting with bodily fluids. This research has conducted a comprehensive assessment of various bioactive coatings for implants that have multifunctional qualities and are beneficial to patients.

References

1. Joudeh, N. and Linke, D., Nanoparticle Classification, Physicochemical Properties, Characterization, and Applications: A Comprehensive Review for Biologists. *J. Nanobiotechnol.*, 20, 262, 2022.
2. Lane, W.A., Some remarks on the treatment of fractures. *Br. Med. J.*, 1, 861, 1895.
3. Sherman, W.O., Vanadium steel bone plates and screws. *Surg. Gynecol. Obstet.*, 14, 629, 1912.
4. Sansone, V., Pagani, D., Melato, M., The Effects on Bone Cells of Metal Ions Released from Orthopaedic Implants. A Review. *Clin. Cases Miner. Bone Metab.*, 10, 34–40, 2013.
5. Subbiahdoss, G., Kuijer, R., Grijpma, D.W., van der Mei, H.C., Busscher, H.J., Microbial Biofilm Growth vs. Tissue Integration: "The Race for the Surface" Experimentally Studied. *Acta Biomater.*, 5, 1399–1404, 2009.
6. Eisenbarth, E., Velten, D., Müller, M., Thull, R., Breme, J., Biocompatibility of Stabilizing Elements of Titanium Alloys. *Biomaterials*, 25, 5705–5713, 2004.
7. Gautam, S., Bhatnagar, D., Bansal, D., Batra, H., Goyal, N., Recent Advancements in Nanomaterials for Biomedical Implants. *Biomed. Eng. Adv.*, 3, 100029, 2022.
8. Zakaria, O., Madi, M., Kasugai, S., A Novel Osteogenesis Technique: The Expansible Guided Bone Regeneration. *J. Tissue Eng.*, 3, 2041731412441194, 2012.
9. Pires, P.C., Mascarenhas-Melo, F., Pedrosa, K., Lopes, D., Lopes, J., Macário-Soares, A., Peixoto, D., Giram, P.S., Veiga, F., Paiva-Santos, A.C., Polymer-Based

Biomaterials for Pharmaceutical and Biomedical Applications: A Focus on Topical Drug Administration. *Eur. Polym. J.*, 187, 111868, 2023.

10. Alipour, M., Baneshi, M., Hosseinkhani, S., Mahmoudi, R., Jabari Arabzadeh, A., Akrami, M., Mehrzad, J., Bardania, H., Recent Progress in Biomedical Applications of RGD-Based Ligand: From Precise Cancer Theranostics to Biomaterial Engineering: A Systematic Review. *J. Biomed. Mater. Res. A*, 108, 839–850, 2020.
11. Dhavalikar, P., Robinson, A., Lan, Z., Jenkins, D., Chwatko, M., Salhadar, K., Jose, A., Kar, R., Shoga, E., Kannapiran, A., *et al.*, Review of Integrin-Targeting Biomaterials in Tissue Engineering. *Adv. Healthcare Mater.*, 9, e2000795, 2020.
12. Chatterjee, S., Saxena, M., Padmanabhan, D., Jayachandra, M., Pandya, H.J., Futuristic Medical Implants Using Bioresorbable Materials and Devices. *Biosens. Bioelectron.*, 142, 111489, 2019.
13. Jiao, G.H., Miao, J.J., Wu, K., Recent Advance of Stainless Steel Used In Non-Active Surgical Implantable Medical Device and Regulatory Perspective. *Zhongguo Yi Liao Qi Xie Za Zhi*, 46, 312–317, 2022.
14. Oate, J.I., Comin, M., Braceras, I., Garcia, A., Viviente, J.L., Brizuela, M., Garagorri, N., Peris, J.L., Alava, J.I., Wear Reduction Effect on Ultra-High-Molecular-Weight Polyethylene by Application of Hard Coatings and Ion Implanation on Cobalt Chromium Ally, as Measured in a Knee Wear Simulation Machine. *Surf. Coat. Technol.*, 142–144, 1056–1062, 2001.
15. Gigante, A., Bottegoni, C., Ragone, V., Banci, L., Effectiveness of Vitamin-E-Doped Polyethylene in Joint Replacement: A Literature Review. *J. Funct. Biomater.*, 6, 889, 2015.
16. Mathew, A.S., Sreenivasan, K., Mohanan, P.V., Kumary, T.V., Mohanty, M., Polyurethane Degradation in the Biological Milieu. *Trends Biomater. Artif. Organs*, 19, 115–121, 2006.
17. Pawelec, K.M., Tu, E., Chakravarty, S., Hix, J.M.L., Buchanan, L., Kenney, L., Buchanan, F., Chatterjee, N., Das, S., Alessio, A., *et al.*, Incorporating Tantalum Oxide Nanoparticles into Implantable Polymeric Biomedical Devices for Radiological Monitoring. *Adv. Health Mater.*, 12, e2203167, 2023.
18. Laing, T.J., Schottenfeld, D., Lacey, J.V., Gillespie, B.W., Garabrant, D.H., Cooper, B.C., Heeringa, S.G., Alcser, K.H., Mayes, M.D., Potential Risk Factors for Undifferentiated Connective Tissue Disease among Women: Implanted Medical Devices. *Am. J. Epidemiol.*, 154, 610–617, 2001.
19. Dunne, N., Tzagiollari, A., Sahebalzamani, M., Dunne, T.J., Acrylic Cements for Bone Fixation in Joint Replacement, in: *Joint Replacement Technology*, Woodhead Publishing, Sawston, UK, 2021.
20. Pidhatika, B., Widyaya, V.T., Nalam, P.C., Swasono, Y.A., Ardhani, R., Surface Modifications of High-Performance Polymer Polyetheretherketone (PEEK) to Improve Its Biological Performance in Dentistry. *Polymers*, 14, 5526, 2022.
21. Piconi, C. and Sprio, S., Oxide Bioceramic Composites in Orthopedics and Dentistry. *J. Compos. Sci.*, 5, 206, 2021.

22. Leroy, A., Ribeiro, S., Grossiord, C., Alves, A., Vestberg, R.H., Salles, V., Brunon, C., Gritsch, K., Grosgogeat, B., Bayon, Y., FTIR Microscopy Contribution for Comprehension of Degradation Mechanisms in PLA-Based Implantable Medical Devices. *J. Mater. Sci. - Mater. Med.*, 28, 87, 2017.
23. Wu, S., Wu, S., Zhang, X., Feng, T., Wu, L., Chitosan-Based Hydrogels for Bioelectronic Sensing: Recent Advances and Applications in Biomedicine and Food Safety. *Biosensors*, 13, 93, 2023.
24. Rosenbloom, S.N., Kumar, P., Lasley, C., The Role of Surface Oxide Thickness and Structure on the Corrosion and Nickel Elution Behavior of Nitinol Biomedical Implants. *J. Biomed. Mater. Res. B Appl. Biomater.*, 109, 1334–1343, 2021.
25. Zhou, H., Yang, L., Gbureck, U., Bhaduri, S.B., Sikder, P., Monetite, an Important Calcium Phosphate Compound—Its Synthesis, Properties and Applications in Orthopedics. *Acta Biomater.*, 127, 41–55, 2021.
26. Manzoor, K., Ahmad, S., Soundarajan, A., Ikram, S., Ahmed, S., Chitosan based nanomaterials for biomedical applications, in: *Handbook of Nanomaterials for Industrial Applications*, pp. 543–562, Elsevier, London, UK, 2018, ISBN 9780128133514.
27. Mutsuzaki, H., Yokoyama, Y., Ito, A., Oyane, A., Formation of Apatite Coatings on an Artificial Ligament Using a Plasma- and Precursor-Assisted Biomimetic Process. *Int. J. Mol. Sci.*, 14, 19155–19168, 2013.
28. Onoki, T., Porous apatite coating on various titanium metallic materials via low temperature processing, in: *Biomaterials Science and Engineering*, InTech, Philadelphia, PA, USA, 2011.
29. Yeo, I.-S., Reality of Dental Implant Surface Modification: A Short Literature Review. *Open Biomed. Eng. J.*, 8, 114–119, 2014.
30. Matsuura, T., Hosokawa, R., Okamoto, K., Kimoto, T., Akagawa, Y., Diverse Mechanisms of Osteoblast Spreading on Hydroxyapatite and Titanium. *Biomaterials*, 21, 1121–1127, 2000.
31. Raczkowska, J., Ohar, M., Stetsyshyn, Y., Zemła, J., Awsiuk, K., Rysz, J., Fornal, K., Bernasik, A., Ohar, H., Fedorova, S., *et al.*, Temperature-Responsive Peptide-Mimetic Coating Based on Poly(N-Methacryloyl-l-Leucine): Properties, Protein Adsorption and Cell Growth. *Colloids Surf., B*, 118, 270–279, 2014.
32. Almehmadi, A.H., Effect of Magnesium-Based Coatings on Titanium or Zirconia Substrates on Bone Regeneration and Implant Osseointegration—A Systematic Review. *Front. Mater.*, 8, 482, 2021.
33. Pardun, K., Treccani, L., Volkmann, E., Streckbein, P., Heiss, C., Gerlach, J.W., Maendl, S., Rezwan, K., Magnesium-Containing Mixed Coatings on Zirconia for Dental Implants: Mechanical Characterization and *in Vitro* Behavior. *J. Biomater. Appl.*, 30, 104–118, 2015.
34. Tong, P., Sheng, Y., Hou, R., Iqbal, M., Chen, L., Li, J., Recent Progress on Coatings of Biomedical Magnesium Alloy. *Smart Mater. Med.*, 3, 104–116, 2022.

35. Puspasari, V., Ridhova, A., Hermawan, A., Amal, M.I., Khan, M.M., ZnO-Based Antimicrobial Coatings for Biomedical Applications. *Bioprocess. Biosyst. Eng.*, 45, 1421–1445, 2022.
36. Hou, S., Solvothermal Fabrication of TiO2 Nanosheet Films on Degradable Mg–Zn Alloys. *Surf. Eng.*, 32, 745–749, 2016.
37. Hou, S., Yu, W., Yang, Z., Li, Y., Yang, L., Lang, S., Properties of Titanium Oxide Coating on MgZn Alloy by Magnetron Sputtering for Stent Application. *Coatings*, 10, 999, 2020.
38. Yu, S., Guo, D., Han, J., Sun, L., Zhu, H., Yu, Z., Dargusch, M., Wang, G., Enhancing Antibacterial Performance and Biocompatibility of Pure Titanium by a Two-Step Electrochemical Surface Coating. *ACS Appl. Mater. Interfaces*, 12, 44433–44446, 2020.
39. Peng, J., Yang, M., Bi, J., Sheng, R., Li, L., Hydrogen Existence State of a Hydrogenated Amorphous Carbon Coating and Its Thermal Stability. *Diamond Relat. Mater.*, 99, 107535, 2019.
40. Petersen, M., Bandorf, R., Bräuer, G., Klages, C.P., Diamond-like Carbon Films as Piezoresistors in Highly Sensitive Force Sensors. *Diamond Relat. Mater.*, 26, 50–54, 2012.
41. Schwarz, F.P., Hauser-Gerspach, I., Waltimo, T., Stritzker, B., Antibacterial Properties of Silver Containing Diamond like Carbon Coatings Produced by Ion Induced Polymer Densification. *Surf. Coat. Technol.*, 205, 4850–4854, 2011.
42. Kaliaraj, G.S., Siva, T., Ramadoss, A., Surface Functionalized Bioceramics Coated on Metallic Implants for Biomedical and Anticorrosion Performance—A Review. *J. Mater. Chem. B*, 9, 9433–9460, 2021.
43. Gobbi, S.J., Orthopedic Implants: Coating with TiN. *Biomed. J. Sci. Tech. Res.*, 16, 11740–11742, 2019.
44. van Hove, R.P., Sierevelt, I.N., van Royen, B.J., Nolte, P.A., Titanium-Nitride Coating of Orthopaedic Implants: A Review of the Literature. *Biomed Res. Int.*, 2015, 485975, 2015.
45. Dion, I., Baquey, C., Candelon, B., Monties, J.R., Hemocompatibility of Titanium Nitride. *Int. J. Artif. Organs*, 15, 617–621, 1992.
46. Cogan, S.F., Neural Stimulation and Recording Electrodes. *Annu. Rev. Biomed. Eng.*, 10, 275–309, 2008.
47. Sauvet, G., Dupond, S., Kazmierski, K., Chojnowski, J., Biocidal Polymers Active by Contact. V. Synthesis of Polysiloxanes with Biocidal Activity. *J. Appl. Polym. Sci.*, 75, 1005–1012, 2000.
48. Kenawy, E.R., Worley, S.D., Broughton, R., The Chemistry and Applications of Antimicrobial Polymers: A State-of-the-Art Review. *Biomacromolecules*, 8, 1359–1384, 2007.
49. Belbekhouche, S., Bousserrhine, N., Alphonse, V., Carbonnier, B., From Beta-Cyclodextrin Polyelectrolyte to Layer-by-Layer Self-Assembly Microcapsules: From Inhibition of Bacterial Growth to Bactericidal Effect. *Food Hydrocolloids*, 95, 219–227, 2019.

50. Stetsyshyn, Y., Raczkowska, J., Harhay, K., Gajos, K., Melnyk, Y., Dąbczyński, P., Shevtsova, T., Budkowski, A., Temperature-Responsive and Multi-Responsive Grafted Polymer Brushes with Transitions Based on Critical Solution Temperature: Synthesis, Properties, and Applications. *Colloid. Polym. Sci.*, 299, 363–383, 2021.
51. Gunatillake, P., Mayadunne, R., Adhikari, R., Recent Developments in Biodegradable Synthetic Polymers. *Biotechnol. Annu. Rev.*, 12, 301–347, 2006.
52. Tachibana, A., Furuta, Y., Takeshima, H., Tanabe, T., Yamauchi, K., Fabrication of Wool Keratin Sponge Scaffolds for Long-Term Cell Cultivation. *J. Biotechnol.*, 93, 165–170, 2002.
53. Ruszczak, Z., Collagen as a Carrier for On-Site Delivery of Antibacterial Drugs. *Adv. Drug Delivery Rev.*, 55, 1679–1698, 2003.
54. Webb, J.C.J. and Spencer, R.F., The Role of Polymethylmethacrylate Bone Cement in Modern Orthopaedic Surgery. *J. Bone Joint Surg. Ser. B*, 89, 851–857, 2007.
55. Kaliaraj, G.S., Muthaiah, B., Alagarsamy, K., Vishwakarma, V., Kirubaharan, A.M.K., Role of Bovine Serum Albumin in the Degradation of Zirconia and Its Allotropes Coated 316L SS for Potential Bioimplants. *Mater. Chem. Phys.*, 258, 123859, 2021.
56. Lishchynskyi, O., Stetsyshyn, Y., Raczkowska, J., Awsiuk, K., Orzechowska, B., Abalymov, A., Skirtach, A.G., Bernasik, A., Nastyshyn, S., Budkowski, A., Fabrication and Impact of Fouling-Reducing Temperature-Responsive Poegma Coatings with Embedded CaCo3 Nanoparticles on Different Cell Lines. *Materials*, 14, 1417, 2021.
57. Yamamuro, T., Bioceramics, in: *Biomech. Biomater. Orthop.*, pp. 22–33, 2004.
58. Shen, J.Z. and Fäldt, J., Requirements of Bioactive Ceramics for Dental Implants and Scaffolds, in: *Adv. Ceram. Dent.*, pp. 279–300, 2014.
59. Piconi, C. and Porporati, A.A., Bioinert Ceramics: Zirconia and Alumina, in: *Handb. Bioceram. Biocomposites*, pp. 59–89, 2016.
60. Whitehouse, M.R. and Blom, A.W., The Use of Ceramics as Bone Substitutes in Revision Hip Arthroplasty. *Materials*, 2, 1895, 2009.
61. Kim, M., An, S., Huh, C., Kim, C., Development of Zirconium-Based Alloys with Low Elastic Modulus for Dental Implant Materials. *Appl. Sci.*, 9, 5281, 2019.
62. Ridzwan, M.I.Z., Shuib, S., Hassan, A.Y., Shokri, A.A., Mohammad Ibrahim, M.N., Problem of Stress Shielding and Improvement to the Hip Implant Designs: A Review. *J. Med. Sci.*, 7, 460–467, 2007.
63. Sankarkumar, R., Bhuvaneshwar, G.S., Magotra, R., Muralidharan, S., Rajan, R.S., Saha, D., Subba Rao, K.S.V.K., Valiathan, M.S., Radhakrishna, S., Ramani, A.V., Chitra Heart Valve: Results of a Multicenter Clinical Study. *J. Heart Valve Dis.*, 10, 619–627, 2001.

64. Kozakiewicz, M., Gmyrek, T., Zajdel, R., Konieczny, B., Custom-Made Zirconium Dioxide Implants for Craniofacial Bone Reconstruction. *Materials*, 14, 840, 2021.
65. Cassagnol, M., *Maha Saad US Pharmacist*, p. 37, Jobson Medical Information LLC, New York, NY, USA, 2010.
66. Gradinaru, S., Popescu, V., Leasu, C., Pricopie, S., Yasin, S., Ciuluvica, R., Ungureanu, E., Hydroxyapatite Ocular Implant and Non-Integrated Implants in Eviscerated Patients. *J. Med. Life*, 8, 90–93, 2015.
67. Prakasam, M., Locs, J., Salma-Ancane, K., Loca, D., Largeteau, A., Berzina-Cimdina, L., Biodegradable Materials and Metallic Implants—A Review. *J. Funct. Biomater.*, 8, 44, 2017.
68. Lam, M., Migonney, V., Falentin-Daudre, C., Review of Silicone Surface Modification Techniques and Coatings for Antibacterial/Antimicrobial Applications to Improve Breast Implant Surfaces. *Acta Biomater.*, 121, 68–88, 2021.
69. Accioni, F., Vázquez, J., Merinero, M., Begines, B., Alcudia, A., Latest Trends in Surface Modification for Dental Implantology: Innovative Developments and Analytical Applications. *Pharmaceutics*, 14, 455, 2022.
70. Sikder, P., Ren, Y., Bhaduri, S.B., Synthesis and Evaluation of Protective Poly(Lactic Acid) and Fluorine-Doped Hydroxyapatite–Based Composite Coatings on AZ31 Magnesium Alloy. *J. Mater. Res.*, 34, 3766–3776, 2019.
71. Bekmurzayeva, A., Duncanson, W.J., Azevedo, H.S., Kanayeva, D., Surface Modification of Stainless Steel for Biomedical Applications: Revisiting a Century-Old Material. *Mater. Sci. Eng., C*, 93, 1073–1089, 2018.
72. Nastyshyn, S., Stetsyshyn, Y., Raczkowska, J., Nastishin, Y., Melnyk, Y., Panchenko, Y., Budkowski, A., Temperature-Responsive Polymer Brush Coatings for Advanced Biomedical Applications. *Polymers*, 14, 4245, 2022.

14

Antifouling Nano Filtration Membranes

Divya D. Achari[1], Nandini A. Pattanashetti[2] and Mahadevappa Y. Kariduraganavar[3]*

[1]Department of Chemistry, Karnatak Science College, Dharwad, India
[2]Department of Chemistry, Dayananda Sagar Academy of Technology & Management, Bangalore, India
[3]Department of Chemistry, Karnatak University, Dharwad, India

Abstract
Availability of fresh water is becoming more and more scarce as a result of population increase, industrialization, and globalization, which has prompted the development of alternate fresh water delivery methods. Membrane-mediated wastewater reclamation and other industrial separations, as well as brackish and seawater desalination are the methods which depends on nanofiltration technology. Despite this, membrane fouling has limited the use of these systems by lowering filtration efficiency. Membrane technology's efficiency is greatly decreased by the non-specific interaction between foulants and the membrane surface. Thus, the current chapter briefly discusses nanofiltration membranes, fouling and antifouling mechanisms, fouling factors, antifouling strategy, and various antifouling methods. The chapter is consistent with the findings and prospects of antifouling membranes.

***Keywords*:** Membrane, nanofiltration, fouling, antifouling, foulants, desalination, wastewater, separations

14.1 Introduction

The growing global population and industrialization have resulted in a massive depletion of natural resources due to pollution. Water, as one of the most necessities for life, is in short supply as a result of global industrialization.

**Corresponding author*: mahadevappayk@gmail.com; kariduraganavarmy@kud.ac.in

Inamuddin, Tariq Altalhi and Mohammad Abu Jafar Mazumder (eds.) Nanostructured Hard Materials: Industrial Applications, (415–446) © 2026 Scrivener Publishing LLC

The scarcity of fresh water resources has become an urgent concern. The amount of water used for drinking, sanitation, agriculture, and industry has all increased. Due to rising industrial use, contamination of water resources, the outflow of residential wastewater, and a decline in ground water level, the fresh water supplies are becoming more and more contaminated. Furthermore, the researchers predict that droughts will become five times more likely by 2050 as a result of climatic changes [1, 2]. Therefore, it is critical to look for alternative sources of fresh water supply. To meet the need for fresh and clean water, water filtration and desalination are being more and more employed. As a result, classical techniques for water purification include membrane filtration, ion exchange, and distillation [2]. Although pollutants like germs and water hardeners like calcium, magnesium, lead, and mercury can be effectively removed by distillation, the mineral-free water that is produced is detrimental to human health due to its acidic nature [3]. As a result, membrane-based water purification technologies have surpassed all other methods of producing drinking water.

Membrane filtration is a versatile and emerging water reuse and wastewater treatment technology. Behold first used ultra-filtration to introduce the technology in 1907 [4, 5]. It is regarded as a dependable and promising approach due to its low energy consumption, simple and easy-to-use equipment, high efficiency, low operation and maintenance costs, consistent sewage, and low chemical usage [6, 7]. Membrane-based processes are increasingly being used in the pharmaceutical, food, and biotechnology industries, as well as in desalination and separation processes and waste-water treatment [8–10]. As a result, membranes have come to be recognized as an essential component in both biological and chemical industrial applications.

Microfiltration (MF), ultrafiltration (UF), nanofiltration (NF), and reverse osmosis (RO) are the membrane-based technologies that are categorized according to their driving force and membrane pore size [11–13]. The pore size of MF ranges from 100 to 10,000 nm, with the largest pore size being the most open. It uses a sieving separation mechanism to reject particles larger than the membrane pores. This technique is employed for bacterial eradication, pre-treatment procedures, and solution clarification [14]. Ultrafiltration membranes use the same sieving technique as MF, but its holes are smaller, measuring 2 and 100 nm. A pressure-driven membrane technology with tiny pores that range in size from 0.5 to 2 nm is known as nanofiltration. Both sieving and the charge effect are used. It is frequently used to reject tiny organic particles and multivalent salts that MF and UF cannot remove [15]. The second most popular method of desalinating water in use today is reverse osmosis. They have the smallest and tightest pores, which work *via* the solution-diffusion mechanism [16, 17].

These membranes are critical in the production of drinking water in our modern lives. In summary, the membranes can treat various feed compositions and salinities of river water, brine water, and salt water. These membranes are also used to clean effluent from a variety of industries, including dairy, paper and pulp, pharmaceuticals, textiles, food, pigment, and colors. Furthermore, the membranes are commercially accepted for the recovery of valuable chemicals from industries [2, 7, 18]. Among the various membrane-processes mentioned, the current chapter addresses the use of nanofiltration membranes, their limitations, and potential solutions.

14.2 Nanofiltration Membranes and Membrane Fouling

Nanofiltration membranes are reverse osmosis membranes that operate at low pressure. NF membranes have a separation characteristic that is intermediate between RO and UF processes. These membranes rely on the sieving and diffusion transport mechanisms that operate at the interface of porous and non-porous membranes, respectively [15]. This method is usually employed for water purification, desalination of seawater, removal of heavy metals, textile coloration etc [19]. The rejection percentages of NF membranes are: 30-40% NaCl, 85-90% $MgSO_4$, 98% sucrose, and 99% raffinose [20, 21].

Based on the NF membranes' characteristics, which are neither entirely dense nor entirely porous, the retention mechanisms of nanofiltration membranes can be assessed using size exclusion (for porous membranes) and sorption-diffusion mechanisms (for dense membranes) [22]. Because of this, the method can successfully retain salts, bacteria, viruses, pesticides, and other organic pollutants from ground water and wastewater streams. NF produces premium permeate for industrial and agricultural use. NF has the potential to take the place of RO in the production of high-quality water due to its inherent benefits [23]. Its use in the creation of drinking water, the treatment of waste water, the recovery of small organic compositions, the filtration and purification of biotechnological and pharmaceuticals, the extraction and filtration of food, the processing of petroleum materials, and other applications makes it incredibly versatile. Despite the excellent performance of NF membranes, the process is associated with the limitations in its performance and the post-treatment process. The major drawbacks of these membranes is found to be the fouling of the membranes since the inception of membrane-based separation technology [24, 25].

Understanding the fouling mechanism is a difficult task. This is due to the presence of multiple types of foulants. A non-specific contact with the membrane surface is produced by bacteria, proteins, biopolymers, natural organic matter, oils, salts, and other organic components in feed streams [26]. Fouling results in a temporary flux decrease or pressure rise, as well as a decrease in salt rejection, which leads to a loss of membrane performance. This is because trapped particles, macromolecules, colloids, and salts are deposited inside the pore wall and at the membrane surface. The primary membrane properties affecting the fouling mechanisms includes, surface roughness, surface hydrophilicity, and electrostatic surface charge [26–28].

14.3 Types of Membrane Fouling

Membrane fouling is the plugging of pores during filtration in a membrane. To do this, a combination of sieving and adsorption of particles and chemicals onto the membrane surface and inside its pores is used. Membrane fouling is also influenced by some process variables, including flow, temperature, pressure, feed concentration, and equipment design [22]. Membrane properties and interactions between solute-solute and solute-membrane are hypothesized to be responsible for the time-dependent and irreversible phenomenon known as membrane fouling. This pore occlusion slows down the formation of permeate and makes membrane filtration more challenging [22, 29]. For instance, scaling and contaminant deposition on the membrane surface layer are the two main causes of membrane fouling in NF membranes. Based on the presence of foulants, they are classified as colloidal, inorganic, organic, or biological fouling [29, 30]. The types of membrane fouling are shown in Figure 14.1 [31].

The colloidal particles like, graphite, silica, sulfur, iron, inorganic materials, and other organic substances results into colloidal fouling. Inorganic fouling is generated by the oxidation-induced precipitation of some dissolved ions and

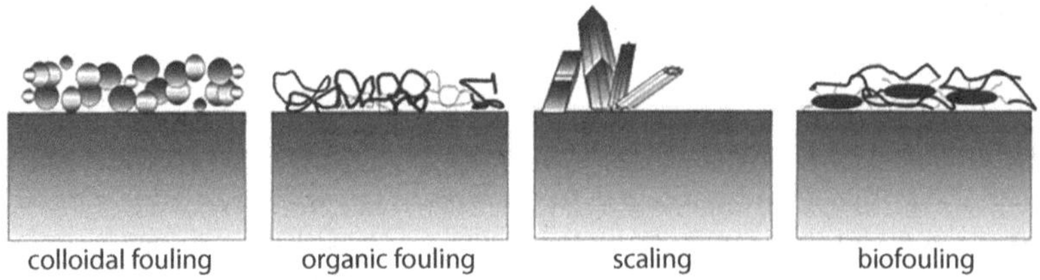

Figure 14.1 Pictorial representation of the types of membrane fouling. (Adapted with consent from Creative Commons) [31].

salts onto the membrane surface, such as carbonates, metal chlorides, hydroxides, and sulfates. Because of their low solubility, inorganic salts produce scales on the surfaces which include barium sulfate, calcium carbonate, silica, and calcium sulfate which can easily deposit and crystallize. Furthermore, organic fouling is the outcome of organic pollutants deposition. Plant materials, humic acid, cationic surfactants, oil, and other hydrocarbons are examples of non-biological substances. Biofouling is the process of biologically active organisms, such as microbial cells of bacteria, fungus, and other eukaryotic microorganisms, adhering to membrane surfaces and developing a biofilm [32–34].

Based on the degree of particle adhesion to the membrane surface, fouling can also be classified as back-washable or non-back-washable, as well as reversible or irreversible. At the conclusion of the separation procedure, back washable fouling on the membrane surface can be eliminated by switching the permeate flow direction. But normal hydraulic backwashing cannot get rid of fouling that is non-back washable. It can, however, be cleaned using a chemical cleaning procedure. Furthermore, no procedure, including flushing, backwashing, or chemical cleaning, will eliminate the irreversible fouling [30].

14.4 Mechanism of Fouling

Fouling can generally be caused by three main factors: the properties of the membrane material, the properties of the solute, and the operating conditions [35, 36]. According to the definition given earlier, fouling is the deposition or adsorption of particles on the membrane's surface or inside its pores, which causes a partial or complete obstruction of the pores and a steady decrease in flow. A key element in pore blockage is the size and shape of the solute in relation to the membrane's distribution of pore sizes. For instance, the pores become partially occluded when isolated macromolecules cover them, generating a coating on the membrane surface and increasing permeation resistance. The volume that may be used for permeate flow is constrained when particles deposit inside the membrane's pores, which leads to an internal obstruction of the pores [22, 37]. The only way to restore the permeate flow that has been irreversibly reduced by these pore blockages is to chemically clean the membrane. Figure 14.2 shows the forms of membrane pore blockages [22].

Transmembrane pressure and permeate flux are the two most essential metrics of membrane fouling. Figure 14.3 depicts the usual flux-time curve of a membrane, which comprises three distinct stages.

i. Stage (I) is defined by a quick and abrupt decline in permeate flux.

ii. Stage (II), which is defined by a progressive flux decline over a long period.
iii. Stage (III), which concludes with a steady-state flux.

Due to pore blockage and the development of a cake layer on the membrane's surface, a decrease in flow in membrane filtration is brought on

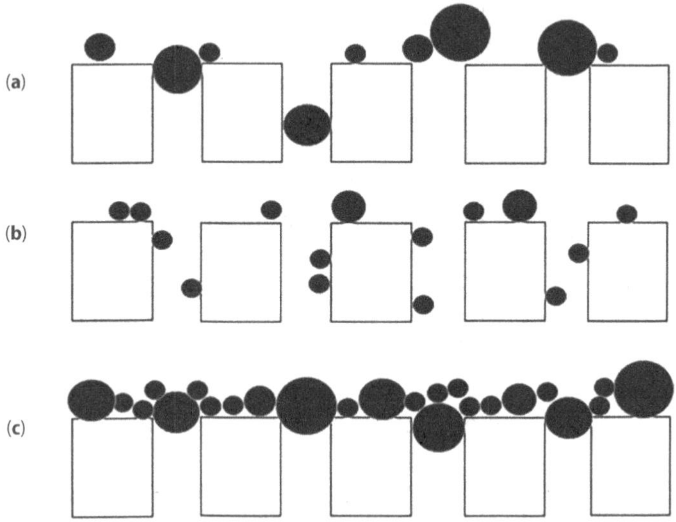

Figure 14.2 Schematic representation of different types of pore blockages: (a) a complete pores blockage; (b) the partial (internal) pore blockage; (c) the polarized layer (Adapted with consent from Intech Open) [22].

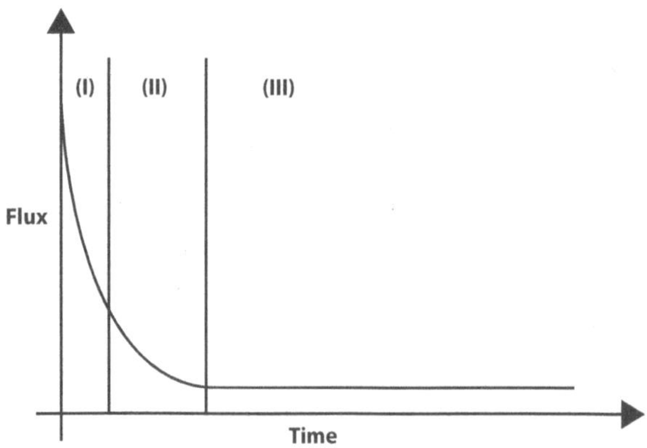

Figure 14.3 Graphical representation of Time – Flux curve. (Adapted with Consent from Intech Open) [30].

by an increase in membrane resistance. Cake formation increases permeate flow resistance whereas pore blockages increase membrane resistance. Hence, it is possible to consider cake formation and pore blocking as two significant membrane fouling mechanisms [30].

According to Figure 14.3, the initial abrupt decrease in permeate flux is caused by the rapid deposition of foulants on the pore. This happens as a result of the solute concentration on the membrane surface. The largest permeation flow always happened at the commencement of filtration because membrane pores are clear and open. When trapped particles close the membrane's perforations, the flux gradually decreases. The size, shape, and foulant of the membranes affect how much of the pores are blocked. When the pores and foulants have the same form and size, the blockage is deemed complete [38, 39].

Although fewer than one layer of particles is necessary for complete pore blocking, pore blocking is frequently a quicker process than cake formation [39–40]. The creation of a cake layer on the membrane's surface causes the flux to decrease even more after complete pore blockage. A cake layer develops on the membrane surface as the number of trapped particles rises, increasing the resistance to permeate flow. The permeate flux thus keeps getting smaller over time [30].

Because fouling results in a decrease in flux under steady-state operating circumstances, it is commonly referred to as a deteriorating performance. The membrane fouling process in NF technology includes depositing foulants on the membrane surface, which lowers the membrane's efficiency. Moreover, chemical cleaning will cause membrane surface damage, increasing the cost of membrane process maintenance. Because of this, it's critical to provide antifouling membrane surfaces that promote the release of absorbed foulants or lessen their adsorption.

14.5 Controlling Factors of Membrane Fouling

Several intrinsic membrane features influence membrane fouling. Pore size, surface structure, charge, as well as hydrophilicity and hydrophobicity, are factors that affect membrane fouling.

14.5.1 Effect of Surface Chemistry (Hydrophilicity/Hydrophobicity)

The membrane in an aqueous system can either attract or repel water. Moisture diffusion is governed by the membrane's architecture and the

Figure 14.4 The surface chemistry of the membranes (a) hydrophilic membrane; (b) hydrophobic membrane. (Adapted with consent from Creative Commons) [30].

surface chemistry that surrounds it, which has an impact on its hydrophilicity [29]. By measuring the contact angle between the water molecules and the membrane's surface, the hydrophilicity of the membrane was ascertained.

The greater the theta (θ) value, the more hydrophobic the membrane is. These factors are connected to the membrane's pore size and surface form. The antifouling procedure is significantly impacted by the membrane's surface chemistry [41]. Wettability is a characteristic of hydrophilic membranes, which are believed to occur as a result of active groups with the capacity to create stronger hydrogen bonds with water, as can be shown in Figure 14.4(a). Because hydrophobic membranes lack the ability to absorb water, water droplets appear as a distinct bead on their surface, as shown in Figure 14.4(b). As a result, unlike hydrophobic membranes, hydrophilic membranes are less impacted by the sorption process, have a higher membrane flow, and have excellent antifouling capabilities.

For instance, there has been a lot of interest in the use of polyether sulfone (PES) in waste water treatment and water purification activities. The hydrophobicity of PES membranes, which results in a strong interaction between the membrane surface and hydrophobic foulants, makes them particularly susceptible to fouling. This can lead to both internal membrane fouling (pore blocking) and external membrane fouling (deposition on the membrane surface during nanofiltration) [42]. As a result, a successful and environmentally friendly modification process is used in the production of antifouling membranes. Therefore, fouling can be decreased by utilising hydrophilic membranes with the right surface chemistry.

14.5.2 Effect of Pore Size and Pore Size Distribution

Because of the pore size and pore size distribution on the membrane surface, the membrane morphology also controls membrane fouling. The membrane surface or pores can easily turn into sites of adsorption and deposition during the nanofiltration process; in addition, they are hindered by tiny particles or macromolecules present in the water [43]. Larger porous membranes are more prone to irreversible fouling, allowing macromolecules to permeate more easily [44]. To achieve a good size-sieving effect and avoid "typical pore blocking," the pores should be smaller in size. Practically speaking, the holes should not be too small because it would cause the membrane resistance to be too high and prevent permeate flux. As a result, pore blockage is known to be a significant mechanism for membrane fouling. Therefore, to monitor membrane fouling, low porosity nanofiltration membranes with adequate pore size and narrow dispersion are proposed [5, 30, 45].

14.5.3 Effect of Surface Structure

The level of fouling is significantly influenced by the membrane's surface structure (surface roughness), which can be discovered using atomic force microscopy. In comparison to the smooth membrane surface, the rough membrane surface accumulates more particles, and it turns out that particles gather preferentially in its valley, causing "valley clogging" and a higher drop in flux [46].

14.5.4 Effect of Surface Charge

Surface charge is a crucial factor that controls how the membrane interacts with its environment and has a significant effect on the rejection of charged species like salts, dyes, and amino acids, among others. Because it affects the electrostatic repulsion between ions or charged molecules and the membrane surface, the membrane surface charge is crucial for membrane performance [47].

A membrane's surface charge can be changed to promote or inhibit interactions with feed water constituents that directly impact membrane fouling. With the same surface charge, the membrane materials are known to repel each other. As a result, negatively charged particles are less likely to attach to negatively charged surfaces, and fouling is reduced [48]. If the surface charge is opposing the species, it will generate attractive forces and readily spoil the membrane, reducing efficiency.

14.5.5 Effect of Salts

Permeability, variations in permeation flux, solute rejection rate, and membrane fouling behaviour are the factors that have the biggest impact on salts' ability to affect nanofiltration. In three different ways, raising the ionic strength of solutions increases the fouling behaviour of nanofiltration membranes:

1. Ion screening reduces the electrostatic attraction between charged species and the membrane's surface [49].
2. In higher ionic strength solutions, organic solutes are more likely to change a smaller apparent macromolecular structure. A smaller solute size led to the creation of a cake layer with less porosity.
3. Bridges can be built between the -COOH group of proteins and calcium and magnesium-like divalent ions at the membrane's surface. As a result, the fouling process is sped up by the foulants' strong attachment to the membrane surface [50, 51].

According to the preceding discussion, greater ionic strength solutions containing divalent cations induce the dense fouling layer on the membrane surface. As a result, it is obvious that using lower salt concentration solutions with the NF membrane can significantly reduce fouling behaviour.

14.5.6 Effect of pH

According to the literature review, as pH increases, the surfaces of nanofiltration membranes acquire a negative charge and become very hydrophilic in nature. Since foulants have a weak affinity for the membrane surface due to their high hydrophilic nature when pH increases, membrane fouling has decreased [49, 52]. If the nanofiltration procedure is carried out in an acidic pH environment, the electrostatic interaction and adsorption of hydrophobic molecules would cause significant fouling. In contrast, increasing the pH in the presence of anionic surfactants considerably enhances the flow due to less fouling [53, 54]. Moreover, the foulant-foulant interaction is impacted by the pH of the solution [55], as well as the foulant-membrane contact.

14.5.7 Effect of Operational Conditions

Temperature cross flow velocity (CFV), membrane flux, aeration, and trans-membrane pressure all have a direct impact on membrane fouling. The fouling and operating expenses of membrane filtration systems that are subjected to filtration requirements can be decreased by optimizing these operating parameters.

There are two common operating modes for membrane filtration systems: (i) constant transmembrane pressure and (ii) constant permeate flux. During filtration of a predetermined permeate volume, the constant flux operation resulted in decreased fouling. Fouling was decreased in specific temperature ranges with constant transmembrane pressure activity as a result of surface water treatment [56].

Under high temperatures, the feed solution's viscosity decreases, lowering the flow resistance. Moreover, as temperature rises, so does diffusivity, speeding up the transfer of solutes from the membrane surface to the bulk stream [57].

Cross-flow velocity is the name given to the feed stream's superficial velocity when it travels parallel to the membrane surface (CFV). The coefficient of mass transfer of the solute as well as the amount of mixing close to the membrane surface both benefit from an increase in CFV [58]. In addition, the permeation rate is significantly influenced by the transmembrane pressure, which is the pressure differential between the feed and permeate streams. With rising transmembrane pressure and correspondingly higher flow, the intensity of penetration towards the membrane increases.

To reduce membrane fouling, aeration is essential. Along with giving the activated sludge metabolism the oxygen it needs, it also cleans the membrane surface, stops foulant build-up, and slows down the membrane fouling process. As a result, increased aeration is frequently used in the membrane filtration process [59].

14.6 Antifouling Methods

Many methods and techniques can be used to treat membrane fouling. A few techniques were employed to reduce membrane fouling and to restore and maintain membrane effectiveness [26]. The primary fouling control strategy is the production of membranes with strong antifouling mechanism. Fouling is significantly influenced by the membrane's surface characteristics, as mentioned under fouling influencing variables. As a result, the inherent physical and chemical properties can be tailored and altered to stop the fouling process. As the fouling mechanism differs depending on the foulants, several approaches and mechanisms are used to fabricate antifouling membranes. There are two types of antifouling techniques:

(i) Passive antifouling method
(ii) Active antifouling method

14.6.1 Passive Antifouling Method

The objective of the passive antifouling technique is to reduce the primary deposition of foulants on the surface without altering their fundamental characteristics. This approach relies on altering the membrane's physicochemical and topographical structure to diminish the surface interactions of foulant-membrane, hence reducing foulant adsorption. Further, based on the defence mechanism it is classified into fouling resistance technique and fouling release method [60].

14.6.1.1 Fouling Resistance Method and Mechanism

The hydrophilic surface of the membrane, also known as "fouling resistance," serves to prevent foulants from adhering to its surface. Fouling resistance is the most used way for developing antifouling surfaces. To introduce fouling resistance qualities on surfaces, many hydrophilic materials are utilized, such as poly(ethylene glycol)-based, glycomimetic polymers, zwitterionic materials, peptides, and peptides polymers [61–65]. The hydrophilic materials produce a hydrophilic membrane surface on top of a fouling membrane, allowing it to tolerate fouling by biological molecules and organic chemicals.

The two main fouling resistance mechanisms are frequently noticed. As shown in Figure 14.5, the first is the steric repulsion effect, and the second is the development of a hydration layer. The adsorption of foulants (e.g., proteins) on one side compresses the expanded polymer brushes, impeding polymer chain movement. This causes an unfavorable loss of entropy, making protein adsorption unfavorable entropically [66]. Although limiting non-selective sorption of foulants through free energy variation brought on by entropic dehydration effects, hydrophilic polymer brushes will provide a dense hydration layer *via* hydrogen bonding or ionic solvation.

The literature claims that neutral PEG chains create the hydration layer through hydrogen bonding, whereas zwitterionic molecules can join water molecules through potent electrostatic interactions [67, 68]. Further research showed that zwitterionic substances, including sulfobetaine (SB), not only connect to more water molecules (about 7-8 H_2O molecules for each SB molecule), but also obtain more mobility on the first hydration layer for unbound water molecules [69]. As a result, zwitterionic compounds are often more fouling resistant than PEG-based materials.

Nanomaterials, in addition to hydrophilic polymers, contribute to the fouling resistance of membrane surfaces. Due to their exceptional abilities

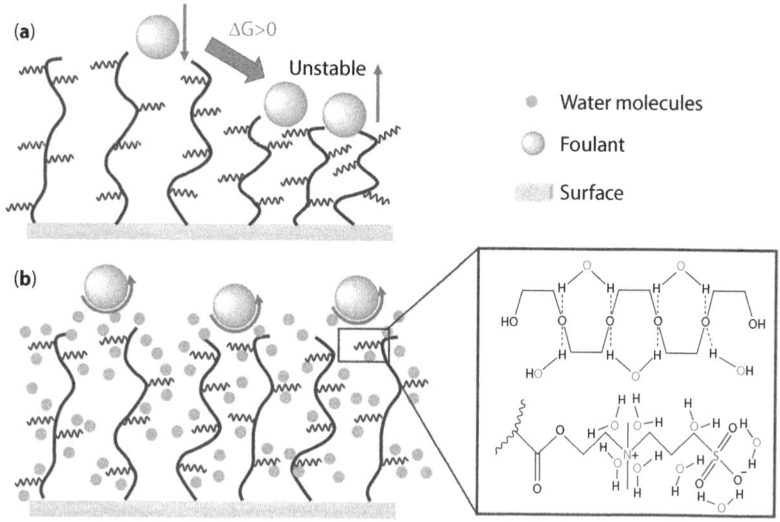

Figure 14.5 The fouling resistance mechanisms are as follows: (a) steric repulsion; (b) creation of a hydration layer on the membrane surface. (Adapted with consent from Royal Society of Chemistry) [60].

to improve fouling resistance, inorganic nanoparticles (iNPs) such as metals, metal oxides, and carbon-based nanomaterials (CBNs), particularly single and multi-walled carbon nanotubes, and graphene oxide (GO), have gained considerable interest.

These nanoparticles are typically found to have a high hydrophilicity. They could increase the overall quantity of hydrogen bonds on the membrane's surface, leading to the development of an interfacial layer of strongly orientated, slowly moving, tightly bound water molecules. For the first time in the literature, Wang *et al.* [70], used graphene oxide (GO) as an additive for the fabrication of a PVDF antifouling ultrafiltration (UF) membrane by physical blending. The GO-PVDF blend membrane beat a virgin PVDF membrane in terms of fouling resistance to bovine serum albumin and reached about 96.4% of water flux recovery ratio. This is because the graphene oxide layer contains polar -COOH and -OH, which exhibits extraordinary hydrophilicity in membranes. According to the literature, the wettability and steric hindrance of membrane surfaces must be utilised through the elemental composition and physical properties of surface modifiers to provide improved fouling resistance ability.

14.6.1.2 Fouling Release Method and Mechanism

Elastomers, poly(dimethyl siloxane), and fluoropolymers are examples of hydrophobic polymers with low surface energy that are utilized to produce the hydrophobic surface in nanofiltration membranes. Because of its hydrophobic nature, polar molecules are not bonded to the hydrophobic surface, and no hydrogen bonds or polar interactions are formed. This hydrophobic surface is designed to reduce interfacial bonds, allowing for "fouling release," or the removal of attached foulants by hydraulic shear forces while inhibiting their adhesion. The hydrophobic membrane's surface is excellent for minimizing bacteria and fouling of hydrocarbons. A novel strategy to produce amphiphilic surfaces composed of hydrophilic groups (exhibiting fouling-resistant characteristics) and hydrophobic moieties (exhibiting fouling release characteristics) has recently been presented [71, 72].

The amphiphilic surface is a combination of dynamic and compositional surfaces that prevents foulant deposition and minimizes foulant contact with the membrane surface. The primary goal of developing an amphiphilic surface is to combine the foulant repellent qualities of the hydrophilic portion with the reduced surface energy properties of the hydrophobic zone. The amphiphilic membrane's combinational features not only diminish foulant interactions with the membrane surface, but also foulant adsorption entropy and enthalpy. Microorganisms, biomacromolecules, and organic compounds are especially avoided by the amphiphilic surface. Amphiphilic membranes for nanofiltration are thus the most promising antifouling membrane surface for a separation system encompassing a varied range of foulants. For instance, Jiang and colleagues' creation of heterogeneous antifouling membranes [73–76], involved grafting non-polar, hydrophobic polymers with low surface energy onto hydrophilic membrane surfaces. The developed membranes showed good fouling resistance and release qualities for various contaminants such as proteins, oil, and non-migratory (NOM) foulants, with only a minor reduction in total flux and maximum flux recovery. Due to the combined effect of hydrophilic and hydrophobic domains on the surfaces during the filtration process, various foulants could be avoided and eliminated from the surface of amphiphilic (heterogeneous) membranes (Figure 14.6); at low hydrodynamic shear force, they can be dispersed back to the bulk feed stream, maintaining low permeation water flux and high-water flux recovery.

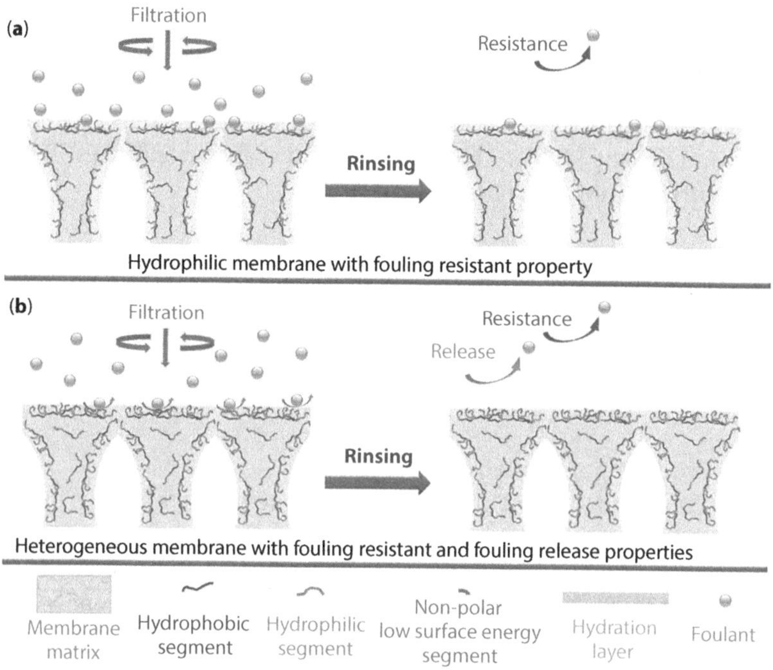

Figure 14.6 Antifouling mechanism of hydrophilic and heterogenous membrane surfaces: (a). Hydrophilic membranes possess property of fouling resistance; (b). Heterogenous (both hydrophilic and hydrophobic) membranes possess both the properties of fouling resistance and release. (Adapted with consent from Royal Society of Chemistry) [60].

14.6.2 Active Antifouling Method

The passive antifouling approach, which was ineffectual in the treatment of proliferative foulants, might reduce foulant adsorption. For instance, the intricate two-stage fouling mechanisms, adsorption, and growth of microorganisms have made them difficult to control [76]. As a result of various physical interactions on the membrane surface, microbes were first reversibly adsorbed; however, this gradually evolved into irreversible adsorbed states with polysaccharide anchors. The final formation of the biofilm on the membrane surfaces was achieved by the bacteria replicating the growth and proliferation processes. Non-migratory foulants' microorganisms and dead microorganisms might be eliminated by antibacterial agents using active antibacterial methods, which could be defeated by active anti-fouling methods.

Antibacterial agents were classified into two groups based on their active antibacterial methods and mechanisms: (i) Release-based antibacterial agents, such as Ag and Cu based [77–81] with off-surface antibacterial processes, and (ii) Release-based antibacterial agents, such as quaternary ammonium groups [82–84], GO [85, 86], and carbon nanotubes [87, 88] with on-surface antibacterial mechanisms.

14.6.2.1 Off-Surface Antibacterial Mechanisms

By leaching antibacterial chemicals from the membrane's surface, the active off-surface antifouling mechanism intends to target bacterial cells, planktonic cells, and organic materials. The mechanisms function in two ways:

1. Producing reactive oxygen species.
2. Catalytic reactions are caused on the membrane's surface by inducing nano-catalysts [89, 90].

Figure 14.7 depicts the active antifouling process proposed for nanoscale Ag membrane surfaces against microorganisms. The absorption of free silver ions (Ag^+) released is a typical technique used to research the function of silver nanoparticles in the process of active off-surface. The

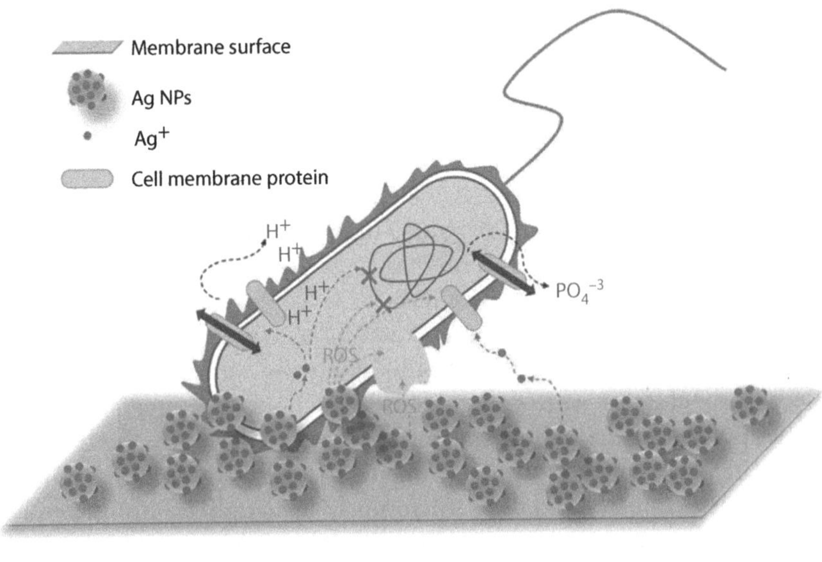

Figure 14.7 Hypothesised active antifouling mechanism for nanoscale Ag membrane surfaces against microorganisms. (Adapted with consent from Royal Society of Chemistry) [60].

process is very intricate and involves primarily the cessation of adenosine triphosphate (ATP) synthesis and DNA replication. Silver ion (Ag^+) has been widely claimed to have antibacterial effects. Silver ion interacts with respiratory chain reaction enzymes, such as NADH dehydrogenase, at sub-micromolar doses, but it also successfully inhibits respiration involved in ATP synthesis [91–93]. At millimolar concentrations, Ag+ produces morphological and metabolic alterations in prokaryotic cells. As a result, they are active, but not in a cluster. Following that, prokaryotic cells experience oxidative stress as a result of an imbalance in the synthesis and accumulation of oxygen reactive species. The cell membrane deteriorates and bacterial DNA is damaged as a result of oxidative stress [94, 95].

14.6.2.2 On-Surface Antibacterial Mechanisms

Active on-surface antimicrobial treatments, which use non-releasable antibacterial chemicals to kill germs in contact with the membrane surface, may be a feasible option. Direct integration of substances onto the membrane's surface allows for a variety of physical or chemical strategies to target the related microbial cells.

Using antimicrobial cationic groups such as quats (NR^{4+}), phosphorane (PH^{4+}), and diaminomethaniminium (CH_6N^{3+}) in cationic polymeric materials (CPMs), the membrane surface is changed with contact-active antimicrobial characteristics [96, 97]. Figure 14.8 depicts the cationic modified groups.

The antimicrobial mode of action is a dynamic activity process in which the following actions take place in a continuous and sequential manner:

(i) Initially, degradation of cell wall by binding to cell wall polymers or by a multi-valent cation substitution process.

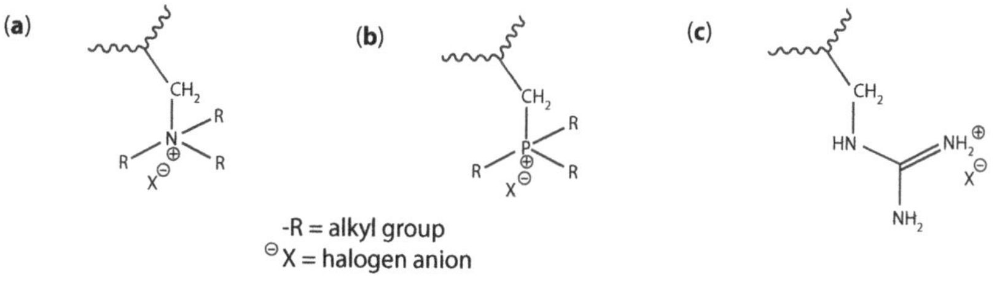

Figure 14.8 The antimicrobial cationic groups: (a) quats (NR_4^+); (b) phosphorane (PH_4^+); and (c) diaminomethaniminium ($CH_6N_3^+$) (Adapted with consent from Royal Society of Chemistry) [60].

(ii) In addition, intracellular matrix leakage and cell death were caused by electrostatic interactions with negatively charged phospholipid bilayers and cellular proteins, as well as by the destabilization and eventual disintegration of the exposed cell membrane.

14.7 Surface Modification of the Nanofiltration Membranes

Enhancing antifouling properties by surface modification is one efficient way to lower membrane fouling. Two fundamental approaches were frequently investigated: (i) thin film surface coating of the membrane surface (Figure 14.9a); and (ii) surface grafting of polymer chains (Figure 14.9b) [25, 98]. Because the coating or grafting materials have no affinity for foulants, foulant-membrane interactions are avoided.

14.7.1 Surface Modification by Surface Coating

Surface coating is an effective approach to provide antifouling qualities to a membrane surface. The non-porous layer coating also prevents foulants that pass through the skin layer from fouling inside. There are two types of surface coating methods;

1. The coating-to method
2. The coating-from method

1. The coating – to method: Coating-to methods include dip-coating and spin-coating. These techniques entail altering nanofiltration membranes by using anti-fouling polymeric substances including polyelectrolytes,

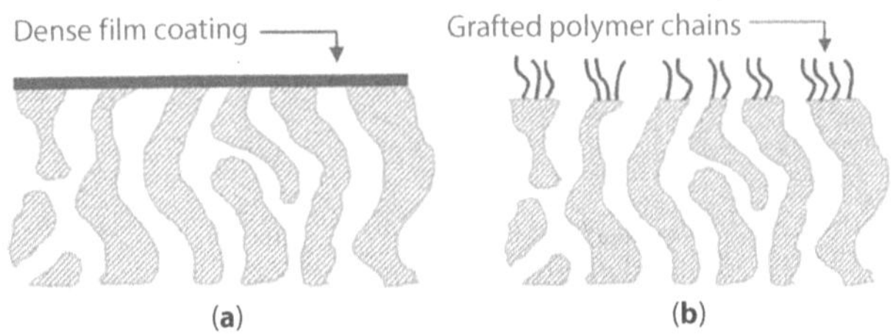

Figure 14.9 Surface modification of nanofiltration membranes: (a) membrane coating; (b) polymer chain surface grafting. (Adapted with consent from Creative Commons) [98].

poly(vinyl alcohol), and poly(ethylene glycol), as well as nanomaterials like silica and graphene oxide [99, 100]. Lee and his colleagues employed PVA as a coating layer in this procedure to improve the anti-fouling efficacy of nanofiltration and reverse osmosis membranes [101, 102]. In the literature, we discovered that the surfaces of thin-film composite NF membranes are negatively charged, and that NH_2/NH_4^+ groups are typically given to nanomaterial layers to facilitate layer coating. The surface of the membrane must be tailored to initiate effective non-covalent interactions, notably electrostatic interactions, to enable the coating-to technique. Dung *et al.* [103], investigated the surface coating of TiO_2 nanoparticles on a polyamide thin film composite nanofiltration membrane with subsequent UV irradiation in the literature. TiO_2-coated UV-irradiated membranes' antifouling properties were improved over uncoated membranes, with increased retention flux and reduced irreversible fouling factors. This is due to TiO_2's two-way action on the polymer surface containing COOH and COOH groups. TiO_2 can be linked to Ti^{4+} cations using oxygen atoms in one way. The alternative way is to form a hydrogen bond between the COOH groups and the TiO_2 hydroxyl groups. Figure 14.10 depicts the two-way activity of TiO_2 nanoparticles.

2. The coating-from method: *In-situ* deposition of polymers or inorganic compounds on a membrane's surface is referred to as the coating-from technique. The membrane's surface is typically pre-treated before being coated with inorganic nanoparticles to promote the adsorption of the inorganic precursors. To improve the coating of inorganic nanomaterials using an *in-situ* method, special techniques such the sol-gel method, biomineralization, and chemical reduction were used [104–107].

The chemical vapor deposition (CVD) method can be conducted without solvents at lower temperature and pressure, and shown tremendous potential as a technique for surface coating. The *in-situ* "coating-from" method is easy to control compared with the "coating-to" technique; the physicochemical characteristics of the antifouling coating, such as

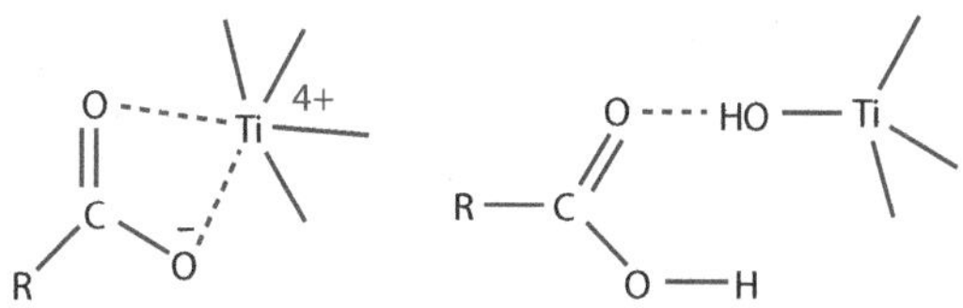

Figure 14.10 The two-ways action of TiO_2 nanoparticles (Adapted with consent from Creative Commons) [103].

thickness, roughness and chemical composition, can be particularly analyzed to obtain optimized efficiency.

14.7.2 Surface Modification *via* Surface Grafting

Surface grafting is the process of altering a thin layer on the surface of a membrane without changing the material's bulk properties so that the membrane surface and the grafted polymer chain form a relatively stable chemical bond [108]. Antifouling chemicals grafted onto the membrane and from the membrane's surface were both employed to chemically graft the membrane surface. Surface grafting procedures are classified as "grafting-to" and "grafting-from" on this basis.

1. Grafting-to method: This method involves grafting modifying polymers containing initiator sites onto the membrane's surface.
2. Grafting-from approach: This technique forms alternating polymer chains from functional monomers on the membrane surface through a surface-facilitated polymerization process.

According to the description of Zhao *et al.* [109], surface-initiated atom transfers radical polymerization (SI-ATRP) and surface graft polymerization were used to graft poly(sulfobetaine methacrylate, or PSBMA), onto a polypropylene membrane. In this grafting procedure, 2-Hydroxyethyl methacrylate (HEMA) was first immobilized on the PP membrane surface *via* UV-induced graft polymerization, and after coupling with 2-bromopropionyl groups, each of the grafted poly(HEMA) chains offered several starting sites for subsequent ATRP of SBMA. The high-density membrane that resulted had a high fouling efficiency.

Photochemical grafting techniques, especially UV-initiated grafting, have been used extensively due to their low operating costs, mild reaction conditions, selectivity to absorb UV light without affecting bulk polymer, and ease of incorporation into the final stages of a membrane manufacturing process [110]. Several hydrophilic monomers have been used to modify the membrane surface through UV-initiated graft polymerization, such as 2-hydroxylethyl methacrylate (HEMA), acrylic acid (AA), N-vinyl-2-pyrrolidinone (NVP), and 2-acrylamidoglycolic acid (AAG). To determine their ability to desalinate Gaza's groundwater, two commercial NF membranes, NF-1 and ASP30, were evaluated for effectiveness and humic acid (HA) fouling potential in the literature. Due to the high concentrations of chloride-based salts in Gaza groundwater, both membranes initially displayed good efficiency in terms of flow and $MgSO_4$ rejection but showed weakness in terms of very low NaCl rejection. As a result, Abuhabib *et al.* [111], used UV-grafting to modify the surface and improve the membrane's effectiveness as well as fouling resistance for the application. Both membranes could reject $MgSO_4$ and NaCl at specific monomer concentrations; however, the percentages of rejection varied. The NF-1 membranes were effectively modified to improve salt rejection. Additionally, only specific modification parameters related to both monomer concentration and UV exposure duration resulted in improved rejection.

14.7.3 Surface Modification *via* Physical Blending

Typically, physical blending entails physically merging a precise amount of the membrane material with the modifying ingredient [112, 113]. Several chemicals, organic compounds, and inorganic nanoparticles were incorporated into the membrane matrix using a modification approach to give anti-fouling capabilities.

A copolymer and hydrophilic building blocks in a polymer matrix enable intermolecular interactions to favorably boost the blending process [114]. As an example, for the desalination process, a polymer blending method was employed to modify a polyvinyl alcohol membrane by adding Arabic gum. Such blending provides superior permeability and biofouling resistance, which improves membrane performance. Because of the presence of -OH groups, the addition of Arabic gum increased the membrane's hydrophilicity. The surface with higher hydrophilicity has higher water affinity, which minimizes foulant adsorption. The amphiphilic polyimide NF membrane reported by Ruan *et al.* [115], was used for the filtration of bovine serum albumin, humic acid, and sodium alginate solutions. The two stage mechanisms were fouling resistance and fouling release protection. Our modified membrane outperformed the clean polyamide NF membrane in terms of antifouling properties.

14.7.4 Surface Modification *via* Hydrophilic Materials

Hydrophilic polymeric materials form a surface hydration layer that acts as a physical barrier, preventing foulants from adhering to the surface and, thus, reducing fouling [116]. As a result, membrane surfaces can be grafted or coated with hydrophilic materials such as poly (ethylene glycol) (PEG), polydopamine (PDA), and zwitter ions to improve hydrophilicity, and hence, antifouling properties.

PEG-based chemicals have commonly been used to modify the membrane surface in the following two approaches to improve antifouling capabilities:

i. Method-I: Make insoluble in water PEG-based polymers and spray them directly on top of membranes.
ii. Method II: PEG chains are grafted to the membrane's surface. The functional groups of poly(ethylene glycol)-based polymers can be covalently bonded to the membrane's surface.

Dopamine is also widely utilized to improve membrane surface hydrophilicity, bonding ability with a wide range of polymers, and simple coatings with water soluble solutions to reduce fouling [117, 118]. Polydopamine can also be utilized to graft amine functional derivatives onto polyether sulfone, such as poly(ethyleneimine) (PEI). The surface of a polysulfone membrane was altered by polydopamine deposition and subsequently grafted with amine-functionalized PEG (PEG-NH_2) by atom transfer radical polymerization (ATRP) to improve fouling resistance [119, 120]. Apart from that, the surface of the immobilized zwitterion membrane was grafted with polydopamine, which improves durability and antibiofouling efficiency [121]. Zwitterions are advanced materials with outstanding fouling resistance. Zwitterions can form electrostatic interactions with up to eight water molecules that are stronger than hydrogen bonding and improve the membrane's surface hydrophilicity and antifouling potential [122].

14.8 Conclusions and Future Perspectives

Nanofiltration methods have been used successfully to treat polluted water and desalinate saltwater. At the same time, due to their fouling mechanism over time, these membranes have performance constraints. Fouling leads in a significant loss of efficiency and greater maintenance expenses. Bacteria, organic foulants, inorganic foulants, proteins, oil foulants, bio macromolecules, and biopolymers that deposit on the membrane's surface reduce its effectiveness. Modifying the membrane to be hydrophilic, pore size and distribution across the membrane, surface structure and charge, salt content and pH, and so on can all be used to affect the fouling mechanism. A great deal of effort has been devoted towards understanding the fouling of nanofiltration membranes at the molecular level and improving membrane performance in the literature. To increase the antifouling capabilities of the nanofiltration membrane, many researchers devised surface coating or grafting approaches. A variety of procedures are used to change the physical and chemical properties of antifouling membrane surfaces to develop antifouling techniques. As a result, proper material selection is critical in order to generate high-performance membranes with increased antifouling resistance. Despite this, much research has been performed to improve the antifouling properties of membrane surfaces. Combining active and passive antifouling approaches could be one solution to the fouling problem. However, it is required to improve antifouling approaches that are widely used, low-cost, and environmentally benign. Many two-dimensional modification procedures have been described in the literature up to this point, including surface grafting and surface coating; nevertheless, more modification process refinement is required to address the trade-off between permeability and antifouling properties.

Acknowledgement

The authors sincerely acknowledge the financial support extended by the DST-PURSE-Phase-II Program of the Department of Science and Technology, New Delhi [Grant No. SR/PURSE PHASE-2/13(G)].

References

1. Walsh, C.L., Blenkinsop, S., Fowler, H.J., Burton, A., Dawson, R.J., Glenis, V., Manning, L.J., Jahanshahi, G., Kilsby, C.G., Adaptation of water resource systems to an uncertain future. *Hydrol. Earth Syst. Sci.*, 20, 1869, 2016.
2. Anand, A., Unnikrishnan, B., Mao, J.Y., Lin, H.J., Huang, C.C., Graphene-based nanofiltration membranes for improving salt rejection, water flux and antifouling–A review. *Desalination*, 429, 119, 2018.
3. Rosborg, I., Nihlgård, B., Ferrante, M., *Drinking Water Minerals and Mineral Balance*, vol. 25, Springer International Publishing, Switzerland, 2015.
4. Liu, B., Wang, D., Yu, G., Meng, X., Giraldo, J.D., Thakur, V.K., Gutiérrez, E., The History and State of Art in Membrane Technologies Tarragona. *J. Membr. Sci.*, 16, 1, 2013.
5. Romero, D.P., Garcia-Arce, Z.P., Ramírez, A.Z., Castro-Muñoz, R., Current Advances in Biofouling Mitigation in Membranes for Water Treatment: An Overview. *Processes*, 8, 182, 2020.
6. Ji, Y., Qian, W., Yu, Y., An, Q., Liu, L., Zhou, Y., Gao, C., Recent developments in nanofiltration membranes based on nanomaterials. *Chin. J. Chem. Eng.*, 25, 1639, 2017.

7. Lee, A., Elam, J.W., Darling, S.B., Membrane materials for water purification: design, development, and application. *Environ. Sci.: Water Res. Technol.*, 2, 17, 2016.
8. Yamjala, K., Nainar, M.S., Ramisetti, N.R., Methods for the analysis of azo dyes employed in food industry - A review. *Food Chem.*, 192, 813, 2016.
9. Hansen, E., Rodrigues, M.A.S., Aragão, M.E., de Aquim, P.M., Water and wastewater minimization in a petrochemical industry through mathematical programming. *J. Clean. Prod.*, 172, 1814, 2018.
10. Lively, R.P. and Sholl, D.S., From water to organics in membrane separations. *Nat. Mater.*, 16, 276, 2017.
11. Moslehyani, A. and Goh, P.S., Recent Progresses of Ultrafiltration (UF) Membranes and Processes in Water Treatment, in: *Membrane Separation Principles and Applications*, pp. 85–110, Elsevier, Amsterdam, The Netherlands, 2019.
12. de Sousa dos Santos, E.L.V., Cruz, J.N., da Costa, G.V., de Sá, E.M.F., da Silva, A.K.P., Fernandes, C.P., de Faria Mota Oliveira, A.E.M., Duarte, J.L., Bezerra, R.M., Tavares, J.F., *et al.*, Essential Oil of Ocimum basilicum against Aedes aegypti and Culex quinquefasciatus: Larvicidal Activity of a Nanoemulsion and *In Silico* Study. *Separations*, 11, 97, 2024.
13. Zhao, D. and Yu, S., A review of recent advance in fouling mitigation of NF/RO membranes in water treatment: Pretreatment, membrane modification, and chemical cleaning. *Desalin. Water Treat.*, 55, 870, 2015.
14. Van der Bruggen, B., Vandecasteele, C., Van Gestel, T., Doyen, W., Leysen, R., A review of pressure driven membrane processes in wastewater treatment and drinking water production. *Environ. Prog. Sustain. Energy*, 22, 46, 2003.
15. Mohammad, A., Teow, Y., Ang, W., Chung, Y., Oatley-Radcliffe, D., Hilal, N., Nanofiltration membranes review: Recent advances and future prospect. *Desalination*, 356, 226, 2015.
16. Paul, D.R., Reformulation of the solution-diffusion theory of reverse osmosis. *J. Mem. Sci.*, 241, 371, 2004.
17. Castro-Muñoz, R., Barragán-Huerta, B.E., Fíla, V., Denis, P.C., Ruby-Figueroa, R., Current role of membrane technology: From the treatment of agro-industrial by-products up to the valorization of valuable compounds. *Waste Biomass Valorization*, 9, 513, 2018.
18. Bhattacharyya, S., Das, P., *et al.*, Removal of Ranitidine from Pharmaceutical Waste Water Using Activated Carbon (AC) Prepared from Waste Lemon Peel, in: *Waste Water Recycling and Management*, S. Bhattacharyya, P. Das, S. Datta (Eds.), pp. 123–141, Springer, Singapore, 2019.
19. Shon, H.K., Phuntsho, S., Chaudhary, D.S., Vigneswaran, S., Cho, J., Nanofiltration for water and wastewater treatment - a mini review. *Drink. Water Eng. Sci.*, 6, 47, 2013.
20. Borea, L., Naddeo, V., Shalaby, M.S., Zarra, T., Belgiorno, V., Abdalla, H., Shaban, A.M., Wastewater treatment by membrane ultrafiltration enhanced

with ultrasound: Effect of membrane flux and ultrasonic frequency. *Ultrasonics*, 83, 42, 2018.

21. Abdallah, H., Shalaby, M.S., Alanezi, A.A., Shaban, A.M., Polyethersulfone/Polyamide11Cost Effective, Antifouling-Nanofiltration Blend Membrane. *16th International Conference on Environmental Science and Technology Rhodes*, Greece, CEST2019_00162, 2019.
22. Silva, F.C., Fouling of Nanofiltration Membranes, in: *Filtration*, M.G. Brazil (Ed.), pp. 119–131, Intech Open, Brazil, 2018.
23. Shannon, M.A., Bohn, P.W., Elimelech, M., Georgiadis, J.G., Marinas, B.J., Mayes, A.M., Science and Technology for Water Purification in the Coming Decades. *Nature*, 452, 301, 2008.
24. Rana, D. and Matsuura, T., Surface Modifications for Antifouling Membranes. *Chem. Rev.*, 110, 2448, 2010.
25. Geise, G.M., Lee, H.S., Miller, D.J., Freeman, B.D., McGrath, J.E., Paul, D.R., Water Purification by Membranes: The Role of Polymer Science. *J. Polym. Sci., Part B: Polym. Phys.*, 48, 1685, 2010.
26. Choudhury, R.R., Gohil, J.M., Mohanty, S., Nayak, S.K., Antifouling, fouling release and antimicrobial materials for surface modification of reverse osmosis and nanofiltration membranes. *J. Mater. Chem. A*, 6, 313, 2018.
27. Jhaveri, J.H. and Murthy, Z.V.P., A comprehensive review on anti-fouling nanocomposite membranes for pressure driven membrane separation processes. *Desalination*, 379, 137, 2016.
28. Kochkodan, V., Johnson, D.J., Hilal, N., Polymeric membranes: Surface modification for minimizing (bio) colloidal fouling. *Adv. Colloid Interface Sci.*, 206, 116, 2014.
29. Jiang, S., Li, Y., Ladewig, B.P., A review of reverse osmosis membrane fouling and control strategies. *Sci. Total Environ.*, 595, 567, 2017.
30. Abdelrasoul, A., Doan, H., Lohi, A., Fouling in membrane filtration and remediation methods, in: *Mass Transfer - Advances in Sustainable Energy and Environmental Oriented Numerical Modeling*, 195, InTech, London, United Kingdom, 2013.
31. Tiraferri, A., Membrane-based water treatment to increase water supply, Laboratory of Colloid and Surface Chemistry (LCSC) University of Geneva, www.colloid.ch/membranes, 2014.
32. Tang, C.Y., Chong, T.H., Fane, A.G., Colloidal interactions and fouling of NF and RO membranes: a review. *Adv. Colloid Interface Sci.*, 164, 126, 2011.
33. Ang, W.S., Tiraferri, A., Chen, K.L., Elimelech, M., Fouling and cleaning of RO membranes fouled by mixtures of organic foulants simulating wastewater effluent. *J. Membr. Sci.*, 376, 196, 2011.
34. Matin, A., Khan, Z., Zaidi, S.M.J., Boyce, M.C., Biofouling in reverse osmosis membranes for seawater desalination: Phenomena and prevention. *Desalination*, 281, 1, 2011.
35. Broeckmann, A., Wintgens, T., Schätfer, A.I., Removal and fouling mechanisms in nanofiltration of polysaccharide solutions. *Desalination*, 178, 149, 2005.

36. Schäfer, A.I., Fane, A.G., Waite, T.D. (Eds.), *Nanofiltration—Principles and Applications*, Elsevier, Great Britain, 2006.
37. Mulder, M.H.V., Polarization phenomena and membrane fouling, in: *Membrane Separations Technology: Principles and Applications*, R.D. Noble and S.A. Stern (Eds.), pp. 45–84, Elsevier Science, Amsterdan, 1995.
38. Belfort, G.R.H. and Zydney, A.L., The behavior of suspensions and macromolecular solutions in crossflow microfiltration. *J. Mem. Sci.*, 96, 1, 1994.
39. Javacek, M.H. and Bouchet, F., Constant flowrate blocking laws and an example of their application to dead-end microfiltration of protein solutions. *J. Mem. Sci.*, 82, 285, 1993.
40. Granger, J., Leclerc, D., Dodds, J.A., Filtration of dilute suspensions of latexes. *Filtr. Sep.*, 22, 58, 1985.
41. Blandin, G., Gautier, C., Toran, S.M., Monclus, H., Rodriguez-Roda, I., Comas, J., Retrofitting membrane bioreactor (MBR) into osmotic membrane bioreactor (OMBR): A pilot scale study. *Chem. Eng. J.*, 339, 268, 2019.
42. Moradi, G., Zinadini, S., R., L., Development of high flux nanofiltration membrane using para-amino benzoate ferroxane nanoparticle for enhanced antifouling behavior and dye removal. *Process Saf. Environ. Prot.*, 144, 65, 2020.
43. Qiu, H.Y., Xiao, T.H., Hu, N.E., The microporous membrane with different pore sizes was used to study the treatment of micro-polluted water in MBR. *Technol. Water Treat.*, 44, 94, 2018.
44. Costa, A.R., de Pinho, M.N., Elimelech, M., Mechanisms of colloidal natural organic matter fouling in ultrafiltration. *J. Membr. Sci.*, 281, 716, 2006.
45. Zheng, X., Major Organic Foulants in Ultrafiltration of Treated Domestic Wastewater and their Removal by Bio-Filtration as Pre-Treatment (Doctoral Thesis), Technische Universität Berlin, Berlin, Germany, 2010.
46. Vrijenhoek, E.M., Hong, S., Elimelech, M., Influence of membrane surface properties on initial rate of colloidal fouling of reverse osmosis and nanofiltration membranes. *J. Membr. Sci.*, 188, 115, 2001.
47. Williams, P.M., Drioli, E., Giorno, L., Membrane charge (Zeta Potential) effect, in: *Encyclopedia of Membranes*, vol. 1, 2016.
48. Membranes and membrane fouling: Surface charge analysis in filtration, https://wiki.anton-paar.com/in-en/surface-charge-analysis-in-filtration/.
49. Al-Amoudi, A.S., Factors affecting natural organic matter (NOM) and scaling fouling in NF membranes: a review. *Desalination*, 259, 1, 2010.
50. Luo, J. and Wan, Y., Effects of pH and salt on nanofiltration—a critical review. *J. Membr. Sci.*, 438, 18, 2013.
51. Luo, J.Q., Ding, L.H., Wan, Y.H., Jaffrin, M.Y., Flux decline control in nanofiltration of detergent wastewater by a shear-enhanced filtration system. *Chem. Eng. J.*, 181, 397, 2012.
52. Manttari, M., Pihlajamaki, A., Nystrom, M., Effect of pH on hydrophilicity and charge and their effect on the filtration efficiency of NF membranes at different pH. *J. Membr. Sci.*, 280, 311, 2006.

53. Childress, A.E. and Elimelech, M., Relating nanofiltration membrane performance to membrane charge (electrokinetic) characteristics. *Environ. Sci. Technol.*, 34, 3710, 2000.
54. Luo, J. and Ding, L., Influence of pH on treatment of dairy wastewater by nanofiltration using shear-enhanced filtration system. *Desalination*, 278, 150, 2011.
55. Wang, Y.N. and Tang, C.Y., Protein fouling of nanofiltration, reverse osmosis, and ultrafiltration membranes—the role of hydrodynamic conditions, solution chemistry, and membrane properties. *J. Membr. Sci.*, 376, 275, 2011.
56. Guo, X., Zhang, Z., Fang, L., Su, L., Study on ultrafiltration for surface water by a polyvinylchloride hollow fiber membrane. *Desalination*, 238, 183, 2009.
57. Cassano, A., Conidi, C., Ruby-Figueroa, R., Castro-Muñoz, R., Nanofiltration and tight ultrafiltration membranes for the recovery of polyphenols from agro-food by-products. *Int. J. Mol. Sci.*, 19, 351, 2018.
58. Salahi, A., Mohammedi, T., Pour, A., Rekabdar, F., Oily wastewater treatment using ultrafiltration. *Desalination*, 6, 289, 2000.
59. Lin, H.J., Lu, X.F., Duan, W., Filtration characteristics and mechanism of membrane fouling in a membrane bioreactor for municipal wastewater treatment. *Environ. Sci.*, 27, 2511, 2006.
60. Zhang, R., Liu, Y., He, M., Su, Y., Zhao, X., Elimelech, M., Jiang, Z., Antifouling membranes for sustainable water purification: strategies and mechanisms. *Chem. Soc. Rev.*, 45, 5888, 2016.
61. Hucknall, A., Rangarajan, S., Chilkoti, A., In pursuit of zero: polymer brushes that resist the adsorption of proteins. *Adv. Mater.*, 21, 2441, 2009.
62. Jiang, S. and Cao, Z., Ultralow Fouling, Functionalizable, and hydrolyzable zwitterionic materials and their derivatives for biological applications. *Adv. Mater.*, 22, 920, 2010.
63. Statz., A.R., Meagher., R.J., Barron., A.E., Messersmith., P.B., New Peptidomimetic Polymers for Antifouling Surfaces. *J. Am. Chem. Soc.*, 127, 7972, 2005.
64. Murthy, R., Cox, C.D., Hahn, M.S., Grunlan, M.A., Protein-resistant silicones: incorporation of poly (ethylene oxide) *via* siloxane tethers. *Biomacromolecules*, 8, 3244, 2007.
65. Murthy, R., Shell, C.E., Grunlan, M.A., The influence of poly (ethylene oxide) grafting *via* siloxane tethers on protein adsorption. *Biomaterials*, 30, 2433, 2009.
66. Jeon, S.I., Lee, J.H., Andrade, J.D., De Gennes, P.G., Protein—surface interactions in the presence of polyethylene oxide: I. Simplified theory. *J. Colloid Interface Sci.*, 142, 149, 1991.
67. Zheng, J., Li, L.Y., Chen, S.F., Jiang, S.Y., Molecular simulation study of water interactions with oligo (ethylene glycol)-terminated alkanethiol self-assembled monolayers. *Langmuir*, 20, 8931, 2004.
68. White, A. and Jiang, S.Y., Local and bulk hydration of zwitterionic glycine and its analogues through molecular simulations. *J. Phys. Chem. B*, 115, 660, 2011.

69. Wu, J., Lin, W., Wang, Z., Chen, S., Chang, Y., Investigation of the hydration of nonfouling material poly (sulfobetaine methacrylate) by low-field nuclear magnetic resonance. *Langmuir*, 28, 7436, 2012.
70. Wang, Z., Yu, H., Xia, J., Zhang, F., Li, F., Xia, Y., Li, Y., Novel GO-blended PVDF ultrafiltration membranes. *Desalination*, 299, 50, 2012.
71. Zhao, X.T., Su, Y.L., Li, Y.F., Zhang, R.N., Zhao, J.J., Jiang, Z.Y., Engineering amphiphilic membrane surfaces based on PEO and PDMS segments for improved antifouling performances. *J. Membr. Sci.*, 450, 111, 2014.
72. Sarfraz, M.H., Zubair, M., Aslam, B., Ashraf, A., Siddique, M.H., Hayat, S., Cruz, J.N., Muzammil, S., Khurshid, M., Sarfraz, M.F., Hashem, A., Dawoud, T.M., Avila-Quezada, G.D., Abd_Allah, E.F., Comparative analysis of phyto-fabricated chitosan, copper oxide, and chitosan-based CuO nanoparticles: antibacterial potential against Acinetobacter baumannii isolates and anticancer activity against HepG2 cell lines. *Front. Microbiol.*, 14, 1188743, 2023.
73. Li, Y., Su, Y., Zhao, X., Zhang, R., Zhao, J., Fan, X., Jiang, Z., Surface fluorination of polyamide nanofiltration membrane for enhanced antifouling property. *J. Membr. Sci.*, 455, 15, 2014.
74. Li, Y., Su, Y., Zhao, X., He, X., Zhang, R., Zhao, J., Fan, X., Jiang, Z., Antifouling, High-Flux Nanofiltration Membranes Enabled by Dual Functional Polydopamine. *ACS Appl. Mater. Interfaces*, 6, 5548–5557, 2014.
75. Zhao, X., Su, Y., Chen, W., Peng, J., Jiang, Z., Grafting perfluoroalkyl groups onto polyacrylonitrile membrane surface for improved fouling release property. *J. Membr. Sci.*, 415–416, 824–834, 2012.
76. Liu, Y., Su, Y., Zhao, X., Li, Y., Zhang, R., Jiang, Z., Improved antifouling properties of polyethersulfone membrane by blending the amphiphilic surface modifier with crosslinked hydrophobic segments. *J. Membr. Sci.*, 486, 195, 2015.
77. Nagy, A., Harrison, A., Sabbani, S., Munson, Jr., R.S., Dutta, P.K., Waldman, W.J., Silver nanoparticles embedded in zeolite membranes: release of silver ions and mechanism of antibacterial action. *Int. J. Nanomed.*, 6, 1833, 2011.
78. Mei, L., Lu, Z., Zhang, X., Li, C., Jia, Y., Polymer-Ag nanocomposites with enhanced antimicrobial activity against bacterial infection. *ACS Appl. Mater. Interfaces*, 6, 15813, 2014.
79. Hu, R., Li, G., Jiang, Y., Zhang, Y., Zou, J.J., Wang, L., Zhang, X., Silver-zwitterion organic-inorganic nanocomposite with antimicrobial and antiadhesive capabilities. *Langmuir*, 29, 3773, 2013.
80. Chen, X., Ku, S., Weibel, J.A., Ximenes, E., Liu, X., Ladisch, M., Garimella, S.V., Enhanced antimicrobial efficacy of bimetallic porous CuO microspheres decorated with Ag nanoparticles. *ACS Appl. Mater. Interfaces*, 9, 39165, 2017.
81. Ben-Sasson, M., Zodrow, K.R., Genggeng, Q., Kang, Y., Giannelis, E.P., Elimelech, M., Surface functionalization of thin-film composite membranes with copper nanoparticles for antimicrobial surface properties. *Environ. Sci. Technol.*, 48, 384, 2014.

82. Zhang, X., Ma, J., Tang, C.Y., Wang, Z., Ng, H.Y., Wu, Z., Antibiofouling polyvinylidene fluoride membrane modified by quaternary ammonium compound: direct contactkilling versus induced indirect contact-killing. *Environ. Sci. Technol.*, 50, 5086, 2016.
83. Ji, W., Koepsel, R.R., Murata, H., Zadan, S., Campbell, A.S., Russell, A.J., Bactericidal specificity and resistance profile of poly(quaternary ammonium) polymers and protein-poly(quaternary ammonium) conjugates. *Biomacromolecules*, 18, 2583, 2017.
84. Elena, P. and Miri, K., Formation of contact active antimicrobial surfaces by covalent grafting of quaternary ammonium compounds. *Colloids Surf., B*, 169, 195, 2018.
85. Liu, S.B., Zeng, T.H., Hofmann, M., Burcombe, E., Wei, J., Jiang, R.R., Kong, J., Chen, Y., Antibacterial activity of graphite, graphite oxide, graphene oxide, and reduced graphene oxide: membrane and oxidative stress. *ACS Nano*, 5, 6971, 2011.
86. de Faria, A.F., Perreault, F., Shaulsky, E., Arias Chavez, L.H., Elimelech, M., Antimicrobial electrospun biopolymer nanofiber mats functionalized with grapheme oxide-silver nanocomposites. *ACS Appl. Mater. Interfaces*, 7, 12751, 2015.
87. Baek, Y., Kim, C., Seo, D.K., Kim, T., Lee, J.S., Kim, Y.H., Ahn, K.H., Bae, S.S., Lee, S.C., Lim, J., Lee, K., Yoon, J., High performance and antifouling vertically aligned carbon nanotube membrane for water purification. *J. Membr. Sci.*, 460, 171, 2014.
88. Al Aani, S., Gomez, V., Wright, C.J., Hilal, N., Fabrication of antibacterial mixed matrix nanocomposite membranes using hybrid nanostructure of silver coated multi-walled carbon nanotubes. *Chem. Eng. J.*, 326, 721, 2017.
89. Tang, L., Livi, K.J.T., Chen, K.L., Polysulfone membranes modified with bioinspired polydopamine and silver nanoparticles formed in situ to mitigate biofouling. *Environ. Sci. Technol. Lett.*, 2, 59, 2015.
90. Kim, S.H., Kwak, S.Y., Sohn, B.H., Park, T.H., Design of TiO_2 nanoparticle self-assembled aromatic polyamide thin-film-composite (TFC) membrane as an approach to solve biofouling problem. *J. Membr. Sci.*, 211, 157, 2003.
91. Dibrov, P., Dzioba, J., Gosink, K.K., Hase, C.C., Chemiosmotic mechanism of antimicrobial activity of Ag^+ in Vibrio cholera. *Antimicrob. Agents Chemother.*, 46, 2668, 2002.
92. Shahane, K., Kshirsagar, M., Tambe, S., Jain, D., Rout, S., Ferreira, M.K.M., Mali, S., Amin, P., Srivastav, P.P., Cruz, J., Lima, R.R., An Updated Review on the Multifaceted Therapeutic Potential of Calendula officinalis L. *Pharmaceuticals*, 16, 611, 2023.
93. Lok, C.N., Ho, C.M., Chen, R., He, Q.Y., Yu, W.Y., Sun, H., Tam, P.K.H., Chiu, J.F., Che, C.M., Proteomic analysis of the mode of antibacterial action of silver nanoparticles. *J. Proteome Res.*, 5, 916, 2006.
94. Schreurs, W.J.A. and Rosenberg, H., Effect of silver ions on transport and retention of phosphate by Escherichia coli. *J. Bacteriol.*, 152, 7, 1982.

95. Yang, W., Shen, C., Ji, Q., An, H., Wang, J., Liu, Q., Zhang, Z., Food storage material silver nanoparticles interfere with DNA replication fidelity and bind with DNA. *Nanotechnology*, 20, 085102, 2009.
96. Bromberg, L. and Hatton, T.A., Poly (N-vinylguanidine): Characterization, and catalytic and bactericidal properties. *Polymer*, 48, 7490, 2007.
97. Kenawy, E.R., Worley, S.D., Broughton, R., The chemistry and applications of antimicrobial polymers: a state-of-the-art review. *Biomacromolecules*, 8, 1359, 2007.
98. Shahkaramipour, N., Tran, T.N., Ramanan, S., Lin, H., Membranes with surface-enhanced antifouling properties for water purification. *Membranes*, 7, 13, 2017.
99. Ayyavoo, J., Nguyen, T.P.N., Jun, B.M., Kim, I.C., Kwon, Y.N., Protection of polymeric membranes with antifouling surfacing *via* surface modifications. *Colloids Surf. A: Physicochem. Eng. Asp.*, 506, 190–201, 2016.
100. Sagle, A.C., Ju, H., Freeman, B.D., Sharma, M.M., PEG-based hydrogel membrane coatings. *Polymer*, 50, 756, 2009.
101. Kim, I.C., Ka, Y.H., Park, J.Y., Lee, K.H., Preparation of fouling resistant nanofiltration and reverse osmosis membranes and their use for dyeing wastewater effluent. *J. Ind. Eng. Chem.*, 10, 115, 2004.
102. Myung, S.W., Choi, I.H., Lee, S.H., Kim, I.C., Lee, K.H., Use of fouling resistant nanofiltration and reverse osmosis membranes for dyeing wastewater effluent treatment. *Water Sci. Technol.*, 51, 159, 2005.
103. Hong Anh Ngo, T., Nguyen, D.T., Dinh Do, K., Thi Minh Nguyen, T., Mori, S., Thi Tran, D., Surface modification of polyamide thin film composite membrane by coating of titanium dioxide nanoparticles. *J. Sci.: Adv. Mater. Devices*, 1, 468, 2016.
104. Choi, H., Sofranko, A.C., Dionysiou, D.D., Nanocrystalline TiO_2 photocatalytic membranes with a hierarchical mesoporous multilayer structure: synthesis, characterization, and multifunction. *Adv. Funct. Mater.*, 16, 1067, 2006.
105. Prince, J.A., Bhuvana, S., Boodhoo, K.V.K., Anbharasi, V., Singh, G., Synthesis and characterization of PEG-Ag immobilized PES hollow fiber ultrafiltration membranes with long lasting antifouling properties. *J. Membr. Sci.*, 454, 538, 2014.
106. Yang, H.L., Lin, J.C., Huang, C., Application of nanosilver surface modification to RO membrane and spacer for mitigating biofouling in seawater desalination. *Water Res.*, 43, 3777, 2009.
107. Zhao, X., Su, Y., Cao, J., Li, Y., Zhang, R., Liu, Y., Jiang, Z., Fabrication of antifouling polymer–inorganic hybrid membranes through the synergy of biomimetic mineralization and non-solvent induced phase separation. *J. Mater. Chem. A*, 3, 7287, 2015.
108. Zhang, Y.C., Yang, Q., Chang, Q., Low temperature plasma modification and electrokinetic characteristics of PVDF hollow fiber membrane. *Technol. Water Treat.*, 37, 57, 2011.

109. Zhao, Y.H., Wee, K.H., Bai, R., Highly hydrophilic and low- protein-fouling polypropylene membrane prepared by surface modification with sulfobetaine-based zwitterionic polymer through a combined surface polymerization method. *J. Membr. Sci.*, 362, 326, 2010.
110. Seman, A., Mazrul, N., Johnson, D., Al-Malek, S., Nidal, H., Surface modification of nanofiltration membrane for reduction of membrane fouling. *Desalin. Water Treat.*, 10, 298, 2009.
111. Abuhabib, A.A., Mohammad, A.W., Hilal, N., Rahman, R.A., Shafie, A.H., Nanofiltration membrane modification by UV grafting for salt rejection and fouling resistance improvement for brackish water desalination. *Desalination*, 295, 16, 2012.
112. Mezohegyi, G., Bilad, M.R., Vankelecom, I.F.J., Direct wastewater up-concentration by submerged aerated and vibrated membranes. *Bioresour. Technol.*, 118, 1, 2012.
113. Naidu, G., Jeong, S., Choi, Y., Vigneswaran, S., Membrane distillation for wastewater reverse osmosis concentrate treatment with water reuse potential. *J. Membr. Sci.*, 524, 565, 2017.
114. Carretier, S., Chen, L.A., Venault, A., Yang, Z.R., Aimar, P., Chang, Y., Design of PVDF/PEGMA-b-PS-b-PEGMA membranes by VIPS for improved biofouling mitigation. *J. Membr. Sci.*, 510, 355, 2016.
115. Ruan, H., Li, B., Ji, J., Sotto, A., Van Der Bruggen, B., Shen, J., Gao, C., Preparation and characterization of an amphiphilic polyamide nanofiltration membrane with improved antifouling properties by two-step surface modification method. *RSC Adv.*, 8, 13353, 2018.
116. Rosenbaum, S., Mahon, H.I., Cotton, O., Permeation of water and sodium chloride through cellulose acetate. *J. Appl. Polym. Sci.*, 11, 2041, 1967.
117. Jiang, J., Zhu, L., Zhu, L., Zhu, B., Xu, Y., Surface characteristics of a self-polymerized dopamine coating deposited on hydrophobic polymer films. *Langmuir*, 27, 14180, 2011.
118. Li, X., Yang, L., Torres, J., Wang, R., Engineering ultra-permeable and antifouling water channel-based biomimetic membranes toward sustainable water purification. *J. Membr. Sci.* Lett., 3, 2, 100049, 2023.
119. Liu, Y., Han, C., Wei, T., Chang, Y., Surface-initiated atom transfer radical polymerization from porous poly(tetrafluoroethylene) membranes using the C–F groups as initiators. *J. Polym. Sci. Part A Polym. Chem.*, 48, 2076, 2010.
120. McCloskey, B.D., Park, H.B., Ju, H., Rowe, B.W., Miller, D.J., Freeman, B.D., A bioinspired fouling-resistant surface modification for water purification membranes. *J. Membr. Sci.*, 413, 82, 2012.
121. Karkhanechi, H., Takagi, R., Matsuyama, H., Enhanced antibiofouling of RO membranes *via* polydopamine coating and polyzwitterion immobilization. *Desalination*, 337, 23, 2014.
122. Chen, S., Li, L., Zhao, C., Zheng, J., Surface hydration: Principles and applications toward low-fouling/nonfouling biomaterials. *Polymer*, 51, 5283, 2010.

15

Nanocatalysts for Chemical Processes: Materials for Catalysis, Hydrogen Production, and Pollution Control

Shruti Mishra[1], Mustafa Aamir Hussain[1], Nisha V. Bora[2] and Leena V. Bora[1]*

[1]*Chemical Engineering Department, Institute of Technology, Nirma University, Ahmedabad, Gujarat, India*
[2]*Mechanical Engineering Department, L.D. College of Engineering, Ahmedabad, Gujarat, India*

Abstract

Nanocatalysis has revolutionized chemical processes for biodiesel production, biomass utilization, electrosynthesis, air and water pollution control, hydrogen production, and several other applications. This study covers recent nanocatalytic material advances and their wide applications in numerous major sectors. The paper begins with nanocatalysis basics and its role in improving reaction kinetics and selectivity. It focuses on nanocatalyst creation and characterization using metal nanoparticles, metal oxides, and other novel nanostructures. In biodiesel manufacturing and biomass conversion, nanocatalysts offer cleaner and more sustainable energy applications. The use of nanocatalysts in electrosynthesis processes for value-added chemical production highlights their potential to change chemical transformations. This study also discusses nanocatalysts and nanoadsorbents for air and water pollution reduction. It describes how photocatalytic nanocatalysts degrade pollutants to improve air quality. Additionally, nanocatalysts' involvement in hydrogen production is highlighted, revealing their ability to unlock hydrogen as a clean and renewable energy carrier. The review of spinel-structured nanocatalysts and their applications is unique. The use of these materials in catalytic processes and their unique structural features demonstrate their potential to change the execution method and space.

**Corresponding author*: leena.bora@nirmauni.ac.in

Inamuddin, Tariq Altalhi and Mohammad Abu Jafar Mazumder (eds.) Nanostructured Hard Materials: Industrial Applications, (447–470) © 2026 Scrivener Publishing LLC

Keywords: Nanocatalysts, pollution control, hydrogen production, spinel structures, photoelectrocatalyst

15.1 Introduction

Nanocatalysts, a product of nanoscience and nanotechnology, have revolutionized the field of catalysis and material structure. These catalysts, frequently developed as nanoparticles, parade unknown properties and performances that surpass those of traditional homogeneous and miscellaneous catalysts. The mechanisms behind these nano-goods remain unclear, furnishing ample room for further disquisition and enhancement in the design of nanocatalysts [1]. Current design strategies for nanocatalysts are grounded on the conflation of large active spots at the nanoscale level and tuning the micro-environment by hosting them in confined spaces in previous nanomaterials. This leads to structural changes, which can alter the physical, electronic, and optical properties of nanomaterials.

Nanomaterials have emerged as promising solutions for water quality management in the field of pollution control, owing to their distinctive physicochemical properties, significant economic advantages, efficient waste disposal capabilities, and environmentally friendly nature [2]. Nanocatalysts also play a significant part in hydrogen production. Hydrogen is primarily produced by the steam reforming of natural gas, with other similar styles like gasification of coal/biomass/waste and water splitting by electrolysis. All these styles use nanocatalysts to achieve high effectiveness in hydrogen production [3].

There are two types of nanocatalysts used in chemical industries: Homogeneous catalysts have a well-defined coordinate structure, and active sites can be designed around the centers of coordination. The two major drawbacks of these catalysts are reusability and catalysts with products, which make them costly. On the other hand, heterogeneous catalysts are made up of solid nanoparticles, which make them structurally more stable than homogeneous nanocatalysts. Additionally, it is easy to recover after the reaction and can be used for the next round. Lu *et al.* [4], discovered that heteroatom-doped carbon material with active sites has shown excellent properties of reusability and high atom utilization. Nanoparticles are generally synthesized by chemical and physical processes which require energy. Figure 15.1 depicts the bottom – down approach of different nanoparticles [5].

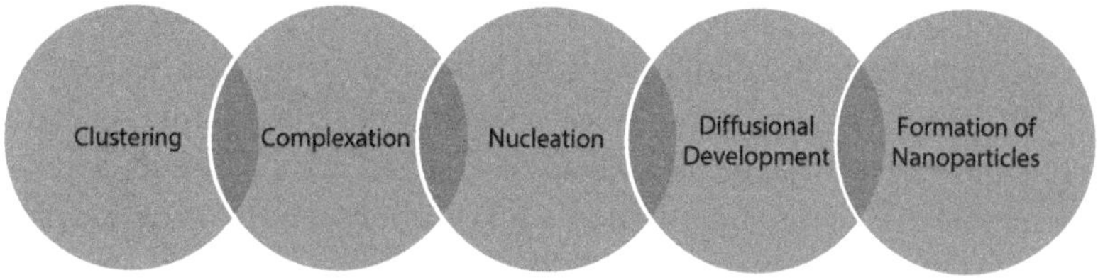

Figure 15.1 A classic way for synthesizing these nanoparticles consists of five steps.

15.2 Applications of Nanocatalysts

15.2.1 Biodiesel Production

Fossil fuel is made up of decomposed plant and animal matter. More than 85% of the world's energy needs are met by fossil fuels. Biodiesel is a renewable alternative to these fossil fuels. It has a better cetane number and lubrication and the emission of greenhouse gases is less than that of petroleum products. Biodiesel is made up of 11% oxygen which results in fewer CO (carbon monoxide) emissions [6].

Several ways can be retained for the conflation of biodiesel, including micro-emulsion, pyrolysis, transesterification, and *in-situ* transesterification. The transesterification synthesis involving methanol and triolein yields methyl oleate, which is a type of biodiesel, as well as glycerol, which is a useful by-product [7]. Figure 15.2 shows the four main steps for production of biodiesel. To enhance the efficiency of this reaction, an appropriate catalyst is required.

To enhance the pace and yield of a response, it's imperative to employ catalysts that are suitable for the specific response conditions. The suitability of nanocatalysts for biodiesel generation was first demonstrated by Yan *et al.* in a patent published in 2009 [8]. Research findings indicate that the

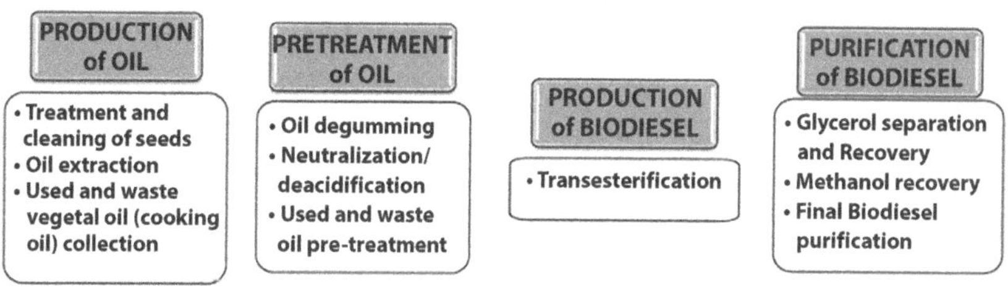

Figure 15.2 Biodiesel production from vegetable/waste oil [80].

utilization of alkaline catalysts significantly accelerates the process, exhibiting a speed enhancement of up to 4,000 times in comparison to acid catalysts [9]. The acid catalysts commonly employed in the production process are HCl and H_2SO_4. Conversely, alkaline catalysts exhibit greater solubility in methanol when compared to their acid counterparts. In order to address these issues, heterogeneous catalysts are employed.

Nanocatalysts are a specific category of heterogeneous catalysts that have been observed to have enhanced surface area, leading to accelerated reaction rates. Additionally, these catalysts have demonstrated resistance to conversion in soap formulations [10]. The application of nanocatalysts has been set up to enhance the effectiveness of biodiesel conflation, particularly in cases involving both low and high situations of free fatty acids (FFA), making it a suitable option. Similarly, nanocatalysts have collected attention owing to their capability to be "reused" in discrepancy to indispensable catalysts [11].

Nanoparticles have the earliest application in catalyst production in spherical (quasi-sphere) balls of less than 100 nm. Compared to nanocatalysts, bulk catalysts are less catalytically active. For instance, Au particles in bulk form are, in general, catalytically inactive while Au nanoparticles are extremely active.

Smaller particles increase the surface-to-volume ratio, which increases the reactivity and decreases the catalyst required. Biodiesel production after nanocatalysts takes place under ambient conditions without the requirement of any other solvent. It indirectly contributes to decreasing global warming by decreasing the activation energy [10].

15.2.2 Biomass Production Using Nanocatalysts

Biomass can be converted into bioenergy using the thermochemical processes. Thermal conversion: it involves the burning of biomass feedstock, removing water or stabilizing the base nature of the feedstock. Pyrolysis: It is a method of burning biomass in the absence of oxygen at 200°C to 300°C. Depending upon the degree of burning it can be classified into slow, fast and flash pyrolysis. Because of its ability to quickly turn biomass feedstock into liquid biofuel, fast pyrolysis has drawn a lot of attention [12]. Figure 15.3 depicts the major findings about biomass production with the help of different types of nanocatalysts.

Conventional catalysts have several drawbacks such as deactivation, coke formation, etc. These drawbacks can be addressed with the help of nanocatalysts. It reduces energy consumption, enhances the surface area of reaction and increases the lifespan of the catalyst.

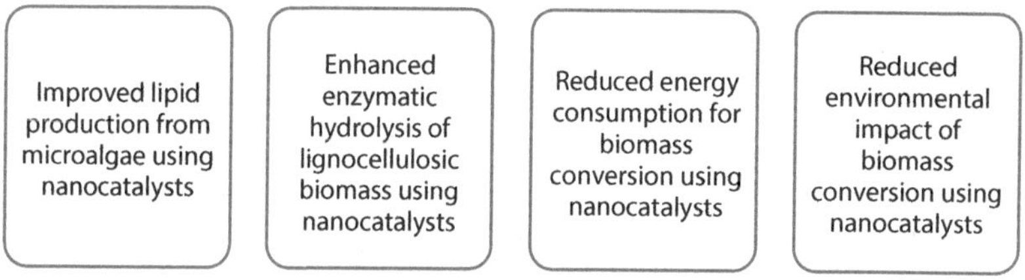

Figure 15.3 Major advantages of biomass production using nanocatalyst.

15.2.3 Nanocatalyst in Electrosynthesis

Electrosynthesis refers to the process of producing chemical compounds using an electrolytic cell. Electrochemical synthesis takes place by electrochemical reduction on the cathode and electrochemical oxidation on the anode, where both the electrodes are separated by an ion conducting membrane. In industrial applications, notable instances of electrosynthesis include the chlor-alkali process for the production of chlorine gas (Cl_2) and sodium hydroxide (NaOH), as well as the Hall-Héroult process utilized to obtain elemental aluminum (Al) [13].

Catalysts have always played a crucial role in increasing the rate and yield of electrochemical reactions. Schaidle *et al.* [14], suggested that in industries more than 85% of chemical reactions rely on catalysts.

Electro production of hydrogen by water splitting using renewable energy sources has become an alternative for fuel. Metal based catalysts such as Pd, Pt, and Ru have been found to outperform the non-noble metals such as V, Ni, Cu etc. [15]. The performance of these catalysts can be increased by doping and creating active sites using hetero atoms. Another popular technique to increase catalyst stability is to encase catalysts made of non-noble metals in carbon shells. Through chemical vapor deposition, Nanoparticles can acquire a carbon shell, the process of high-temperature co-annealing, in conjunction with an appropriate carbon source [16], or the utilization of pyrolysis techniques applied to Metal-organic frameworks (MOFs) [17], can be employed.

15.2.4 Air Pollution

Air pollutants are a serious threat to human life and have always demanded special attention. The detrimental effects of poor air quality extend beyond humans, having an effect on the natural system, which includes vegetation and living things. The following is a discussion of some notable developments in the application of nanomaterials to air pollution:

15.2.4.1 Nanoadsorbents

Based on the principles of nanotechnology, it is suggested that the implementation of nanoscale adsorbents, generally known as nanoadsorbents, might potentially lead to substantial mitigation of several prevailing challenges, such as substandard air quality. The nanoparticles exhibit the potential to efficiently collect a wide range of air pollutants, including atmospheric NOx, volatile organic compounds (VOCs), and microorganisms. The following discussion will delve into the different methodologies and configurations employed by researchers in their endeavors to combat and mitigate the dissemination of harmful substances and air pollutants, utilizing these vibrant nanoadsorbent tools.

1. Carbon Nanotubes (CNTs)

Numerous scholarly investigations have been conducted to examine the usage of carbon nanotubes as nanoadsorbents for the adsorption of various adsorbates. These accessories possess exceptional physical characteristics such as average separation volume, separation perimeter, and surface area, rendering them appropriate for commercial applications [18]. CNTs are employed for the purpose of regulating the emission of methane and CO_2 from synthetic chimney stack exhausts, and vehicular exhausts. Additionally, they serve to capture greenhouse gases released from coal mining operations and power generation facilities. CNTs can be utilized for the purpose of separating adulterants in the atmosphere, owing to their improved properties that enable selective adsorption of these adulterants [19]. The adsorptive properties of these nanoparticles can be modified by the introduction of new functional groups. In their study, Gui *et al.* [20], conducted modifications on multi-walled carbon nanotubes (MWCNTs) utilizing (3- amino propyl) triethoxysilane (APTES). The findings of their research revealed the chemisorption of CO_2 through the carbamate conformation.

The CO_2 adsorption packets from air aqueducts were investigated by Lu *et al.* [21], using both pure and 3 -aminopropyltriethoxysilane (APTS) -modified CNTs. The carbon nanotubes that were transformed had a higher capacity for adsorbing CO_2 when compared to other adsorbents that were modified with APTS and grounded silica. This indicates that the carbon nanotubes modified with APTS are promising adsorbents for capturing CO_2 from air ducts.

2. Zinc Oxide

Zinc oxide nanostructures present a challenge because they are resistant to non-catalytic fixing reactions, but they can be an effective tool for reducing

H_2S. Huy *et al.* [22], conducted a study employing an ultrasonic-assisted approach to synthesize zinc oxide nanoparticles (U-ZnO) in a single step, resulting in an enhanced H_2S adsorption capacity of approximately 29.51 mg/g. In a separate investigation, Ghosh *et al.* [23], utilized a detonation technique to fabricate ZnO nanoparticles for the purpose of studying carbon dioxide adsorption. The results demonstrated an enhanced adsorption capacity of approximately 0.4 mmol/g at bar and 283 K. Thermodynamic analysis confirmed the regenerability of carbon dioxide adsorbed on ZnO nanoparticles. Carbon nanotubes, along with nanoadsorbents based on zinc oxide, exhibit favorable efficiency for the removal of toxic gas emissions. Furthermore, modified versions of CNT have exhibited superior adsorption capabilities compared to unmodified counterparts.

15.2.4.2 Photocatalytic Nanocatalysts

Through photocatalytic remediation, in which the material is exposed to light and energy to create electron-hole dyads, nanomaterials can manage air impurities. These nanocatalysts are largely promising for air remediation, as they enable quick and precise chemical metamorphosis with superior product yield and catalyst recovery compared to standard catalysts [24].

Titanium dioxide is a popular semiconductor photocatalyst due to its affordability, vacuity, stability, and non-toxicity. To break down atmospheric pollutants including NOx and VOCs, nanoparticles of titanium dioxide are used in self-cleaning coatings. The surface absorbs UV light and degrades air pollutants when it is covered underneath a photocatalytic TiO_2 layer [25].

Face hydroxyl groups play a pivotal part in adulterant adsorption and hydroxyl radical generation, making the fabrication system's influence on the hydroxyl group population pivotal for air sanctification. Face-fluorinated titanium dioxide nanosheets have shown better exertion in gaseous ammonia separation, compared to aeroxide, mixed anatase and rutile TiO_2. Still, pristine TiO_2 have lower adsorption affinity towards air pollutants and lower face areas, reducing response rate and separation effectiveness [26]. Graphitic carbon nitrides, graphene oxide, CNTs, carbon filaments and actuated carbon are a few examples of carbonaceous accessories that have been looked at as TiO_2 supports. TiO_2 photocatalytic efficiency is frequently improved by using carbon-grounded nanostructures such as graphene nanosheets and carbon nanotubes [27]. TiO_2-CNT composites provide efficient electron transport, preventing electron-hole recombination. Long-term exposure to the substantial adulterant formaldehyde can cause health problems such as skin irritation and nasal excrescences.

15.2.5 Water Pollution

While there are advancements in materials and technologies of water treatment, they often result in residues that have an impact on the environment. Nanomaterials are utilized for purifying dirty groundwater, and wastewater, filtration of heavy metals and hydrocarbon-polluted sediments, and separating organic or inorganic compounds from the soil, as discussed below:

15.2.5.1 Carbon Nanotubes

The interaction between CNTs and metal cations, through electrostatic attraction and chemical bonding, removes salt and pollutants with high efficiency [28]. Plasma-modified ultra-long CNTs have a large surface area and variable surface chemistry, but their production cost is a major barrier to industrial-scale applications. They exhibit strong antibacterial activity and have shown potential in removing oil pollutants from oil spills [29].

15.2.5.2 Polymeric Nanoadsorbents

Due to their large face areas, adjustable face chemistry and optimal mechanical severity, polymeric nano-adsorbents are more practical than conventional designs [30]. They are divided into four categories: dendrimer-grounded nanocomposites, essence and essence oxide polymer nanocomposites, carbon polymer nanocomposites, and graphene/polymer nanocomposites [31]. The unique fanning patch structure of dendrimer nanocomposite, which eliminates heavy essence and organic and inorganic contaminants, makes it perfect for catalysis [32].

15.2.5.3 Zeolites

Zeolites are aluminosilicate compounds that have microscopic pores and a high surface area. They are supported by reversible dehydration and ion exchange [33, 34]. Using cutting-edge technologies, nano-zeolites with silver conjugation have been decreased to 10-500 nm and have a higher affinity for pollutant adsorption. These selective zeolites improve ion exchange and reversible dehydration, which makes them efficient at preventing microbial development [35]. Figure 15.4 shows zeolites in a simplified crystal structure.

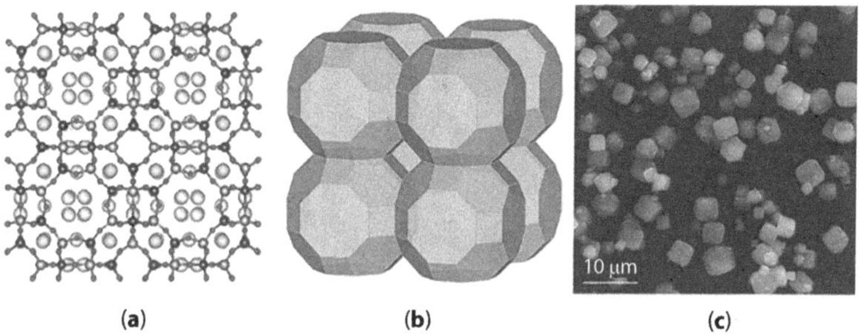

Figure 15.4 (a) Model structure of zeolite Na-A; (b) A simplified illustration of the crystal structure of zeolite; (c) Zeolite A's SEM image [81].

15.2.5.4 Metal Based Nanoadsorbent

Heavy metals and radioactive elements can be removed more successfully using metal-based nanoadsorbents than with traditional adsorbents like activated carbon [36]. Due to their high surface area, these adsorbents have short intraparticle diffusion distances and can be compressed without losing surface area. Nano magnetite, nano iron hydroxide, nano-sized Ag, nano-sized TiO_2, and nano-sized zero-valent Fe are a few examples of metal based nanoadsorbents [37, 38]. Nano hydroxyapatite and nano clay are examples of highly efficient adsorbents that can effectively remove both inorganic and organic pollutants. These metal-oxide nanoparticulate adsorbents are primarily used to remove radioactive and heavy metal contaminants from toxic water sources [39].

15.2.6 Hydrogen Production

Hydrogen is a promising future energy source due to its eventuality to reduce the greenhouse effect, excluding reactionary energy dependence, and facilitating energy metamorphosis in inner combustive apparatuses and polymer electrolyte energy cells [40]. It offers substantial efficacy in energetic metamorphoses, water as a starting media, ample vacuity, and colorful storehouse systems [41].

15.2.6.1 Hydrogen Production by Water Splitting Method

"Water splitting" is the process by which H_2 is created from water by excluding oxygen. Hydrogen production can be achieved by utilizing energy sources, including both non-renewable options such, as biomass

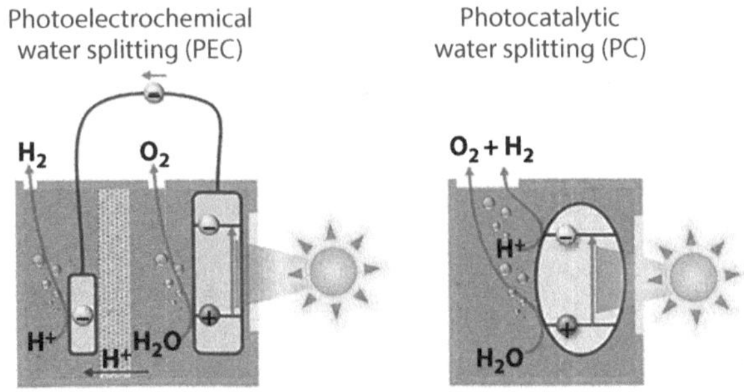

Figure 15.5 Two main methods of producing green hydrogen represented schematically [82].

gasification [42], pyrolysis, and solvothermal processes [43]. These methods help break down the H-O-H bond to obtain hydrogen. For water splitting, electrocatalysts, photoelectrocatalysts, and photocatalysts are frequently used (Figure 15.5). To produce green hydrogen, there are two main methods: photoelectrochemical (PEC) and photocatalytic (PC) water splitting. These two methods vary greatly in terms of their underlying theories and technology readiness levels (TRL). TRL are a method for understanding the technical maturity of a technology during its acquisition phase.

One of the obstacles faced is the occurrence of electron transfers, in oxygen evolution. This issue significantly impacts the rate limiting step. Another challenge encountered is the amount of level drop resulting from continuous gas purification, which causes degradation of the electrode materials. It is crucial to develop catalysts that can effectively reduce over potential. A well - designed electrocatalyst should naturally exhibit an appropriate over potential; it should also possess an electrochemically reactive surface, demonstrate identifiable charge conductance, exhibit chemical robustness in the underlying electrolytic environment and display thermomechanical durability [44]. Although rare materials have proven effective as electrocatalysts, they present challenges due to their porosity, cost and inefficiency in low pH environments. Therefore, there is a growing interest in reducing or replacing metals with noble metal nanocatalysts [45].

Nanoscience reveals that the surface properties of nanomaterials differ significantly from those of bulk materials because surface atoms increase as the scale decreases. 2 nm quantum dots have 70-80% more accessible atoms. Dangling atoms and crystal structures of loosely bound atoms, enhance reactivity and have different opto-bio-physico-chemical properties.

Using nanocatalysts, the following techniques can be utilized to improve water splitting for the generation of hydrogen:

I. Pointed edges may be added to the catalyst's upper surface to produce enough charge species at the catalyst-electrolyte interface. The most desirable charge quantity is possible in dispersed nanoparticles when the particle magnitude is less than the mean free dispersion extent. While smooth surface features result in ineffective photon assimilation, nanoscale surface structuring increases photon absorption [46].

II. By widening the band gap, dropping the LUMO (Least unoccupied molecular orbital) level, and oxidizing the HUMO (Highest occupied molecular orbital), nanocrystals can be effectively used for electro-catalysis. Due to the intense electron confinement, external stimuli will most likely influence this effect. 0D nano confinement Quantum Dots (QDs) exhibit effective low-energy photon absorption, producing the desired photocatalytic H_2 evolution rate [47].

III. Rapid charge conductance and the suppression of charge annihilation are well-known properties of 1D nanoarchitecture, including nanorods and nanotubes. Due to its significant aspect ratio and exceptional surface reactivity, tubular nanoconfinement is desired for improved electrocatalysis [48]. At the edges of clearly defined lattice symmetry, these structures enable quick ion conduction pathways. Certain features of 2D plenary structures, like ultra-large surface area and atomic thickness, are beneficial for catalyzing hydrogen redox behavior. Furthermore, via limited diffusion pathways, 2D nanomaterials provide more dynamic sites and improved electrical conductivity [49].

IV. Due to its increased surface area, noticeable ionic species mobility and robustness, graphene, a 2D carbon nanosheet, continues to retain its distinctive properties. A TiO_2-reduced Graphene Oxide (composite) primed using a sol-gel method can increase hydrogen production by up to 8.6 $molh^{-1}$ [50]. The hydrazine method may not be as effective as the hydrothermal method, which employs composites containing r-GO as the finest material for the Hydrogen Evolution Reaction (HER) process. An analysis of the HER activity of a Fe_2O_3/TiO_2 composite electrocatalyst coated with rGO revealed that it increased stability and the surface area [51].

V. 3D nanostructures are characterized by high phonon assimilation and restricted diffusive path for photo-triggered h^+ to migrate through the interface between the electrolytic

> solution and the electrode. These micro-nano characteristics have piqued a great deal of interest in the field of water splitting because they are hierarchical nanostructures that facilitate the split of H-OH bond photo-electrocatalytically [52].

There are two methods used to create electrodes, viz; using powdered catalysts and growing expensive materials on conductive substrates. However, one drawback of these methods is the use of a binder. This binder reduces the surface area of the catalyst that comes into contact with electrolytes and blocks sites [53].

The development of artificial principles for inexpensive 3D electrodes is crucial for effective photoelectrochemical hydrogen production. Hierarchical scaffolds have gained attention because of their lowered ion diffusive paths, which allow simpler and more surface area that is readily acceptable to electronic and ionic conductance. The less expensive nickel foam (NF), which has a large surface area, a 3D porous scaffold, and exceptional electronic conductivity, serves primarily as a foundation for electrode materials. Effective mass movement is provided by its zigzag flow channels, micro holes and exterior space. In various energy gathering processes, particularly in electrochemical H-OH bond cleavage, a variety of base surfaces, with different conductive bedrock with permeable textures, metallic thin coatings, and meshes, have been studied, when it comes to improving the bulk conduction of electrolytic species [54]. Figure 15.6 visually represents an insightful overview of the most common graphitic nanostructures, highlighting their various forms and applications in nanotechnology. The graphitic nanostructures shown in the image are graphene, carbon nanotubes (CNTs), graphene oxide, graphite, rGO, and carbon nano-onions.

15.2.7 Hydrogen Production Using Spinel-Structured Nanocatalysts

Spinel nanocatalysts seem pretty promising for taking carbon dioxide (CO_2) and hydrogenating it into useful chemicals like methanol, methane, and formic acid. With their tetrahedral and octahedral metal ions, the unique structure of spinels lets scientists engineer active sites with special electrical and geometric properties to boost the CO_2 hydrogenation reaction. One major example is copper-based spinels that are remarkable at selectively converting CO_2 and H_2 into methanol. The high reactivity of the copper spinels relates to how strongly the copper ions interact with the spinel support material, which holds the active copper parts in place and helps CO_2

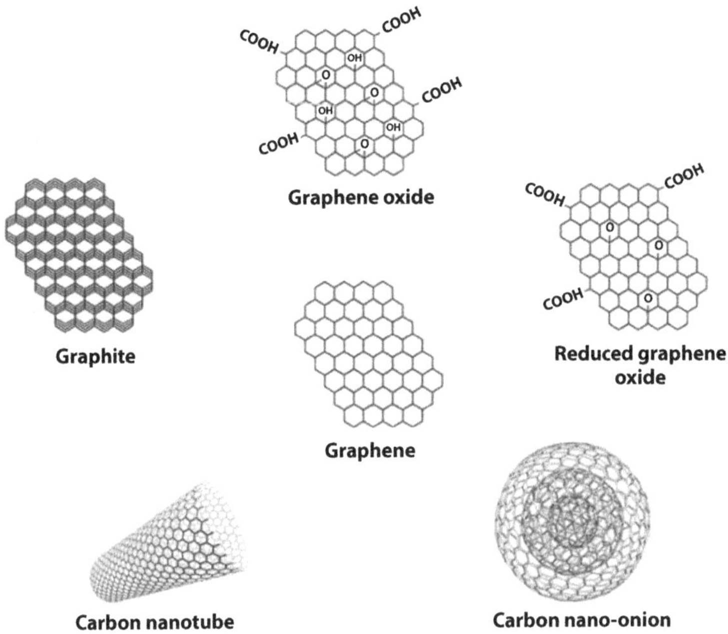

Figure 15.6 Most prevalent graphitic nanostructures [83].

stick to the surface and get activated. Other metal spinels besides copper have also been checked out for CO_2 hydrogenation, like iron, cobalt, nickel, and mixtures thereof. Further improvements as to how well the spinel nanocatalysts work could come from adding extra metals or tweaking their size and shape [55]. Figure 15.7 shows the unique structure of the spine, where tetrahedral (sites A) and octahedral (sites B) units with cations, are

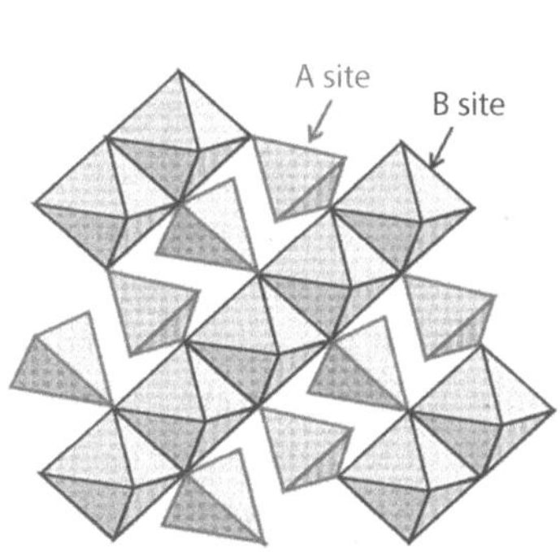

Figure 15.7 Structure of a spine. Tetrahedron (sites A) and octahedron (sites B) have cations at their centers oxygen anions are at the intersection of two octahedron and one tetrahedron are oxygen anions [84].

found at their centers. Two octahedra and one tetrahedron are present at the intersections in this framework. Cations are at the centers of tetrahedral and octahedral sites. Oxygen anions at the meeting point of two octahedral and one tetrahedral unit stabilize and shape the spine structure. This cross-reference emphasizes the spine's intricate organization, including tetrahedral and octahedral units and oxygen anions at specific junctions.

15.2.7.1 Fabrication Process

Spinel structured shells play a role in their effectiveness and reactivity in CO_2-H_2 channels. Techniques to control the conformation of spinel catalysts have evolved from temperature solid-state methods to outcome-based or chemical-based approaches [56]. Factors such as forerunner selection, surfactants, and temperature of treatment affect the development of the spinel structure. By utilizing precursor substances like nitrates, chlorides, acetates, and sulfates it becomes possible to create spinel catalysts with nanoparticulate morphologies. Colloidal synthetic chemistry techniques offer opportunities for synthesizing spinel nanoparticles with control over their size, shape, and composition. In settings, these crafted spinel catalysts have been successfully employed in CO_2 hydrogenation processes that yield diverse products. Various methods have been utilized for nanofabrication purposes including co-precipitation [57], combustion system [58], oxalate co-precipitation [59], decay of thermal complexes [60], honey spray pyrolysis [61], solvothermal method [62], hydrothermal approaches (including assisted techniques) [63], and electrospinning process [64]. Furthermore, combined methods like rush/deposition and impregnation/citrate complexation have also been used to create reagents with spinel structures. [55].

15.2.7.1.1 Co-Precipitation and Co-Precipitation of Oxalate

Co-precipitation is a widely used method for adding metallic ions to catalytic precursors to generate spinel catalysts for CO_2 hydrogenation [65]. It offers low cost, scalability, and straightforward medication [66]. Controlling the size and form of the accompanying spinel catalyst is crucial since it might be challenging to obtain homogenous catalyst flyspeck structures [67]. In bimetallic or ternary oxide catalysts, pH, growth period, and temperature can induce the conformation of the spinel phase [68]. As a result, the performance of CO_2 hydrogenation in methanol conflation is improved due to an increase in the number of introduction places. Oxalate precipitation is a different type of precipitation system that employs oxalic acid as the precipitating agent. The thermal deterioration of hydrotalcite can also result in co-precipitated spinel catalysts [69].

15.2.7.1.2 Self-Combustion

The auto-combustion system uses chemical ligands with carboxylic, carboxylate, or amine functional groups that serve as energies for petroleum and biomedical products. It's easy, inexpensive, and doesn't call for any specialized equipment or clothing. To improve the performance of hetero-atom-unravel Fe catalysts in CO_2 hydrogenation, Guo *et al.* [58], prepared them using the urea tone-combustion method. The quick reaction allows for the production of spinel catalyst nanopowders that are clearly defined, less agglomerated, thicker, and smaller than flyspecks.

15.2.7.1.3 Thermal Decomposition

Spinel nanocrystals with great face area and small flyspeck size can be synthesized using solvo- and hydro-thermal methods. These styles are flexible, environmentally friendly, and can yield high-chastity products with large mass effectiveness. Still, hydrothermal slurries can be corrosive and can create accidental explosions. Choi *et al.* [63], established an evenly distributed spinel ZnFe with great effectiveness for hefty hydrocarbons using microwave oven-supported hydrothermal conflation, an enhancement on the conventional hydrothermal system. In the same way, Zhang *et al.* [70], produced a Fe-Zn-K nanocatalyst with high olefin selectivity. Conrad *et al.* [71], synthesized spinel $Cu_xZn_{1-x}Al_2O_4$ using the microwave oven-supported hydrothermal system as a precursor for methanol conflation.

15.2.7.1.4 Electrospinning

Electrospinning is a fast and important approach for producing nanomaterials, including spinel oxides in nanofiber shapes. The fabrication process is firm and requires a diminutive heating time to achieve the required crystal phase. These nanofibers are generally nanobelts, concave nanofibers and lamellar such as mesoporous nanofibers, and dimensional nanostructures [72]. Controlled heat treatment is applied to untreated electrospun filaments to produce ceramic nanostructures with larger nanocrystals. Elishav *et al.* [64], developed electrospun infrastructures using the inner heat released due to the oxidation of organic factors. In comparison to the co-precipitated spinel powder, the carburized electrospun-made ceramic K/Fe-Al-O spinel nanobelts demonstrated greater CO_2 conversion and $C_2 - C_5$ selectivity. Figure 15.8 shows the schematics of the above process.

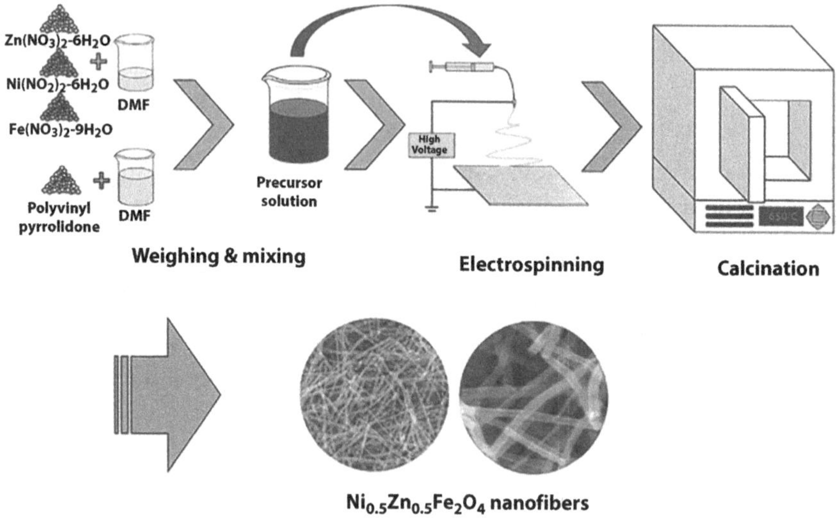

Figure 15.8 Electrospinning and calcination fabrication schematic [85].

15.3 Challenges and Future Perspectives

Significant improvements in the design of electrocatalysts for water splitting have resulted in improvements in hydrogen production. Hydrogen's efficiency is restricted by delayed kinetics in electrocatalysts for Hydrogen Evolution Reaction (HER) and Oxygen Evolution Reaction (OER), hydrogen is a clean and sustainable energy source [73, 74]. The development of non-noble substance OER electrocatalysts with high effort and long-term stability performance in acidic conditions still faces obstacles. Additionally, nothing is known about the specific catalytic mechanisms for HER and OER electrocatalysts that are predicted by transition substances. Catalysts with the best performance require both in-situ characterization techniques and theoretical modeling, which must be integrated. To fully comprehend catalyst behavior, the creation of intermediate products, and their decomposition, cutting-edge equipment including X-ray absorption spectrometers, FT-IR spectrometers, Raman spectrometers, and Scanning electrochemical microscopy (SECM) is required. By using fewer chemicals and extreme physicochemical conditions, these developments can be used to guide the nanocatalyst design and increase the hardness and reusability of nanocrystals [75].

Since the operation of a catalyst in actual circumstances (as opposed to lab-grounded conditions) it necessitates a stable and provident catalyst; the performance of nanocatalysts is a crucial quality that helps to apply a catalyst successfully on a bigger scale for practical operation. A nanocatalyst must be stable when used in oil residue and wastewater to be used in practical applications. To improve the recycling effectiveness of a catalyst, it is suggested that suitable catalyst recovery techniques should be identified for a specific catalyst [76]. Each type of waste has a variable composition and elements that can influence the catalyst performance. Additionally, investigation on the stability of nanocatalysts ought to be carried out at various pH values, temperatures, agitation rates, and time periods [77]. An extensive investigation of reaction circumstances and active areas is required. An understanding of the catalytic decline of pollutants is required to maximize the catalytic efficacy by discussing the reactive regions in the nanocatalysts [78]. It is crucial to produce low-cost, high-performance nanocatalysts. However, the expenses connected to similar catalytic reactions (nanocatalysts-grounded advanced organic processes AOPs) are not taken into account by most research. Also provided in various findings were electrical energy requirements, chemical expenses, and functional costs; however, catalyst prices were not included in comparable estimates. Future improvements to the functionality of AOPs can be encouraged by interpreting and improving the oxidation technology. Considerations for artificially using AOPs include minimizing the treatment system's budget and finding a solution to the erosion difficulty of the response structure [79].

15.4 Conclusion

Nanocatalysts are really making a difference in a lot of chemical processes. What makes them special is their enormous surface area and the fact that their properties can be favorably modified. They're particularly good at speeding up the process of transesterification in biodiesel production. Plus, nanocatalysts can simplify complex organic compounds into easier-to-handle molecules in biomass conversion. These simpler molecules can then be used to create chemicals or biofuels. Nanocatalysts can boost efficiency in electron transfers during electrosynthesis. This facilitates the creation of beneficial chemicals using renewable power and feedstock. Regarding pollution control in our air and water sources, nanocatalysts help by accelerating the decomposition of pollutants, leading to cleaner air and water. Also, nanocatalysts can quicken the hydrogen evolution reaction, allowing fuel cells to gradually accumulate hydrogen. It is necessary to broaden the scope of nanocatalyst exploration in a few key areas. Firstly, they should focus on enhancing catalytic performance by creating new nanomaterials that have specific structures and functionalities. Secondly, it's important to understand the kinetics and mechanisms of catalysis at the nanoscale level for the design of more efficient catalysts. Further, tweaking reaction conditions and parameters will ensure the maximum productivity of the catalyst. Finally, the environmental and economic impact of nanocatalysis must be assessed to ensure its sustainability in the long run.

References

1. Prinsen, P. and Luque, R., *Nanoparticle Design and Characterization for Catalytic Applications in Sustainable Chemistry*, R. Luque and P. Prinsen (Eds.), ch. 1, pp. 1–36, The Royal Society of Chemistry, 2019, doi: 10.1039/9781788016292-00001.
2. Olveira, S., Forster, S.P., Seeger, S., Nanocatalysis: Academic Discipline and Industrial Realities. *J. Nanotechnol.*, 2014, 324089, 2014, doi: 10.1155/2014/324089.
3. Roh, H.-S., Nanocatalysts for Hydrogen Production. *Catalysts*, 11, 2, 288, 2021, doi: 10.3390/catal11020288.
4. Lu, J., *et al.*, Preparation of Ag nanoparticles by spark ablation in gas as catalysts for electrocatalytic hydrogen production. *RSC Adv.*, 10, 63, 38583–38587, 2020, doi: 10.1039/D0RA06682F.
5. Ghosh, N. and Halder, G., Current progress and perspective of heterogeneous nanocatalytic transesterification towards biodiesel production from edible and inedible feedstock: A review. *Energy Convers. Manage.*, 270, 116292, Oct. 2022, doi: 10.1016/j.enconman.2022.116292.
6. Hasan, M.M. and Rahman, M.M., Performance and emission characteristics of biodiesel–diesel blend and environmental and economic impacts of biodiesel production: A review. *Renew. Sustain. Energy Rev.*, 74, 938–948, Jul. 2017, doi: 10.1016/j.rser.2017.03.045.
7. Ahmed, M., *et al.*, Feedstocks, catalysts, process variables and techniques for biodiesel production by one-pot extraction-transesterification: a review. *Environ. Chem. Lett.*, 20, 1, 335–378, Feb. 2022, doi: 10.1007/s10311-021-01358-w.
8. Yan, S., Salley, S.O., Ng, K.Y.S., ZnO nanoparticle catalysts for use in biodiesel production and method of making, US8895764B2, 2009.
9. Zhang, Y., Duan, L., Esmaeili, H., A review on biodiesel production using various heterogeneous nanocatalysts: Operation mechanisms and performances. *Biomass Bioenergy*, 158, 106356, Mar. 2022, doi: 10.1016/j.biombioe.2022.106356.
10. Mofijur, M., *et al.*, Effect of nanocatalysts on the transesterification reaction of first, second and third generation biodiesel sources- A mini-review. *Chemosphere*, 270, 128642, May 2021, doi: 10.1016/j.chemosphere.2020.128642.

11. Bora, A.P., Konda, L. D. N. V. V., Pasupuleti, S., Durbha, K.S., Synthesis of MgO/MgSO4 nanocatalyst by thiourea–nitrate solution combustion for biodiesel production from waste cooking oil. *Renew. Energy*, 190, 474–486, May 2022, doi: 10.1016/j.renene.2022.03.127.
12. Xue, Y. and Bai, X., Synergistic enhancement of product quality through fast co-pyrolysis of acid pretreated biomass and waste plastic. *Energy Convers. Manage.*, 164, 629–638, May 2018, doi: 10.1016/j.enconman.2018.03.036.
13. Botte, G.G., Electrochemical Manufacturing in the Chemical Industry. *Interface Mag.*, 23, 3, 49–55, Jan. 2014, doi: 10.1149/2.F04143if.
14. Schaidle, J., Habas, S., Baddour, F., Farberow, C., Ruddy, D., Hensley, J., Malmstadt, N., Robota, H., *Transitioning rationally designed catalytic materials to real "working" catalysts produced at commercial scale: nanoparticle materials*, volume 29, pp. 213–281, 2017. 10.1039/9781788010634-00213.
15. Millet, P., *et al.*, PEM water electrolyzers: From electrocatalysis to stack development. *Int. J. Hydrogen Energy*, 35, 10, 5043–5052, May 2010, doi: 10.1016/j.ijhydene.2009.09.015.
16. Yang, Y., *et al.*, Encapsulating Co_2P@C Core–Shell Nanoparticles in a Porous Carbon Sandwich as Dual-Doped Electrocatalyst for Hydrogen Evolution. *ChemSusChem*, 11, 2, 376–388, Jan. 2018, doi: 10.1002/cssc.201701705.
17. Wang, R., Dong, X., Du, J., Zhao, J., Zang, S., MOF-Derived Bifunctional Cu_3P Nanoparticles Coated by a N,P-Codoped Carbon Shell for Hydrogen Evolution and Oxygen Reduction. *Adv. Mater.*, 30, 6, 1703711, Feb. 2018, doi: 10.1002/adma.201703711.
18. Gupta, V.K. and Saleh, T.A., Sorption of pollutants by porous carbon, carbon nanotubes and fullerene- An overview. *Environ. Sci. Pollut. Res.*, 20, 5, 2828–2843, May 2013, doi: 10.1007/s11356-013-1524-1.
19. Gabal, E., Chatterjee, S., Ahmed, F.K., Abd-Elsalam, K.A., Carbon nanomaterial applications in air pollution remediation, in: *Carbon Nanomaterials for Agri-Food and Environmental Applications*, pp. 133–153, Elsevier, 2020, doi: 10.1016/B978-0-12-819786-8.00007-4.
20. Gui, M.M., Yap, Y.X., Chai, S.-P., Mohamed, A.R., Multi-walled carbon nanotubes modified with (3-aminopropyl)triethoxysilane for effective carbon dioxide adsorption. *Int. J. Greenhouse Gas Control*, 14, 65–73, May 2013, doi: 10.1016/j.ijggc.2013.01.004.
21. Lu, C., Wu, B., Chen, W., Lin, Y.-K., Bai, H., Capture of Carbon Dioxide by Modified Multiwalled Carbon Nanotubes, in: *Environanotechnology*, pp. 55–69, Elsevier, 2010, doi: 10.1016/B978-0-08-054820-3.00004-6.
22. Huy, N.N., *et al.*, Facile one-step synthesis of zinc oxide nanoparticles by ultrasonic-assisted precipitation method and its application for H2S adsorption in air. *J. Phys. Chem. Solids*, 132, 99–103, Sep. 2019, doi: 10.1016/j.jpcs.2019.04.018.
23. Ghosh, S., Ranjan, P., Ramaprabhu, S., Sarathi, R., Carbon Dioxide Adsorption of Zinc Oxide Nanoparticles Synthesized by Wire Explosion Technique. *INAE Lett.*, 3, 4, 197–202, Dec. 2018, doi: 10.1007/s41403-018-0049-9.

24. Li, X., Zhang, H., Lü, H., Zuo, S., Zhang, Y., Yao, C., Photo-assisted SCR removal of NO by upconversion CeO2/Pr3+/attapulgite nanocatalyst. *Environ. Sci. Pollut. Res.*, 26, 13, 12842–12850, May 2019, doi: 10.1007/s11356-019-04802-1.
25. Shen, W., Zhang, C., Li, Q., Zhang, W., Cao, L., Ye, J., Preparation of titanium dioxide nano particle modified photocatalytic self-cleaning concrete. *J. Clean. Prod.*, 87, 762–765, Jan. 2015, doi: 10.1016/j.jclepro.2014.10.014.
26. Mamaghani, A.H., Haghighat, F., Lee, C.-S., Gas phase adsorption of volatile organic compounds onto titanium dioxide photocatalysts. *Chem. Eng. J.*, 337, 60–73, Apr. 2018, doi: 10.1016/j.cej.2017.12.082.
27. Wu, H., Ma, J., Li, Y., Zhang, C., He, H., Photocatalytic oxidation of gaseous ammonia over fluorinated TiO2 with exposed (001) facets. *Appl. Catal., B*, 152–153, 82–87, Jun. 2014, doi: 10.1016/j.apcatb.2014.01.021.
28. Anjum, H., *et al.*, A review on adsorptive removal of oil pollutants (BTEX) from wastewater using carbon nanotubes. *J. Mol. Liq.*, 277, 1005–1025, Mar. 2019, doi: 10.1016/j.molliq.2018.10.105.
29. Aslam, M.M.-A., Kuo, H.-W., Den, W., Usman, M., Sultan, M., Ashraf, H., Functionalized Carbon Nanotubes (CNTs) for Water and Wastewater Treatment: Preparation to Application. *Sustainability*, 13, 10, 5717, May 2021, doi: 10.3390/su13105717.
30. Agboola, O., *et al.*, A Review on Polymer Nanocomposites and Their Effective Applications in Membranes and Adsorbents for Water Treatment and Gas Separation. *Membranes (Basel)*, 11, 2, 139, Feb. 2021, doi: 10.3390/membranes11020139.
31. Bushra, R., Nanoadsorbents-based polymer nanocomposite for environmental remediation, in: *New Polymer Nanocomposites for Environmental Remediation*, pp. 243–260, Elsevier, 2018, doi: 10.1016/B978-0-12-811033-1.00011-1.
32. Goebel, J.F., Stemmer, J., Belitz, F., Gooßen, L.J., Manganese(I) Catalyzed *ortho* C–H Allylation of Benzoic Acids. *Angew. Chem. Int. Ed.*, 62, 24, e202301839, Jun. 2023, doi: 10.1002/anie.202301839.
33. Singh, R., Singh, M., Kumari, N., Janak, Maharana, S., Maharana, P., A Comprehensive Review of Polymeric Wastewater Purification Membranes. *J. Compos. Sci.*, 5, 6, 162, Jun. 2021, doi: 10.3390/jcs5060162.
34. Chmielewská, E., Tylus, W., Bujdoš, M., Study of Mono- and Bimetallic Fe and Mn Oxide-Supported Clinoptilolite for Improved Pb(II) Removal. *Molecules*, 26, 14, 4143, Jul. 2021, doi: 10.3390/molecules26144143.
35. Schwanke, A.J., Balzer, R., Pergher, S., Microporous and Mesoporous Materials from Natural and Inexpensive Sources, in: *Handbook of Ecomaterials*, pp. 1–22, Springer International Publishing, Cham, 2017, doi: 10.1007/978-3-319-48281-1_43-1.
36. R., J., Gurunathan, B., K, S., Varjani, S., Ngo, H.H., Gnansounou, E., Advancements in heavy metals removal from effluents employing nano-adsorbents: Way towards cleaner production. *Environ. Res.*, 203, 111815, Jan. 2022, doi: 10.1016/j.envres.2021.111815.

37. Pooyamanesh, M., Ahari, H., Anvar, A.A., Karim, G., Synthesis and Characterization of Silver Nanocomposite as a Food Packaging. *J. Food Biosci. Technol.*, 09, 2, 73–82, 2019, [Online]. Available: https://jfbt.srbiau.ac.ir/article_14382.html.
38. Jo, Y., *et al.*, Characterization and antibacterial properties of nanosilver-applied polyethylene and polypropylene composite films for food packaging applications. *Food Biosci.*, 23, 83–90, Jun. 2018, doi: 10.1016/j.fbio.2018.03.008.
39. Tian, Z., Lu, C., Zhou, Y., Zhang, Y., Wei, W., Phosphoric acid-induced activation of sepiolite for enhanced As(III) adsorption: role of *in situ* deposition of nano-hydroxyapatite. *J. Dispersion Sci. Technol.*, 44, 3, 406–418, Mar. 2023, doi: 10.1080/01932691.2021.1948424.
40. Niemann, M.U., Srinivasan, S.S., Phani, A.R., Kumar, A., Goswami, D.Y., Stefanakos, E.K., Nanomaterials for Hydrogen Storage Applications: A Review. *J. Nanomater.*, 2008, 950967, 2008, doi: 10.1155/2008/950967.
41. Kumaravel, V. and Abdel-Wahab, A., A Short Review on Hydrogen, Biofuel, and Electricity Production Using Seawater as a Medium. *Energy Fuels*, 32, 6, 6423–6437, Jun. 2018, doi: 10.1021/acs.energyfuels.8b00995.
42. Jian, Y., Meng, Y., Li, H., Selectivity Control of C-O Bond Cleavage for Catalytic Biomass Valorization. *Front. Energy Res.*, 9, Jan. 2022, doi: 10.3389/fenrg.2021.827680.
43. Den, W., Sharma, V.K., Lee, M., Nadadur, G., Varma, R.S., Lignocellulosic Biomass Transformations via Greener Oxidative Pretreatment Processes: Access to Energy and Value-Added Chemicals. *Front. Chem.*, 6, 141, Apr. 2018, doi: 10.3389/fchem.2018.00141.
44. Tian, B., Wu, Y., Lu, G., Metal-free plasmonic boron phosphide/graphitic carbon nitride with core-shell structure photocatalysts for overall water splitting. *Appl. Catal., B*, 280, 119410, Jan. 2021, doi: 10.1016/j.apcatb.2020.119410.
45. Yu, S., Wu, Y., Xue, Q., Zhu, J.-J., Zhou, Y., A novel multi-walled carbon nanotube-coupled CoNi MOF composite enhances the oxygen evolution reaction through synergistic effects. *J. Mater. Chem. A Mater.*, 10, 9, 4936–4943, 2022, doi: 10.1039/D1TA10681C.
46. Dat, V.D. and Vu, T.V., Janus monolayer HfSO with improved optical properties as a novel material for photovoltaic and photocatalyst applications. *New J. Chem.*, 46, 4, 1557–1568, 2022, doi: 10.1039/D1NJ05096F.
47. Lu, H., Huang, Z., Martinez, M.S., Johnson, J.C., Luther, J.M., Beard, M.C., Transforming energy using quantum dots. *Energy Environ. Sci.*, 13, 5, 1347–1376, 2020, doi: 10.1039/C9EE03930A.
48. Grewe, T., Meggouh, M., Tüysüz, H., Nanocatalysts for Solar Water Splitting and a Perspective on Hydrogen Economy. *Chem. Asian J.*, 11, 1, 22–42, Jan. 2016, doi: 10.1002/asia.201500723.
49. Zhu, J., Hu, L., Zhao, P., Lee, L.Y.S., Wong, K.-Y., Recent Advances in Electrocatalytic Hydrogen Evolution Using Nanoparticles. *Chem. Rev.*, 120, 2, 851–918, Jan. 2020, doi: 10.1021/acs.chemrev.9b00248.

50. Luo, F., *et al.*, Robust and Stable Acidic Overall Water Splitting on Ir Single Atoms. *Nano Lett.*, 20, 3, 2120–2128, Mar. 2020, doi: 10.1021/acs.nanolett.0c00127.
51. Sumi, V.S., Meera, M.S., Sha, M.A., Shibli, S.M.A., Effect of rGO on Fe2O3–TiO2 composite incorporated NiP coating for boosting hydrogen evolution reaction in alkaline solution. *Int. J. Hydrogen Energy*, 45, 4, 2460–2477, Jan. 2020, doi: 10.1016/j.ijhydene.2019.11.167.
52. Cai, M., *et al.*, Large-Scale Tunable 3D Self-Supporting WO_3 Micro-Nano Architectures as Direct Photoanodes for Efficient Photoelectrochemical Water Splitting. *ACS Appl. Mater. Interfaces*, 9, 21, 17856–17864, May 2017, doi: 10.1021/acsami.7b02386.
53. Xu, S., *et al.*, The bifunctional 3D-on-2D FeCo/Ni(OH)2 hierarchical nanocatalyst for industrial-level electrocatalytic water splitting. *Int. J. Hydrogen Energy*, 48, 47, 17882–17893, Jun. 2023, doi: 10.1016/j.ijhydene.2023.01.289.
54. Chaudhari, N.K., Jin, H., Kim, B., Lee, K., Nanostructured materials on 3D nickel foam as electrocatalysts for water splitting. *Nanoscale*, 9, 34, 12231–12247, 2017, doi: 10.1039/C7NR04187J.
55. Orege, J.I., Wei, J., Ge, Q., Sun, J., Spinel-structured nanocatalysts: New opportunities for CO2 hydrogenation to value-added chemicals. *Nano Today*, 51, 101914, Aug. 2023, doi: 10.1016/j.nantod.2023.101914.
56. Witoon, T., Lapkeatseree, V., Numpilai, T., Kui Cheng, C., Limtrakul, J., CO2 hydrogenation to light olefins over mixed Fe-Co-K-Al oxides catalysts prepared via precipitation and reduction methods. *Chem. Eng. J.*, 428, 131389, Jan. 2022, doi: 10.1016/j.cej.2021.131389.
57. Li, X., Ren, P., Zhang, J., Zhang, L., Liu, G., Preparation and magnetic properties of Mn-Zn ferrites by the Co-precipitation method. *J. Wuhan Univ. Technol. Mater. Sci. Ed.*, 24, 6, 875–878, Dec. 2009, doi: 10.1007/s11595-009-6875-6.
58. Guo, L., *et al.*, Heteroatom doped iron-based catalysts prepared by urea self-combustion method for efficient CO2 hydrogenation. *Fuel*, 276, 118102, Sep. 2020, doi: 10.1016/j.fuel.2020.118102.
59. Gnanamani, M.K., *et al.*, Hydrogenation of Carbon Dioxide over Co–Fe Bimetallic Catalysts. *ACS Catal.*, 6, 2, 913–927, Feb. 2016, doi: 10.1021/acscatal.5b01346.
60. Kim, K.Y., *et al.*, Cobalt Ferrite Nanoparticles to Form a Catalytic Co–Fe Alloy Carbide Phase for Selective CO_2 Hydrogenation to Light Olefins. *ACS Catal.*, 10, 15, 8660–8671, Aug. 2020, doi: 10.1021/acscatal.0c01417.
61. Piriyasurawong, K., Panpranot, J., Mekasuwandumrong, O., Praserthdam, P., CO2 hydrogenation over FSP-made iron supported on cerium modified alumina catalyst. *Catal. Today*, 375, 307–313, Sep. 2021, doi: 10.1016/j.cattod.2020.03.007.
62. Shi, B., Zhang, Z., Zha, B., Liu, D., Structure evolution of spinel Fe-MII (M=Mn, Fe, Co, Ni) ferrite in CO hydrogeneration. *Mol. Catal.*, 456, 31–37, Sep. 2018, doi: 10.1016/j.mcat.2018.06.019.

63. Choi, Y.H., *et al.*, Sodium-Containing Spinel Zinc Ferrite as a Catalyst Precursor for the Selective Synthesis of Liquid Hydrocarbon Fuels. *ChemSusChem*, 10, 23, 4764–4770, Dec. 2017, doi: 10.1002/cssc.201701437.
64. Elishav, O., *et al.*, Electrospun Fe–Al–O Nanobelts for Selective CO_2 Hydrogenation to Light Olefins. *ACS Appl. Mater. Interfaces*, 12, 22, 24855–24867, Jun. 2020, doi: 10.1021/acsami.0c05765.
65. Bahmanpour, A.M., Le Monnier, B.P., Du, Y.-P., Héroguel, F., Luterbacher, J.S., Kröcher, O., Increasing the activity of the $Cu/CuAl_2O_4/Al_2O_3$ catalyst for the RWGS through preserving the Cu^{2+} ions. *Chem. Commun.*, 57, 9, 1153–1156, 2021, doi: 10.1039/D0CC07142K.
66. Liu, X., *et al.*, Tandem Catalysis for Hydrogenation of CO and CO_2 to Lower Olefins with Bifunctional Catalysts Composed of Spinel Oxide and SAPO-34. *ACS Catal.*, 10, 15, 8303–8314, Aug. 2020, doi: 10.1021/acscatal.0c01579.
67. Gao, W., *et al.*, Selective Conversion of CO_2 into *para* -Xylene over a $ZnCr_2O_4$-ZSM-5 Catalyst. *ChemSusChem*, 13, 24, 6541–6545, Dec. 2020, doi: 10.1002/cssc.202002305.
68. Liu, Z., Jia, G., Zhao, C., Xing, Y., Efficient Fischer-Tropsch to light olefins over iron-based catalyst with low methane selectivity and high olefin/paraffin ratio. *Fuel*, 288, 119572, Mar. 2021, doi: 10.1016/j.fuel.2020.119572.
69. Zhang, J., Zheng, K., Wang, X., Zhang, Q., Zeng, C., Tan, Y., Promotional Effect of Dispersant Modification to ZnCr on CO_2 Hydrogenation into Aromatics over Hybrid Catalysts. *Ind. Eng. Chem. Res.*, 61, 30, 10766–10774, Aug. 2022, doi: 10.1021/acs.iecr.2c01408.
70. Zhang, J., Lu, S., Su, X., Fan, S., Ma, Q., Zhao, T., Selective formation of light olefins from CO2 hydrogenation over Fe–Zn–K catalysts. *J. CO2 Util.*, 12, 95–100, Dec. 2015, doi: 10.1016/j.jcou.2015.05.004.
71. Conrad, F., *et al.*, Microwave-hydrothermal synthesis and characterization of nanostructured copper substituted ZnM2O4 (M = Al, Ga) spinels as precursors for thermally stable Cu catalysts. *Nanoscale*, 4, 6, 2018, 2012, doi: 10.1039/c2nr11804a.
72. Khajavi, R. and Abbasipour, M., Electrospinning as a versatile method for fabricating coreshell, hollow and porous nanofibers. *Sci. Iran.*, 19, 6, 2029–2034, Dec. 2012, doi: 10.1016/j.scient.2012.10.037.
73. Skúlason, E., *et al.*, Density functional theory calculations for the hydrogen evolution reaction in an electrochemical double layer on the Pt(111) electrode. *PCCP*, 9, 25, 3241–3250, 2007, doi: 10.1039/B700099E.
74. Nørskov, J.K., *et al.*, Origin of the Overpotential for Oxygen Reduction at a Fuel-Cell Cathode. *J. Phys. Chem. B*, 108, 46, 17886–17892, Nov. 2004, doi: 10.1021/jp047349j.
75. Wang, H.-Y., Hsu, Y.-Y., Chen, R., Chan, T.-S., Chen, H.M., Liu, B., Ni^{3+}-Induced Formation of Active NiOOH on the Spinel Ni-Co Oxide Surface for Efficient Oxygen Evolution Reaction. *Adv. Energy Mater.*, 5, 10, 1500091, May 2015, doi: 10.1002/aenm.201500091.

76. Buthiyappan, A., Abdul Aziz, A.R., Wan Daud, W.M.A., Recent advances and prospects of catalytic advanced oxidation process in treating textile effluents. *Rev. Chem. Eng.*, 32, 1, 1–47, 2016, doi:10.1515/revce-2015-0034.
77. Sajjad, S., Ikhlaq, A., Javed, F., Ahmad, S.W., Qi, F., A study on the influence of pH changes during catalytic ozonation process on alumina, zeolites and activated carbons for the decolorization of Reactive Red-241. *Water Sci. Technol.*, 83, 3, 727–738, Feb. 2021, doi: 10.2166/wst.2020.609.
78. Nawrocki, J. and Kasprzyk-Hordern, B., The efficiency and mechanisms of catalytic ozonation. *Appl. Catal., B*, 99, 1–2, 27–42, Aug. 2010, doi: 10.1016/j.apcatb.2010.06.033.
79. Masood, Z., Ikhlaq, A., Akram, A., Qazi, U.Y., Rizvi, O.S., Javaid, R., Alazmi, A., Madkour, M., Qi, F., Application of Nanocatalysts in Advanced Oxidation Processes for Wastewater Purification: Challenges and Future Prospects. *Catalysts*, 12, 741, 2022, https://doi.org/10.3390/catal12070741.
80. De Paola, M.G., Mazza, I., Paletta, R., Lopresto, C.G., Calabrò, V., Small-Scale Biodiesel Production Plants—An Overview. *Energies (Basel)*, 14, 7, 1901, Mar. 2021, doi: 10.3390/en14071901.
81. Król, M.K. and Jeleń, P., The Effect of Heat Treatment on the Structure of Zeolite A. *Materials*, 14, 16, 4642, Aug. 2021, doi: 10.3390/ma14164642.
82. Bollmann, J., Pitchaimuthu, S., Kühnel, M.F., Challenges of Industrial-Scale Testing Infrastructure for Green Hydrogen Technologies. *Energies (Basel)*, 16, 8, 3604, Apr. 2023, doi: 10.3390/en16083604.
83. Plachá, D. and Jampilek, J., Graphenic Materials for Biomedical Applications. *Nanomaterials*, 9, 12, 1758, Dec. 2019, doi: 10.3390/nano9121758.
84. Vozniuk, O., *et al.*, Mixed-Oxide Catalysts with Spinel Structure for the Valorization of Biomass: The Chemical-Loop Reforming of Bioethanol. *Catalysts*, 8, 8, 332, Aug. 2018, doi: 10.3390/catal8080332.
85. Na, K.-H., Jang, K.-P., Kim, S.-W., Choi, W.-Y., Fabrication of Electrospun Ni0.5Zn0.5Fe2O4 Nanofibers Using Polyvinyl Pyrrolidone Precursors and Electromagnetic Wave Absorption Performance Improvement. *Polymers (Basel)*, 13, 23, 4247, Dec. 2021, doi: 10.3390/polym13234247.

Index

Abrasive machining, 185
Abrasive processing, 333
Abrasive wear, 229–230, 376, 380
Active antifouling, 425, 429, 430
Additive materials, 39
Adhesion, 225, 270, 271, 274, 277
Adhesive wear, 359
Adsorption, 243
Adsorption and breakdown (degradation), 246
Adsorption and chelation, 256
Adsorption capability, 252
Advanced functionalities, 262
Advanced sensing technologies, 262
Aerospace applications, 150, 163, 200, 285, 294, 297, 298
Aerospace industry, 272, 274, 279, 280, 326
Aerospace materials, 150–151
AFM, 130
Aircraft component, 298
Air remediation, 240–241
AISI 52100 steel, 318, 321, 323, 324, 326
Al_2O_3 composites, 216
Alloy steels, 318
AlTiN (Alumina-Titanium Nitride), 273
Alumina, 273, 281
Alumina (Al_2O_3), 157–158
Aluminium oxide (Al_2O_3), 273, 279, 280
Aluminium titanate, 158–159
Aluminum alloys, 176, 364, 375
Aluminum nitride (AlN), 344
Antibacterial activity, 156, 162
Antibacterial properties, 246
Antibiotic-resistant biofilm, 403
Antifouling mechanism, 425, 429, 430
Antifouling methods, 425
Application of nanocomposite in environmental remediation, 240–242
Applications of composites, 362–363, 380–381
Applications of nanocatalysts, 449
  biodiesel production, 449
  biomass production using nanocatalysts, 450
  hydrogen production, 455
  hydrogen production using spinel-structured nanocatalysts, 458
  nanocatalyst in electrosynthesis, 451
  water pollution, 454
Artificial intelligence (AI), 115
Atomic layer deposition (ALD), 113, 124
Au and Ag quantum dots, 113, 132
Automotive industries, 299
Automotive industry, 272, 273, 274, 278, 279, 280, 319, 326
Automotive sector, 178–179

Ball milling, 352–354
Ballistic protection, 160–161
Barium titanate ($BaTiO_3$), 160–161
Bearing steel (AISI 52100), 183

Bearing steel, 329
Binary liquid crystal composites, 113, 132, 137
Binder-free carbides, 349–356
Binderless carbides, 214–219
Bioactive ceramics, 405
Biocompatibility, 276
Bioengineering, 326
Biomedical applications, 161–162
Biomedical gadgets, 406
Bohr radius, 114
Borate nanoceramics, 159
Boring, 318, 328, 330
Born–Oppenheimer (BO) approximation, 11–12
Boron carbide (B_4C), 160
Boron nitride (BN), 156–157
Boron nitride (PCBN), 178–179, 183
Borophene, 118
Bottom-up approach, 119–120
  liquid-phase synthesis, 120
  self-assembly, 113, 120, 124–125
  sol-gel, 120
  spinning methods, 120
Broaching, 332, 333

Carbide alloys, 209, 214
Carbide cutters, 329, 332, 335
Carbide tools, 320, 332, 336, 337
Carbides (tungsten, fine-grain), 178–179, 182
Carbon allotropes in TiCN, 229
Carbon composites, 157
Carbon nanotubes (CNTs), 114, 118, 362, 364–367, 370–374, 376–379
Carbonitride of Titanium (TiCN), 210–213, 219–221
Case-hardened steel, 178, 318, 329
Cast iron, 176
Catalysis, 244
Cathodic arc deposition, 122
CBN (Cubic Boron Nitride) tools, 322, 334
Ceramic composites, 149–151
Ceramic cutting tools, 178–179, 204, 226–229
Ceramic matrix composites, 363–364, 368–369
Ceramic nanocomposites, 154–156
Ceramic tools, 317, 318, 323, 330
Ceramic–metal composites (cermets), 207–209, 222–223
Cermet, 323, 330, 344–345, 349, 357
Challenges and future perspectives, 261–262
Chemical precipitation, 196
Chemical vapor deposition (CVD), 39, 70, 113, 119, 123, 190–191, 284–286
Chip formation, 329
Chiral dopants, 113, 135, 137
Cholesteric, 126–128, 129, 135
Chromium nitride (CrN), 274, 280
CNC (computer numerical control), 331
Coated tools, 317, 333, 337
Coating deposition, 317, 321
Coating materials, 317, 319, 320
Coating methods, 39–40, 69–70
Coating technology, 336, 337
Coatings (multilayer, electrophoretic), 157
Coatings (nanocoatings, TiN, TiAlN, AlTiN) 182–183, 202–204
Coatings (nanoscale), 269–282
Coatings (TiN, TiCN), 211–213, 229–230, 347–348
Coatings, 362, 369, 372–374, 377
Cold sprayed Ni-based coatings, 59, 78–79, 104, 109–110
Combustion synthesis (SHS), 151–152
Composite coatings, 52, 54, 59, 62, 66–67, 74, 94, 96, 98, 102–111
Composites, 113, 116, 133, 135, 137
Compressive strength, 370, 375–376
Contrast ratio, 132
Coolant/lubrication, 184
Coolant, 321, 330

Corrosion resistance, 50–51, 55, 57–63, 65, 68–70, 80, 89, 91, 93–94, 96, 98, 104–109, 111, 273, 274, 275, 276, 278, 208–209, 222–223, 317, 344–345, 357, 359
Crystallography, 126
Cutting edge geometry, 180–181
Cutting fluids, 184
Cutting force, 322, 323, 324, 325, 326
Cutting speed ranges, 176–177
Cutting speed, 319, 322, 332, 333
Cutting tools, 270, 275, 278, 280, 207–208, 211–213, 226–228, 317, 319, 320, 326, 337, 338, 343–345, 347–348, 350, 355–356
CVD (chemical vapor deposition) coatings, 336, 337

Defense materials, 163–165
Deformation, 327, 328
Degradation of organic pollutants, 244–246
Degradation of pollutants, 258
Density functional theory (DFT), 18–19
Dental abutments, 407
Desalination, 415, 416, 417, 435
- antibacterial, 430, 431
- antimicrobial cationic groups, 431
- catalytic reactions, 430
- membrane flux, 424
- permeability, 423, 435, 437
- polymerization, 434, 436
- pore blockages, 419, 420
- steric repulsion, 426, 427
- valley clogging, 423

Desalination and ion removal, 247–248
Detection of pollutants, 249–250
Diamond-like carbon (DLC), 47–48, 55, 95, 102, 110, 273, 279
Dielectric ceramics, 160–161
Dies and molds, 175
Diffusion kinetics, 216, 349
Dimensional accuracy, 185–186
Dispersion strengthening, 366, 368–369
Display technology, 113, 133, 135, 137
Drill inserts, 209
Drilling applications, 343–344, 347
Drilling hardened steels, 182–183
Drilling, 328, 330
Dry cutting/milling, 321, 330, 337, 338
Dry machining, 184
Durability, 273, 274, 275, 277, 278

Economic viability, 261
Elasticity, 327
Electrical conductivity, 370, 373
Electrical contacts, 207
Electrochemical methods, 195–196
Electrodeposition, 57, 59, 70–71, 74–75, 84, 89, 96, 104–105, 111
Electroless coating, 39, 71–72, 98, 105, 111
Electronics, 272, 276
Electronics applications, 160–161
Electroplating, 39, 58, 60, 70, 77, 85, 96
Electrostatic capture, 251–252
Elongation, 323, 324, 326
Energy sector, 272
Engine components, 272, 273, 274, 279, 299
Enhanced bioremediation, 242
Enhanced filters, 251
Environmental applications, 162–163
Environmental impact, 184, 203, 261
Erosion wear, 359

Fabrication challenges, 277
Fabrication techniques, 113, 119, 121, 123–124, 137
Fatigue resistance, 376
FDTD, 21–22
Feed rate, 176, 182
Ferroelectric 0D materials, 135, 137
Ferromagnetic properties, 306, 307
Feynman, 114

Filtering of particulate matter, 250–252
Finite element method (FEM), 21
First-principle methods, 20–21
Flame synthesis (FSP), 193–194
Fluorapatites and HA, 405
Foulants, 415, 418, 421, 424–426, 428–429, 432, 436
Fouling release, 428, 429, 435
Fouling resistance, 426–429, 435, 436
Fracture toughness, 209, 213, 219, 223, 285–287, 362, 364, 370–372, 374–376
Friction, 39–49, 51–52, 54, 56–57, 59, 61–62, 65, 67–68, 71–73, 76–77, 79–89, 91, 93–97, 99–102, 108–110, 269–281
Friction discs, 180–181
Friction, 326, 328, 333, 336, 337
Fuel cells, 158–159
Fullerene, 117
Furnace linings, 348
Future perspectives, 262

Gear manufacturing, 333, 334, 335
Gel casting, 154
Glass ceramics, 159
Global collaboration and regulation, 262
Gradient coatings, 109
Grain boundary effects, 223
Grain microstructure, 218–221
Grain size effects, 348–349, 355–357
Graphene, 49, 59, 76, 85, 91, 95, 97, 104, 108–109, 116, 118
Graphene nanoplatelets, 362, 367–369, 372–373, 376
Green cutting technology, 184
Grinding operations, 317, 318, 329, 333

Hafnium boride (HfB_2), 300
Hard disk drives, 276
Hard machining, 226–229, 318, 321, 322, 323, 324, 328
Hard machining (hard turning, drilling, milling), 177–183
Hard materials, 397
Hard steels (> 60 HRC), 178–179
Hard turning, 319, 325, 329, 330, 334
Hardness, 39, 46–47, 51, 53–55, 59–63, 65–70, 77, 79–86, 88–89, 91–98, 103, 106–108, 207–209, 223–224, 269, 270, 271, 273, 274, 278, 279, 280, 317, 318, 319, 322, 323, 334, 335, 336, 337, 343–344, 346, 348, 355, 359, 362, 364, 368–370, 373–375, 377–378
Hartree–Fock (HF) approximation, 13
Health risks, 261
Heat treatment, 330, 333, 334
Heavy metal immobilization, 256–257
High entropy alloys, 214, 350
High-speed machining, 212, 228
High-speed steel, 333, 335, 337
Hobbing, 333, 335
Hohenberg–Kohn theorems, 15
Hologram applications, 136–137
Honing, 185–187, 330
Hybrid composites, 363, 370, 372, 378–380
Hydrogen evolution reaction, 103, 108
Hydrophilicity, 421, 422, 427, 435
Hydrothermal synthesis, 151–152, 199–200
Hydrothermal synthesis

Integrated systems, 262
Interface bonding, 366–367, 370–373, 376–378

Kohn–Sham formalism, 15–16
Kurzweil's singularity, 114

Laser ablation, 194–195
Laser cladding, 39, 59, 79, 99, 104, 108–110
Laser heat, 326
Lightweight armor, 300

Lightweight structures, 150–151, 163
Liquid crystal composites, 113, 133–134, 137
Liquid crystals (LCs), 113, 125–127, 131, 133, 135, 137
Lithography, 113, 119, 121
Longevity, 273–277
Lubrication, 39–40, 52, 80, 82–83, 86, 93, 95, 99, 109–110, 269, 270, 271, 275, 281, 282
Lycurgus cup, 114
Lyotropic, 126–129

Machine tools (HSM requirements), 177
Machining, 343–344, 346–348, 355
Magnesium alloys, 362–364, 366–369, 374–378
Magnetic nanocomposites, 244
Manufacturing industry, 272, 275, 277, 281
Mechanical properties, 223–225, 348, 358–360, 363–364, 369–377, 380
Medical devices, 272, 276
Medical implants, 398
Membrane fouling, 418–424, 432
Membrane technology, 248
MEMS/NEMS (nanoelectronics), 156
Mesogenic, 127
Metal matrix composites (MMCs), 52, 102, 362–364, 367, 370–371, 374–380
Metallic implantation devices, 396–397
Microhardness, 217
Microstructure, 59, 63, 73, 75, 79–81, 101–106, 109–111, 345, 349, 352, 355–357, 364, 366–368, 372, 374, 377–379
Milling, 318, 321, 331, 332, 333
Minimum quantity lubrication (MQL), 184
Mining applications, 343–344
Molecular dynamics (MD), 16–17
Monte Carlo (MC), 19–20
Nanoceramics (general), 151–157
Nanocoatings (applications, properties), 202–204
Nanocomposite coatings, 98, 106, 108–110
Nanocomposite filters, 250
Nanocomposite recycling and reuse, 262
Nanocomposite-based sensors, 249
Nanocomposites (types, structures), 154–156
Nanocomposites in air purification, 250–255
Nanocomposites in soil remediation, 256–260
Nanocomposites in water remediation, 242–250
Nanocomposites, 362, 365–370, 372–375, 377–380
Nanocrystalline materials, 326
Nanofabrication, 119, 125
Nanofiltration, 416, 417, 422–424, 428, 432, 433
Nanolayers, 319
Nanomaterials, 116, 118, 120, 122, 131
Nanoparticles, 277, 281, 282
Nanoparticles (synthesis methods), 193–202
Nanorods, 117–118
Nanoscale coatings, 269–281
Nanosheets, 118
Nanostructure
  alloy, 7
  carbon-based, 6
  composites, 7
  magnetic, 4
  metal, 3
  semiconductor, 4
Nanostructured hard materials, 3
Nanostructured materials, 113, 115–116, 118–120, 122, 131, 133, 137
Nanostructures, 319
Nanotechnology, 114, 115, 117–119, 131, 137, 275, 276

Nanotribology, 40–43, 45, 83, 95, 100–101
Nanotubes, 116, 118
Nanowires, 117, 118, 121, 122
Nematic, 127, 131–135
Neutron capture properties, 303, 304
Ni3Al binders, 344
Ni-B coating, 65–69, 91–92, 95–96, 106–108
Ni-based surfaces, 40, 79
Nickel-based alloys, 39, 70–71, 84, 103, 108–110
Niobium carbide (NbC), 221–222, 357–359
Ni-P coating, 60–65, 71, 77, 88, 92–94, 96, 104–106
Ni-W coating, 74, 84
NOx and SOx removal, 254
Nuclear reactors, 302–304

Oil spill cleanup, 241
Optical properties (nanomaterials), 201–202
Organic contaminant breakdown, 257–258
Orthopaedic implants, 276, 279
Osteoconductive coatings, 399
Oxidation, 273
Oxidation resistance, 212, 229, 294–296, 301, 343–344, 348, 362, 364, 369, 377

Passive antifouling, 425, 426, 429, 437
Pathogen elimination, 246–247
PCBN (Polycrystalline Cubic Boron Nitride), 330
Peclet criterion, 324
Photocatalysis, 156, 162
Photocatalytic applications, 253–254
Photocatalytic degradation, 241, 245, 253
Photochromic/photoresponsive polymers, 113, 135, 137
Physical blending, 427, 435
Physical vapor deposition (PVD), 39, 47, 110, 113, 119, 122
Phytoremediation enhancement, 257
Plasma arc cutting, 191–192
Plasma spraying, 103, 110
Polycrystalline cubic boron nitride (PCBN) 178–179, 183
Polymer-based ceramics, 149
Pore size, 416, 422, 423
Porosity in nanoceramics, 154–155
Powder metallurgy, 209, 230, 344, 362
Precision machining, 185
Processing techniques, 363–364, 366–368, 370–372, 374–376
Public perception, 261
Pulsed Laser Deposition (PLD), 122
PVD (physical vapor deposition) coatings, 319, 336, 338

Quantum computers, 115
Quantum dots (QDs), 114, 116, 117, 132, 134, 137

Radiation resistance, 298–303
Radioactive contaminant removal, 259
Radioactive waste treatment, 259–260
Reaming, 330
Refractory carbides/nitrides, 345–347
Regulatory framework, 261
Reinforcement distribution, 366–368, 372–373
Reinitzer F., 125
Removal of heavy metals, 243–244
Residual stresses, 182
Response time, 132–133
Rigidity, 319, 329
Rocket motors, 299

Saladin's Damascus sword, 114
Sawtooth chips, 322, 325
Scramjet engines, 299
Secondary carbides, 222–223, 230, 359
Selective adsorption, 253
Selective ion exchange, 248

Self-cleaning surfaces, 203, 254
Self-diffusion, 344
Self-healing properties, 271
Self-lubricating oxide layer, 39
Self-lubricating properties, 373–374, 378
Self-propagating high-temperature synthesis, 289, 291
Semi-empirical methods, 22–23
Sensing and detection, 241
Sensing and monitoring, 249–250
Sensing and monitoring of air quality, 255
Sensing and monitoring of soil quality, 260
Silicate nanoceramics, 157
Silicon carbide (SiC), 159
Silicon dioxide quantum dots, 113, 132, 137
Silicon nitride (Si_3N_4), 204–205
Simulations
  atomistic, 9
  continuum, 9
  multiscale modeling, 10
  quantum mechanical, 10
Sintering techniques, 213–218, 230
Sintering, 344–345, 348–350, 351–356
Sintering, 366–368, 370, 373, 376–377
Skiving, 334, 335
Smart nanocomposites, 262
Smectic, 126–129
Soil remediation, 240
Sol-gel method, 152–154, 197–199
Solid lubricants, 39, 82, 85, 88, 97, 109
Solution liquid solid (SLS) approach, 113, 122
Sorption and destruction, 258
Spark plasma sintering (SPS), 213, 216, 349
Spindle, 318, 331
Spindles (HSM), 176–177
Stabilization of contaminated soil, 258–259
Steel, 318, 319, 321, 323, 324, 326, 327, 329, 333, 335, 337
Strengthening mechanisms, 344, 362–363, 368–370, 374, 377–378
Substrate compatibility, 277, 278
Substrate, 320, 336
Superalloys, 318
Superfinishing, 185–186
Surface charge, 418, 423
Surface coating, 432, 433, 437, 347–348, 357
Surface engineering, 328
Surface fatigue, 225
Surface finish/quality, 319, 321, 329, 330, 332, 333
Surface grafting, 432, 434, 437
Surface integrity, 182, 185–186
Surface modification, 432, 434, 435
Surface roughness, 271
Surface-to-volume ratio, 114
Sustainable development, 262

Taniguchi, 115
Technological challenges, 261
Template for chapter index, 130
Ternary composite materials, 113, 135, 137
Thermal conductivity, 301, 364, 368, 370, 373–374
Thermal conductivity (Si_3N_4 ceramics), 205
Thermal management, 157–159, 163, 207
Thermal spraying, 59, 70, 104
Thermal stability, 212, 221, 229, 273, 275, 317, 347–348, 355
Thermotropic, 126–128
Thin films, 269, 280
Thomas-Fermi Theory, 13–15
Threshold, 114, 132–136
Tissue engineering, 398
Titania, 113, 132, 137
Titanium carbide (TiC), 208–210, 220–221, 344, 346–347, 355–356

Titanium carbonitride (TiCN), 343, 346–348, 350–356
Titanium diboride, 298, 300
Titanium dioxide (TiO_2, titania), 158–159
Titanium implants, 400
Titanium nitride (TiN), 273, 278, 209–211, 219–221, 319, 320, 336, 343, 345–348, 352–356
Titanium, 324, 325, 326
Tool life, 177–183, 319, 333, 337, 338
Tool steel coatings, 208, 211
Tool steel, 318, 329
Tool wear, 182–184, 321, 337
Top-down approach, 119–120
Toughness, 318, 336, 337, 338, 343–344, 348, 359
Toxicity, 277
Transition metal dichalcogenides (TMDs), 118
Tribo-corrosion, 109
Tribo-layer, 42
Tribological behavior, 363, 366–368, 372–374, 376–378, 380
Tribological properties, 40, 44–46, 79–80, 83, 88, 95, 100, 102, 104–105, 109–111, 185–186, 208, 224–225, 343, 348
Tribology, 269–281
Tungsten carbide (WC), 179, 182–183
Turbine blades, 275
Turning, 318, 319, 329, 330, 334

Ultra-high-temperature ceramics, 284
Uniformity issues, 277

Vapor liquid solid (VLS) technique, 113, 121
Vibration, 319, 329, 331
Virus removal, 247
VOCs and hazardous gas adsorption, 252–253
Volume-specific surface area, 11, 114

Water and wastewater treatment, 156, 162
Water remediation, 240–241
WC + TiC particles, 108, 110
Wear, 39–41, 43, 45–55, 57–63, 65–77, 79–99, 101–103, 105, 107–111, 269–282
Wear mechanisms, 223–225, 345, 359–360
Wear resistance, 39–40, 43–47, 49–55, 57–58, 60–62, 66, 69, 71, 73–77, 79–81, 83, 85–86, 88–98, 101, 107–109, 111, 150, 156, 208–209, 212, 220–223, 273–280, 336, 337, 338, 343–344, 346–348, 350, 355–356, 359, 362–363, 366–367, 370, 372–374, 376–378, 380
Wear-resistant coatings, 203
Workpiece, 321, 322, 323, 326, 329, 330, 333, 334, 336

X-ray, 119, 121, 129–130

Zero-dimensional (0D) nanostructures, 116, 117, 132, 135
  nanocrystals, 117
  nanodots, 117
  nanoparticles, 117, 132–133, 135
Zirconium diboride (ZrB_2), 298, 300
Zirconium dioxide (ZrO_2, zirconia), 157–158

Printed in the USA/Agawam, MA
June 30, 2026

907143.083